Relationship between ionic charge and ionic radius for the commoner elements

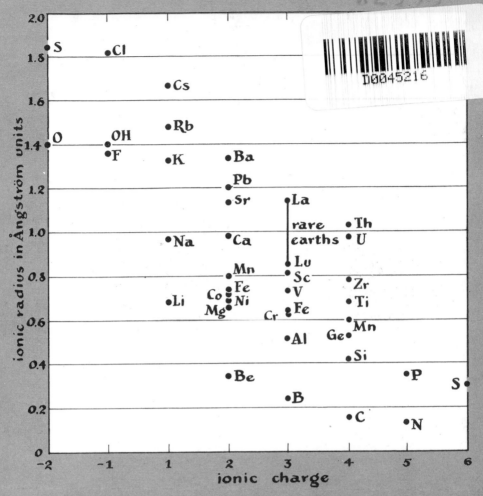

The major chemical elements in the earth's crust

	WEIGHT PERCENT	ATOM PERCENT	IONIC RADIUS (Å)	VOLUME PERCENT
O	46.60	62.55	1.40	93.77
Si	27.72	21.22	0.42	0.86
Al	8.13	6.47	0.51	0.47
Fe	5.00	1.92	0.74	0.43
Mg	2.09	1.84	0.66	0.29
Ca	3.63	1.94	0.99	1.03
Na	2.83	2.64	0.97	1.32
K	2.59	1.42	1.33	1.83

Elements of Mineralogy

BRIAN MASON   *U.S. National Museum, Washington, D.C.*

L. G. BERRY   *Queen's University, Kingston, Ontario*

W. H. Freeman and Company

SAN FRANCISCO

Library of Congress Catalogue Card Number: 68-13311

Standard Book Number: 7167 0235-5

8 9

Preface

This book is a modified and revised version of our *Mineralogy: Concepts, Descriptions, Determinations*, published in 1959 and still available, this alternate edition is designed to satisfy the requirements of a first course in mineralogy, and also to be useful to the student as a permanent reference. It is intended not only for the student who will later take more advanced work in the subject, but also for the student for whom this may be his only formal course in mineralogy. Emphasis is therefore given to general principles and to the significance of mineralogical data in interpreting geological phenomena, especially in the fields of petrology and economic geology.

A guiding principle has been that each individual mineral—a phase in the earth's crust—documents the chemical and physical conditions that caused it to form at a specific place and at a particular time. Minerals are products of geological processes; from these products we can draw conclusions about the nature of the processes themselves. An attractive feature of mineralogy is that we can use the laboratory to help us understand what we observe in the field. It is possible to reproduce most minerals in the laboratory under controlled conditions of temperature, pressure, and chemical environment. This enables us to elucidate the stability range of different minerals, and hence the conditions under which the rocks containing them were formed.

These aims have placed special emphasis on certain aspects of the subject as it is treated here. The consideration of minerals as crystalline solids involves giving attention to both external form and internal structure and to their interdependence. The chemistry of minerals is regarded as a specialized part of solid state chemistry, requiring consideration of the relative sizes of atoms, the types of bonding, and the significance of these bonds in determining crystal structure; this leads to discussions of isomorphism, atomic substitution, and polymorphism in minerals. In the treatment of the physics of minerals we have emphasized the interrelations of physical properties, crystal structure, and chemical composition. The chapter on genesis (Chapter 5) goes beyond the occurrence and associations of minerals to a consideration of their origin; here we have placed emphasis on the correlation of geological observations with physicochemical principles and experimental data. Chapters 8 through 15 describe some 200 minerals in detail, and many others are mentioned incidentally. We believe that this number is adequate for a course in general mineralogy, although some uncommon minerals of great local significance may have been omitted. The determinative tables provide the student with a logical scheme for the examination of an unknown specimen and its ultimate identification.

Selected bibliographies appended to some chapters will, it is hoped, guide the student to the most useful literature on the subject and encourage him to read more widely on topics of special interest.

A course in mineralogy generally has two aims. One is to develop in the student the facility for identifying minerals, especially the more common ones; the other is to provide him with an understanding of the physics, chemistry, and genesis of minerals, so that he will be able to go beyond the identification of the minerals to consideration of their geological significance and the evidence they bear for the interpretation of earth history. It is hoped that this book is an adequate guide for such a course. As far as possible, the text develops the subject from first principles; it is assumed, however, that the student is conversant with the basic principles of physics and chemistry, and with elementary plane trigonometry.

Elements of Mineralogy differs from our previous book in several ways. Chapter 2, on crystallography, has been considerably shortened, and the discussion is less mathematical. A section has been added on refractive index determination as a means of identifying nonopaque minerals, and relevant optical data have been incorporated into the mineral descriptions and the determinative tables. Diagrams have been included that relate refractive indices to chemical composition for the major mineral groups. Information on production and uses of economically important minerals has been revised in accordance with current data.

The text has benefited greatly from suggestions by many of our colleagues, and from the critical readings and comments of Professor A. O. Woodford, Dr. P. Leavens, and, especially, Professor Adolf Pabst. To all these persons we express our appreciation and thanks. We are also greatly indebted to Roger Hayward and Evan Gillespie, who have contributed both artistic merit and scientific understanding to the task of illustrating this book.

Brian Mason

October 1967   L. G. Berry

Diedra Bohn

Contents

Part I CONCEPTS

Notations and Abbreviations

CHEMICAL

Chemical elements	Na
Anions	O^{-2}, SO_4^{-2}
Cations	Fe^2, K^1
Isotopes	U^{238}, Pb^{207}, He^4
Rare-earth elements	$\mathbf{R}$
Smaller atoms (cations) in oxides, sulfides, halides	A, B
Larger atoms (anions) in sulfides, halides	X
Larger cations in sulfates, phosphates	A, B
Smaller cations in sulfates (S), phosphates (P, As, V)	X
Anion (OH, Cl, F) in sulfates, phosphates	Z
Large cation in silicates	W
Medium-sized bivalent cations (and lithium) in sixfold coordination in silicates	X
Medium-sized trivalent and quadrivalent cations in sixfold coordination in silicates	Y
Small cations in fourfold coordination in silicates	Z
Total weight of atoms in a unit cell	M
Number of formula units in a unit cell	Z

CRYSTALLOGRAPHIC

Crystal axes, unit cell dimensions	
Orthorhombic, monoclinic, triclinic	$a\ b\ c$
Tetragonal	$a_1\ a_2\ c$
Hexagonal	$a_1\ a_2\ a_3\ c$
Rhombohedral	a_r
Isometric	$a_1\ a_2\ a_3$

Unit cell interaxial angles
    Triclinic $\alpha\ \beta\ \gamma$
    Monoclinic β
    Rhombohedral α

Interplanar spacing d

Unit cell volume V

Crystal face indices (hkl)

Crystal form indices $\{hkl\}$

Crystal zone axis (lattice row) indices $[u\ v\ w]$

Reflection plane of symmetry m

Rotation axes of symmetry   2, 3, 4, 6

Center of symmetry $\bar{1}$

Rotary inversion axes of symmetry $\bar{2}, \bar{3}, \bar{4}, \bar{6}$

Rotation axes perpendicular to mirror planes $2/m, 3/m, 4/m, 6/m$

PHYSICAL

Ångstrom unit (10^{-8} cm)   Å

Specific gravity (density) G

Hardness H

Fusibility F

Index of refraction (isotropic media), or mean index
    (anisotropic media) n

Index of refraction (anisotropic, uniaxial) $\omega,\ \epsilon$

Index of refraction (anisotropic, biaxial) $\alpha,\ \beta,\ \gamma$

Part I
CONCEPTS

1

Introduction

THE SUBJECT OF MINERALOGY

Mineralogy is the science of minerals, which are the naturally occurring elements and compounds making up the solid parts of the universe. We generally think of mineralogy in terms of the materials of the earth's crust. However, meteorites provide us with samples of minerals from outside the earth, and geophysical measurements furnish some indication of the nature of the minerals below the accessible crust. Mineralogy is a subdivision of geology, since minerals constitute the rocks of the earth's crust. Chemistry and mineralogy are also closely related, since minerals are chemical compounds; indeed, the terms *mineral chemistry* and *inorganic chemistry* have sometimes been used interchangeably. Crystallography began as a subdivision of mineralogy, being the study of the external form of naturally occurring crystals, but has developed into the study of all crystals, natural and artificial, and the investigation not only of external form but also of internal structure.

Inasmuch as mineralogy is the scentific study of minerals, we should start with a clear understanding of what the term "mineral" connotes. It is quite difficult to formulate a precise definition of a mineral, and, in fact, there is no general agreement on any such definition. We prefer to say: *A mineral is a naturally occurring homogeneous solid, inorganically formed, with a definite chemical composition and an ordered atomic arrangement.*

It may help our understanding to scrutinize this definition point by point. The qualification *naturally occurring* is essential because it is possible to reproduce most minerals in the chemical laboratory. For example, evaporating a solution of sodium chloride produces crystals indistinguishable from those of the mineral halite, but such laboratory-produced crystals are not minerals. Replicas of most gemstones are made commercially, but they are referred to as synthetic ruby, synthetic spinel, and so on, to distinguish them from the natural minerals. On the other hand, substances that have formed naturally from man-made materials *are* considered to be minerals. An interesting case in point concerns a group of lead oxychloride minerals from Laurium, a town on the Greek coast some miles south of Athens. During the fourth and fifth centuries B.C. lead-bearing minerals were mined and smelted at Laurium and the slag dumped into the sea. These slags were reworked during the nineteenth century, and beautiful crystals of lead oxychlorides were found in cavities in the slag, evidently formed by the long-continued action of sea water on the lead compounds therein. The compounds in the slag itself are not considered to be minerals, but these lead oxychlorides are.

A mineral is a *homogeneous solid;* that is to say, it consists of a single solid phase—one kind of material only, which cannot be separated into simpler compounds by any physical method. The requirement that a mineral be solid eliminates liquids and gases from consideration. This may seem rather arbitrary; ice is a mineral (a very common one, especially at high altitudes and latitudes) but water is not. Some mineralogists dispute this restriction and would include water (and native mercury, which is sometimes found associated with cinnabar) as a mineral.

The restriction of minerals to *inorganically formed* substances eliminates those homogeneous solids produced by animals and plants. Thus the shell of an oyster (and the pearl inside), though consisting of calcium carbonate indistinguishable chemically or physically from the mineral aragonite, is not usually considered to be a mineral. The human body, unfortunately, sometimes produces homogeneous solids in the form of stones in certain internal organs. The stones are often identical in all respects to natural minerals, but are not classified as minerals (although the Riksmuseum in Stockholm has on display in its mineral gallery the gallstone of a famous Swedish admiral of the seventeenth century). The restriction of "inorganically formed" does not, however, eliminate the possibility of organic compounds being minerals. A few such substances (solid hydrocarbons, calcium oxalates, and other compounds) have been found in peat and coal, and distilled from the products of natural combustion.

The requirement of a *definite chemical composition* implies that a mineral is a chemical compound, and chemical compounds have a definite composition which is readily expressible by a formula. Mineral formulas may be simple or complex, depending upon the number of elements present and the proportions in which they are combined. It is important to distinguish between a *definite* and a *fixed* chemical composition; many minerals vary in composition (i.e., the composition is not fixed), but this variation is within definite limits.

An *ordered atomic arrangement* is the criterion of the crystalline state; another way of expressing this is to say that minerals are crystalline solids. Under favorable conditions of formation the ordered atomic arrangement may be expressed in the external crystal form—in fact, the presence of an ordered atomic arrangement in crystalline solids was deduced from the external regularity of crystals long before x-rays provided the means of demonstrating it. There are, however, a few exceptions to this requirement. Some minerals are *metamict;* when they were formed they had an ordered atomic arrangement which has since been partly or completely destroyed by radiation from uranium or thorium. A few minerals, the commonest being opal, are formed by the solidification of a colloidal gel and are noncrystalline initially; many such minerals become crystalline during geologic time.

THE HISTORY OF MINERALOGY

The origins of mineralogy lie far back in prehistory. Long before the art of writing was developed men recognized natural pigments such as hematite (red) and manganese oxides (black) and used them in their cave paintings. Stone Age Man was well aware that the hardness and toughness of fibrous actinolite (nephrite jade) made it a superior material for adzes, and the distribution of nephrite tools indicates that a lively trade must have developed in this material, since these tools are found in places far removed from any possible source of the raw material. The mining and smelting of metallic minerals to produce iron, copper, bronze, lead, and silver probably dates back 4,000 years or more, although we have no written records to verify this.

One of the earliest writings on minerals was the book *On Stones* by the Greek philosopher Theophrastus (*ca.* 372–287 B.C.). Pliny, in the first century A.D., recorded a great deal of natural history as it was understood by the Romans, and described a number of minerals that were mined as gemstones, as pigments, or as metallic ores. Little is known of developments during the next millennium, but the Renaissance in Europe brought with it an upswing of interest in science and technology. The classic works on minerals from this period were written by a German mining expert, Agricola, who published *De Re Metallica* (1556), and *De Natura Fossilium* (1546), in which he recorded the state of geology, mineralogy, mining, and metallurgy at that time. These works have been translated into English and are available in many libraries. After Agricola the next step in the development of mineralogy was provided by the Dane Niels Stensen (better known by the Latinized version of his name—Nicolaus Steno), who in 1669 showed that the interfacial angles of quartz crystals were constant, no matter what the shape and size of the crystals. He thereby drew attention to the significance of crystal form, which was to lead to the development of the whole science of crystallography.

Throughout the eighteenth century slow but steady progress in mineralogy was recorded. New minerals were recognized and described, and various attempts were made to achieve a rational classification of them. Most of the active workers in this endeavor were in Sweden and Germany, and the greatest teacher of the time was A. G. Werner (1750–1817), Professor at the Mining Academy in Freiberg, who attracted students from all parts of Europe. At this time mineralogy and chemistry were closely linked, since the chemists of the day worked largely with minerals as their raw materials. This resulted in the recognition of many new elements—cobalt, nickel, manganese, tungsten, molybdenum, uranium, and others. The significance of crystallography in the study of minerals was brought out largely by the work of the French scientist Haüy (1743–1822).

The early years of the nineteenth century saw rapid advances in mineralogy, following the enunciation of the atomic theory and the realization that minerals were chemical compounds with a definite composition. The invention of the reflecting goniometer also provided the means for much more precise measurements on crystals and an acceptable classification of crystal forms and crystal systems. The Swedish chemist Berzelius (1779–1848) and his pupils, especially Mitscherlich (1794–1863), studied the chemistry of minerals and enunciated the principles of their chemical classification.

Throughout the nineteenth century many new minerals were discovered and described, often as a result of the opening up of new mining districts in previously unexplored territories. The development of the polarizing microscope, and its application to the determination of the optical properties of minerals from about 1870 on, placed a new and powerful tool in the hands of the mineralogist.

The greatest development within the present century has been the demonstration by von Laue in 1912 that crystals diffract x-rays and that the diffraction patterns can be interpreted to give the actual positions of the atoms in the crystal. The mineralogist, previously limited to examining the external form and optical properties of his minerals, and to determining their chemical composition, could now proceed to investigate their internal structure. Since 1912 the structures of hundreds of minerals have been determined, but many still remain to be analyzed. In addition, improvements in technique have made it possible to detect extremely subtle differences between similar crystal structures. This whole field of research continues to be an exceedingly active branch of mineralogy.

THE LITERATURE OF MINERALOGY

The basic literature of mineralogy consists of individual papers describing original researches on minerals and published in scientific journals all over the world. The most important journals in the English-speaking countries are the *American Mineralogist*, published by the Mineralogical Society of America, and the *Mineralogical Magazine*, published by the Mineralogical Society in Great Britain. The latter society has also been responsible for a journal devoted to the abstracting of mineralogical papers wherever they appear; this journal, *Mineralogical Abstracts*, is now expanded in its size and scope through the cooperative efforts of mineralogical societies in several countries. Another important journal of abstracts is the *Zentralblatt für Mineralogie*, published in Germany. *Chemical Abstracts*, published by the American Chemical Society, carries a section entitled "Mineralogical and Geological Chemistry."

A standard reference work in mineralogy has long been *Dana's System of Mineralogy*. The first edition of this book was written by J. D. Dana and published in 1837, and it aimed at a complete account of all minerals described up to that time. It was revised and brought up to date by successive editions in 1844, 1850, 1854, 1868, and 1892, and by supplements issued at intervals between each edition. Three supplements to the sixth edition (1892) have been published, in 1899, in 1909, and in 1915. Some years ago the colossal task of preparing a seventh edition was begun. Since it is impossible to publish such a comprehensive work in one volume, the seventh edition is appearing in four volumes, of which three have been published, in 1944, 1951, and 1962; the fourth volume, dealing with the silicate minerals, has not yet appeared.

Two very comprehensive treatises on mineralogy, both in German, have been published. In some respects these are more detailed than *Dana's System of Mineralogy*. One is *Handbuch der Mineralchemie*, by C. Doelter and co-workers, published 1911–1931. The other is *Handbuch der Mineralogie*, begun by C. Hintze and carried on after his death by other workers; the first volume appeared in 1897, the last in 1933, and supplementary parts covering new minerals and new data on previously described minerals are published from time to time.

THE IMPORTANCE OF MINERALS

From earliest times man has found important uses for minerals, and these uses have expanded tremendously with the expansion of science and industry. At first minerals were used as they were found: clay for bricks and pottery; flint, quartz, and jade for weapons or implements; oxides of iron and manganese as paints; turquois, garnet, amethyst, and other colored stones for ornaments; and native gold, silver, and copper for ornaments and utensils. The art of smelting, by which metals are extracted from minerals, dates back to prehistoric times. The Bible mentions six metals: gold, silver, copper, tin, lead, and iron. A specimen of mercury was found in an amulet from an Egyptian tomb dating from the fifteenth or sixteenth century B.C., and Theophrastus describes its extraction from cinnabar. However, the discovery of new metals advanced little until the eighteenth century, when advances in chemistry resulted in the isolation of previously unknown metals from their naturally occurring compounds. Many of these metals were, at first, treated as curiosities. Even in more recent times some metal-bearing minerals have been discarded in mining, to become valuable at a later date. Extensive uses for many such metals were developed only after the costs of production had been reduced by the discovery of adequate mineral deposits or cheap smelting methods. Nickel was discovered in niccolite from the mines of Saxony in the eighteenth century, but little use was made of the metal until after the discovery of large deposits of pentlandite at Sudbury, Ontario, at the close of the nineteenth century. Aluminum metal was a curiosity, and the ore, bauxite, was of little use until the simultaneous discovery of an electrolytic reduction process in 1886 by Hall in America and Héroult in France. Many minerals will gain greater commercial importance in the future with the development of new uses for their constituents, more economical methods of extraction, or the discovery of more concentrated deposits.

The development of the polarizing microscope in the nineteenth century, which made possible the study of even the fine-grained rocks of the earth's crust, established the fact that virtually all the rocks of the crust are aggregates of one or more minerals. The more powerful tools of x-ray diffraction and electron microscopy have extended the study, leaving only the volcanic glasses, petroleum, and coal as important rock bodies which do not consist of minerals according to our definition. (It should be noted in passing that petroleum and coal are commonly referred to as "mineral fuels," and their production is usually included in statistics on mineral production.) The detailed study of rock masses and veins in the crust has greatly increased our knowledge about the processes of formation, both of the mass as a whole and of each individual mineral.

An important field of investigation in mineralogy in recent years has been the synthesis of minerals. These researches define the physicochemical conditions necessary for the formation of certain minerals, and, in turn, throw a great deal of light on the genesis of rocks and minerals. Many new compounds not previously known as minerals have been produced during investigations of mineral synthesis; some of these have later been recognized in nature.

Minerals are natural objects, and their description, naming, and classification is the fundamental task of the science of mineralogy. Minerals are given distinctive names, and each is described in terms of its physical and chemical properties. As the science progresses, other properties unknown at the time of the original description become important, and it is therefore necessary to re-study many minerals. This would often be impossible if it were not for the mineralogical collections housed in numerous public and private museums throughout the world. These museum collections thus have an importance far above that of displaying the mineral specimens to the general public. It is a duty of mankind to maintain these mineral collections for posterity. A few minerals are of common occurrence and are available in the earth's crust to anyone at any time, but many minerals are found only in a few isolated localities and others may be found only during the lifetime of a particular mine. Wherever possible, authentic original specimens and other specimens from the original localities must be preserved, as well as representative minerals from other localities as they are discovered Many interesting mineral occurrences have become inaccessible for further study, due to the advance of civilization, or have been completely destroyed by mining and conversion into valuable industrial products.

ECONOMIC MINERALOGY

This aspect of the study of minerals is seldom treated as a distinct subject. The scientific study of the identity and relationship of the minerals in an ore deposit, which may lead to useful information about the possible origin of the minerals in the ore, is usually included in the broader subject of economic geology. In this book it is appropriately combined with a study of the geological processes that were responsible for the physical conditions favorable to the formation of ore minerals. *Ore minerals* are usually considered to be those minerals which can be utilized as the source of a metal. Some authorities would include all metallic elements, sulfides, and oxides which occur in nature, and in addition some other minerals of metals such as copper, tungsten, vanadium, and uranium. Others would restrict the term to those minerals from which a metal can be profitably extracted. According to the latter interpretation, a particular mineral might be an ore mineral in only a few localities, whereas in other localities, due to remote location or to the small amount present it would not be so designated. This designation of an ore mineral will also change with time in many cases—improved technology, better transportation, and similar factors all influence the possibility of profitable exploitation. The term *gangue mineral* is usually applied to those minerals in an ore deposit that have no value and which must be removed in treating the ore. In Chapters 8–15, where the individual minerals are described, those minerals which commonly serve as sources of metals are so indicated.

The term *industrial mineral* embraces another group of minerals of economic importance. This term may be defined as covering those minerals which are raw materials for industry other than as a source of a particular element. These include the minerals from which are manufactured electrical and thermal insulators, refractories, ceramics, glass, abrasives, fertilizers, fluxes used in metallurgical processes, and cement and other building materials. Gemstones should also be included, since many minerals that provide beautiful precious and semi-precious stones also have numerous industrial uses, mainly because of their hardness.

The unequal and erratic distribution of ore and industrial minerals and the variable quality of the deposits containing these minerals in different parts of the earth's crust have far-reaching economic and political effects. A considerable portion of ocean and railway freight consists of mineral material either as raw ore or in a concentrated form ready for smelting. It is difficult to obtain an accurate figure on the percentage in terms of value or tonnage, since most statistics include in this category the mineral fuels, coal and petroleum. The unequal distribution of raw materials over the earth has caused numerous crises in many countries, particularly during the two World Wars, and the term *strategic mineral* has come into wide usage. "Mineral," as used here, included metals as well as the ore minerals required to produce the metals. In some cases the more appropriate word "material" was used instead of mineral. In March 1944 a new definition of strategic and critical materials was announced by the Army and Navy Munitions Board of the United States of America (DeMille, 1947, p. 3):

"*Strategic and Critical Materials* are those materials required for essential uses in war emergency, the procurement of which in adequate quantities, quality, and time is sufficiently uncertain for any reason to require prior provision for the supply thereof."

The shortage of materials in this category in North America led to extensive prospecting for new mineral deposits, to government assistance in the development and mining of deposits which would be uneconomical under normal peacetime conditions, and to expansion of existing mines to meet the increased demands. Since the end of World War II strategic materials have been stockpiled in considerable quantities, and government assistance to uneconomical mining ventures has been greatly reduced or terminated; such mines can be considered as being held in reserve for the future.

Selected Readings

Agricola, G., 1546, *De Natura Fossilium:* New York, Geological Society of America Special Paper 63 (1955). English translation from the first Latin edition, by M. C. and J. A. Bandy.

————, 1556, *De Re Metallica:* New York, Dover (1950). Reprint of 1912 English translation from the first Latin edition, by H. C. and L. H. Hoover.

Bateman, A. M., 1950, *Economic mineral deposits* (2nd ed.): New York, Wiley.

Caley, E. R., and J. F. C. Richards, 1956, *Theophrastus on stones* (Introduction, Greek text, English translation, and Commentary): Columbus, Ohio State University.

Dana, J. D., *A system of mineralogy* (6th ed.): New York, Wiley. Rewritten by E. S. Dana, with supplements, 1899, 1909, 1915.

————, *A system of mineralogy* (7th ed.), vol. 1, 1944; vol. 2, 1951; vol. 3, 1962: Wiley, New York. Rewritten by C. Palache, H. Berman, and C. Frondel.

DeMille, J. B., 1947, *Strategic minerals:* New York, McGraw-Hill.

Doelter, C., et al., 1911–1931, *Handbuch der Mineralchemie:* Dresden, Steinkopff.

Flawn, P. T. 1966, *Mineral Resources:* Chicago, Rand McNally.

Haüy, R. J., 1801, *Traité de minéralogie* (2nd ed., 1822): Paris, Louis.

Hintze, C. et al., 1897–1939, with supplements, *Handbuch der Mineralogie:* Berlin, De Gruyter (early volumes, Veit, Leipzig).

Voskuil, W. H., 1955, *Minerals in world industry:* New York, McGraw-Hill.

Werner, A. G., 1774, *On the external characters of minerals:* Urbana, Univ. Illinois Press (1962). English translation by A. V. Carozzi.

2
Crystallography

THE FORMATION OF CRYSTALS

A crystal is a solid body bounded by plane natural surfaces, which are the external expression of a regular internal arrangement of constituent atoms or ions. Excellent crystals may be grown in the laboratory by slowly cooling, or evaporating, a saturated solution of a salt such as common salt, alum, cupric sulfate, or ammonium dihydrogen phosphate. If a small seed crystal is suspended in such a saturated solution and the solution then cooled at a slow steady rate, a large single crystal bounded by smooth plane surfaces will grow as long as the cooling rate and the supply of material can be maintained. Crystals may also be grown at constant temperature by the use of equipment which permits addition of material to the solution in order to maintain saturated or very slightly supersaturated conditions. Commercial growth of large crystals requires very careful control of temperature. Numerous methods which have been used for growing crystals are described in detail by Buckley (1951). The quality of the plane surfaces varies greatly between different crystals, or even between different faces of a single crystal, from mirror-like to rough and pitted or deeply striated.

Crystals of minerals are found in nature wherever constituent atoms or ions were free to come together in the correct proportions to form a certain mineral and under conditions which permitted formation or "growth" of the mineral at a reasonably slow and steady rate. Well-developed crystals of minerals are commonly found lining the walls of open fractures, solution cavities, or vesicles in rocks. They may grow from water solutions descending through the surface rocks of the crust or ascending from sources at depth (hydrothermal solutions).

They may also form late in the crystallization of an igneous rock in cavities (miarolitic cavities) kept open by accumulations of gases and vapors. However, if the supply of material is such that the space is completely filled, the natural plane surfaces that are characteristic of a crystal may not develop.

Also, when solid material forms from a solution which is cooled or evaporated quickly, or from a vapor which is cooled quickly, many minute nuclei form. Under the microscope these nuclei may exhibit the geometrical shapes characteristic of crystals. As crystallization proceeds the nuclei grow until they impinge on one another, and eventually form a solid mass of grains bounded by irregular interfaces. This mass, in which each grain has the same internal structure and physical properties as well-formed crystals of the same mineral, chemical compound, or metal, is called a crystalline aggregate.

Under conditions of very sudden cooling some molten magmas freeze to volcanic glass or obsidian—in this case the atoms or ions do not have time to come together in a regular arrangement. Similarly, artificial melts of many types, such as mixtures of oxides, may form glasses if cooled quickly. In glasses there is virtually no ordered arrangement of constituents; therefore they lack crystal structure. An insoluble solid such as $BaSO_4$, formed quickly from the mixing of two solutions, will possess an ordered arrangement but only in extremely small crystallites.

In general, when a solid substance is formed slowly from a solution, a melt, or a vapor, under certain conditions of temperature and pressure, that substance consists of one or more chemical compounds or phases in each of which the constituent atoms or ions take on a regular orderly arrangement. The size of the domain of a specific orderly arrangement, or the size of a single crystal grain, depends on the rate of formation of nuclei of the new solid phase and the growth rate of the crystallites developing from these nuclei. If the velocity of nucleation is high, many small crystallites form simultaneously and the domains will then be small. On the other hand, if the velocity of nucleation is low, a few crystallites may continue to grow and use all the material separating from solution. If the supply of material is large, each regular domain develops until it impinges on the next, usually resulting in an aggregate of irregular grains. Regardless of such differences in formation, the well-formed crystal and the irregular grain of the same substance have the same orderly internal arrangement of constituents.

THE DEVELOPMENT OF CRYSTALLOGRAPHY

The term "crystal" is the Anglicized Greek word for "ice," and was generally employed throughout the Middle Ages to designate rock-crystal or quartz. The term was eventually applied to all solid objects of natural origin possessing polyhedral form. However, the use of the word crystal for any quartz clear enough for vessels or ornaments has also persisted, and we still hear the word used in reference to crystal ware (now made of specially designed lead-glass), or to crystal balls, which also were once made of clear quartz.

The earliest known comments on crystal form and the quality of crystal faces are found in Volume XXXVII of Pliny's *Natural History*. The first fundamental law of crystallography—The Constancy of Interfacial Angles—was announced by Steno of Florence in 1669. His early work was based on very rough measurements on quartz crystals, but in 1783 it was confirmed as a general law of nature by Romé de l'Isle (1736–1790), in Paris, who used a simple contact goniometer for measuring interfacial angles (see Fig. 2-5). As we shall see later, this law is a natural result of the regular internal arrangement in crystals. In 1611 the great astronomer Johannes Kepler wrote a small pamphlet

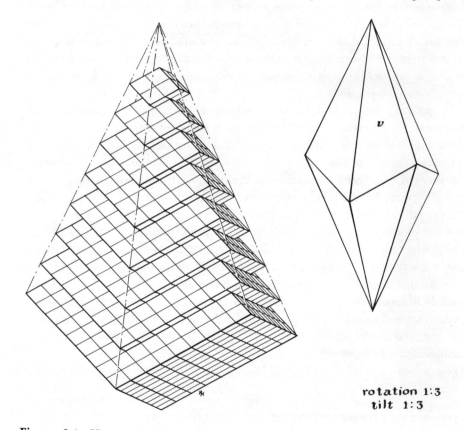

rotation 1:3
tilt 1:3

Figure 2-1 Haüy's conception of structural units, with the shape of the cleavage rhombohedron, building up a crystal of calcite in the form of a scalenohedron $v\{21\bar{3}1\}$. [Left-hand drawing after Haüy.]

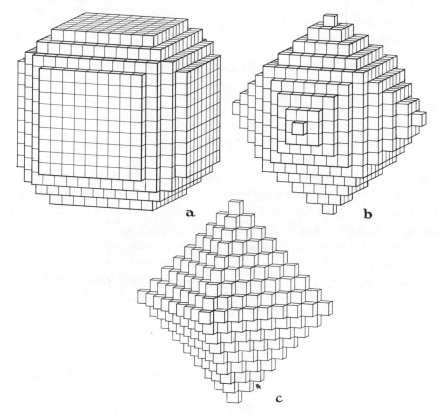

Figure 2-2 Haüy's conception of structural units, with the form of cleavage cubes of galena or halite, building up (a) a cube modified by dodecahedron faces, (b) a dodecahedron, (c) an octahedron.

on "hexagonal snow," in which he suggested that the regularity of crystal form is probably due to a regular geometrical arrangement of minute building units. Guglielmini (1655–1710) also suggested a theory of crystal structure, based on the constancy of cleavage directions in crystals. However, these two works are little known and apparently had little effect on later developments.

In 1784, Haüy (1743–1822) published, in Paris, an essay on a theory of crystal structure, and this was followed by his *Traité de Minéralogie* in 1801. A study of the cleavage of calcite led him to suggest that all crystals are made up of small polyhedral units, the unit for each mineral having a characteristic shape. He illustrated this with calcite and some cubic crystals (Figs. 2-1, 2-2). Haüy also defined axes of reference for crystals, and recognized that all faces on crystals of a single substance cut these axes at simple rational multiples of certain lengths; i.e., a simple mathematical relation exists between all planes which are possible faces on crystals of the same substance. This is the second fundamental law of crystallography, known as Haüy's Law, or the Law of Simple Rational Indices.

THE CLASSIFICATION OF CRYSTALS

Thousands of different crystals, each showing characteristic morphology, have been described. The significance of crystal morphology can only be understood if some property common to all crystals can be used as a basis of classification. This basic property is *symmetry*. The characteristic feature of symmetry is repetition. A crystal may show repetition with respect to a point, in which case it has a center of symmetry; with respect to a line, in which case it has an axis of symmetry; and with respect to a plane, in which case it has a plane of symmetry. Every crystal is characterized by a specific combination of symmetry elements (or, for a few substances, by the complete absence of symmetry). The nature of the various symmetry elements is best appreciated by the study of natural crystals and crystal models. The formal definitions of these elements are:

Plane of symmetry: A plane of symmetry divides a crystal into two parts in such a way that one part is the mirror image of the other part.

Axis of symmetry: If a crystal can be rotated about an axis in such a way that it repeats itself two or more times in a revolution, it is said to have an axis of symmetry. Symmetry axes in crystals may be 2-fold, or diad (repeat after 180° rotation); 3-fold, or triad (repeat after 120° rotation); 4-fold, or tetrad (repeat after 90° rotation); and six-fold, or hexad (repeat after 60° rotation).

Center of symmetry: A crystal has a center of symmetry if every face has a similar face parallel to it, repeated by inversion through the center.

The significance of symmetry as a basis for the classification of crystals was first realized early in the nineteenth century. Through studies of the external form of crystals and the angular relationships between crystal faces, the early investigators established the existence of these symmetry elements. The next step was to determine how many different ways in which these symmetry elements could be combined in crystals. In 1830 Hessel, a German crystallographer, showed by mathematical reasoning that there could be altogether 32 different classes of crystals, each class characterized by a specific combination of symmetry elements (Table 2-I); at that time only a few of these classes were known to be represented by minerals.

It will be noticed that two classes, the sphenoidal and the tetragonal disphenoidal, are apparently identical, each being characterized by a single diad axis. They are actually quite distinct. Here it is necessary to introduce an additional symmetry element, a rotatory-inversion axis (hereafter called an inversion axis). The diad axis in the tetragonal disphenoidal class is not a simple diad axis as in the sphenoidal class, but is a tetrad inversion axis (Fig. 2-3). This axis differs from a simple tetrad axis in that is requires inversion

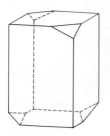

Figure 2-3 Crystal with a tetrad inversion axis (vertical). Note that crystals with a $\bar{4}$ axis do not have a center of symmetry.

after a 90° rotation in order to produce repetition. There are inversion axes corresponding to each of the simple rotation axes, but all except the tetrad axis can be described in terms of the other symmetry elements:

One-fold inversion axis $\equiv$ center of symmetry
Two-fold inversion axis $\equiv$ a plane of symmetry
Three-fold inversion axis $\equiv$ triad axis plus a center of symmetry
Six-fold inversion axis $\equiv$ triad axis normal to a plane of symmetry

Some of the class names are rather complex, and different authorites use different systems of naming; the class names in Table 2-I are derived from those of the general form (p. 23) in each class.

It is convenient to have short, self-explanatory symbols with which to refer to the individual crystal classes. The Hermann-Mauguin symbols (Table 2-I) are not only extremely concise, but also self-explanatory in that they present the essential symmetries of the crystal classes. Two-, three-, four-, and six-fold rotation axes of symmetry are represented by the numbers 2, 3, 4, and 6, whereas three-, four-, and six-fold inversion axes have the symbols $\bar{3}$, $\bar{4}$, and $\bar{6}$. In conformity with this scheme, asymmetry is represented by the figure 1 (only one repetition in a complete rotation), and a center of symmetry, or inversion through a point, by $\bar{1}$. A plane of symmetry is represented by the letter m (mirror). In arranging the symbols to denote the symmetry of a specific class, the convention is to give the symmetry of the principal axis first—for instance, 4 or $\bar{4}$ for tetragonal classes. If there is a plane of symmetry parallel to the principal axis, the two symbols are associated thus: $4m$; if there is a plane of symmetry perpendicular to the principal axis, the two symbols are associated thus: $\frac{4}{m}$, or, more conveniently for printing, $4/m$. Then follow the symbols for

Table 2-I *The thirty-two crystal classes*

#	AXES 2-FOLD	3-FOLD	4-FOLD	6-FOLD	PLANES	CENTER	HERMANN-MAUGIN SYMBOLS	CLASS NAME	SYSTEM
1	—	—	—	—	—	—	1	Pedial	Triclinic
2	—	—	—	—	—	yes	$\bar{1}$	Pinacoidal	
3	1	—	—	—	1	—	m	Domatic	Monoclinic
4	1	—	—	—	—	—	2	Sphenoidal	
5	1	—	—	—	1	yes	$2/m$	Prismatic	
6	1	—	—	—	2	—	$mm2$	Orthorhombic pyramidal	Orthorhombic
7	3	—	—	—	—	—	222	Orthorhombic disphenoidal	
8	3	—	—	—	3	yes	$2/m\ 2/m\ 2/m$	Orthorhombic dipyramidal	
9	—	1	—	—	—	—	3	Trigonal pyramidal	Trigonal
10	—	1	—	—	—	yes	$\bar{3}$	Rhombohedral	
11	—	1	—	—	3	—	$3m$	Ditrigonal pyramidal	
12	3	1	—	—	—	—	32	Trigonal trapezohedral	
13	3	1	—	—	3	yes	$\bar{3}2/m$	Hexagonal scalenohedral	
14	—	1	—	—	1	—	$\bar{6}$	Trigonal dipyramidal	Hexagonal
15	—	—	—	1	—	—	6	Hexagonal pyramidal	
16	—	—	—	1	1	yes	$6/m$	Hexagonal dipyramidal	
17	3	1	—	—	4	—	$\bar{6}m2$	Ditrigonal dipyramidal	
18	—	—	—	1	6	—	$6mm$	Dihexagonal pyramidal	
19	6	—	—	1	—	—	622	Hexagonal trapezohedral	
20	6	—	—	1	7	yes	$6/m\ 2/m\ 2/m$	Dihexagonal dipyramidal	
21	1	—	1	—	—	—	$\bar{4}$	Tetragonal disphenoidal	Tetragonal
22	—	—	1	—	—	—	4	Tetragonal pyramidal	
23	—	—	1	—	1	yes	$4/m$	Tetragonal dipyramidal	
24	3	—	—	—	2	—	$\bar{4}2m$	Tetragonal scalenohedral	
25	—	—	1	—	4	—	$4mm$	Ditetragonal pyramidal	
26	4	—	1	—	—	—	422	Tetragonal trapezohedral	
27	4	—	1	—	5	yes	$4/m\ 2/m\ 2/m$	Ditetragonal dipyramidal	
28	3	4	—	—	—	—	23	Tetartoidal	Isometric
29	3	4	—	—	3	yes	$2/m\bar{3}$	Diploidal	
30	3	4	—	—	6	—	$\bar{4}3m$	Hextetrahedral	
31	6	4	3	—	—	—	432	Gyroidal	
32	6	4	3	—	9	yes	$4/m\ \bar{3}\ 2/m$	Hexoctahedral	

the secondary axes, if any, and then any other symmetry planes. (Note that $4/mmm$ means $\frac{4}{m}\, mm$; that is, the second and third m's refer to planes of symmetry parallel to the fourfold axis.)

Abbreviations are sometimes made, the principle being to give only those elements which are necessary to describe the class symmetry uniquely. Thus Class 8 has three nonequivalent, mutually perpendicular diad axes, with a plane of symmetry perpendicular to each; the full class symbol is $2/m\, 2/m\, 2/m$, but it is rather common practice to leave out the axial symbols and simply write mmm; the three planes of symmetry automatically call into being the three diad axes. Similarly, Class 32 has the full symbol $4/m\, \bar{3}\, 2/m$, but $m3m$ is adequate to describe uniquely the class symmetry. (Note that $3m$ is quite a different class—a trigonal class with planes of symmetry parallel to the triad axis.)

A study of these symbols indicates why it is general practice to group the 32 crystal classes into six or seven crystal systems. Thus the triclinic system comprises the two classes with neither axes nor planes of symmetry; the monoclinic system the three classes with one diad axis and/or one plane of symmetry; the orthorhombic system, three classes with three diad axes or one diad axis and two planes; the trigonal system, five classes each with one triad axis (rotation or inversion); the hexagonal system, seven classes each with one hexad axis (rotation or inversion); the tetragonal system, seven classes each with one tetrad axis (rotation or inversion); the isometric system, five classes each with four triad axes. The trigonal system is commonly referred to as the rhombohedral division of the hexagonal system.

The number of minerals crystallizing in each of the 32 crystal classes varies markedly; in fact, for Class 14 no mineral has yet been recognized. In any system the class with the highest symmetry is represented by the largest number of minerals. The percentages of minerals crystallizing in the different systems are: triclinic, 7.4; monoclinic, 31.7; orthorhombic, 22.3; trigonal, 9.0; hexagonal, 7.6; tetragonal, 9.8; isometric, 12.2. Thus more than 50 percent of all known minerals crystallize in two systems, monoclinic and orthorhombic.

THE DESCRIPTION OF CRYSTALS

Two crystals of the same substance may differ considerably in appearance—that is, in the number, size, and shape of the individual faces. This was recognized at an early stage in the development of crystallography, and in order to describe the external form of crystals it was clearly necessary to devise some way of identifying the individual faces and relating different faces to one another. This is done by the mathematical method of relating planes to certain imaginary lines in space.

The position of any plane can be uniquely fixed by the intercepts it makes on three intersecting lines, the axes of reference (Fig. 2-4). Let us denote the axes of reference as a, b, and c. The intercept of the plane ABC on axis a is OA, on axis b OB, and on axis c OC. These intercepts, OA, OB, OC, uniquely fix the position of the plane ABC with reference to the axes a, b, and c. Similarly, another plane XYZ has an intercepts OX on a, OY on b, and OZ on c. These intercepts fix its position in space with reference to axes a, b, and c.

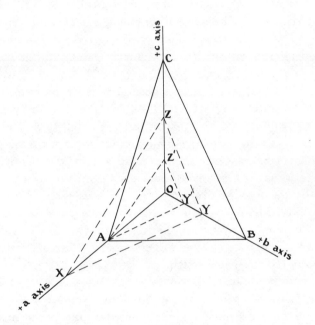

Figure 2-4   Intercepts of crystal faces on axes of reference.

Now in applying this technique to crystallography, we are interested not so much in the actual position of an individual face as we are in the position of one face in relation to all other faces on the crystal. For this purpose a face which cuts all three axes of reference is chosen as a unit or reference face, and all other faces are related to this face by parameters. Thus if we consider the planes ABC and XYZ as being different faces on the same crystal, and ABC is the unit or reference face, the parameters of XYZ are $\dfrac{OX}{OA}$, $\dfrac{OY}{OB}$, and $\dfrac{OZ}{OC}$.

That is, the position of XYZ with reference to ABC is defined by the ratio of the intercepts of XYZ on the axes of reference a, b, c to the intercepts of ABC, the unit or reference face, on the same axes.

This method was developed to provide a means of establishing the positions of faces with respect to axes of reference, and their positions in relation to one another. Soon after it was first applied, however, the method was found to be more than just a useful mathematical device for describing the external form of crystals; its application revealed that the method had fundamental significance in the science of crystallography. It was observed that for any crystal the parameters of the different faces with respect to the one chosen as a unit face were always relatively simple fractions. The intercepts of ABC on the three axes might be, for example, 15, 22, 36; the intercepts of XYZ, 30, 11, 24. The parameters of XYZ relative to ABC as the unit face are then $\frac{30}{15}$, $\frac{11}{22}$, $\frac{24}{36}$, or $\frac{2}{1}$, $\frac{1}{2}$, $\frac{2}{3}$. This is the sort of relation that was found to hold true for all crystals and led to the Second Law of Crystallography, which can be stated as follows: *A simple mathematical relation exists between all planes which are possible faces on crystals of the same substance.* The axial ratio—the ratio of the intercepts of the unit face on the axes of reference—is a fundamental property for each crystalline substance. If ABC is the unit face of a crystal, the axial ratio is 15:22:36, or, following the standard convention of setting *b* equal to unity, 0.682:1:1.636.

Parameters could be used directly as symbols for the faces of a crystal, and were so used in the Weiss system, but because fractions and infinity symbols (a face parallel to an axis intercepts it at infinity) are awkward to work with, parameters are usually converted to Miller indices. To convert parameters to Miller indices one simply inverts the parameters and multiplies through in order to clear fractions. In this way ∞ becomes $1/\infty$, i.e., 0, and fractions become whole numbers. The Miller indices of XYZ are, for example, (143). The parallel plane AY'Z' cuts OA at 1/1, OB at 1/4, and OC at 1/3. Miller indices are always expressed in order of the three axes, the index on *a* first, on *b* second, and on *c* last. The Miller indices of the unit face are, of course, (111). Since the origin (O in Fig. 2-4) of the axes of reference is the center of the crystal, it is necessary to distinguish opposite ends of the axes as positive and negative. For faces intersecting the negative ends of the axes, the Miller indices are expressed with a bar over the digit; thus the face parallel with and opposite to (101) is expressed as ($\bar{1}0\bar{1}$).

The use of Miller indices gives rise to another, often used statement of the Second Law of Crystallography, or the Law of Simple Rational Indices: *the Miller indices of any crystal face are either small whole numbers or zero.*

The Second Law of Crystallography, which was first stated by Haüy in 1784, before the atomic theory had been established, is in fact the natural consequence of the atomic structure of crystals. A crystal is a regular arrangement of atoms that repeats itself indefinitely in three dimensions. The repeating group of atoms consists of a small parallelepiped, called the *unit cell*. The edges of the unit cell are parallel to the axes of reference, and provided the unit face

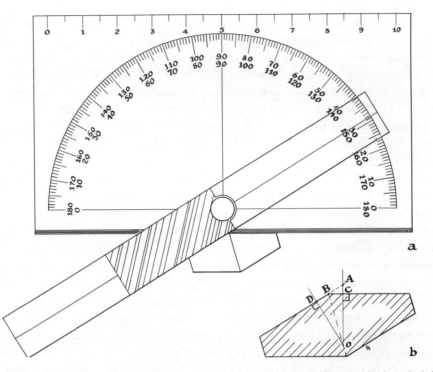

Figure 2-5   (a) A contact goniometer on which may be read directly the interfacial angle $CBD = 148\frac{1}{2}°$ or the polar angle $COD = ABC = 31\frac{1}{2}°$, for the example shown at (b).

has been correctly selected, the ratio between the lengths of the edges of the unit cell will be the same as the axial ratio derived from measurements of the crystal faces. Thus in the nineteenth century the axial ratio of sulfur was determined to be 0.813:1:1.903; after the introduction of x-ray determination of unit cell dimensions in 1912 the unit cell of sulfur was found to have the dimensions $a = 10.44$ Å, $b = 12.84$ Å, $c = 24.37$ Å, which gives an axial ratio of 0.811:1:1.900, in excellent agreement with the earlier determination, within the margin of error of the two determinations.

The investigation of a crystal requires the measurement of the angles between the different faces. Interfacial angles must be measured in a plane perpendicular to both of the crystal faces concerned. This may be done with a simple contact goniometer* consisting of a printed protractor with a straight strip of celluloid that is pivoted at the center and which has a hairline scratch that is read against the scale (Fig. 2-5). The goniometer is held with the straightedge of the protrac-

---

* A convenient model for general class use is available from Ward's Natural Science Establishment, Rochester, N. Y.

tor in contact with one face and the straightedge of the celluloid strip in contact with the other face. The plane surface of the protractor and strip must be perpendicular to both crystal faces. Two values of the interfacial angle, which total 180°, may be read from the protractor. One is the internal solid angle DBC (Fig. 2-5b), the other is the external angle ABC between one face and the extension of the other. The latter angle, which is equal to the angle COD between the perpendiculars to the two faces (since $ODA = OCB = 90°$), is usually called the polar angle. It is this angle COD that is quoted in this book, and in most other textbooks, in describing crystals. In Part II of this book the interfacial angles (polar angles) between two faces are given as (001) $\wedge$ (111) = 54° 44′ (for any cubic mineral).

With small crystals, the interfacial angles may be measured more conveniently and accurately with a reflecting goniometer. This instrument has wider application, since small crystals occur more commonly than large ones for most minerals; in fact many minerals are very rarely found in crystals large enough for contact goniometry.

Forms

The similar faces of a crystal constitute a *form*, the term being used in a special sense. A form comprises all those faces on a crystal required by the class symmetry if one such face is present. For example, in a triclinic crystal with a center of symmetry each form will comprise two faces, parallel to each other and on opposite sides of the crystal. This is an example of an *open form*, so called because it does not enclose space. If the faces of a single form are sufficient to enclose space, it is known as a *closed form*.

The Miller indices for a single face are designated by a symbol enclosed in parentheses, such as (010), (111), (hkl). Miller indices are also used as form symbols, in which case they are enclosed in braces: $\{010\}$, $\{111\}$, $\{hkl\}$. It is customary in describing crystals to designate each form by a letter, in addition to the Miller indices, for example, $m\{110\}$.

A crystal form is named according to the number of its faces and their mutual relations. Thus a *pedion* is a single plane; a *pinacoid* is a pair of parallel planes; a *prism* is a set of equivalent planes, more than two in number, all parallel to a common axis; a *pyramid* is a set of three or more planes, all equally inclined to a common axis. The position of a form in relation to the crystallographic axes may be indicated by placing the Miller symbol of the form in front of the name: $\{hk0\}$ prism, $\{100\}$ pinacoid, $\{111\}$ pyramid, etc.

In all crystal classes, any crystal form that intersects each of the three axes of reference in different multiples of the unit lengths a, b, c—for example, $\{321\}$, or $\{432\}$—is known as a *general form*. In each class a general form has the largest number of equivalent faces possible in that class. Crystal forms comprising faces that cut the axes of reference in special ways, such as {101}, {011}, {011}, etc., have fewer equivalent faces in most crystal classes. In each crystal class the general form is diagnostic of the symmetry of that class, whereas special forms may be identical in two or more classes of a system. The name of the general form in each class is used also as the name of the class; these are given in Table 2-I.

Any two faces of a crystal either intersect or are parallel. A group of faces whose lines of intersection are mutually parallel constitute a *zone*. The line of intersection of any two faces defines the direction of the zone axis (Fig. 2-6).

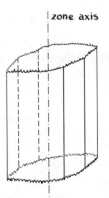

Figure 2-6   Crystal faces lying in a zone.

If the indices of two nonparallel faces are added together, their sum represents the indices of a face lying between the two faces and in the same zone. This is a useful relationship for deciding whether a face lies in a specific zone. This remains true if the indices of one or both faces are multiplied by a simple whole number, i.e., (110) + (001) = (111), (110) + (002) = (112), (210) + (001) = (211), and so on.

The first step in a crystallographic investigation is to make a sketch of the crystal. In this sketch each face is identified by a number or a letter. The angles between the faces in a prominent zone are measured in succession and recorded in the following manner:

$$1 \wedge 2 = 27°$$
$$2 \wedge 3 = 43°$$

and so on, and the same procedure is followed for all other zones until each interfacial angle on the crystal has been measured.

THE STEREOGRAPHIC PROJECTION

In crystallography, as in other fields of mineralogy and geology, we are constantly confronted with the problem of representing a three-dimensional object

—in this case, the crystal—on a two-dimensional surface, a sheet of paper. To achieve this we utilize various kinds of projections. One of the most useful of these is the *stereographic projection*.

In this method of projection the center of the crystal coincides with the center of a sphere which surrounds the crystal. Lines from the center of the sphere and perpendicular to each face are then extended to the surface of the sphere (Fig. 2-7). The points where these lines penetrate the spherical surface are the

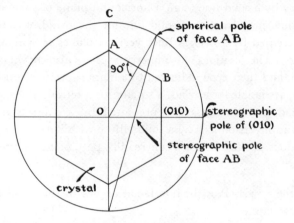

Figure 2-7   Section of the spherical projection and stereographic projection of a crystal.

spherical poles of the faces. The assembly of spherical poles is called the spherical projection of the crystal. The stereographic projection is derived from the spherical projection by projecting the spherical poles to the south pole of the sphere; the points where lines of projection intersect the equatorial plane of the sphere are the stereographic poles of the faces, and the complete pattern is the stereographic projection of the crystal.

In preparing a stereographic projection the standard orientation of a crystal is with the vertical crystallographic axis as the vertical axis of the sphere (OAC in Fig. 2-7), the positive end being up and the negative end down. With this orientation the poles of all faces parallel to the vertical axis (i.e., those with indices of the type $(hk0)$) fall on the circumference of the stereographic projection (the circumference is often referred to as the primitive circle of the stereographic projection). The crystal is oriented so that the pole of the face (010) falls at the east point of the primitive circle.

The projection of faces on the lower half of the crystal (i.e., those with negative indices on the *c* axis) will fall outside the circumference of the stereographic projection if the spherical poles of these faces are projected to the south pole. By projecting them to the north pole, however, their stereographic poles then fall within the circumference of the stereographic projection. Their poles are usually represented by small circles to distinguish them from faces projected from the northern hemisphere, which are represented by small crosses. In all crystals with a horizontal plane of symmetry the poles of faces projected from the southern hemisphere coincide with the poles of corresponding faces projected from the northern hemisphere.

The stereographic projection of a crystal represents the actual symmetry of the crystal, as will be seen when studying the stereographic projections for the different crystal classes. On a stereographic projection it is customary to indicate a plane of symmetry by a solid line (e.g., a horizontal plane of symmetry appears on a stereographic projection as a solid line for the primitive circle, which is otherwise represented by a broken line; vertical planes of symmetry appear as solid diameters). The positions of symmetry axes are also shown on a stereographic projection, and their type indicated by an appropriate symbol—a solid ellipse for a diad, a triangle for a triad, a square for a tetrad, a hexagon for a hexad axis. A triad inversion axis is represented by a solid triangle within a hexagon, a tetrad inversion axis by a solid ellipse within a square.

Stereographic projections of most crystals can readily be plotted by simple constructions, as follows:

1. Faces parallel to the c axis: Plot the interfacial angles in this zone directly around the primitive circle.

2. Faces in zones forming vertical great circles—i.e., those whose poles fall on diameters in the stereographic projection: Let us take, for example, the faces (110), (111), (112), (001), ($\bar{1}\bar{1}2$), ($\bar{1}\bar{1}1$), ($\bar{1}\bar{1}0$) in an orthorhombic crystal. Draw a circle whose diameter is equal to that of the primitive circle; this represents a vertical section through the spherical projection for this specific zone. An east-west diameter represents the plane of the stereographic projection, the eastern end being the pole of (110) and the western end that of ($\bar{1}\bar{1}0$). From

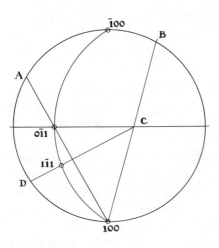

Figure 2-8 Stereographic projection of faces lying on oblique great circles.

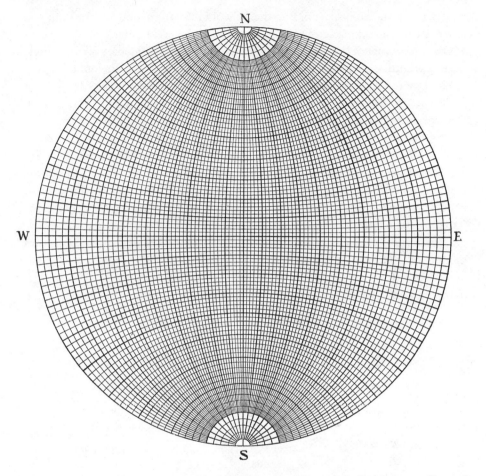

Figure 2-9 A Wulff net. Stereographically projected great circles appear as lines of equal longitude, and small circles centered about the north and south poles appear as lines of equal latitude at two-degree intervals.

the angles (110) ∧ (111), (111) ∧ (112), etc., the spherical poles of these faces are laid off on the circumference of the circle, and these poles are joined to the south pole. The respective stereographic poles are located where these joins cut the east-west diameter. This east-west diameter and the located poles are then transferred to the appropriate diameter of the projection.

3. Faces in zones forming oblique great circles: It is first necessary to plot the pole of the great circle. This is a point on the diameter normal to the great circle, and 90° from its intersection with the great circle. This pole is found as indicated in Fig. 2-8. Assume we are dealing with the zone (100)(0$\bar{1}$1)($\bar{1}$00) in an orthorhombic crystal. Join the pole of (100) to that of (0$\bar{1}$1) and extend it until it cuts the primitive circle at A. Lay off 90° from A on the primitive

circle, giving the point B; join B to the pole of (100). This join cuts the diameter normal to the great circle at C; C is the pole of the great circle.

Given the angle (100) $\wedge$ (1$\bar{1}$1), we can now plot (1$\bar{1}$1). Lay off this angle on the primitive circle ((100) − D). Join D to C, the pole of the oblique great circle. Where DC cuts the oblique great circle is the stereographic pole of (1$\bar{1}$1).

Construction of stereographic projections is expedited by the use of the Wulff stereographic net (Fig. 2-9).* This is simply the stereographic projection of a sphere marked off at 2° intervals of latitude and longitude. For plotting data, a piece of tracing paper is placed over the Wulff net and fastened by a pin through the center of the net; the pin then serves as an axis around which the tracing can be rotated. Using the Wulff net, interfacial angles in the vertical zone—i.e., faces parallel to the *c* axis—can be plotted directly from the graduated circumference of the net. Faces in zones forming vertical great circles—i.e., zones represented by diameters in the stereographic projection—are plotted by rotating the tracing so that the zone coincides with the E-W diameter of the net, and then laying off the angles from the graduations on this diameter. Any face Q whose interfacial angles with two vertical zone faces P and R are known can be plotted as follows (Fig. 2-10). Locate the pole of P

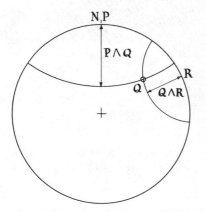

Figure 2-10 Locating the pole of face Q on a stereographic projection from measured angles P $\wedge$ R, P $\wedge$ Q, and Q $\wedge$ R, using the Wulff net.

on the tracing paper at N on the Wulff net. Lay off the angle P $\wedge$ Q on the diameter from N toward the center, and trace the small circle through this angle. Turn the tracing through the angle P $\wedge$ R so that R falls at N on the net. Lay off the angle Q $\wedge$ R along the diameter, and trace the small circle

* Printed Wulff nets may be obtained in various sizes: nets 10 cm and 14 cm in diameter may be obtained from Ward's Natural Science Establishment, Rochester, N. Y.; nets 40 cm in diameter may be obtained from the U. S. Navy Hydrographic Office—it is designated H.O. Misc. No. 7736-1; 5 inch and 30 cm diameter nets may be obtained from Polycrystal Book Service, P.O. Box 11567, Pittsburgh, Pa. 15238, or from the Institute of Physics, 40 Belgrave Sq., London S.W. 1, England.

through this angle. The intersection of these two small circles is the stereographic pole of Q.

THE SEVEN CRYSTAL SYSTEMS

In this section we shall describe the general features of each of the seven crystal systems and give a detailed account of those classes which are well represented by mineral species. In any system the class of highest symmetry always has far more natural representatives than any of the classes of lower symmetry. Therefore, except for the isometric system, we shall discuss in detail only the class of highest symmetry in each system, although we shall mention minerals crystallizing in other classes.

Triclinic System

Axes of reference (Fig. 2-11): The only symmetry element possessed by crystals in the triclinic system is a center of symmetry (in the pinacoidal class), and thus the position of the axes of reference is not controlled by the positions of

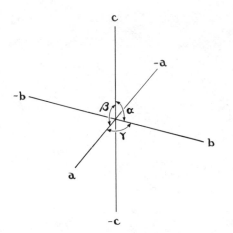

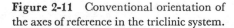

Figure 2-11   Conventional orientation of the axes of reference in the triclinic system.

axes and planes of symmetry, as in the other systems. Usually the direction of a prominent zone is selected as the *c* axis; a prominent face in this zone is chosen as (010), and its intersection with (001) defines the *a* direction (Fig. (2-12); a second face in this zone can be chosen as (100), thereby defining the direction of the *b* axis. The angle between *a* and *b* is denoted γ, that between *c* and *a* β, and that between *c* and *b* α. These choices are usually consistent with $c < a < b$, α and β obtuse, and γ either acute or obtuse. The possibility of choice in selecting the axes of reference in triclinic crystals has resulted in

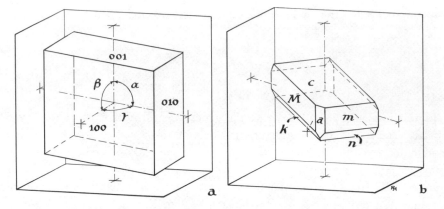

Figure 2-12 (a) A hypothetical triclinic crystal with the three pinacoids {100}, {010}, {001}. (b) A crystal of rhodonite showing the forms $a\{100\}$, $c\{001\}$, $m\{110\}$, $M\{1\bar{1}0\}$, $n\{22\bar{1}\}$, $k\{2\bar{2}\bar{1}\}$, all pinacoids.

various settings being chosen for a single mineral, each setting giving a different axial ratio and different interaxial angles. Thanks to the availability of x-ray methods for determining the actual unit cell of any crystal, it is now possible

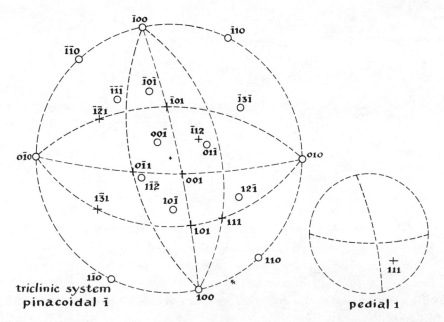

Figure 2-13 Stereographic projection of some forms in the pinacoidal class $\bar{1}$. The small stereogram shows the pedion {111} in the pedial class 1. Here and in later figures a face with l positive is shown by $+$, one with l negative by $\bigcirc$, and one with l zero by $\bigcirc$ on the primitive.

to decide on a setting consistent with this—i.e., a setting in which the axes of reference are parallel to the edges of the unit cell.

Forms: As implied by the name, all forms in this class are pinacoids. The presence of a center of symmetry, and no other symmetry element, means that each form must consist of two faces, parallel to each other on opposite sides of the center of symmetry. There are seven distinct types of pinacoids, distinguished by their positions relative to the axes of reference, as follows:

{100}, front or *a* pinacoid: parallel to *b* and *c*, cuts *a* axis
{010}, side or *b* pinacoid: parallel to *a* and *c*, cuts *b* axis
{001}, basal or *c* pinacoid: parallel to *a* and *b*, cuts *c* axis
{*hk*0} pinacoid: parallel to *c*, cuts *a* and *b* axes
{*h0l*} pinacoid: parallel to *b*, cuts *a* and *c* axes
{0*kl*} pinacoid: parallel to *a*, cuts *b* and *c* axes
{*hkl*} pinacoid: cuts all three axes

In the stereographic projection (Fig. 2-13) several forms are plotted, each comprising two poles.

Minerals crystallizing in the pinacoidal class include the plagioclase feldspars, microcline, rhodonite, amblygonite, pectolite, wollastonite, and turquois.

Monoclinic System

Axes of reference (Fig. 2-14): The prismatic class has a diad symmetry axis, and a plane of symmetry normal to this axis. Symmetry requires $\alpha = \gamma = 90°$. The *b* axis coincides with the diad axis; *a* and *c* lie in the plane of symmetry, the angle between them, β, being obtuse toward the front. The positions of

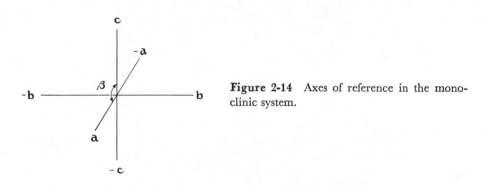

Figure 2-14 Axes of reference in the monoclinic system.

axes *a* and *c* in this plane are chosen to provide the simplest indices for the faces; when the crystal has a direction of elongation in the symmetry plane, this direction is usually, but not always, *c* (Fig. 2-15). The positions of the *a* and *c*

axes chosen on the basis of external morphology may have to be modified
when the unit cell is determined by x-ray diffraction.

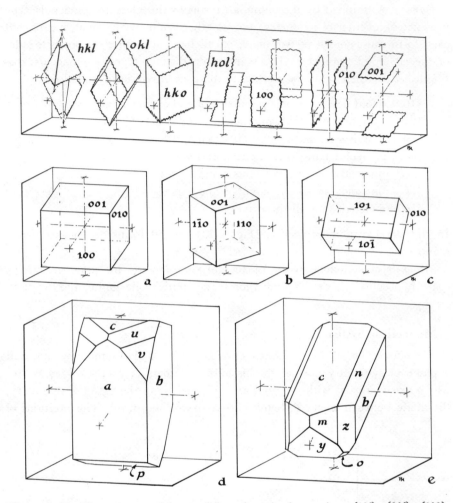

Figure 2-15 The type crystal forms of the prismatic class: prisms $\{hkl\}$, $\{0kl\}$, $\{hk0\}$;
pinacoids $\{h0l\}$, $\{100\}$, $\{010\}$, $\{001\}$. (a) A combination of three pinacoids: $\{100\}$,
$\{010\}$, $\{001\}$. (b) A combination of a prism $\{110\}$ and a pinacoid $\{001\}$. (c) A combination of three pinacoids: $\{101\}$, $\{\bar{1}01\}$, $\{010\}$. (d) A crystal of pyroxene with the
following forms: pinacoids $a\{100\}$, $b\{010\}$, $c\{001\}$, $p\{\bar{1}01\}$; prisms $u\{111\}$, $v\{221\}$.
(e) A crystal of orthoclase elongated along a with the following forms: pinacoids $b\{010\}$,
$c\{001\}$, $y\{\bar{2}01\}$; prisms $m\{110\}$, $z\{130\}$, $n\{021\}$, $o\{\bar{1}11\}$.

The axial ratio for a monoclinic crystal is expressed in the form $a:1:c$, the
length b being taken as unity. The axial ratio can be calculated from interfacial
angles by means of the following formulae:

$$a = \frac{\tan(100 \wedge 110)}{\sin \beta} \qquad (1)$$

$$c = \frac{\tan(001 \wedge 011)}{\sin \beta} \qquad (2)$$

or generally

$$a = \frac{\tan(100 \wedge hk0)}{\sin \beta} \times \frac{h}{k}$$

$$c = \frac{\tan(001 \wedge 0kl)}{\sin \beta} \times \frac{l}{k}$$

The derivation of these formulae is illustrated in Fig. 2-16. Axial ratios can be determined by calculation or by graphical construction.

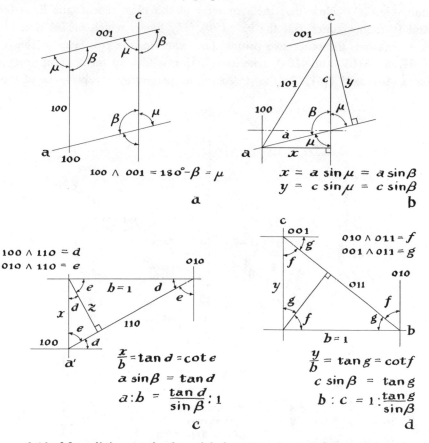

Figure 2-16 Monoclinic crystals, the axial elements $a:b:c$ and β derived from interfacial angles. (a) $\beta = (001) \wedge (\bar{1}00)$ or $(00\bar{1}) \wedge (100)$, $\mu = 180° - \beta = (100) \wedge (001)$ or $(\bar{1}00) \wedge (00\bar{1})$. (b) $x =$ projected length of a normal to c, and $y =$ projected length of c normal to a. (c) $a:b$ from $d = (100) \wedge (110)$ or $e = (010) \wedge (110)$, (d) $b:c$ from $g = (001) \wedge (011)$ or $f = (010) \wedge (011)$.

Forms: There are only two types of forms in this class, pinacoids and prisms. They are as follows:

NAME	NUMBER OF FACES
{001}, basal or *c* pinacoid	2
{010}, side or *b* pinacoid	2
{100}, front or *a* pinacoid	2
{0*kl*} prism	4
{*h*0*l*} pinacoid	2
{*hk*0} prism	4
{*hkl*} prism	4

The different forms are illustrated in Fig. 2-15 and in the stereographic projection, Fig. 2-17. Note that the symmetry of this class repeats the faces (*hk*0) and (0*kl*) four times, but the face (*h*0*l*) only twice, which means that {*h*0*l*} is a pinacoid, the other two prisms. For example, the two faces in the form {101} are (101) and ($\bar{1}$0$\bar{1}$), whereas {10$\bar{1}$} is a distinct form, comprising the faces (10$\bar{1}$) and ($\bar{1}$01). In the stereographic projection the presence of the *ac*

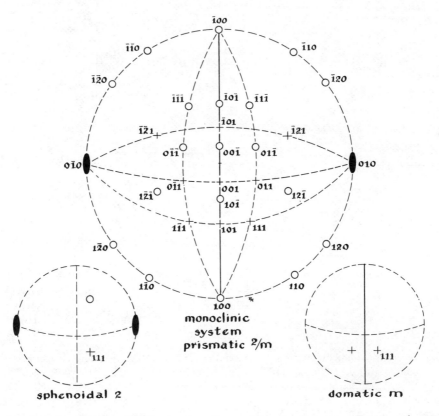

Figure 2-17 Stereographic projection of some forms in the prismatic class 2/*m*. The small stereograms show a sphenoid {111} in the class 2, and a dome {111} in the class *m*.

symmetry plane is indicated by drawing the N-S diameter as a solid line, and the presence of the diad symmetry axis is indicated by the appropriate symbols at the poles of (010) and (0$\bar{1}$0).

The {100} and {$h0l$} pinacoids are sometimes referred to as orthopinacoids, since they are parallel to b, which has been called the ortho-axis; similarly, the {010} pinacoid is sometimes referred to as a clinopinacoid, being parallel to a, the clino-axis.

More minerals—more than 300 in all—crystallize in the monoclinic prismatic class than in any other single class. Some of the more common are gypsum, orthoclase, hornblende, augite, chlorite, the micas, the clay minerals, epidote, heulandite, talc, pyrophyllite, malachite, azurite, borax, sphene, realgar, and orpiment.

The sphenoidal class, which has a single diad axis but no plane or center of symmetry, is not represented to any extent in minerals, but is represented by a great many organic compounds. Many of these substances are of scientific (and economic) interest because they rotate the plane of polarized light and are pyroelectric and piezoelectric (Chapter 4). Crystals of these substances (and all substances in classes having neither planes nor a center of symmetry) may show *enantiomorphism*—that is, they may form two varieties of crystals which are mirror images, related to each other as a left-hand glove to a right-hand glove. Tartaric acid, which crystallizes in the sphenoidal class, occupies an important place in the history of science on this account; in 1844 Pasteur showed that right-hand and left-hand crystals of this substance rotated the plane of polarized light in opposite directions, and that this difference of rotation persisted in solutions, thereby indicating a fundamental asymmetry of the tartaric acid molecule.

Orthorhombic System

Axes of reference (*Fig. 2-18*): The dipyramidal class has three diad symmetry axes at right angles to each other, and these are used as the axes of reference. Symmetry requires $\alpha = \beta = \gamma = 90°$. The standard orientation for an orthorhombic crystal is for the unit lengths on the three axes to be $c < a < b$; for

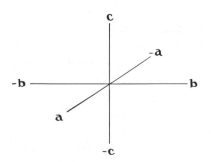

Figure 2-18 Axes of reference in the orthorhombic system.

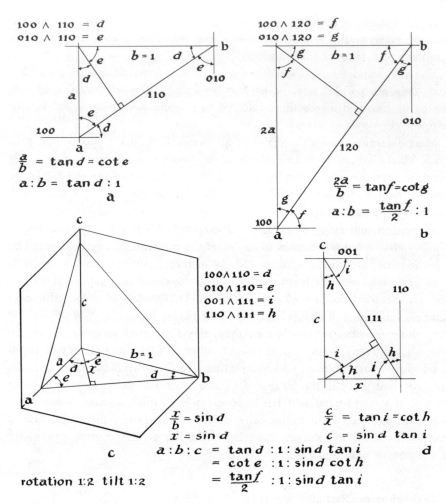

Figure 2-19 Orthorhombic crystals, the axial ratio $a:b:c$ derived from interfacial angles. (a) $a:b$ from $d = (100) \wedge (110)$ or $e = (010) \wedge (110)$. (b) $a:b$ from $f = (100) \wedge (120)$ or $g = (010) \wedge (120)$. (c) and (d) $a:b:c$ from $a:b$ and $i = (001) \wedge (111)$ or $h = (110) \wedge (111)$.

example, the axial ratio of sillimanite is $0.975:1:0.752$. However, this convention is not consistently followed. If an orthorhombic crystal shows a pronounced direction of elongation, this direction is normally chosen as the c axis, and this usually corresponds to the shortest of the three edges of the unit cell. (It is a rather general rule that the direction of elongation in a crystal corresponds to the shortest edge of the unit cell, which suggests that crystal growth is fastest in the direction of the shortest repeat in the structure.)

    The axial ratio for an orthorhombic crystal can be calculated from interfacial angles by the following formulae:

$$a = \tan (100 \wedge 110)$$
$$c = \tan (001 \wedge 011)$$

or generally

$$a = \tan (100 \wedge hk0) \times h/k$$
$$c = \tan (001 \wedge 0kl) \times l/k$$

The derivation of these formulae is illustrated in Fig. 2-19.

Forms The forms in the orthorhombic dipyramidal class are as follows:

FORM	NAME	NUMBER OF FACES
{001}	basal or *c* pinacoid	2
{010}	side or *b* pinacoid	2
{100}	front or *a* pinacoid	2
{0*kl*}	prism	4
{*h*0*l*}	prism	4
{*hk*0}	prism	4
{*hkl*}	dipyramid	8

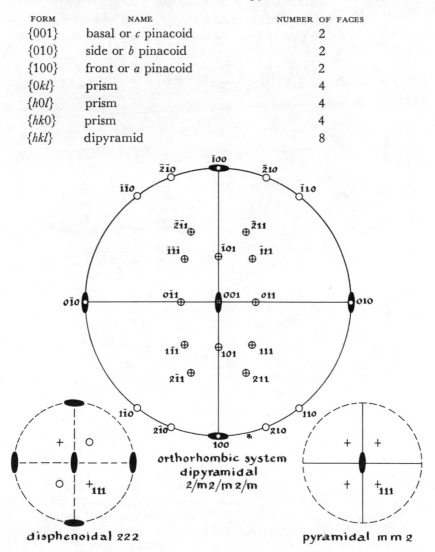

orthorhombic system
dipyramidal
2/m 2/m 2/m

disphenoidal 222

pyramidal m m 2

Figure 2-20 Stereographic projection of some forms in the rhombic dipyramidal class 2/m 2/m 2/m. The small stereograms show the rhombic disphenoid {111} in class 222, and the rhombic pyramid {111} in class *mm*2.

The symmetry is illustrated by the stereographic projection (Fig. 2-20), and the forms are shown in Fig. 2-21. The symmetry requires that the forms {100}, {010}, and {001} consist of two parallel opposite faces, and they are therefore pinacoids. Similarly, the symmetry requires that the forms {0*kl*}, {*h*0*l*}, and

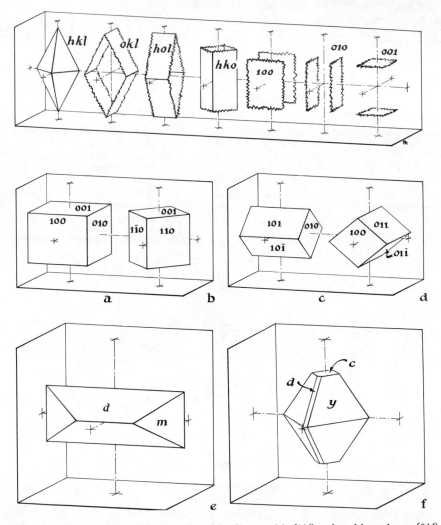

Figure 2-21 Type crystal forms: rhombic dipyramid {*hkl*}; rhombic prisms {0*kl*}, {*h*0*l*}, {*hk*0}; pinacoids {100}, {010}, {001}. (a) A combination of three pinacoids. (b), (c), (d) Combinations of a prism, with one pinacoid. (e), (f) Typical crystals of barite with rhombic prisms *m*{210}, *d*{101}, *o*{011}, basal pinacoid *c*{001}. (See also Figs. 13-1, 13-2).

{hk0} consist of four equivalent faces, or prisms; they are sometimes referred to as rhombic prisms, because the cross section is a rhombus. The form {hkl} comprises eight faces, and is called a dipyramid, because each end of the crystal is terminated by equivalent four-faced pyramids.

The orthorhombic dipyramidal class is second only to the monoclinic prismatic class in the number of minerals it includes. Among the more common are the barite group, the aragonite group, enstatite, anthophyllite, the olivine group, sillimanite, andalusite, sulfur, cordierite, marcasite, columbite, and topaz. The other two classes of the orthorhombic system are represented by relatively few minerals; enargite, hemimorphite, natrolite, and prehnite crystallize in the pyramidal class, epsomite in the disphenoidal class.

Tetragonal System

Axes of reference (Fig. 2-22):   In this system the c axis is fixed as coinciding with the tetrad symmetry axis. The ditetragonal dipyramidal class has four diad

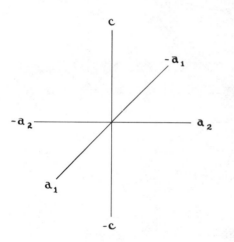

Figure 2-22 Axes of reference in the tetragonal system.

axes normal to the tetrad axis, at 45° to each other; two of these, at right angles to each other, are selected as the horizontal axes of reference. In the tetragonal system the unit distances on the horizontal axes are required by symmetry to be equal, and these axes are therefore referred to as a_1 (front and back) and a_2 (side, or left and right), rather than as a and b, as in the systems with lower symmetry. Of the five planes of symmetry four are vertical (each containing the tetrad axis and one of diad axes) and one horizontal (the plane of the diad axes). The symmetry is illustrated in the stereographic projection (Fig. 2-23). The determination of the axial ratio is illustrated in Fig. 2-24.

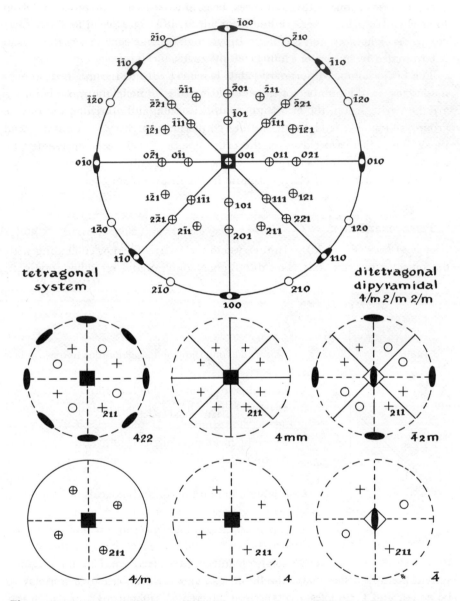

Figure 2-23 Stereographic projection of some forms in the ditetragonal dipyramidal class $4/m\ 2/m\ 2/m$. The small stereograms show one general form $\{211\}$ in classes 422, $4mm$, $\bar{4}2m$, $4/m$, $\bar{4}$, 4.

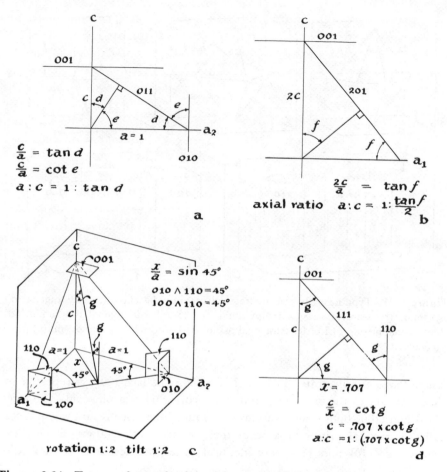

Figure 2-24 Tetragonal crystals, the axial ratio $a:c$ derived from interfacial angles. (a) From $d = (001) \wedge (011)$ or $e = (010) \wedge (011)$. (b) From $f = (101) \wedge (201)$. (c) and (d) From $g = (110) \wedge (111)$.

Forms: The forms in the ditetragonal dipyramidal class are as follows:

FORM	NAME	NUMBER OF FACES
$\{001\}$	basal or c pinacoid	2
$\{110\}$	I-order prism	4
$\{100\}$	II-order prism	4
$\{hk0\}$	ditetragonal prism	8
$\{hhl\}$	I-order dipyramid	8
$\{h0l\}$	II-order dipyramid	8
$\{hkl\}$	ditetragonal dipyramid	16

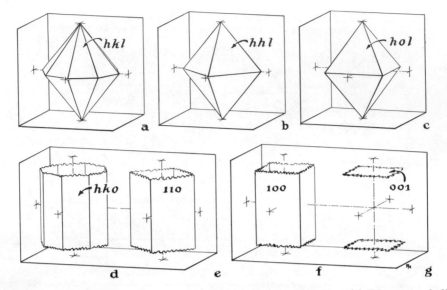

Figure 2-25 Type forms in the ditetragonal dipyramidal class. (a) Ditetragonal dipyramid, (b) I-order tetragonal dipyramid, (c) II-order tetragonal dipyramid, (d) ditetragonal prism, (e) I-order tetragonal prism, (f) II-order tetragonal prism, (g) basal pinacoid.

These forms are illustrated in Figs. 2-25, 2-26. The face (001) is repeated by the symmetry in the face (00$\bar{1}$), thus the form {001} is a pinacoid. The faces (110) and (100) both result in forms with four symmetrically equivalent faces, called tetragonal prisms because of their square cross sections. Since these two forms {110} and {100} are identical in appearance, it is usual to dis-

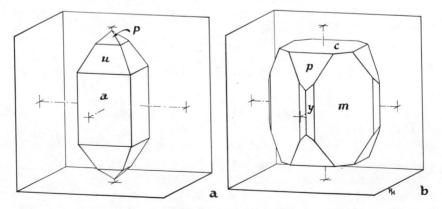

Figure 2-26 (a) Typical zircon crystal with a{100}, p{101}, u{301}. (b) Typical crystal of apophyllite with c{001}, p{101}, m{110}, y{310}.

tinguish them by referring to them as I-order and II-order, respectively. Alternatively, the indices clearly define each form. The face (210) requires seven other equivalent faces, resulting in a ditetragonal prism {210}, or generally {$hk0$}. The ditetragonal prism is not a regular octagon in cross section. Alternate angles are equal, but adjacent angles are unequal; for example, in apophyllite $(3\bar{1}0) \wedge (310) = 36°52'$, $(310) \wedge (130) = 53°08'$, $(130) \wedge (\bar{1}30) = 36°52'$. The face (111) results in the form {111} with eight faces, a tetragonal dipyramid. Any face with indices of the type (hhl) is also a tetragonal dipyramid, differing from (111) only in the slope of the pyramid face. The face (101) or any face ($h0l$) also results in a tetragonal dipyramid; it can be referred to as a II-order dipyramid to distinguish it from (hhl), a I-order dipyramid. The form {hkl} is a ditetragonal dipyramid, with sixteen faces; any cross section perpendicular to c is ditetragonal, similar to the ditetragonal prism {$hk0$}.

Important minerals crystallizing in the ditetragonal dipyramidal class include apophyllite, autunite, cassiterite, pyrolusite, rutile, idocrase, and zircon.

There are six other classes of lower symmetry in the tetragonal system. Comparatively few minerals crystallize in any of these classes, but some are important and widespread. Scapolite and scheelite crystallize in class $4/m$, chalcopyrite in class $\bar{4}2m$, and wulfenite in class 4. Of these chalcopyrite is the only mineral which commonly shows forms characteristic of its specific class (Fig. 9-9).

Hexagonal System

Axes of reference (Fig. 2-27): It is possible to index hexagonal (and trigonal) crystals on the basis of three axes, as in the other systems, but because of the symmetry a method of indexing on the basis of one vertical and three horizontal

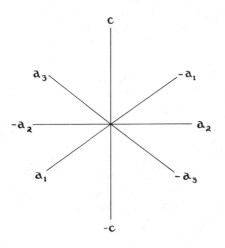

Figure 2-27 Axes of reference in the hexagonal and trigonal systems.

axes was devised by the French crystallographer Bravais, and has been generally adopted. In the hexagonal system the hexad symmetry axis is the c axis; perpendicular to it are the three horizontal axes, at 60° to each other. Since the unit lengths on all three horizontal axes are the same (required by the six-

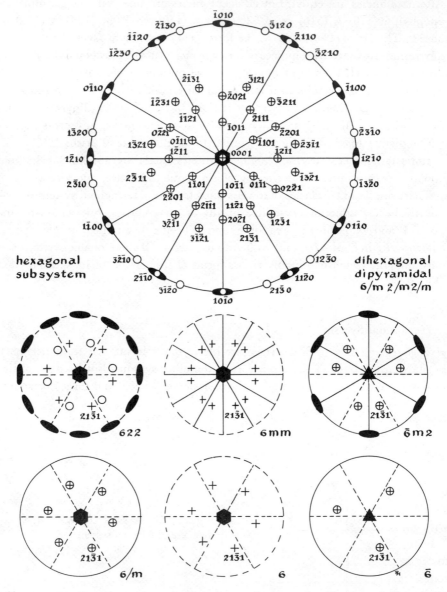

Figure 2-28 Stereographic projection of some forms on the dihexagonal dipyramidal class $6/m\ 2/m\ 2/m$. Small stereograms show one general form $\{21\bar{3}1\}$ for classes 622, 6mm, $\bar{6}m2$, $6/m$, 6, $\bar{6}$.

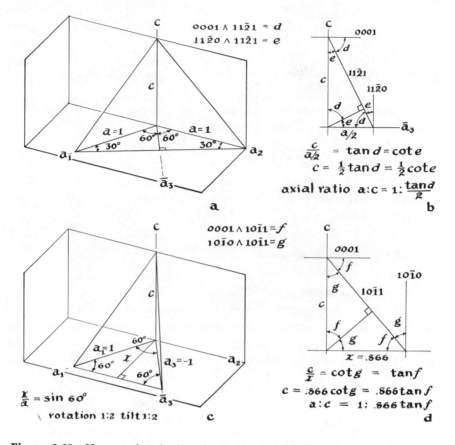

Figure 2-29 Hexagonal and trigonal crystals, the axial ratio $a:c$ derived from interfacial angles. (a) and (b) From $d = (0001) \wedge (11\bar{2}1)$ or $e = (11\bar{2}0) \wedge (11\bar{2}1)$. (c) and (d) From $f = (0001) \wedge (10\bar{1}1)$ or $g = (10\bar{1}0) \wedge (10\bar{1}1)$.

fold axis), they are designated a_1, a_2, a_3. Axis a_2 is placed left and right, and a_1 and a_3 make angles of 30° left and right of a line perpendicular to it. Positive and negative ends of the horizontal axes alternate; a_1 is positive to the front, a_3 negative to the front, a_2 positive to the right. With this convention the general indices of any face are written $(hk\bar{i}l)$, the order of the indices being on a_1, a_2, a_3, and c; $h + k = i$; i.e., the third digit of the index is the sum of the first two with changed sign.

In the dihexagonal dipyramidal class there are six diad axes at right angles to the hexad axis. Three of these, at 60° to each other, are chosen as the horizontal axes of reference. The symmetry of this class is illustrated in the stereographic projection (Fig. 2-28).

The axial ratio for hexagonal and trigonal crystals can be calculated from interfacial angles by means of the following formulae:

$$c = \tan (0001 \wedge 10\bar{1}1) \times \tfrac{1}{2}\sqrt{3} = \tan (0001 \wedge 11\bar{2}2)$$

or generally:

$$c = \tan (0001 \wedge h0\bar{h}l) \times \tfrac{1}{2}\sqrt{3} \times l/h$$
$$= \tan (0001 \wedge hh\bar{2}hl) \times 1/2h$$

The derivation of these formulae is illustrated in Fig. 2-29.

Forms: The forms in this class are as follows:

FORM	NAME	NUMBER OF FACES
{0001}	basal pinacoid	2
{10$\bar{1}$0)	I-order prism	6
{11$\bar{2}$0}	II-order prism	6
{$hk\bar{i}$0}	dihexagonal prism	12
{$h0\bar{h}l$}	I-order dipyramid	12
{$hh\overline{2hl}$}	II-order dipyramid	12
{$hk\bar{i}l$}	dihexagonal dipyramid	24

These forms are illustrated in Fig. 2-30.

The face (0001) has one other equivalent face, (000$\bar{1}$), resulting in a form again referred to as the basal pinacoid. The faces (10$\bar{1}$0) and (11$\bar{2}$0) give rise to hexagonal prisms with six faces each; these are commonly distinguished as I-order {10$\bar{1}$0} and II-order {11$\bar{2}$0}. The faces (10$\bar{1}$1) or any face ($h0\bar{h}l$) will

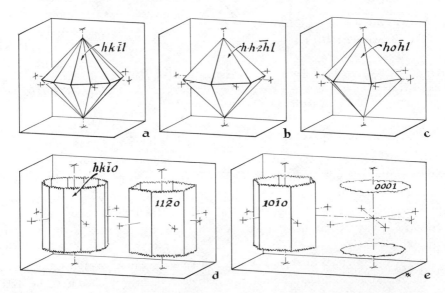

Figure 2-30 Type forms in the dihexagonal dipyramidal class. (a) Dihexagonal dipyramid, (b) II-order hexagonal dipyramid, (c) I-order hexagonal dipyramid, (d) dihexagonal prism and II-order hexagonal prism, (e) I-order hexagonal prism and basal pinacoid.

generate a I-order hexagonal dipyramid with twelve faces, and $(11\bar{2}1)$ or $(hh\bar{2}hl)$ a II-order hexagonal dipyramid. All these prisms and pyramids have a six-sided cross-section. The faces $(21\bar{3}0)$, or $(hk\bar{i}0)$, and $(21\bar{3}1)$, or $(hk\bar{i}l)$, will generate a dihexagonal prism with twelve faces and a dihexagonal dipyramid with 24 faces, respectively. The dihexagonal prism outline in a plane perpendicular to c is twelve-sided, but in this form no combination of whole numbers for h and k' will give a regular dodecagon with interfacial angles of 30° between every pair of faces. In the dihexagonal outline alternate interfacial angles are equal, but not equal to the other set of alternate angles. Beryl, a typical crystal of this class, is shown in Fig. 2-31.

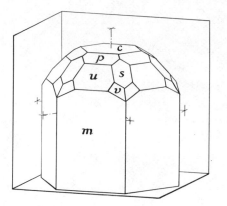

Figure 2-31 Crystal of beryl, dihexagonal dipyramidal, showing the following forms: hexagonal prism $m\{10\bar{1}0\}$; basal pinacoid $c\{0001\}$; hexagonal dipyramids $p\{10\bar{1}2\}$, $u\{10\bar{1}1\}$, $s\{11\bar{2}2\}$; dihexagonal dipyramid $v\{21\bar{3}2\}$.

Important minerals crystallizing in the dihexagonal dipyramidal class include beryl, cancrinite, covellite, graphite, ice, molybdenite, niccolite, and pyrrhotite. Some of the classes of lower symmetry in the hexagonal system also have important mineral representatives. The apatite group crystallizes in class $6/m$, nepheline in class 6, zincite and wurtzite in class $6mm$; however, it is rare to find specimens of these minerals showing the diagnostic forms of the specific class.

Trigonal System

Axes of reference: The trigonal system comprises five classes, all with a single triad axis of symmetry. This system includes both rhombohedral and trigonal crystals. These crystals are indexed on the same set of four axes as hexagonal crystals. The triad axis is c. In the trigonal scalenohedral class the three diad axes, which are perpendicular to c, are used as a_1, a_2, and a_3. The three planes of symmetry in this class are vertical and bisect the angles between the diad axes, as shown on the stereographic projection (Fig. 2-32).

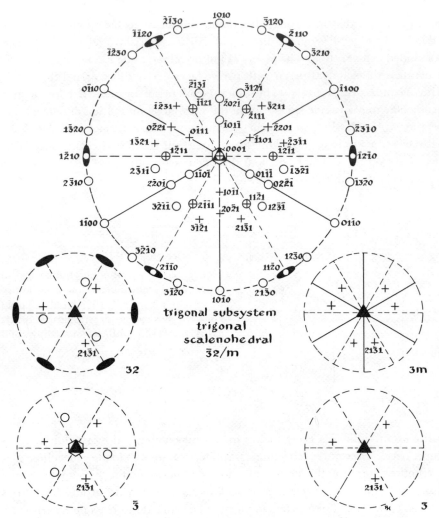

Figure 2-32 Stereographic projection of some forms in the hexagonal scalenohedral class $\bar{3}\,2/m$. The small stereograms show the symmetry, and a general form {$21\bar{3}1$} in each class, 32, 3*m*, $\bar{3}$, 3.

Forms: The forms in this class are as follows:

FORM	NAME	NUMBER OF FACES
{0001}	basal pinacoid	2
{10$\bar{1}$0}	I-order prism	6
{11$\bar{2}$0}	II-order prism	6
{$hk\bar{i}$0}	dihexagonal prism	12
{$h0\bar{h}l$}	rhombohedron	6
{$hh\overline{2h}l$}	hexagonal dipyramid	12
{$hk\bar{i}l$}	scalenohedron	12

These forms are illustrated in Fig. 2-33. It is immediately apparent that many of the forms in this class are the same as those in the dihexagonal dipyramidal class, the only distinctive ones being the rhombohedron and the scalenohedron. In the absence of these distinctive forms on a crystal, it is not possible from morphology to decide whether it belongs to the hexagonal or the trigonal system. For example, corundum is frequently found as hexagonal prisms; however, its trigonal nature is usually manifested by a rhombohedral parting.

The rhombohedron is a form consisting of six rhomb-shaped faces, cor-

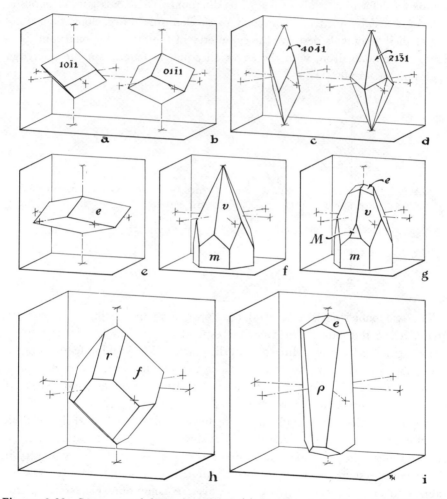

Figure 2-33 Some crystal forms in calcite. (a) Positive rhombohedron $\{10\bar{1}1\}$; (b) negative rhombohedron $\{01\bar{1}1\}$; (c) positive rhombohedron $\{40\bar{4}1\}$; (d) hexagonal scalenohedron $\{21\bar{3}1\}$; (e) negative rhombohedron $e\{01\bar{1}2\}$; (f) combination of hexagonal prism $m\{10\bar{1}0\}$ and trigonal scalenohedron $v\{21\bar{3}1\}$; (g) combination of $m\{10\bar{1}0\}$, $M\{40\bar{4}1\}$, $v\{21\bar{3}1\}$, $e\{01\bar{1}2\}$; (h) combination of $r\{10\bar{1}1\}$ with negative rhombohedron $f\{20\bar{2}1\}$; (i) combination of $e\{01\bar{1}2\}$ and positive rhombohedron $\rho\{16\cdot0\cdot\bar{16}\cdot1\}$.

responding in position and indices to alternate faces of a I-order hexagonal dipyramid. Two sets of rhombohedrons are thus possible with indices corresponding to those of the I-order hexagonal dipyramid, and the form $\{h0\bar{h}l\}$ is distinguished as the positive rhombohedron from $\{0h\bar{h}l\}$, the negative rhombohedron. Frequently both sets of rhombohedrons are present on a single crystal, giving an appearance of a hexagonal pyramid when equally developed; however, usually one set is more strongly developed than the other, so that large and small faces alternate. The rhombohedron can also be derived from a cube by deforming it along one of its diagonals. Diagonals joining opposite corners of a cube are triad symmetry axes. Compression or elongation along one of these diagonals destroys the symmetry of the other diagonals and produces a rhombohedron, with the edges all equal in length but the interedge angles no longer 90°.

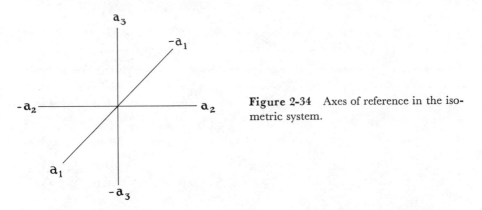

Figure 2-34   Axes of reference in the isometric system.

The scalenohedron is a twelve-faced form, each face being a scalene triangle, hence the name. The faces correspond in position and indices to alternate pairs of faces of a dihexagonal dipyramid. The striking features of the scalenohedron are the zigzag appearance of the middle edges, and its ditrigonal cross section (of the six angles of the cross section three alternate ones are equal, but different from the other three).

Important minerals crystallizing in the hexagonal scalenohedral class include brucite, calcite, magnesite, siderite, rhodochrosite, smithsonite, chabazite, corundum, hematite, and soda niter. Some of the classes of lower symmetry in the trigonal system also have important mineral representatives. Quartz crystallizes in class 32 below 573°C. Crystals are commonly hexagonal prisms capped by positive and negative rhombohedrons, but occasionally show forms characteristic of class 32—trigonal pyramids $\{11\bar{2}1\}$ and trigonal trapezohedrons $\{51\bar{6}1\}$. Since this class has neither planes nor a center of symmetry, quartz is enantiomorphic, occurring in both right- and left-hand crystals

(Fig. 15-6). Tourmaline crystallizes in class $3m$; since this class possesses neither axes nor a plane of symmetry perpendicular to c, crystals of tourmaline, when sufficiently developed, will display different forms at opposite ends of the c axis (Fig. 15-41). The form $\{10\bar{1}0\}$ is a trigonal (three-faced) prism in this class, and this can be observed in the triangular cross section common in tourmaline crystals. Dolomite, ankerite, willemite, and ilmenite crystallize in class $\bar{3}$; the lower symmetry of dolomite and ankerite compared to calcite, and of ilminite compared to hematite, is the result of the cation positions in the structure being occupied by two different elements in an ordered fashion.

Isometric System

Axes of reference (*Fig. 2-34*): All crystals in this system have four triad axes, and either three tetrad or three diad axes at right angles to each other; these tetrad or diad axes are used as the axes of reference. Since symmetry requires that the unit lengths on all three axes be the same, the axes are denoted a_1 (horizontal, front and back), a_2 (horizontal, right and left), and a_3 (vertical).

The class of highest symmetry in the isometric system is the hexoctahedral class, and includes the largest number of minerals in this system. Two other classes, the hextetrahedral and the diploidal, are also well represented among minerals, and will be discussed in detail. Stereographic projections of these classes are given in Fig. 2-35.

Forms: The forms in these three classes are as follows (number of faces in each form is given in brackets):

INDICES	HEXOCTAHEDRAL	HEXTETRAHEDRAL	DIPLOIDAL
$\{100\}$	cube (6)	cube (6)	cube (6)
$\{110\}$	dodecahedron (12)	dodecahedron (12)	dodecahedron (12)
$\{111\}$	octahedron (8)	tetrahedron (4)	octahedron (8)
$\{hk0\}$	tetrahexahedron (24)	tetrahexahedron (24)	pyritohedron (12)
$\{hhl\}$ $h > l$	trisoctahedron (24)	deltohedron (12)	trisoctahedron (24)
$\{hkk\}$ $h > k$	trapezohedron (24)	tristetrahedron (12)	trapezohedron (24)
$\{hkl\}$	hexoctahedron (48)	hextetrahedron (24)	diploid (24)

The forms in the hexoctahedral class are illustrated in Figs. 2-36 and 2-37. The first three—cube or hexahedron, octahedron, and dodecahedron—are familiar geometric figures. The dodecahedron is sometimes referred to as the rhombic dodecahedron, to distinguish it from the pyritohedron in the diploidal class, which is a pentagonal dodecahedron. The tetrahexahedron is so named because it can be derived from a cube (hexahedron) by replacing each cube face with a four-faced pyramid. The distinction between the trisoctahedron and the trapezohedron is obvious in simple crystals, but when present in combinations with other forms they may be confused. Both forms can be considered

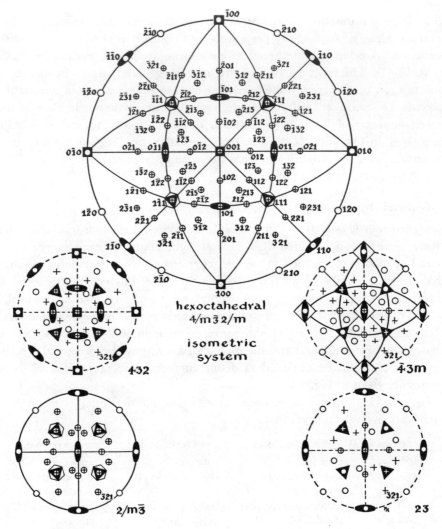

Figure 2-35 Stereographic projection of seven type forms in the hexoctahedral class $4/m\ 3\ 2/m$. The small stereograms show one general form {321} and the forms {100}, {110}, {111} in classes 432, $\bar{4}3m$, $2/m\bar{3}$, and 23.

as derived from an octahedron by the replacement of each octahedron face with three faces. Trisoctahedron faces intersect two axes equally, the third at a *greater* distance, giving indices of the type (221); trapezohedron faces intersect two axes equally, the third at a *lesser* distance, giving indices of the type (211). The hexoctahedron, as the name implies, can be regarded as an octahedron in which each face has been replaced by a group of six faces; note that this is the only fully diagnostic form for this class—all the other forms are present also in classes of lower symmetry.

Many important minerals, and hundreds of inorganic compounds, crystallize in the hexoctahedral class. Among the minerals we find the garnet group, analcime, the spinel group, copper, silver, gold, diamond, galena, halite, sylvite, pentlandite, fluorite, and uraninite. Many metals and alloys crystallize in this class; many artificial compounds with the general formula AX crystallize in the halite structure, and many compounds with the formula AX_2 in the fluorite structure. A large number of synthetic compounds with the spinel structure have been made, and synthetic garnet-type compounds are important industrially.

The forms characteristic of the hextetrahedral class (Fig. 2-38) are the tetrahedron and the forms derived from it—deltohedron and tristetrahedron (three faces replacing each tetrahedron face), and hextetrahedron (six faces replacing each tetrahedron face). It should be noted that the diad symmetry

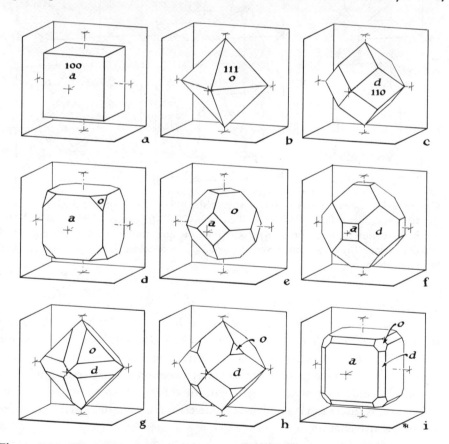

Figure 2-36 The fixed isometric forms: cube $a\{100\}$, dodecahedron $d\{110\}$, octahedron $o\{111\}$, and various combinations of two or three of them, as found in the hexoctahedral class.

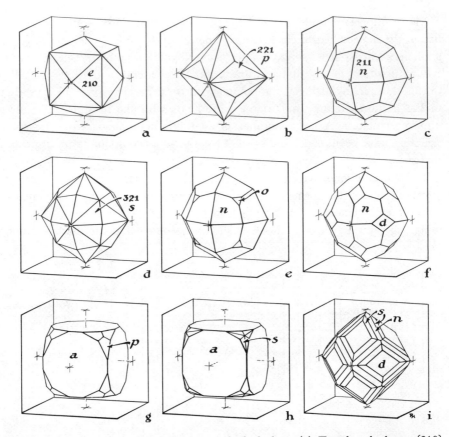

Figure 2-37 Crystal forms of the hexoctahedral class. (a) Tetrahexahedron $e\{210\}$, (b) trisoctahedron $p\{221\}$, (c) trapezohedron $n\{211\}$, (d) hexoctahedron $s\{321\}$, (e) octahedron and trapezohedron, (f) dodecahedron and trapezohedron, (g) cube and trisoctahedron, (h) cube and hexoctahedron, (i) dodecahedron, trapezohedron, and hexoctahedron.

axes in the hextetrahedral class are actually tetrad inversion axes. A tetrahedron can be derived from an octahedron by the development of alternate faces of the latter. Depending on which set of alternate faces is developed, the resultant form is either a positive tetrahedron $\{111\}$ or a negative tetrahedron $\{1\bar{1}1\}$. Sometimes both of these are present on a single crystal, in which case one set of faces is usually larger than the other.

Important minerals crystallizing in this class are sodalite, sphalerite, tetrahedrite, and tennantite.

The diploidal class is represented by the common mineral pyrite, and hence is frequently referred to as the pyritohedral class. The characteristic forms

(Fig. 2-39) are the pyritohedron and the diploid. Each form can be derived from the corresponding form in the hexoctahedral class by development of alternate faces, which provides two possibilities. The form {210} is known as the positive pyritohedron, the form {120} as the negative pyritohedron; similarly the positive diploid is {321}, the negative diploid {231}.

Important minerals in the diploidal class are pyrite and several other sulfides with the general formula AS_2, cobaltite, skutterudite, and sperrylite. Pyrite frequently crystallizes in cubes, apparently indistinguishable from this form in the hexoctahedral class; however, it is often possible to detect the distinctive symmetry of the diploidal class by the striations on the faces, whose directions show a two-fold symmetry around the cube axes (Figs. 9-21, 9-22).

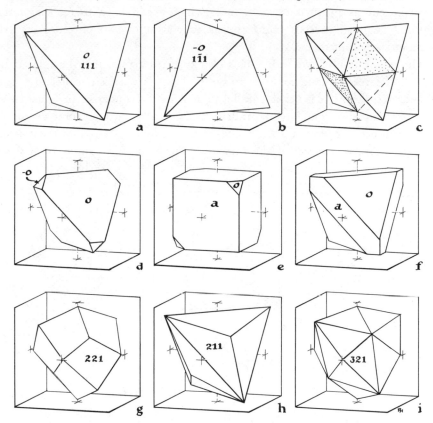

Figure 2-38 Crystal forms of the hextetrahedral class $\bar{4}3m$. (a) Positive tetrahedron $o\{111\}$; (b) negative tetrahedron $-o\{1\bar{1}1\}$; (c) relation of positive tetrahedron to octahedron; (d) positive and negative tetrahedrons; (e), (f) cube and positive tetrahedron; (g) positive deltohedron $\{221\}$; (h) positive tristetrahedron $\{211\}$; (i) positive hextetrahedron $\{321\}$.

TWINNED CRYSTALS

Composite crystals of a single substance, in which the individual parts are related to one another in a definite crystallographic manner, are known as *twinned crystals*. The nature of the relation between the parts of the twinned crystal is expressed in a *twin law*. Twin laws are often given specific names, which are related to the characteristic shape of the twin or to a certain locality where such twin crystals were first found; others are named for the mineral which commonly displays the particular twin law. Many important rock-forming minerals such as orthoclase, microcline, plagioclase, and calcite commonly occur as twinned crystals, and twinning can be a useful diagnostic feature. Twinned crystals often appear to consist of two or more crystals symmetrically united.

The orientation of two individuals of a twinned crystal may be related by reflection across a plane common to both; this *twin plane* is parallel to a possible crystal face, but is never parallel to a plane of symmetry. In other twins one part may appear to have been derived by rotation around some crystallographic direction common to both; the rotation is usually 180°, and the crystallographic direction is called the *twin axis*. The twin axis cannot be a diad,

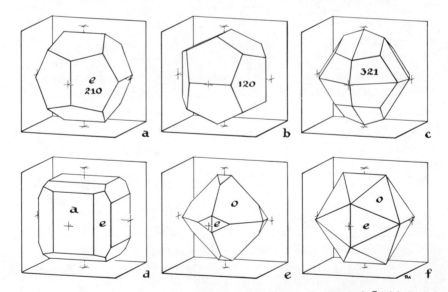

Figure 2-39 Crystal forms of the diploidal (pyritohedral) class $2/m\bar{3}$. (a) Positive pyritohedron $e\{210\}$; (b) negative pyritohedron $\{120\}$; (c) positive diploid $\{321\}$; (d) cube and pyritohedron; (e), (f) octahedron and pyritohedron. Form (f) approximates a regular 20-faced form; o and e faces are not identical in shape.

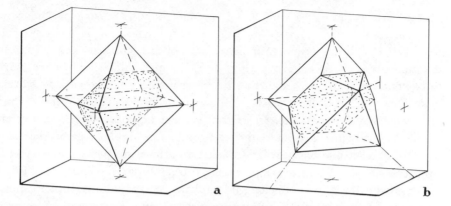

Figure 2-40 (a) Octahedron, showing the plane ($\bar{1}\bar{1}1$), which may act as a twin plane. (b) Octahedron twinned on ($\bar{1}\bar{1}1$) (stippled plane).

tetrad, or hexad axis of symmetry in the single crystal. If two individuals of a twin meet in a plane surface, this surface is known as the *composition plane;* it is commonly, but not invariably, the twin plane.

Twinned crystals are described as *simple twins* if composed of two parts in related orientation; *multiple twins* if more than two orientations are present; *contact twins* if a definite composition plane is present; *penetration twins* if two or more parts of a crystal appear to interpenetrate each other, the surface between the parts being indefinable and irregular. Both contact and penetration twins may be multiple as well as simple. If three or more individuals are repeated alternately on the same twin plane (that is, if all twin planes are parallel in the twinned crystal), the result is a *polysynthetic twin*. If the individuals of the polysynthetic twin are thin plates, the twinning is called lamellar. If successive composition planes are not parallel, a *cyclic twin* results.

On many twin crystals corresponding crystal faces on opposite parts of the twin meet in re-entrant angles, i.e., interior angles greater than 180° (Fig. 2-40b). If the twinning is polysynthetic and lamellar, these re-entrant angles across an edge alternate with a normal angle over the next edge, resulting in striations (Fig. 15-16). The striations on the {001} cleavage of plagioclase are due to lamellar twinning with {010} as the twin plane. Such striations on cleavage faces of calcite (Figs. 12-5, 12-6), dolomite, galena, and sphalerite are also indicative of lamellar twinning.

Examples of Twin Laws

In the remainder of this chapter some examples of twinning in common minerals of different crystal classes are described.

ISOMETRIC SYSTEM. Twinning is commonly found in crystals of the hexoctahedral class ($4/m\ \bar{3}\ 2/m$) with {111} as twin plane or [111] as twin axis
(*Spinel law*). This law occurs in many crystals of this symmetry in addition to
the minerals of the spinel group (Fig. 2-49). Simple contact twins have (111)
as composition plane. In magnetite, twinning is sometimes lamellar, resulting
in striations on {111} faces; such twinning is possibly the cause of the parting
often observed in magnetite. In galena, contact or penetration twins occur
with {111} as twin plane, and lamellar twinning on {114} results in striations
that are often visible on cleavage surfaces. In sphalerite, twins are described,
with reference to [111] as twin axis, as simple or multiple contact twins or interpenetration twins, also as lamellar gliding twins on {111} due to directed pressure. Fluorite twin crystals commonly consist of interpenetrating cubes with
[111] as twin axis (Fig. 11-3b).

In the diploidal class pyrite commonly twins with [110] as twin axis, and
two interpenetrating pyritohedrons result in the iron-cross twin (Fig. 9-23).
In this class the [110] directions are not symmetry axes and therefore may act
as twin axes.

TETRAGONAL SYSTEM. The most common examples of twinning among
crystals in this system are seen in cassiterite and rutile; particularly in the
latter, twinning is quite varied and complex. In almost all cases {011} is the
twin plane. In cassiterite simple contact (Fig. 10-14) or interpenetration twins
are usually found. In rutile repeated twinning on {011} may produce geniculate

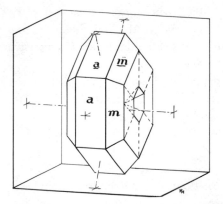

Figure 2-41 Cyclic twin of rutile, twin
plane {101}; forms a{100}, m{110}.

(knee-shaped) forms (Fig. 10-12) or polysynthetic twinning. Repeated twinning on the different equivalent faces of {011} may result in complex sixlings
or eightlings (Fig. 2-41). Twinning is so common that the sharp knee-shaped
offsets in columnar or thin prismatic crystals of rutile may be utilized as a
diagnostic feature for identification.

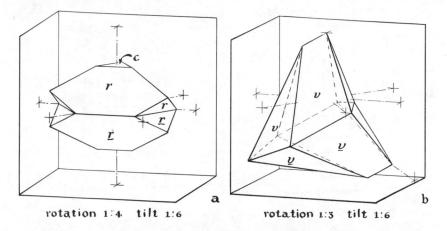

rotation 1:4 tilt 1:6   rotation 1:3 tilt 1:6

Figure 2-42 (a) Calcite twin, twin plane and composition plane (0001). (b) Calcite twin, twin plane and composition plane ($1\bar{1}02$); crystal forms $v\{21\bar{3}1\}$, $r\{10\bar{1}1\}$, $c\{0001\}$.

HEXAGONAL AND TRIGONAL SYSTEMS. Twinning is of rare occurrence in the common minerals of the hexagonal system. However, it is very common in calcite, quartz, and other minerals of the trigonal system.

Calcite forms twin crystals by several laws:

The twin plane is $\{0001\}$ (not a symmetry plane in calcite), and this is also the composition plane (Fig. 2-42a). Re-entrant angles are present about the equator of the crystal except when it is bounded by $\{10\bar{1}0\}$. In that case the twinning may be revealed by cleavage or by the apparent horizontal plane of symmetry.

More commonly the twin plane and composition face are $\{01\bar{1}2\}$ (Fig. 2-42b), sometimes repeated, also lamellar (Fig. 12-6), and produced by pressure, as in marble, or artificially. Two other planes have been recognized as twin planes in rare cases.

Quartz twins in several ways. Two of the more important are Brazil twins and Dauphiné twins. Brazil twins (optical twins), with $\{11\bar{2}0\}$ as twin plane, combine right- and left-handed crystals in a complex penetration twin usually with plane composition surfaces (Fig. 2-43b). Complex twinning of this type renders quartz crystals completely useless for optical or electrical work. This twinning may be detected in polarized light, since the two parts of the twin rotate the plane of polarization oppositely. Dauphiné twins (electrical twins), with c the twin axis, combine two right- or two left-handed individuals with very irregular composition surfaces (Fig. 2-43a). The horizontal striations commonly found on the prism faces of Dauphiné twins are interrupted at an irregular line that marks the boundary between the two crystals (Fig. 2-44a). The positive rhombohedron faces $r\{10\bar{1}1\}$ of one individual and the negative rhombohedron faces $z\{01\bar{1}1\}$ of the other will coincide on a Dauphiné twin,

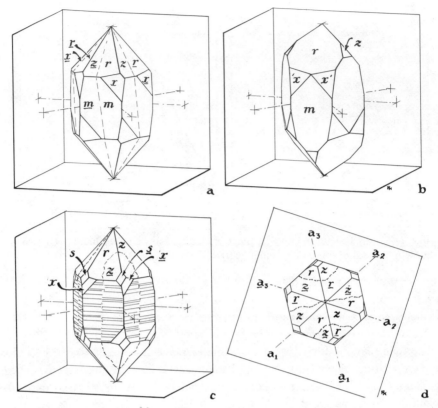

Figure 2-43 Low-temperature quartz. (a) Dauphiné twin, twin axis c, interpenetration of two right-hand crystals. (b) Brazil twin, twin plane $\{11\bar{2}0\}$, interpenetration of a right- and a left-hand crystal. (c) Dauphiné twin, twin axis c, interpenetration of two left-hand crystals; note interruption of striations on prism faces at irregular composition surface and parallelism of r and z. (d) Plan view of crystal in (c); forms: $m\{10\bar{1}0\}$, $r\{10\bar{1}1\}$, $z\{01\bar{1}1\}$, $s\{2\bar{1}\bar{1}1\}$, $x'\{51\bar{6}1\}$, $'x\{6\bar{1}\bar{5}1\}$; underlined letters indicate twin positions.

resulting in an irregular line on most rhombohedron faces that delimits the two parts of the twin. When the trigonal trapezohedron is well-developed the right or left form will be repeated by twinning, giving twelve faces which simulate a hexagonal trapezohedron. This type of twinning also renders crystals useless for electrical work because it reverses the direction of the a axes in the two parts of the twin, and since it combines two crystal orientations with identical optical properties it cannot be recognized in polarized light. It can be recognized in cut plates by etching in hydrofluoric acid or ammonium bifluoride and viewing the etched surface in a spotlight.

In the Japanese law two individuals are combined in a contact twin with $\{11\bar{2}2\}$ as twin plane and composition plane; the c axes intersect at 84° 33′, and one pair of prism faces is common to both parts of the twin (Fig. 15-8a). This twin law is named from the prevalence of these twins at a locality in Kai province, Japan, although such twins have also been recognized at Dauphiné, France.

ORTHORHOMBIC SYSTEM. All minerals of the aragonite group show twinning with {110} as twin plane, resulting in contact and, less often, penetration twins (Fig. 12-11) which are often cyclic with three or more individuals, or polysynthetic. The twinned crystals of the cyclic type simulate hexagonal symmetry. Contact twins, simple or repeated in cyclic fashion, are also found in marcasite and arsenopyrite (Fig. 9-28) with {110} as twin plane. Staurolite commonly forms cruciform twin crystals, which are penetration twins with twin plane {031} (Fig. 15-50) giving a nearly right-angle cross, or with twin plane {231} giving a sawhorse twin with an angle of about 60° (Fig. 2-44).

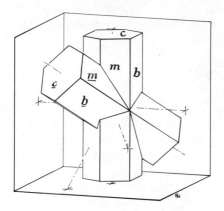

Figure 2-44 Interpenetration twin of staurolite (sawhorse twin), twin plane ($\bar{2}31$); forms $c\{001\}$, $b\{010\}$, $m\{110\}$.

MONOCLINIC SYSTEM. Gypsum, pyroxene, and hornblende commonly form simple contact twins with twin plane {100}. In gypsum these twin crystals are called swallow-tail twins (Figs. 2-45, 13-7). Pyroxene very commonly shows lamellar twinning with {001} as twin plane, resulting in striations on the vertical faces {hk0} and parting on {001} (Fig. 15-37). This lamellar twinning may be produced by shear stress.

Orthoclase crystals may twin according to several laws, the most clearly recognizable being the Carlsbad, Baveno, and Manebach laws. In the Carlsbad law, the c axis is the twin axis, and the twins usually have (010) as composition plane and are interpenetrating (Fig. 15-12c). In these twin crystals {001} (cleavage) of one individual and {$\bar{1}$01} of the second individual very nearly coincide, but they are distinguishable by luster and cleavage (Fig. 15-12). In the Baveno law the twin and composition plane is {021}. Simple contact twins in crystals elongated along a are nearly square prisms, since (001) ∧ (021) = 44° 56$\frac{1}{2}$′

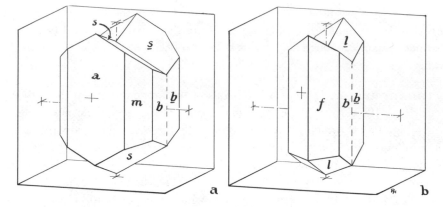

Figure 2-45   (a) Augite twin crystal, twin and composition plane (100); forms $a\{100\}$, $b\{010\}$, $m\{110\}$, $s\{\bar{1}11\}$. (b) Gypsum twin crystal, twin and composition plane (100), forms $b\{010\}$, $f\{120\}$, $l\{\bar{1}11\}$.

(Fig. 15-12d). In the Manebach law the twin and composition plane is $\{001\}$, and the crystals are usually simple contact twins (Fig. 15-12e).

TRICLINIC SYSTEM.   There are no symmetry planes or symmetry axes in triclinic crystals; therefore, any lattice row or lattice plane of simple indices may act as a twin axis or twin plane.

The plagioclase feldspars provide striking examples of twinning in this system, and the two commonest twin laws result from the lower symmetry of plagioclase and microcline compared with monoclinic orthoclase. The *albite law*, with twin plane $\{010\}$, is represented as lamellar twins with composition surface (010), resulting in abundant striations on the $\{001\}$ cleavage of almost all specimens of the plagioclase series (Fig. 15-16). The composition surface is parallel to the other perfect cleavage $\{010\}$ of feldspar. The *pericline law*, with twin axis $b[010]$, also produces lamellar twinning in which the composition surface is a rhombic section parallel to b. The orientation of the rhombic section varies substantially with the composition of the plagioclase; from 21° on one side of the a axis (between positive a and c) through a position parallel to a for andesine to 18° on the other side of a in anorthite (between positive a and negative c). This lamellar twinning, though rarely observed in hand specimens, results in a series of fine striations on $\{010\}$ cleavage which will be inclined to the cleavage edge $\{001\}:\{010\}$ by this small angle, depending on the plagioclase composition. These are the most common types of twinning in plagioclase; they may occur simultaneously, and the simple twinning of the Carlsbad, Baveno, or Manebach laws may be surperimposed on the albite and/or pericline laws.

In microcline the albite and pericline laws combine to give two sets of polysynthetic twin lamellae which intersect at nearly right angles and form a grating, or grid structure. The pericline lamellae are again parallel to *b*, but in microcline are also nearly parallel to *c*; thus the grid twinning is seen in sections which are nearly perpendicular to *c*. This structure is very striking when seen in the polarizing microscope, and can often be detected on the crystal faces {001} and {$\bar{1}$01} with a good hand lens.

Exercises

1. Two lots of crystals of iron disulfide show the following symmetry and interfacial angles:

   2/m $\bar{3}$
   (001) $\wedge$ (102) = 26°34′
   (001) $\wedge$ (111) = 54°44′
   (001) $\wedge$ (112) = 35°16′

   2/m 2/m 2/m
   (010) $\wedge$ (130) = 22°08′
   (101) $\wedge$ ($\bar{1}$01) = 74°38′
   (001) $\wedge$ (111) = 44°34′

   Calculate the axial ratio for each. Plot an accurate stereogram for each, showing all the faces that should be present on well-developed crystals which could yield the above measurements.

2. Crystals of zircon with symmetry 4/m 2/m 2/m show the crystal forms {101}, {110}, {100}, {301}, {211} and the following interfacial angles:

   (100) $\wedge$ (301) = 20°12½′   (011) $\wedge$ (0$\bar{1}$1) = 84°20′

   Plot a stereographic projection of zircon, showing the faces normally expected on a crystal with the above forms. Derive the axial ratio for zircon.

3. Show by means of stereograms the crystal forms developed by the planes (101) and (121) when acted upon by the symmetry of the following classes: (a) orthorhombic system, 2/m 2/m 2/m (b) tetragonal system, 4/m 2/m 2/m, (c) monoclinic system 2/m.

4. Crystals of scapolite (4/m) and of pyroxene (2/m) may be easily confused because both have prismatic cleavage parallel to {110}. What crystallographic features should distinguish them? Contrast the crystal forms {001}, {010}, {111} on each.

5. In an orthorhombic dipyramidal crystal the following angles were measured:

   (110) $\wedge$ (1$\bar{1}$0) = 68°10′   (001) $\wedge$ (102) = 39°30′
   (001) $\wedge$ (011) = 53°40′

   Draw a stereographic projection showing the positions of all the faces of each of the above forms. Determine the axial ratio. Plot the faces (101), (111), (120), and (410), and measure the angles between these faces and (100).

6. In a crystal belonging to the class of highest symmetry in its system the following angles were measured:

   (100) $\wedge$ (110) = 40°30′   (100) $\wedge$ (101) = 19°20′
   (110) $\wedge$ (010) = 49°30′   (101) $\wedge$ (001) = 25°40′

   What system does the crystal belong to? Give the names of the above forms. Draw a stereographic projection showing all the faces of the above forms with positive indices on *c*. Determine the angles (110) $\wedge$ (001), (101) $\wedge$ (111), and (010) $\wedge$ (011).

7. Draw stereographic projections for crystals of enargite, sanidine, cassiterite, calcite, and topaz, using the crystallographic data given in their descriptions.

8. Draw stereographic projections illustrating the symmetry of each of the following crystal classes: diploidal, ditetragonal dipyramidal, prismatic, and pinacoidal. On each projection show the approximate positions of all the faces of the form {210}, and insert the indices of each face thus shown. Give the name of this form for each of these classes.

9. In a crystal the following angles were measured:

(010) ∧ (140) = 14°	(140) ∧ (110) = 31°
(110) ∧ (410) = 31°	(100) ∧ (101) = 38°
(410) ∧ (100) = 14°	(101) ∧ (001) = 52°

Plot these faces on a stereographic projection. Determine the axial ratio. What system does this crystal belong to? Assuming that this crystal belongs to the class of highest symmetry in its system, indicate on the projection the positions of the axes and planes of symmetry. Add to the projection the position and indices of all the faces of each of the above forms with positive indices on *c*. Give the names of these forms. Show the position of (111), and determine the angle (100) ∧ (111).

Selected Readings

Buckley, H. E., 1951, *Crystal growth:* New York, Wiley.

De Jong, W. F., 1959, *General crystallography:* San Francisco, W. H. Freeman and Company.

Holden, A., and P. Singer, 1960, *Crystals and crystal growing:* New York, Doubleday.

Phillips, F. C., 1964, *An introduction to crystallography:* New York, Longmans, Green.

Terpstra, P., and L. W. Codd, 1961, *Crystallometry:* London, Longmans, Green.

Tunell, G., and J. Murdoch, 1959, *Introduction to crystallography:* San Francisco, W. H. Freeman and Company.

Wolfe, C. W., 1953, *Manual for geometrical crystallography:* Ann Arbor, Michigan, Edwards.

Wood, E. A., 1964, *Crystals and light:* Princeton, Van Nostrand. (Momentum Book No. 5.)

3

The Chemistry of Minerals

Mineral chemistry as a science was established in the early years of the nineteenth century, following the proposal of the Law of Constant Composition by Proust in 1799, the enunciation of the atomic theory by Dalton in 1805, and the development of accurate methods of quantitative chemical analysis. The theoretical developments provided a key to interpreting the data resulting from experimental determinations of the proportions of the elements composing each compound.

Because the science of mineral chemistry is based on a knowledge of the composition of minerals, the possibilities and limitations of the chemical analysis of minerals should be understood. A quantitative chemical analysis aims at the identification of the elements present in a substance and the determination of their relative amounts. An analysis should be complete—all the elements in the mineral should be determined—and it should be accurate—the amounts determined should correspond to the amounts actually present. Accuracy depends on the quality of the determinative methods and on the quality of the analyst's work. Even the best methods have some margin of error, though it may be insignificant; and it may truly be said that an analysis is no better than the analyst who makes it.

In the statement of an analysis, amounts are expressed in percentages by weight, and in the complete analysis of a mineral the percentage total should be 100. In practice, as a result of the limitations on accuracy, a summation of exactly 100 is fortuitous; generally a summation between 99.5 and 100.5 is considered a good analysis. However, an analysis that shows a total which lies between these figures may not necessarily be accurate, since the good total may result from a balancing of plus and minus errors, or the analyst may have overlooked or misidentified one or more elements. For example, the rare mineral bavenite, discovered in 1901, was described as a hydrated calcium aluminum silicate, on the basis of an analysis with the good total of 99.72; re-examination thirty years later showed that bavenite also contains beryllium, the beryllium having been overlooked and precipitated and weighed together with aluminum in the original analysis.

INTERPRETATION OF ANALYSES

In our definition of a mineral we said that a mineral has a characteristic chemical composition. A characteristic chemical composition can be expressed by a formula that indicates the elements present in the mineral and the proportions in which they are combined. Thus, the characteristic chemical composition of halite is expressed by the formula NaCl, which indicates that in halite equal numbers of sodium ions and chloride ions are combined; similarly, the formula for brucite, a compound of one magnesium with two hydroxyls, is $Mg(OH)_2$. Formulas may be simple or complex, depending upon the number of elements present and the way in which they are combined. The basic evidence for assigning the correct chemical formula to a mineral is provided by its chemical analysis. However, this may not always be sufficient. The chemical analysis shows what elements are present, and how much of each, but not how they are combined in the structure of the mineral. This last uncertainty is well exemplified by the role of water in an analysis. The water may have been adsorbed in the mineral powder, when it constitutes an impurity, or as water of crystallization, or it may have been formed from hydroxyl groups or hydrogen ions in the structure. As a general rule, small quantities of water lost at low temperatures (below 105°C) are assumed to be adsorbed, but this is not always valid. A correct decision as to how the water shown in the analysis is combined in the mineral may require extensive laboratory investigation. A knowledge of the dimensions of the unit cell and the density of a mineral enables one to check the correctness of a suggested formula, as described later in this chapter.

The results of chemical analyses are expressed in weight percentages, and in order to determine the formula of a mineral these weight percentages must be converted to atomic proportions. This is done by dividing the weight percentage of each element in the mineral by the atomic weight of that element. Here, for a simple example, is the analysis of a specimen of marcasite from Jasper County, Missouri.

	WEIGHT PERCENT	ATOMIC WEIGHT	ATOMIC PROPORTIONS
Fe	46.55	55.85	.834 = 1
S	53.05	32.07	1.654 = 1.988
Total	99.60		

Within the experimental error of the analysis, the formula of marcasite is FeS_2.

The reverse procedure, that of calculating the percentage composition from the formula, is carried out as follows. The formula for marcasite is FeS_2; since the atomic weight of Fe is 55.85, and of sulfur 32.07, the gram-formula weight of marcasite is $55.85 + 2 \times 32.07 = 119.99$. Then

$$\% \text{ Fe} = \frac{55.85}{119.99} \times 100 = 46.54$$

$$\% \text{ S} = \frac{64.14}{119.99} \times 100 = 53.46$$

It is interesting to compare analyses of different specimens of marcasite with the theoretical composition calculated from the formula.

	1	*2*	*3*	*4*	*5*
Fe	46.54	46.55	46.53	47.22	46.56
S	53.46	53.05	53.30	52.61	53.40
	100.00	99.60	99.83	99.83	99.96

1: Calculated for FeS_2; *2:* Jasper County, Missouri; *3:* Joplin, Missouri; *4:* Osnabrück, Germany; *5:* Loughborough Township, Ontario.

All the analyses are good in terms of addition within the limits 99.5 and 100.5, but all deviate slightly from the composition calculated from the formula. These deviations can probably be attributed to the imperfections of analytical procedures and not to any deviation of marcasite from the formula FeS_2.

Within the experimental error, these analyses show that marcasite, whatever its source, has a fixed chemical composition. In this respect marcasite is rather exceptional among minerals. Most minerals are not *fixed* in composition, but they do have a *characteristic* composition, one that can be expressed by a formula. This is illustrated by the following analyses of sphalerite.

	1	*2*	*3*	*4*
Fe	0.15	7.99	11.05	18.25
Mn	—	—	—	2.66
Cd	—	1.23	0.30	0.28
Zn	66.98	57.38	55.89	44.67
S	32.78	32.99	32.63	33.57
	99.91	99.59	99.87	99.43

1: Sonora, Mexico; *2:* Gadoni, Sardinia; *3:* Bodenmais, Germany; *4:* Isère, France.

These analyses show that sphalerite may be nearly pure zinc sulfide or may contain considerable quantities of iron and minor amounts of manganese and cadmium. The situation is clarified when the analyses are recalculated in atomic proportions:

	1	*2*	*3*	*4*
Fe	.003	.143	.198	.327
Mn	—	—	—	.048
Cd	—	.011	.003	.003
Zn	1.026	.879	.856	.684
	1.029	1.033	1.057	1.062
S	1.024	1.032	1.020	1.049

For all the analyses the atomic proportions of metal to sulfur is 1:1, corresponding to the formula ZnS, but with some Zn replaced by Fe, Mn, and Cd. Evidently the formula ZnS is an oversimplification for expressing the composition of sphalerite. Sphalerite may be practically pure ZnS (analysis 1), but it may also have a third or more of the zinc replaced by iron. This phenomenon is common in minerals, and in this example it is indicated by writing the formula as (Zn,Fe)S showing that the total of Zn + Fe is 1 with respect to S = 1, but that the actual amounts of Zn and Fe are variable. When it is desired to express as a formula the composition corresponding to a particular analysis, the atomic proportions of the mutually replacing elements are reduced to decimal fractions of unity. Thus analysis 4 below can be expressed as a formula in this way (omitting the very small amount of cadmium):

	ATOMIC PROPORTIONS	DECIMAL FRACTIONS
Fe	.327	$\dfrac{.327}{1.062} = .31$
Mn	.048	$\dfrac{.048}{1.062} = .05$
Cd	.003	
Zn	.684	$\dfrac{.684}{1.062} = \dfrac{.64}{1.00}$
	1.062	
S	1.049	

Formula: $(Zn_{.64}Mn_{.05}Fe_{.31})S$

The formula $(Zn_{.64}Mn_{.05}Fe_{.31})S$ is a special case of the general formula (Zn,Fe)S, and expresses the composition of a specific analyzed sample of sphalerite.

Another example of this type of relationship is the mineral olivine, for which the following analyses are recorded:

	1	*2*	*3*	*4*
SiO_2	40.99	38.11	33.72	31.85
FeO	8.58	31.48	47.91	58.64
MnO	0.20	0.22	0.41	0.85
MgO	50.00	30.50	18.07	8.49
	99.77	100.31	100.11	99.83

1: Sardinia; *2, 3, 4:* Kangerdlugssuak, Greenland

These analyses illustrate a significant limitation of the analytical technique when it is applied to oxygen-containing compounds. In effect there is no analytical procedure for determining the total amount of oxygen in a compound. As a result, the analyses of such compounds are expressed in terms of the oxides of the individual elements, instead of the elements themselves; and instead of converting weight percentages to atomic proportions they are converted to molecular proportions by dividing the analytical figures by the formula weight of each oxide [e.g., formula weight of $SiO_2 = 28.06 + (2 \times 16.00) = 60.06$].

The above analyses of olivine differ so much one from another that one might well think they represent different minerals. Conversion into molecular proportions, however, shows that they are all variants of the same basic formula.

	1	*2*	*3*	*4*
FeO	.119	.438	.667	.816
MnO	.003	.003	.006	.012
MgO	1.240	.756	.448	.211
	1.362	1.197	1.121	1.039
SiO_2	.680	.635	.561	.530

Despite the wide variation in the amounts of MgO and FeO, the ratio of molecular proportions of FeO + MnO + MgO to SiO_2 is always 2:1, corresponding to the formula $(Mg,Fe)_2SiO_4$. Neglecting the minor quantity of manganese, each analysis can be expressed precisely in the form $(Mg_xFe_{1-x})_2SiO_4$ as follows: *1:* $(Mg_{.91}Fe_{.09})_2SiO_4$; *2:* $(Mg_{.63}Fe_{.37})_2SiO_4$; *3:* $(Mg_{.40}Fe_{.60})_2SiO_4$; *4:* $(Mg_{.21}Fe_{.79})_2SiO_4$.

Another method that is sometimes used for indicating specific compositions in a mineral of variable composition is stating values in terms of percentages of the components. For olivine the components are Mg_2SiO_4 (forsterite, abbreviated Fo) and Fe_2SiO_4 (fayalite, abbreviated Fa). In this terminology the compositions corresponding to the above analyses can be expressed in the following way: *1:* $Fo_{91}Fa_9$; *2:* $Fo_{63}Fa_{37}$; *3:* $Fo_{40}Fa_{60}$; *4:* $Fo_{21}Fa_{79}$. Since the sum of the percentages must be 100, one component in these expressions can be omitted without creating any ambiguity; thus "olivine (Fo_{15})" indicates olivine with 15 percent Mg_2SiO_4 and 85 percent Fe_2SiO_4 in its composition. These are molecular percentages, and the composition by weight of such an olivine can be calculated as follows. The gram-formula weight of Mg_2SiO_4 contains 80.64 grams of MgO and 60.06 grams of SiO_2; that of Fe_2SiO_4 contains 143.70 grams of FeO and 60.06 grams of SiO_2. Hence the gram-formula weight of olivine (Fo_{15}) contains

			WEIGHT PERCENT
MgO	$80.64 \times 0.15 =$	12.10	6.23
FeO	$143.70 \times 0.85 =$	122.14	62.86
SiO_2	$60.06 \times 1 \quad =$	60.06	30.91
		194.30	100.00

Densities for these analyzed olivines are *1:* 3.35; *2:* 3.69; *3:* 3.88; *4:* 4.16. There is evidently a direct relationship between density and iron content, the density increasing with an increase in iron. This relationship can be most clearly shown by a graph of density versus composition, as given in Fig. 3-1. Plotting Fa content on the abscissa and density on the ordinate results in a straight-line graph which shows that the density of an olivine is directly related to its composition. With this graph it is therefore possible to deduce the composition of a particular specimen if its density is known, or its density if its composition is known. For example, from the graph it can be seen that the densities of Mg_2SiO_4 and of Fe_2SiO_4 are 3.22 and 4.41, respectively; these values agree with those determined for the pure compounds.

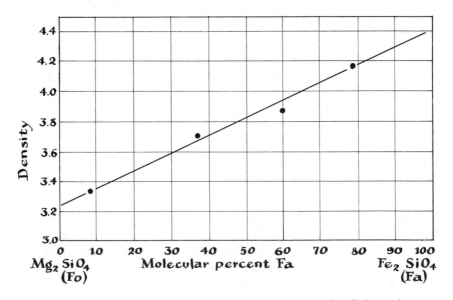

Figure 3-1 Relationship between density and composition in the olivine series.

Such graphs that correlate composition with physical properties are available for many minerals of variable composition, and they are widely used for a rapid determination of composition without the necessity of chemical analysis. A practical example of their usefulness is the determination of niobium and tantalum content in minerals of the columbite-tantalite series from a density measurement; a chemical analysis for niobium and tantalum is difficult, time consuming, and requires a well-equipped laboratory, whereas a density measurement can be made rapidly with simple equipment. For this reason density measurements are especially useful for a quick estimate of niobium and tantalum in samples of columbite and tantalite concentrates.

CHEMICAL COMPOSITION AND UNIT CELL CONTENT

The unit cell of any substance will contain one or an integral multiple of chemical formula units. In the structure of most minerals and inorganic substances the formula unit is not distinguishable as a molecule. In a few minerals— sulfur with S_8 ring molecules—and in most organic substances in the solid state, separate molecules can be recognized. If the dimensions of the unit cell (determined from x-ray measurements) and the density of the substance are known, the unit cell content can be calculated. The dimensions of the unit cell give its volume V as follows.

Isometric:	$V = a^3$
Tetragonal:	$V = a^2c$
Hexagonal and Trigonal:	$V = a^2c \sin 60°$
Orthorhombic:	$V = abc$
Monoclinic:	$V = abc \sin \beta$
Triclinic:	$V = abc (1 - \cos^2 \alpha - \cos^2 \beta - \cos^2 \gamma + 2 \cos \alpha \cos \beta \cos \gamma)^{1/2}$

The total weight (M) of atoms in the unit cell is given by $M = V \times G$, where G is the density. The dimensions of the unit cell are given in Ångstrom units $(1 \text{ Å} = 10^{-8} \text{ cm})$, and to obtain M in grams V in Å^3 must be multiplied by 10^{-24}; i.e. $M = (V \times G \times 10^{-24})$ grams.

Knowing the cell weight M, the number of atoms in the unit cell can be calculated from the chemical analysis. If a mineral contains $P\%$ of an element X of atomic weight N, the weight of X in the unit cell must be $(PM)/100$. The actual weight in grams of an atom of X is $N \times 1.6602 \times 10^{-24}$ (1.6602×10^{-24} is the weight in grams of a hypothetical atom of atomic weight 1.0000). The number of atoms of X in the unit cell is thus

$$\frac{PM}{100} \times \frac{1}{N} \times \frac{1}{1.6602 \times 10^{-24}} = \frac{PVG}{166.02 \ N}$$

For a chemical analysis expressed in percentages of the elements, the number of atoms of each element in the unit cell is obtained by dividing the percentage of the element by the atomic weight—i.e., converting the analysis to atomic proportions—and then multiplying each atomic proportion by the factor $(VG)/166.02$. Since chemical analyses do not generally add up exactly to 100, the figure given for each element is not the true percentage. The analysis figures must be converted to a total of 100.

The procedure is exemplified by the recalculation of analysis 4 of sphalerite on p. 67.

	1	*2*	*3*	*4*	
Fe	18.25	18.36	0.3287	1.23	
Mn	2.66	2.68	0.0488	0.18	4.07 ∼ 4
Cd	0.28	0.28	0.0025	0.09	
Zn	44.67	44.92	0.6875	2.57	
S	33.57	33.76	1.0530	3.93	3.93 ∼ 4
	99.43	100.00			

1: Chemical analysis; *2:* Chemical analysis converted to a total of 100; *3:* Atomic proportions P/N; *4:* Atoms in unit cell [atomic proportions P/N multiplied by $(VG)/166.02$; the unit cell edge being 5.41 Å, and the density 3.92].

In the unit cell of sphalerite there are thus four metal atoms and four sulfur atoms, the small deviations from whole numbers being due to errors in the analysis, the cell dimensions, and the density. The unit cell thus contains four units of the usual formula (Zn,Fe)S. The number of formula units in the unit cell is usually designated as Z.

The calculation of unit cell content from an analysis expressed in terms of oxides follows the same general procedure. The only variation is that the weight percentages of the oxides are converted to molecular proportions by dividing by the molecular weights of the specific oxides; these are then calculated into oxide contents per unit cell, and the oxide contents converted into element and oxygen contents according to the proportions in each oxide. This is illustrated by the recalculation of analysis 1 of olivine on p. 169.

	1	*2*	*3*	*4*	*5*		
SiO_2	40.99	41.08	0.6835	4.01	Si	4.01	
FeO	8.58	8.60	0.1197	0.70	Fe	0.70	
MnO	0.20	0.20	0.0028	0.02	Mn	0.02	8.02
MgO	50.00	50.12	1.2430	7.30	Mg	7.30	
	99.77	100.00			O	8.02 + 0.70 + 0.02	
						+ 7.30 = 16.04	

1: Chemical analysis; *2:* Chemical analysis converted to a total of 100; *3:* Molecular proportions; *4:* Oxide contents per unit cell ($V = 4.76 \times 10.21 \times 5.99$, $G = 3.35$); *5:* Atoms per unit cell.

For the unit cell of olivine Z is clearly 4, i.e., there are four units of $(Mg,Fe)_2SiO_4$ in the unit cell.

The above procedure can be used to check the correctness of the formula assigned to a specific mineral or to check the correctness of a measured density.

COMPONENTS AND PHASES

The discussion of the composition of olivine illustrates the use of the term *component*. This is an important concept in mineralogy. The composition of a material system (i.e., a limited amount of material, whether solid, liquid, or gas, or all three together) can be stated in terms of its components. The components of a system are the smallest number of independent chemical entities (elements or compounds) by which any composition in the system may be expressed. In the example of olivine the components are Mg_2SiO_4 and Fe_2SiO_4; this is known as a two-component, or binary, system. Depending on the number of components, we speak of systems being unary, binary, ternary, quaternary, etc.

Another concept of importance is that of a *phase*. A phase is any portion of a system that is physically homogeneous within itself and is mechanically separable from the other portions. For example, if a flask is partly filled with water in which ice is floating, the contents of the flask are a three-phase system comprised of the ice, the liquid water, and the air. Any pure mineral is a single phase; any rock is a system in which the phases are the individual minerals. In any system the number of phases is related to the number of components by the *phase rule* (Chapter 5).

Diagrams giving the relationship between the components, the phases, and the physical properties of the phases are widely used in mineralogy. A simple example has already been given in Fig. 3-1, which relates density and composition of the phase olivine in the two-component system Mg_2SiO_4–Fe_2SiO_4. In a binary system we can represent any specific composition as a point on the horizontal axis of such a diagram, and any other property along the vertical axis. In ternary systems we can represent any composition as a point in an equilateral triangle (Fig. 3-2). Each apex represents 100 percent of one component; each edge represents the binary system of the components at the ends; any point within the triangle represents a specific composition containing all three components. Physical properties can be shown on such a diagram, either by distances above the plane of the triangle in a solid model or by representing these distances on the plane in the form of contours.

The distinction between a component and a phase is fundamental and must be clearly understood. Unfortunately it has become the practice in mineralogy to use mineral names for components—for example, the Mg_2SiO_4–Fe_2SiO_4 system is often referred to as the forsterite-fayalite system. In this usage forsterite

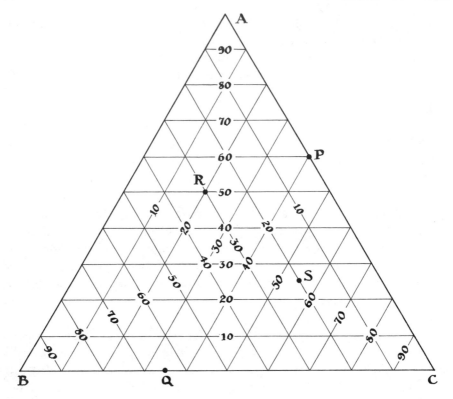

Figure 3-2 The representation of compositions in a three-component system. A represents 100% of component A, B 100% of component B, C 100% of component C. P represents 60% of A, 40% of C; Q represents 65% of B, 35% of C; R represents 50% of A, 20% of C, 30% of B; S represents 25% of A, 20% of B, 55% of C.

and fayalite denote components. However, minerals with approximately the compositions Mg_2SiO_4 and Fe_2SiO_4 are known as forsterite and fayalite, respectively; in this usage forsterite and fayalite are phases. Generally the context makes clear which usage is being employed.

Components are sometimes referred to as *end members*, to emphasize that these are the extreme compositions within the system. Often, too, the term molecule is used in the same sense as component; however, this usage should be avoided, since in most minerals molecules do not exist.

PRINCIPLES OF CRYSTAL CHEMISTRY

At this point it is appropriate to discuss the principles that govern the chemical composition of minerals and determine whether a mineral will have a fixed composition or will vary widely in its composition. Since minerals are crystalline, their chemistry is part of the wider science of crystal chemistry. The idea that in a crystalline substance there is an orderly systematic arrangement of atoms was first suggested by Haüy at the end of the eighteenth century. This concept was developed by later workers, and was finally confirmed in 1912 by the discovery of x-ray diffraction by crystals. X-ray diffraction provided the tool which made possible the determination of the arrangement of atoms in a crystal. The atomic arrangement largely determines the properties of a crystalline substance, and is thus a fundamental feature.

The basic unit in all crystal structures is the atom (in this discussion the term *atom* also implies *ion*, an atom carrying an electric charge), which may, however, be associated with other atoms in a group behaving as a single unit in the structure. Atoms are made up of a very small, positively charged nucleus surrounded by one or more shells of electrons, the whole acting as a sphere whose effective radius is of the order of 1 Å. The radius depends not only on the nature of the element but also on its state of ionization and the manner in which it is linked to adjacent atoms.

THE BONDING OF ATOMS

The Metallic Bond

This bond type is responsible for the cohesion of a metal. Metals are elements whose atoms readily lose their outer electrons, and the crystal structure of a metal is determined by the packing of the positively charged atoms, the detached electrons being dispersed among the atoms and freely mobile. This electron mobility is responsible for the good electrical and thermal conductivity of metals. In minerals metallic bonding is present in native metals and to some extent in a few sulfides and arsenides.

The Covalent or Homopolar Bond

The most stable configuration for an atom is one in which the outer shell of electrons is completely filled. This is the atomic structure of the inert gases, and it accounts for their practically complete lack of reactivity. One way for this configuration to be achieved is by two or more atoms sharing electrons in their outer shells. For example, the atoms in chlorine gas, Cl_2, are always linked in diatomic molecules; each chlorine atom has seven outer shell electrons, and the stable condition of eight electrons in the outer shell is reached by each atom sharing one electron with another atom:

$$: \overset{..}{\underset{..}{Cl}} \cdot \ + \ \cdot \overset{..}{\underset{..}{Cl}} : \ = \ : \overset{..}{\underset{..}{Cl}} : \overset{..}{\underset{..}{Cl}} :$$

This type of bonding is common in organic compounds but is rare, at least in an unmodified form, in minerals. The best mineralogical example is diamond, in which every carbon atom is surrounded by four other carbon atoms, each sharing one electron with the central atom. This pattern is repeated throughout the structure, and the whole crystal is thus a giant molecule (Fig. 3-7).

The Ionic or Polar Bond

Another way for an atom to achieve a completely filled outer shell of electrons is for it to gain or lose a sufficient number of electrons to reach the configuration of the nearest inert gas. Thus a chlorine atom by adding an additional electron becomes a negatively charged ion with the electron configuration of argon. This type of adjustment clearly requires the presence of an atom that can provide the additional electrons—i.e., an atom that attains a stable configuration by losing electrons. Sodium is an element which by losing one electron becomes a positively charged ion with the electron configuration of neon. Sodium and chlorine therefore combine readily to give a structure of oppositely charged ions bonded by electrostatic attraction. Each ion is surrounded with ions of opposite charge, the number being determined by the relative sizes of these ions; there is no pairing of individual positive and negative ions to give discrete molecules, such as we can distinguish in covalent compounds.

Ionic bonding is common in inorganic compounds and is therefore very important in the structure of minerals. Practically all minerals, except the elements and the sulfides, are ionic compounds.

The van der Waals Bond

This bonding is typically present in crystals of the inert gases. One result of the completely filled outer electron shells of such atoms is an inability to form bonds of the metallic, covalent, or ionic type. Consequently, the attractive forces between the atoms are quite weak, a condition reflected in the very low temperatures and high pressures necessary to condense the inert gases to liquids and solids.

These four types of bonding provide a convenient basis for the classification of crystal structures. It must be realized, however, that although each type has well-defined properties the classification is arbitrary, because the bonding in many compounds may be more or less intermediate. The silicon-oxygen bonds in silica and the silicates are neither purely ionic nor purely covalent but are intermediate in nature. The structure assumed by any solid is such that the whole system of atomic nuclei and electrons tends to arrange itself in a form with minimum energy content.

More than one bond type may occur in a single compound. In sulfur, for example, the atoms are covalently linked in rings of eight atoms—in effect, S_8 molecules—and these molecules are bound together in the crystal by van der Waals linkages. Physical properties such as hardness and mechanical strength are determined by the weakest bonds, which are the first to be disrupted under increasing mechanical or thermal strain. In graphite, for example, the carbon atoms are covalently linked in sheets, and the sheets are linked by van der Waals bonds. The latter are weak and easily disrupted, hence the softness of graphite and its ready cleavability parallel to the sheets of carbon atoms.

According to the nature of the bonding, the structures of ionic compounds may be conveniently grouped into two types, the *isodesmic* and *anisodesmic*. In isodesmic compounds the bonds are all of comparatively equal strength; these compounds comprise the oxides, hydroxides, and simple halides. In anisodesmic compounds there is a pronounced difference in the strength of different bonds in the structure. This results in the presence of discrete groups of atoms within the structure. Such groups are the oxyacid anions, such as CO_3^{-2}, SO_4^{-2}, PO_4^{-3}; in these groups the bonding within the groups is stronger than the bonding to the external cations, and the group acts as a unit in the structure.

THE SIZES OF IONS

Among minerals ionic structures are dominant. The total number of mineral species is about 2,000, and of these it has been estimated that about 1,800 can be considered as ionic compounds. Thus the structures of most minerals are determined by the numbers and sizes of the specific ions entering into the composition. Ions can be considered as being approximately spherical with a definite radius characteristic for the element in question and the charge on the ion. The sizes of some of the commoner ions in minerals are given in the appendix and in Fig. 3-3. It will be noticed that hydrogen is not included in Fig. 3-3. The hydrogen ion has unique properties: it consists of a charged nucleus, the proton, which has no orbital electrons associated with it and so is exceedingly small; it acts rather like a dimensionless center of positive charge. The radius of the OH^{-1} ion is essentially the same as that of the O^{-2} ion; the hydrogen is embedded in the oxygen atom, and the OH group is effectively a sphere.

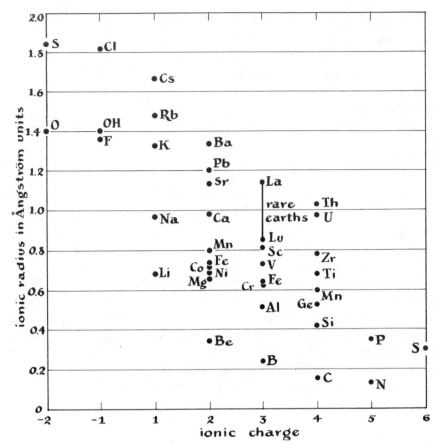

Figure 3-3 Relationship between ionic charge and ionic radius for the commoner elements.

Since the radius of an ion depends upon the atomic structure, it is related to the position of the element in the periodic table (Table 3-I). In this regard the following rules are generally valid.

1. For elements in the same group of the periodic table the ionic radii increase as the atomic number of the elements increases; thus, Be^2: 0.35; Mg^2: 0.66; Ca^2: 0.99; Sr^2: 1.12; Ba^2: 1.34. This is, of course, to be expected, since for elements in the same group of the periodic table the number of electron orbits around the nucleus (and hence the effective radius) increases going down the column.

2. For positive ions of the same electronic structure the radius decreases with increasing charge. As an example we may consider the elements in the second horizontal row in the periodic table, all of which have two electrons in the inner orbit and eight in the outer orbit, thus, Na^1: 0.97; Mg^2: 0.66

Table 3-I   *The periodic table, giving the charge and radius (in Å) of the commoner ions*

I	II	III	IV	V	VI	VII	VIII	VIII	VIII	I	II	III	IV	V	VI	VII	0
H																	He
Li^1 0.68	Be^2 0.35											B^3 0.23	C^4 0.16	N^5 0.13	O^{-2} 1.40	F^- 1.36	Ne
Na^1 0.97	Mg^2 0.66											Al^3 0.51	Si^4 0.42	P^5 0.35	S^6 0.30 / S^{-2} 1.84	Cl^- 1.81	A
K^1 1.33	Ca^2 0.99	Sc^3 0.81	Ti^4 0.68	V^5 0.59 / V^3 0.74	Cr^6 0.52 / Cr^3 0.63	Mn^4 0.60 / Mn^2 0.80	Fe^3 0.64 / Fe^2 0.74	Co^3 0.63 / Co^2 0.72	Ni^2 0.69	Cu^2 0.72 / Cu^1 0.96	Zn^2 0.74	Ga^3 0.62	Ge^4 0.53	As^5 0.46	Se^6 0.42	Br^- 1.95	Kr
Rb^1 1.47	Sr^2 1.12	Y^3 0.92	Zr^4 0.79	Nb^5 0.69	Mo^6 0.62		Ru^4 0.67	Rh^3 0.68	Pd^2 0.80	Ag^1 1.26	Cd^2 0.97	In^3 0.81	Sn^4 0.71	Sb^5 0.62 / Sb^3 0.76	Te^6 0.56	I^- 2.16	Xe
Cs^1 1.67	Ba^2 1.34	La^3 1.14 / Lu^3 0.85	Hf^4 0.78	Ta^5 0.68	W^6 0.62	Re^4 0.72	Os^6 0.69	Ir^4 0.68	Pt^2 0.80	Au^1 1.37	Hg^2 1.10	Tl^1 1.47	Pb^4 0.84 / Pb^2 1.20	Bi^3 0.96			
			Th^4 1.02		U^4 0.97												

Al^3: 0.51; Si^4: 0.42; P^5: 0.35; S^6: 0.30. Thus in going from left to right in rows of the periodic table we find that the radius of the ions, in general, decreases. As electrons are lost the nucleus exerts a greater pull on those remaining, decreasing the effective radius of the ion.

3. For an element that can exist in several valence states—that is, form ions of different charge—the ionic radius decreases with increasing positive charge; e.g., Mn^2: 0.80; Mn^3: 0.66; Mn^4: 0.60. The reason for the previous rule applies here also—the loss of an electron causes the remaining electrons to be more strongly attracted by the nucleus, thus effectively contracting the outer electron orbits and decreasing the ionic radius.

Oxygen is the most abundant element in the earth's crust, and practically all the common minerals are oxygen compounds. Oxygen is therefore the commonest ion, and in this connection the large size of the O^{-2} ion (1.40 Å) is very significant. The earth's crust contains about 47 percent oxygen by weight; however, when weight percentages are converted into volume percentages it is found that oxygen makes up more than 90 percent. In oxygen-containing minerals the amount of oxygen on a volume basis far exceeds all other elements. Thus in oxygen compounds the structure is generally determined by the arrangement of the oxygen ions, the ions of the other elements fitting in the interstices between these large ions.

The structure of an ionic compound is determined by the size of the ions, and the charge on the ions (which is expressed by the valency). In other words, the controlling factors in ionic structures are the demands of geometrical and electrical stability. Geometrical stability implies that the relative ionic sizes and mode of packing must result in the ions being more or less rigidly held in the structure, just as in a house built of blocks each block must support its neighbor. Electrical stability means that the sum of positive and negative charges on the ions must balance—a useful check on the correctness of a formula, especially for complex minerals such as amphiboles and micas.

Table 3-II *Relationship between radius ratio and coordination number for ions acting as rigid spheres*

RADIUS RATIO	ARRANGEMENT OF ANIONS AROUND CATION	COORDINATION NUMBER OF CATION
0.15–0.22	Corners of an equilateral triangle	3
0.22–0.41	Corners of a tetrahedron	4
0.41–0.73	Corners of an octahedron	6
0.73–1	Corners of a cube	8
1	Midpoints of cube edges	12

In an ionic structure each cation tends to surround itself with anions; the number that can be grouped around it will depend upon the relative sizes of the cations and anions. Relative size is most clearly expressed by the *radius ratio*, which is the ratio of the radius of the cation to that of the anion. The number of anions that can fit around each cation is known as the *coordination number* of the cation. Assuming that ions act as rigid spheres of fixed radii, the stable arrangements of cations and anions for particular radius ratios can be calculated from purely geometric considerations (Table 3-II). The different types of coordination are illustrated in Fig. 3-4. Oxygen is the commonest anion, and when the term "coordination number" is used without qualification it refers to the coordination with respect to oxygen. Table 3-III gives the radius ratio and predicted coordination number, with respect to oxygen, for the commoner cations, together with the coordination actually observed in minerals. The close correlation between observation and prediction confirms the assumption that ions do in fact act as spheres of definite radius.

Many cations occur exclusively in a particular coordination. Others—for

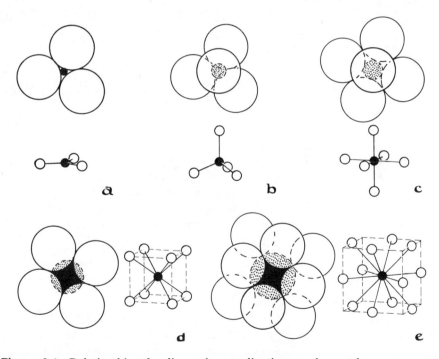

Figure 3-4 Relationship of radius ratio, coordination number, and arrangement of anions around a central cation. (a) Radius ratio 0.15–0.22, coordination number 3. (b) Radius ratio 0.22–0.41, coordination number 4. (c) Radius ratio 0.41–0.73, coordination number 6. (d) Radius ratio 0.73–1, coordination number 8. (c) Radius ratio 1, coordination number 12.

example, aluminum, whose radius ratio lies near the theoretical boundary between *two* types of coordination—may occur in both. The coordination of such cations is to some extent controlled by the temperature and pressure at which crystallization took place. High temperatures and low pressures favor low coordination, and low temperatures and high pressures favor high coordination. Aluminum is a good example; in typical high-temperature minerals

Table 3-III   *Relationship between ionic size and coordination number with oxygen for the common cations*

ION	RADIUS (R)	$R/R_{O^{-2}}$	PREDICTED COORDINATION NUMBER	OBSERVED COORDINATION NUMBER
Cs^1	1.67	1.19	12	12
Rb^1	1.47	1.05	12	8–12
Ba^2	1.34	0.96	8	8–12
K^1	1.33	0.95	8	8–12
Sr^2	1.12	0.80	8	8
Ca^2	0.99	0.71	6	6, 8
Na^1	0.97	0.69	6	6, 8
Mn^2	0.80	0.57	6	6
Fe^2	0.74	0.53	6	6
V^3	0.74	0.53	6	6
Li^1	0.68	0.49	6	6
Ti^4	0.68	0.49	6	6
Mg^2	0.66	0.47	6	6
Fe^3	0.64	0.46	6	6
Cr^3	0.63	0.45	6	6
Al^3	0.51	0.36	4	4, 6
Si^4	0.42	0.30	4	4
P^5	0.35	0.25	4	4
Be^2	0.35	0.25	4	4
S^6	0.30	0.21	4	4
B^3	0.23	0.16	3	3, 4

it tends to assume fourfold coordination and substitute for silicon, whereas in minerals formed at lower temperatures it occurs more often in sixfold coordination.

The foregoing principles are basic for the crystal chemistry of minerals. They express the conditions for low potential energy of the atoms and hence for high stability. Only very stable compounds can occur as minerals; less stable compounds either do not form in nature or soon decompose. Artificial compounds have been made in which these general principles of crystal structure are not closely followed, but such substances are not found as minerals.

ISOMORPHISM

Substances with analogous formulas and in which the relative sizes of cations and anions are similar often have closely related crystal structures; they are then said to be *isomorphous*, and the phenomenon is known as *isomorphism*. As a result of similar internal structures such substances crystallize with similar external forms, and they show the same cleavage. The term isomorphism was introduced in 1819 by Mitscherlich, who prepared crystals of KH_2PO_4, KH_2AsO_4, $(NH_4)H_2PO_4$, and $(NH_4)H_2AsO_4$, and found that they showed the same forms and that the interfacial angles between corresponding faces were very similar. By Mitscherlich's original definition, substances with analogous formulas and similar crystallography were said to be isomorphous. *X*-ray studies have shown that similar crystallography is a reflection of similar internal structure, hence the rewording of his original definition; sometimes the terms *isostructural* or *isotypic* are used instead of isomorphous.

Isomorphism is widespread among minerals and is one of the bases of their classification. Many isomorphous groups are recognized—for example, the spinel group, the garnet group, the amphibole group. The cause of the phenomenon is that anions and cations of the same relative size (i.e., showing the same coordination) and in the same numbers tend to crystallize in the same structure type. This is well-exemplified by some of the carbonate minerals (Table 3-IV). The anhydrous carbonates of the bivalent elements form two isomorphous groups, one orthorhombic and one trigonal. In Table 3-IV the angles between

Table 3-IV   *Isomorphism among the carbonates of bivalent metals*

		ARAGONITE GROUP (ORTHORHOMBIC)	
		$(110) \wedge (1\bar{1}0)$	CATION RADIUS (Å)
$BaCO_3$	(witherite)	62° 38′	1.34
$PbCO_3$	(cerussite)	63° 16′	1.20
$SrCO_3$	(strontianite)	62° 30′	1.12
$CaCO_3$	(aragonite)	63° 48′	0.99
		CALCITE GROUP (TRIGONAL)	
		$(10\bar{1}1) \wedge (\bar{1}101)$	CATION RADIUS (Å)
$CaCO_3$	(calcite)	74° 57′	0.99
$MnCO_3$	(rhodochrosite)	73° 04′	0.80
$FeCO_3$	(siderite)	73° 0′	0.74
$ZnCO_3$	(smithsonite)	72° 12′	0.74
$MgCO_3$	(magnesite)	72° 33′	0.66

two prominent faces are quoted, to show the similarity between isomorphous substances; the angles between other pairs of faces are equally similar. It can be seen that the nature of the structure is determined by the size of the bivalent cation; those minerals with cations larger than calcium crystallize in an orthorhombic structure, those with cations smaller than calcium crystallize in a trigonal structure. Calcium carbonate itself can crystallize in either structure, this phenomenon being known as polymorphism.

Other substances with analogous formulas are isomorphous with these carbonates. Thus soda niter, $NaNO_3$, is isomorphous with calcite, whereas niter, KNO_3, is isomorphous with aragonite, reflecting the similar size of nitrate and carbonate groups and the larger size of potassium ions as compared with sodium ions. Until the development of x-ray techniques for the determination of crystal structures it was somewhat of an enigma why substances as different chemically as calcite and soda niter should show complete similarity in crystal form. Other isomorphous pairs at first sight do not even have analogous formulas. Thus the rare mineral berlinite ($AlPO_4$) is isomorphous with quartz; the true analogy is seen when the formula of quartz is written $SiSiO_4$. Both Al and P are similar in ionic size to Si and hence can exist in a crystal structure in fourfold coordination with oxygen, and thus $AlPO_4$ can crystallize with the same structure as quartz. Similarly, tantalite, $FeTa_2O_6$, is isomorphous with brookite, TiO_2 ($TiTi_2O_6$); the metallic ions are similar in size, and all show sixfold coordination with oxygen.

The important factor in isomorphism, therefore, is similarity in size relations of the different ions, rather than any chemical similarity. This explains many apparently unusual examples of isomorphism, and the absence of isomorphism between chemically similar compounds. Thus, many corresponding calcium and magnesium compounds are not isomorphous, although these elements are similar in chemical behavior; when it is noted that the radius of Ca^2 is 0.99 Å and that of Mg^2 is 0.66 Å it seems natural that the substitution of one for the other without producing a change in structure is improbable.

ATOMIC SUBSTITUTION AND SOLID SOLUTION

We have observed that most minerals are variable in their composition. Substitution of one element by another is the rule rather than the exception. When this phenomenon was first observed it was described in terms of the concept of *solid solution* or *mixed crystals*, which implied the presence, in a single homogeneous crystal, of molecules of two or more substances. For example, common olivine may be described as a solid solution of Mg_2SiO_4(Fo) and Fe_2SiO_4(Fa), and the precise composition of any sample of olivine may be stated in terms of these components, as, for example, $Fo_{85}Fa_{15}$; that is $(Mg_{.85}Fe_{.15})_2SiO_4$. This concept and the related terminology remain in general use, but the light thrown upon the structure of crystals by x-ray investigation has resulted in a revised interpretation. In an ionic structure there are no molecules, the structure being an infinitely extended three-dimensional network. Any ion in the structure may be replaced by another ion of similar radius without causing serious distortion of the structure, just as a bricklayer, running short of red bricks, may incorporate yellow bricks of the same size here and there in his wall. Since minerals usually crystallize from solutions containing many ions other than those essential to the mineral, some foreign ions are often incorporated in the structure.

A solid solution (or mixed crystal) can be simply defined as a homogeneous crystalline solid of variable composition. It was early found that many isomorphous substances have the property of forming solid solutions. There has been a tendency to equate isomorphism and solid solution, in spite of marked inconsistencies. For example, many isomorphous substances show little or no solid solution (e.g., calcite and smithsonite), and extensive solid solution may occur between substances that are not isomorphous (as evidenced by the presence of considerable amounts of iron in sphalerite, although FeS and ZnS have quite different crystal structures). On this account it must be emphasized that isomorphism is neither necessary to nor sufficient for solid solution formation. Isomorphism and solid solution are distinct concepts, and should not be confused.

In atomic substitution the size of the atoms or ions is the governing factor, and it is not essential that the substituting ions have the same charge or valency, provided that electrical neutrality is maintained by concomitant substitution elsewhere in the structure. Thus in the albite $(NaAlSi_3O_8)$–anorthite $(CaAl_2Si_2O_8)$ series, Ca^2 substitutes for Na^1, and electrical neutrality is maintained by the coupled substitution of Al^3 for Si^4; similarly, in diopside $(CaMgSi_2O_6)$, Mg^2–Si^4 may be replaced in part by Al^3–Al^3. Such coupled substitutions are especially common in silicate minerals, and this made the interpretation of their composition exceedingly difficult before the phenomenon was recognized and understood.

As a general rule, little or no atomic substitution takes place when the difference in charge on the ions is greater than 1, even when size is appropriate (e.g., Zr^4 does not substitute for Mn^2, nor does Y^3 replace Na^1, and so on); this may be due in part to the difficulty in balancing the charge requirements by other substitutions.

The extent to which atomic substitution takes place is determined by the nature of the structure, the closeness of correspondence of the ionic radii, and the temperature of formation of the substance. The nature of the structure evidently has considerable influence on the degree of atomic substitution; some structures, such as those of spinel and apatite, are well-known for extensive atomic substitution, whereas others, such as quartz, show very little. To some extent this is due to the lack of foreign ions of suitable size and charge. Ionic size has, of course, a fundamental influence on the degree of substitution, since the substituting ion must be able to occupy the lattice position without causing excessive distortion of the structure. From a study of many mixed crystals it has been found that, provided the radii of substituting and substituted ions do not differ by more than 15 percent, a wide range of substitution may be expected at room temperature. Higher temperatures permit a somewhat greater tolerance; in this respect solid solutions are analogous to solutions of salts in water, solubility increasing with temperature. This property of increased atomic substitution at higher temperatures provides a means of estimating the temperature of mineral deposition (*geological thermometry*). If for a specific mineral the degree of atomic substitution has been determined for different temperatures, the composition of the naturally occurring mineral may indicate the temperature of its formation. Thus, as an example, the amount of iron in solid solution in sphalerite, as a function of temperature, is known from laboratory investigations (Fig. 3-5). Sphalerite is a common ore mineral. Provided the ore-forming solutions contained sufficient iron sulfide to saturate the sphalerite, the iron content of the mineral will indicate the temperature during ore deposition.

Since atomic substitution is generally greater at higher temperatures, it follows that a solid solution formed at high temperature may no longer be stable at lower temperatures. A solid solution in which two different elements, *A* and *B*, are completely interreplaceable at high temperatures but not at lower temperatures will tend to break down on cooling into two separate phases, one rich in *A* and the other rich in *B*. This breakdown of a homogeneous solid solution is known as *exsolution*. For example, in the alkali feldspars potas-

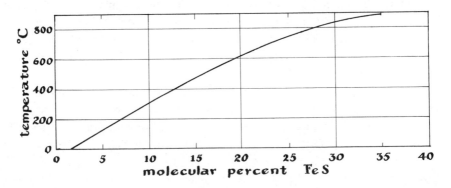

Figure 3-5 Increasing replacement of Zn by Fe in sphalerite with increasing temperature.

sium and sodium are completely interchangeable at high temperatures; for any composition in the system $KAlSi_3O_8$–$NaAlSi_3O_8$ there is a single-phase solid solution $(K,Na)AlSi_3O_8$. At ordinary temperatures the degree of mutual replacement of Na and K in feldspar is quite small. Solid solutions of intermediate composition in this system generally break down, on cooling, into an intergrowth of sodium-rich feldspar and potassium-rich feldspar known as perthite (Fig. 15-15).

The consequence of atomic substitution is that most minerals contain not only the elements characteristic of the particular species but also other elements able to fit into the crystal lattice. For instance, dolomite is theoretically a simple carbonate of magnesium and calcium, but dolomites are found whose analyses show a considerable content of iron and manganese. Traditionally these dolomites were described as solid solutions of the carbonates of all these elements, but it is more illuminating as well as more correct to consider them as products of the substitution of iron and manganese for magnesium. Nevertheless we continue to use the traditional terms solid solution, mixed crystals, and solid solution series, since the terminology of atomic substitution has not yet provided expressions to take their place. The useful term *diadochy* has been introduced to describe the ability of different elements to occupy the same lattice position in a crystal; thus Mg, Fe, and Mn are diadochic in the structure of dolomite. The concept of diadochy, if used rigorously, always applies to a particular structure; two elements may be diadochic in one mineral and not in another.

In interpreting chemical analyses of minerals and deriving their formulas, due consideration must be given to the effects of atomic substitution. In some minerals, such as olivine, substitution is comparatively simple, one element being replaced by another of the same valency, and is readily elucidated by a study of analyses. The situation becomes progressively more complicated when substitution affects several elements and when coupled substitution with elements of different valency takes place. Additional difficulty arises when one element can play a dual structural role and be present in two different coordinations; this is common in silicates, where aluminum may be present in four-coordination replacing silicon, or in six-coordination replacing cations such as Mg^2 and Fe^3. Interpretation of analyses of such minerals requires care and discrimination as well as a real understanding of the principles of crystal structure and the factors governing atomic substitution. The analysis of a pyroxene in Table 3-V is a typical example. When this analysis is converted to molecular proportions in the usual way there is no obvious relationship between the proportions of the different oxides. Conversion to atomic proportions and the arrangement of the elements in order of ionic size allows the elements to be grouped according to ionic size and coordination number. Aluminum evidently

Table 3-V *Interpretation of a pyroxene analysis*

	WEIGHT (PERCENT)	MOLECULAR PROPORTIONS	ATOMIC PROPORTIONS		ATOMIC CONTENT[†] OF UNIT CELL	
SiO_2	48.40	0.806	Si^4 (0.42 Å)	0.806	7.42	$\left.\begin{array}{l}0.58\end{array}\right\}8.00 = 4 \times 2$
Al_2O_3	3.95	0.039	Al^3(0.51 A)	0.078	0.72	$\left\{\begin{array}{l}0.14\end{array}\right.$
TiO_2	0.27	0.004	Ti^4(0.68 A)	0.004	0.04	$4.06 \sim 4 \times 1$
Fe_2O_3	3.90	0.024	Fe^3(0.64 Å)	0.048	0.44	
MgO	8.92	0.221	Mg^2(0.66 Å)	0.221	2.04	
FeO	10.52	0.146	Fe^2(0.74 Å)	0.146	1.34	
MnO	0.39	0.006	Mn^2(0.80 Å)	0.006	0.06	
Na_2O	0.46	0.007	Na^1(0.97 Å)	0.014	0.13	$3.94 \sim 4 \times 1$
CaO	23.20	0.414	Ca^2(0.99 A)	0.414	3.81	
Total	100.01		O^{-2}(1.40 Å)	2.603*	23.96	$23.96 \sim 4 \times 6$

* O = $(2 \times 0.806) + (3 \times 0.039) + (2 \times 0.004) + (3 \times 0.024) + 0.221 + 0.146$
$+ 0.006 + 0.007 + 0.414 = 2.603$
† Atomic proportions multiplied by $M/100 = VG/1.66 \times 100 = 9.2063$, where
$V = 9.73 \times 8.91 \times 5.25 \times \sin 105° 50'$ and $G = 3.49$, giving $Z = 4$.

occurs in both fourfold and sixfold coordination, and the amount in each is adjusted to give the best fit with the pyroxene formula, which for this specimen can be written $(Ca,Na)(Mg,Fe^2,Fe^3,Mn,Al,Ti)(Al,Si)_2O_6$. The specimen is intermediate in composition between diopside, $CaMgSi_2O_6$, and hedenbergite, $CaFeSi_2O_6$, with minor substitutions of aluminum, ferric iron, manganese, titanium, and sodium.

INTERSTITIAL AND DEFECT SOLID SOLUTION

Crystal structure investigations have revealed two other types of solid solution in addition to that due to atomic substitution. One is known as *interstitial solid solution*, whereby foreign atoms or ions do not replace atoms or ions in the structure but fit into *interstices* of the lattice. This type is very common in metals, which take up hydrogen, carbon, boron, and nitrogen—all small atoms—in interstitial solid solution. If a substance has an open structure, interstitial solid solution may take place even with atoms or ions of a considerable size. Thus cristobalite and tridymite, the high-temperature forms of SiO_2, have been found with an appreciable content of sodium and aluminum; the Al^3 replaces Si^4, and the Na^1 needed to maintain electrical neutrality occupies large openings in the cristobalite and tridymite structures. Similarly, in the amphibole structure there are interstices of the right size to accommodate a sodium ion;

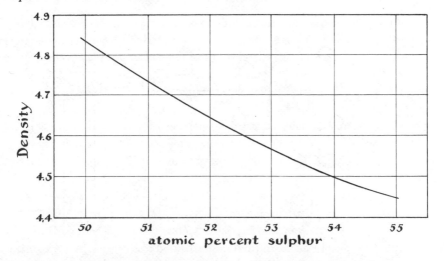

Figure 3-6   Relationship between density and sulfur content in pyrrhotite.

in many amphiboles these interstices are unoccupied, but in hornblende they are partly or completely occupied by sodium.

The other type of solid solution is that associated with *defect lattices*, in which some of the atoms are missing, leaving vacant lattice positions. A good example is the mineral pyrrhotite, whose analyses always show more sulfur than corresponds to the formula FeS. This was, for a long time, described as solid solution of sulfur in FeS. Actually the excess of sulfur shown by analyses is due to the absence of some iron atoms from their places in the lattice; there is a deficiency of Fe, not an excess of S. This was demonstrated by plotting density against sulfur content for different specimens of pyrrhotite (Fig. 3-6); the density decreases regularly with increase in sulfur content, indicating that the cell content must be decreasing in iron rather than increasing in sulfur. Just as in a wall where a brick may be omitted here and there without seriously affecting the stability of the structure, so is it possible to omit some of the Fe atoms in FeS without the lattice collapsing. More and more defect structures are being recognized among minerals, and an explanation is thereby afforded for otherwise puzzling deviations of chemical compositions from those predicted by the Law of Constant Proportions.

POLYMORPHISM

An element or compound that can exist with more than one atomic arrangement is said to be polymorphous. Each arrangement has different physical properties and a distinct crystal structure; that is, the atoms or ions are ar-

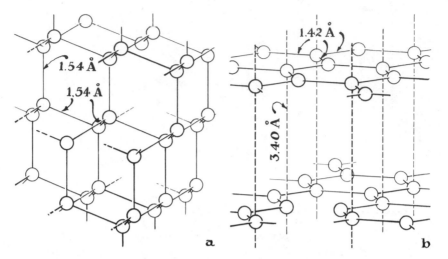

Figure 3-7 Atomic arrangement in (a) diamond, (111) plane horizontal; and (b) graphite, (0001) plane horizontal. Note the differences in the interatomic distances in the two polymorphs; the broken lines in the graphite structure indicate van der Waals linkages, solid lines represent covalent linkages. Small circles indicate locations of centers of carbon atoms and bear no relation to their size.

ranged differently in different polymorphs of the same substance. A polymorphic substance may be described as dimorphic, trimorphic, etc., according to the number of distinct crystalline forms. Polymorphism is an expression of the fact that crystal structure is not exclusively determined by chemical composition, and there is often more than one structure into which the same atoms or ions in the same proportions may be built up. A simple illustration is the relationship between diamond and graphite (Fig. 3-7). In diamond each carbon atom is linked to four other carbon atoms by homopolar bonds, all the linkages being of equal strength and the crystal as a whole a giant molecule. In graphite each carbon atom is linked to three other carbon atoms by homopolar bonds, which results in the formation of planar sheets of carbon atoms; these sheets are joined by weak residual van der Waals forces.

Different polymorphs of the same substance are formed under different conditions of pressure, temperature, and chemical environment; hence the presence of one polymorph in a rock will often tell something about the conditions under which that rock was formed. For example, marcasite is formed from acid solutions at temperatures below 300°C, and the presence of marcasite in a deposit thus puts some limits on the conditions of origin.

Two types of polymorphism are recognized, according to whether the change from one polymorph to another is reversible and takes place at a definite temperature and pressure, or is irreversible and does not take place at a definite temperature and pressure. The first type is known as *enantiotropy*, and is exemplified by the relationship between quartz and tridymite

$$\text{quartz} \xrightleftharpoons{\text{867°C, 1 atm}} \text{tridymite}$$

The second type is known as *monotropy*; an example is the marcasite–pyrite relationship, in which marcasite may invert to pyrite but pyrite does not change to marcasite. With monotropic polymorphs one form is always inherently unstable and the other inherently stable. The unstable form always tends to change into the stable form, but the stable form cannot be changed into the unstable form unless its structure is first completely destroyed by melting, vaporization, or solution.

This distinction between enantiotropic and monotropic polymorphs is useful, but the recognition of monotropic polymorphs is usually based on experimental evidence, and investigation over wide ranges of temperature and pressure or determination of energy relationships of the different polymorphs sometimes indicates that supposedly monotropic polymorphs actually have an enantiotropic relationship under conditions far removed from those usually attainable. Thus studies of the energy relationships between calcite and aragonite (previously thought to be monotropic polymorphs) indicate an enantiotropic transition between them at about −60°C.

The diamond–graphite relationship is particularly interesting in this respect, both from the geological significance of the occurrence of these two polymorphs and from the practical aspect of developing ways to make diamond synthetically. For a long time it was unknown whether diamond and graphite were enantiotropic or monotropic polymorphs; the latter conclusion was favored because under laboratory conditions the transition was always diamond ⟶ graphite, never the reverse. However, it has been established that the relation-

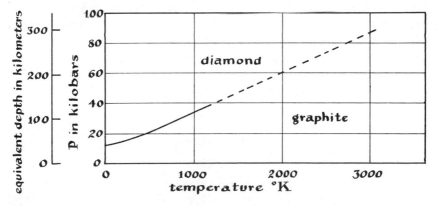

Figure 3-8 Diamond-graphite equilibrium curve, calculated to 1200°K, extrapolated beyond.

ship is enantiotropic, and the actual conditions of the diamond ⇌ graphite equilibrium have been worked out as shown in Fig. 3-8. This figure shows that the practical problem of making diamond synthetically is maintaining pressures within the stability field of diamond at temperatures for which the reaction velocity of its formation is appreciable; this has been achieved by the development of special equipment capable of withstanding great pressures at high temperatures. Figure 3-8 also indicates that the natural occurrence of diamond in igneous rocks implies an origin at considerable depths in the earth, where the combination of temperature and pressure is within the diamond stability field. Diamond is actually unstable under the physical conditions in which it is found (and worn); that it does not change spontaneously into graphite is due solely to the infinitesimal rate of a reaction which the energy relations nevertheless favor.

The rate of change from one polymorph to a more stable one may thus be very slow or very rapid, depending largely on the degree of reconstitution of the structure. Sometimes the change does not involve the breaking of bonds between neighboring atoms or ions but simply their bending, for example, low-quartz ⇌ high-quartz, low-leucite ⇌ high-leucite. Such transformations are practically instantaneous at the transition temperature, and the high-temperature form cannot be preserved at lower temperatures (however, original crystallization as the high-temperature form can often be recognized from the nature of the crystals or from the twinning that so often results from inversions of this type). High-low polymorphs are also characterized by the fact that the high-temperature form has higher symmetry than the corresponding low-temperature form. Transformations other than the high-low type require the breaking of bonds in the structure and the rearrangement of atomic or ionic linkages, are often sluggish, and may require the presence of a solvent in order to attain an appreciable rate of change. These changes have been termed reconstructive transformations, and they are exemplified by the quartz ⇌ tridymite ⇌ cristobalite inversions.

The high-temperature polymorph of a substance generally has a more open structure than a low-temperature form and therefore has lower density. The open character of the structure is dynamically maintained at high temperatures by thermal agitations. It may also be statically maintained by the incorporation of foreign ions into the interstices of the lattice. These foreign ions will buttress the structure and prevent its transformation to a different polymorph when the temperature is lowered. Their complete removal is usually necessary to permit inversion to the close-packed form stable at low temperatures. Thus impure high-temperature polymorphs may be formed far below the normal stability range of pure compounds and may survive indefinitely—a situation that is likely to arise in nature. This phenomenon is probably responsible for the formation and survival of cristobalite and tridymite under conditions in which the stable form of SiO_2 is quartz. As mentioned previously, cristobalite and tridymite have been found with a considerable amount of sodium in interstitial solid solution, and the sodium atoms presumably stabilize the open structure of this mineral. The occurrence of a high-temperature polymorph at ordinary temperatures therefore should not necessarily be interpreted as indicating metastability; the polymorph may be simply a stable impure form.

An interesting transformation that may be considered a variety of polymorphism is the order-disorder type. It has been studied mostly in alloys, since it has important effects on their physical properties, but it is also common in minerals. A simple example is an alloy of 50 percent Cu and 50 percent Zn. Two distinct phases of this alloy exist; in the disordered form the copper atoms and the zinc atoms are randomly distributed over the lattice positions, whereas in the ordered form each element occupies a specific set of positions (Fig. 3-9). The structures of the two forms are related, but the ordered one has lower symmetry than the disordered one. There is no definite transition point between the two forms; perfect order will be achieved only at absolute zero, and with increasing temperature the degree of order gradually decreases to complete disorder above a certain temperature characteristic of the structure and the composition of the crystal. The relationship between microcline and orthoclase is evidently an order-disorder transformation, the one aluminum atom and three silicon atoms in $KAlSi_3O_8$ being disordered in orthoclase but ordered in microcline. This accounts for the monoclinic-triclinic feature of the polymorph-

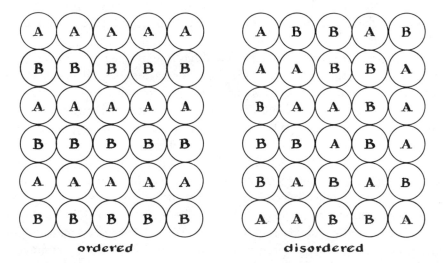

ordered   disordered

Figure 3-9 Theoretical representation in one plane of order-disorder in a binary compound *AB*.

ism, and the typical twinning of microcline is characteristic of the twinning often observed in ordered forms.

Laboratory investigations have shown that most substances show polymorphism if the conditions of temperature and pressure are sufficiently varied. Developments in equipment capable of withstanding both high temperatures and high pressures have led to the discovery of many previously unknown polymorphs. Thus a new polymorph of SiO_2 (coesite) has been made in the laboratory; its density is greater than that of quartz, and it is even more resistant to chemical attack, being insoluble in HF. Such discoveries are of great geological significance, since they suggest that the earth's interior is probably made up of minerals that are unknown at the surface. The earth is solid to a depth of about 2,900 km, and below the crust it is generally believed to consist of magnesium-iron silicates, but at great depths the material probably exists in polymorphic forms different from the olivine and pyroxene of crustal rocks.

PSEUDOMORPHISM

A mineral can be replaced by another mineral without any change in the external form. Such replacements are called *pseudomorphs*, and the phenomenon is known as *pseudomorphism*. There are two types, one in which no change of substance occurs, the other in which there is addition of some element or elements and removal of others. The first type is that observed when one polymorph changes to another without change in external form; this specific type of pseudomorphism is known as *paramorphism*, and the replacing form is a *paramorph* of the replaced form (e.g., paramorphs of calcite after aragonite, of rutile after brookite).

Pseudomorphs in which the later mineral has been formed from the original mineral by a process of chemical alteration may originate by: (1) the loss of a constituent (for example, native copper after cuprite or azurite); (2) the gain of a constituent (gypsum after anhydrite and malachite after cuprite); (3) a partial exchange of constituents [goethite (limonite) after pyrite]; (4) a complete exchange of constituents (quartz after fluorite).

The formation of a pseudomorph implies, of course, that the original mineral was no longer stable under changed physical and chemical conditions and was replaced by another mineral more suited to these conditions. The study of pseudomorphs can therefore provide valuable evidence toward deciphering the geological history of the rock containing them. They may indicate the nature and composition of circulating solutions which added or subtracted certain elements. If the stability fields of the original mineral and of the pseudomorph are known, it may be possible to estimate the temperature and pressure at which the alteration took place.

NONCRYSTALLINE MINERALS

The definition of a mineral usually includes a statement that it is a crystalline solid. There are, however, a few naturally occurring solids that are not crystalline, but which are generally considered as minerals. Two different types may be distinguished: *metamict* minerals, which originally formed as crystalline compounds and in which the crystalline structure has later been destroyed; and *amorphous* minerals, which originally formed in the noncrystalline state either by rapid cooling from the molten state or by slow hardening of gelatinous material.

Metamict minerals may be considered as noncrystalline pseudomorphs after originally crystalline material. They are optically isotropic and do not diffract x-rays, indicating an amorphous condition. They show no cleavage, and are glassy or pitchy in appearance, with a conchoidal fracture. On heating they recrystallize, often with the evolution of so much heat that they become incandescent and glow brightly; recrystallization is accompanied by an increase in density. They are always radioactive from the presence of uranium and/or thorium, although the amount may be quite small—1 percent or less.

The reason for the development of the metamict condition in minerals is the breakdown of the crystal structure through bombardment by alpha particles ejected from the disintegrating radioactive elements. Metamict minerals generally are compounds of weak acids and weak bases, such as zircon, $ZrSiO_4$, and thorite, $ThSiO_4$. The presence of radioactive elements alone is insufficient to induce the metamict state, since thorianite, ThO_2, is apparently never metamict. Some minerals, such as allanite, may be either metamict or nonmetamict.

It has been found that many crystalline substances can be rendered metamict by subjecting them to strong alpha-particle and neutron bombardment in a uranium reactor.

Amorphous minerals include glasses and gels. Glasses may be formed when a melt is rapidly cooled. The only naturally formed glass that can be considered a mineral is lechatelierite, or silica glass, which is often formed in small amounts when lightning strikes in sand. When this occurs the concentrated heat fuses the sand, and the melt immediately solidifies forming a fragile tube of silica glass sometimes several feet long. Such tubes are known as fulgurites.

Gels are formed when colloidal solutions solidify. Colloidal solutions are intermediate between true solutions and suspensions; organic compounds with large molecules often form colloidal solutions, but inorganic compounds, ordinarily insoluble in water, may also form such solutions. The diameter of the particles in a colloidal solution may range from approximately 10^{-3} to 10^{-6} mm. Sometimes these particles are still crystalline, sometimes amorphous. Many colloidal solutions, when they cool or when they lose water, solidify in the form of a gel.

The commonest mineral of this kind is opal, which is formed by the consolidation of colloidal solutions of silica. Opal is hydrated silica with a variable water content, and its formula can be written $SiO_2 \cdot nH_2O$; the amount of water present is usually between 3 and 10 weight percent. Other naturally occurring substances that form colloidal solutions are hydrated aluminum silicates (the clay minerals) and hydrated iron, manganese, and aluminum oxides. Often, however, these substances are crystalline even in the colloidal state and do not form gels on solidification. Even when gels are formed they usually crystallize within a comparatively short time. Minerals that originally solidified as gels can often be recognized by botryoidal surfaces and a fibrous internal structure, the fibers being perpendicular to the surface. Psilomelane and goethite (limonite) frequently appear in this form.

Exercises

1. Calculate the percentage composition by weight, in terms of the constituent oxides, of the following minerals.
   a) ilmenite, $FeTiO_3$
   b) lepidolite, $KLi_2AlS_4O_{10}(OH)_2$
   c) carnotite, $K_2(UO_2)_2(VO_4)_2 \cdot 3H_2O$
   d) cordierite, $(Mg,Fe)_2Al_4Si_5O_{18}$ (Assume equal atomic amounts of Mg and Fe.)
   e) labradorite, $Ab_{50}An_{50}$
2. The formula of pyrrhotite is written $Fe_{1-n}S$. Evaluate n in the following analyses (include atom proportions of Ni and Co with Fe; neglect insoluble).

	1	*2*	*3*	*4*	*5*
Fe	57.49	60.87	61.49	59.91	56.74
Co	1.50	—	—	0.12	0.49
Ni	4.30	—	0.39	0.61	1.10
S	35.71	36.56	37.18	39.69	41.67
Insol.	0.33	2.42	0.46	—	—
	99.33	99.85	99.52	100.33	100.00

3. Determine the formulas and identify the minerals represented by the following analyses.
   a) Cu: 63.3; Fe: 11.1; S: 25.6

b) SiO_2: 43.82; Al_2O_3: 24.60; CaO: 27.33; H_2O: 4.45

c) SiO_2: 37.87; Al_2O_3: 24.13; Fe_2O_3: 12.60; CaO: 23.51; H_2O: 1.89

d) SiO_2: 38.0; Al_2O_3: 21.5; FeO: 26.6; MgO: 6.3; MnO: 7.6

4. Define and discuss the following mineralogical phenomena.
   a) Polymorphism as illustrated by Al_2SiO_5 or TiO_2
   b) Isomorphism as exhibited by the spinel group or the garnet group
   c) Pseudomorphism as exhibited by the common alterations of cuprite, olivine, siderite, and andalusite

5. The following analyses are of different specimens of columbite and tantalite.

	1	*2*	*3*	*4*	*5*
Ta_2O_5	20.00	37.0	44.5	56.3	69.95
Nb_2O_5	60.10	44.2	38.0	26.5	14.47
FeO	6.77	7.1	13.5	1.2	2.68
MnO	13.24	11.2	4.0	15.1	12.54
	100.11	99.5	100.0	99.1	99.64
Density	5.65	6.08	6.33	6.70	7.09

Express each analysis in terms of a formula of the type $(Fe_xMn_{1-x})(Ta_yNb_{1-y})_2O_6$. Plot density against y, and from this plot answer the following questions.
   a) What is the density of $(Fe,Mn)Ta_2O_6$?
   b) What is the density of $(Fe,Mn)Nb_2O_6$?
   c) What is the density of $(Fe,Mn)(Ta_{0.5}Nb_{0.5})_2O_6$?

6. What is the chemical composition of a rock consisting of 40 percent olivine (Fa_{20}) and 60 percent plagioclase (An_{70})?

7. What is the chemical composition of an ore consisting of 17 weight percent quartz, 38 weight percent magnetite, and 45 weight percent ilmenite? (Express in terms of SiO_2, FeO, Fe_2O_3 and TiO_2.)

8. A complex ore consisting of quartz, galena, pyrite, and chalcopyrite gives the following analysis: SiO_2: 12.0; Pb: 28.9; Cu: 13.8; Fe: 19.1; S: 26.3. Calculate the weight percent of each mineral.

9. Give the names and formulas of the minerals that might form in the following systems, and plot their positions in triangular composition diagrams: SiO_2–Al_2O_3–H_2O; Fe–Mn–O; MgO–Al_2O_3–SiO_2; $CaAl_2O_4$–SiO_2–H_2O; MgO–SiO_2–H_2O.

10. Calculate the density of dumortierite from the data given for this mineral.

11. Write equations for the following reactions:
    Muscovite + quartz = sillimanite + orthoclase + water
    Pyrope = spinel + cordierite + forsterite
    Grossularite = anorthite + wollastonite + gehlenite ($Ca_2Al_2SiO_7$)
    Anorthite + enstatite + water = zoisite + chlorite* + quartz
    Tremolite + water + carbon dioxide = serpentine + calcite + quartz
    Chlorite* = forsterite + cordierite + spinel + water

12. Calculate the density of the cordierite of Exercise 1d, given that $Z = 4$ and the unit cell dimensions are $a = 17.13$, $b = 9.80$, $c = 9.35$.

13. The unit cell dimensions of tremolite, $Ca_2Mg_5Si_8O_{22}(OH)_2$, are $a = 9.74$, $b = 17.80$, $c = 5.26$; $\beta = 105°14'$; the density is 2.98. Calculate Z.

* For chlorite use the formula $Mg_5Al_2Si_3O_{10}(OH)_8$.

14. In the system CaO–MgO–Al₂O₃–SiO₂, each of the associations—(1) wollastonite + cordierite, (2) diopside + andalusite + quartz, (3) anorthite + enstatite + quartz —could be equivalent to a single bulk chemical composition. Write equations expressing this equivalence, and calculate the composition.

15. A rock is composed of grossularite, wollastonite, diopside, and quartz; its chemical analysis is SiO_2: 52.75; Al_2O_3: 5.65; Fe_2O_3: 1.23; FeO: 3.71; MgO: 6.33; CaO: 30.20. Calculate the weight percentage of each mineral in the rock. (Assume all FeO is in diopside, all Fe_2O_3 in grossularite.)

16. A rock is composed of orthoclase, plagioclase, olivine, ilmenite, magnetite, and, calcite; its chemical analysis is SiO_2: 53.54; TiO_2: 1.33; Al_2O_3: 16.89; Fe_2O_3; 0.88; FeO: 6.21; MgO: 9.03; CaO: 4.28; Na_2O: 5.60; K_2O: 1.29; CO_2: 0.66. Calculate the weight percentage of each mineral in the rock.

Selected Readings

Bragg, W. L., and G. F. Claringbull, 1965, *Crystal structures of minerals:* New York, Cornell Univ. Press.

Bunn, C. W., 1963, *Chemical crystallography* (2nd ed.): Oxford, Clarendon Press.

Evans, R. C., 1964, *An introduction to crystal chemistry* (3rd ed.): London, Cambridge Univ. Press.

Fyfe, W. S., 1963, *Geochemistry of solids:* New York, McGraw-Hill.

Niggli, P., 1954, *Rocks and mineral deposits:* San Francisco, W. H. Freeman and Company.

Strunz, H., 1966, *Mineralogische Tabellen* (4th ed.): Leipzig, Akademische Verlagsgesellschaft.

Winkler, H. G. F., 1955, *Struktur und Eigenschaften der Kristalle* (2nd ed.): Berlin, Springer-Verlag.

4

The Physics of Minerals

A close connection exists between the physical properties of a mineral, its crystal structure, and its chemical composition. The study of physical properties can thus enable us to make deductions about crystal structure and chemical composition. Physical properties can also be of great technical significance, since a mineral may have important industrial uses that depend on its physical properties alone; for example, the extreme hardness of diamond makes it a highly efficient abrasive, and the piezoelectric nature of quartz is the basis of its use in electronics equipment. Finally, physical properties are of great practical significance, since they provide us with readily determined characteristics for mineral identification; physical properties are, for the most part, more quickly and easily determined than chemical composition or crystal structure and are often uniquely diagnostic of a particular mineral. We must therefore consider the more important physical properties of minerals from three aspects: the scientific, the technical, and the determinative.

DENSITY

The two terms *density* and *specific gravity* are often used interchangeably, although strictly speaking there is a distinction between them. According to the convention in English-speaking countries, the density of a substance is the mass per unit volume, and it is therefore necessary to specify the units used—generally grams per cubic centimeter or pounds per cubic foot. Specific gravity, however, is a pure number—the number of times heavier a body of any volume is than an equal volume of water; in other words, it is the ratio of the density of the substance to the density of water. However, in some countries this convention is reversed; for example, in Germany *Spezifisches Gewicht* is the mass per unit volume, and *Dichte* is the number of times heavier a body is than the same volume of water. In this book the term *density* is used (in the same sense as *Dichte*) in preference to specific gravity, on the grounds of conciseness and convenience.

The density of a substance is primarily determined by its crystal structure and its chemical composition; density will vary somewhat with changes in temperature and pressure, since changes in these factors cause expansion or contraction. Thus the density of a pure substance with a fixed chemical composition and crystallizing in a specific structure should be constant at a stated temperature and pressure. Careful measurements have shown this to be true—quartz, which is practically invariable in composition, has a constant density of 2.65 at ordinary temperature and pressure. The other polymorphs of SiO_2, which crystallize in different structures, have different densities, that of cristobalite being 2.32 and of tridymite 2.26. For a substance of variable composition crystallizing in a specific structure the variation in density will depend essentially on the mass of the individual atoms; for example the density of olivine, $(Mg,Fe)_2SiO_4$, increases with increasing replacement of the light magnesium atoms by the heavier iron from 3.22 for pure Mg_2SiO_4 to 4.41 for pure Fe_2SiO_4. Similarly, in a group of isomorphous compounds the density will show a direct relationship to the mass of the atoms present, as can be seen in the following tabulation for the aragonite group:

	Composition	Density
Aragonite	$CaCO_3$	2.93
Strontianite	$SrCO_3$	3.78
Witherite	$BaCO_3$	4.31
Cerussite	$PbCO_3$	6.58

The density of a substance therefore reflects the nature of the atoms in the structure and the manner in which they are packed together. If the dimensions of the unit cell have been measured, and the number and kinds of atoms in the unit cell are known, it is possible to calculate the density. The procedure was described in Chapter 3, where it was used to determine the number of formula units Z in the unit cell from the cell dimensions, the formula of the substance, and the measured density. The calculation can be used to determine the density, provided Z is known.

A great advantage of the x-ray method of determining densities is that no special care is necessary to prepare the material used, since measurements are not affected by holes or cracks, or by solid, liquid, or gaseous inclusions in the material; also, the amount of material required is extremely small. Densities determined from x-ray measurements provide a most useful independent check on measured densities, and, conversely, measured densities can be used to check the correctness of a proposed structure or chemical formula.

A further application of x-ray crystallography to density and other physical properties is in calculating the factor known as the packing index, which is derived from crystal structure data in the following way:

$$\text{packing index} = \frac{\text{volume of ions}}{\text{volume of unit cell}} \times 10$$

The concept of a packing index assumes that ions behave as spheres which support each other in a crystal structure; however, even with the same ions in the same numbers, different modes of packing are possible, and these have different volume requirements. For ionic compounds, packing indices vary between about 3 and 7; i.e., in actual crystal structures between 30 percent 70 percent of the volume is occupied by the atoms. The relationship between packing index and density is readily seen in the difference between polymorphs of the same substance (Table 4-I).

Table 4-I *Relationship between packing index and density in polymorphs (calculated for ionic radius of $O^{2-} = 1.32$ Å)*

		DENSITY	PACKING INDEX
	rutile	4.25	6.6
TiO_2	brookite	4.14	6.4
	anatase	3.90	6.3
	kyanite	3.63	7.0
Al_2SiO_5	sillimanite	3.24	6.2
	andalusite	3.15	6.0

Determination of Density

The density of a crystalline substance is a fundamental property and is characteristic for that substance. As such it is a valuable diagnostic property, and it should be carefully determined. The accurate determination of the density of a mineral requires considerable care, since there are numerous possibilities of error that must be guarded against. The more serious of these are: errors inherent in the method used, errors arising from inhomogeneity of the sample, and errors arising from the observer himself. Because some determinative methods are inherently more precise than others, it is important to select the technique that is best suited to give accurate results with the material available. The second source of error deserves a great deal more attention than is usually given to it. It is often difficult to obtain large pieces of homogeneous material even of a mineral that is available in considerable amounts, due to inclusions of foreign material. In general, therefore, the best results will be obtained by working with small amounts chosen as carefully as possible (preferably small grains whose purity can be controlled by microscopic examination). Great difficulty arises in working with fine-grained porous materials, such as the clay minerals, since air that is trapped in the pores will give a fictitiously low value for the density. With such materials special precautions must be taken to remove the trapped air; this is usually done by boiling the material in water or in the displacement liquid used. Errors due to the observer himself can be obviated only by care in working.

Of the methods that are available for determining the densities of solids, the following are particularly suitable for minerals.

1. The weight is measured directly, the volume by the Principle of Archimedes.

2. The weight is measured directly, the volume from the weight of liquid displaced in a pycnometer.

3. The density is measured by direct comparison with heavy liquids—the suspension method.

In the first method the volume is determined by measuring the apparent loss of weight when a weighed fragment of the mineral is immersed in a suitable liquid. The fragment displaces an amount of liquid equal to its own volume, and its weight is apparently diminished by the weight of the liquid displaced. If W_1 is the weight of the fragment in air, and W_2 the weight of the fragment in liquid of density L, the density G is

$$G = \frac{W_1}{W_1 - W_2} \times L$$

Water is often used as the displacement liquid because it is readily available, and since its density is 1 or close to 1 the factor L can be eliminated in routine determinations. However, water is not the most suitable liquid for accurate determinations because it has a high surface tension and does not wet solids readily; as a result bubbles are tenaciously held by the solid, thereby giving low figures for the density. Organic liquids of known density and high purity, such as toluene or carbon tetrachloride, are much better in this respect because their surface tension is a third or a quarter that of water.

This method is one of the simplest for determining the density of minerals, and if homogeneous pieces of sufficient size are obtainable it is one of the most accurate. A number of special balances have been developed for direct and rapid determination of densities by the Archimedes method, the best known being the Jolly balance, as improved by Kraus (Fig. 4-1). However, an ordinary chemical balance is quite satisfactory. Torsion balances can readily be adapted for rapid and accurate determinations of density, and an exceedingly useful adaption is the Berman balance, named after its originator, the late Dr. Harry Berman of Harvard University. The Berman balance (Fig. 4-2) is a torsion microbalance weighing up to 25 mg with a precision of 0.01 mg, with an attachment which allows rapid weighing both in air and in a suitable

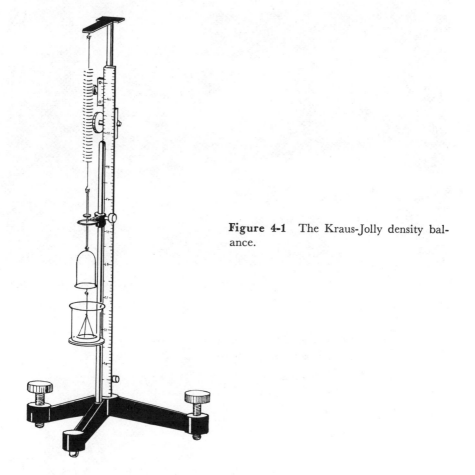

Figure 4-1 The Kraus-Jolly density balance.

liquid. Grains weighing a few milligrams can be controlled for purity by microscopic examination and easily handled with tweezers.

The pycnometer (second method) is simply a small, stoppered glass bottle which holds a definite volume of liquid (determined by measuring the weight of liquid of known density required to fill it). The volume of a known weight of solid is determined from the weight of liquid displaced. If

G = density of the solid

L = density of the liquid used

Figure 4-2   The Berman density balance. [Courtesy Bethlehem Instrument Co., Inc. Bethlehem, Pa.]

W_1 = weight of the pycnometer empty
W_2 = weight of the pycnometer with solid
W_3 = weight of the pycnometer filled with solid and liquid
W_4 = weight of the pycnometer filled with liquid

then

$$G = \frac{L(W_2 - W_1)}{(W_4 - W_1) - (W_3 - W_2)}$$

The pycnometer method requires good technique for accurate results. Care must be taken that air is not trapped in and between the grains of the solid. A serious source of error is often the variation in volume of the pycnometer, which is due to variation in the seating depth of the stopper, which depends on the force with which it is inserted. For accurate results a fair quantity of homogeneous material is required. On the whole it is best to avoid the pycnometer if another technique is available which will yield the accuracy required. However, with certain materials, such as friable fine-grained clays, it may be the only usable method.

The principle of the third method—using heavy liquids—is very simple. The grains of the mineral are immersed in a suitable liquid and it is noticed whether they sink or float. If they float, the liquid is gradually diluted with a miscible liquid of lower density until, when mixing is uniform, they neither sink nor float. If the grains sink initially, the liquid is made more dense by adding a miscible liquid of higher density until equality is reached. The density of the liquid at equality between liquid and solid is determined by one of the standard procedures, generally by using a Westphal balance or by the equivalent procedure of weighing a plumb bob in air, in water, and in the liquid. A particular advantage of the third method is that small grains can be used, and several of them can be compared at the same time; if some are impure this will be immediately revealed by a variation in density from grain to grain.

Suitable liquids for determining the densities of minerals are:

Bromoform, $CHBr_3$; $G = 2.9$

Acetylene tetrabromide (tetrabromethane), $C_2H_2Br_4$; $G = 2.96$

Methylene iodide, CH_2I_2; $G = 3.3$

Clerici solution, a saturated aqueous solution of equal amounts of thallous malonate and thallous formate; $G = 4.2$ (at room temperature)

To vary their densities the organic liquids can be diluted with acetone, Clerici solution with water.

Mineral Separation by Density Differences

Another application of heavy liquids in mineralogy is their use for the separation of individual minerals or groups of minerals from mixtures. This is an important technique in sedimentary petrology, since the so-called heavy minerals in a sediment—those with density greater than the common minerals quartz, the feldspars, calcite, and dolomite—may provide valuable evidence as to the source of the sediments and the conditions under which they were deposited.

Separation of different minerals by differences in density is also an important ore dressing technique for preparing concentrates of the valuable minerals. Sometimes heavy liquids are employed—not the heavy liquids mentioned above, which are far too expensive for large-scale use, but heavy media which are suspensions of finely ground heavy minerals (such as magnetite or galena) in water. More often mechanical devices such as vibrating tables are used to separate the mixture of minerals fed onto the table as a suspension in water.

When dealing with minerals of known density it is possible to determine the composition of a binary mixture from a measurement of its density. Thus if a specimen of vein material consisting of a mixture of x weight percent quartz

$(G = 2.65)$ and $(100 - x)$ weight percent pyrite $(G = 5.01)$ has a density of 3.8, the actual percentages of quartz and pyrite can be determined as follows:

	M	G	V
Pyrite	$100 - x$	5.01	$\dfrac{100 - x}{5.01}$
Quartz	x	2.65	$\dfrac{x}{2.65}$
Vein material	100	3.8	$\dfrac{100}{3.8}$

The sum of the volumes of pyrite and quartz must equal the volume of the mixture; hence

$$\frac{100 - x}{5.01} + \frac{x}{2.65} = \frac{100}{3.8}$$

and $x = 35.8$ percent.

If the compositions of the minerals are known, the composition of the mixture can be calculated. Pyrite contains 46.6 percent Fe and 53.4 percent S; therefore the chemical composition of the above mixture is 35.8 percent SiO_2, 29.9 percent Fe, and 34.3 percent S.

OPTICAL PROPERTIES

The optical properties of minerals comprise a wide variety of phenomena—reflection and refraction, luster, color and streak, and luminescence—and can be dealt with only in a summary fashion in this book.

Reflection and Refraction

When a ray of light in air impinges obliquely on the surface of a nonopaque solid, part of the light is reflected back into the air (the reflected ray) and part enters the solid (the refracted ray) (Fig. 4-3). The direction of the reflected ray is governed by the law of reflection, which states that the angle of reflection r' is equal to the angle of incidence i, and the reflected and incident rays lie in the same plane. The light that passes into the solid is known as the refracted ray, since its path is bent, or refracted, from the path of the incident ray. The relationship between the paths of the incident and refracted rays is known as the Law of Refraction, and as Snell's Law, the law having been discovered by Willebrod Snellius, professor of mathematics at Leyden in Holland, in about 1621. It states that the ratio of the sine of the angle of incidence i to the sine of the angle of refraction r is constant, i.e.

$$\frac{\sin i}{\sin r} = n$$

The constant n is known as the *index of refraction* (assuming $n = 1$ for air). It was later proved that the index of refraction is also the ratio of the velocity of light in air to the velocity of light in the solid, so that if V is the velocity of light in air, and v the velocity in the solid, then

$$n = \frac{V}{v}$$

The velocity of light in air is 300,000 km/sec; if the velocity of light in a substance is 200,000 km/sec, the refractive index is 300,000/200,000, or 1.5. Most solids have refractive indices between 1.4 and 2.0. The refractive index of a substance is related to its chemical composition and crystal structure, just as is the density; in fact, density (G) and refractive index (n) are related by the following approximate equation

$$\frac{n - 1}{G} = K$$

where K is a constant related to the composition of the substance.

There is a close connection between optical properties and the crystal structure of a solid. In isometric and noncrystalline substances the velocity of light

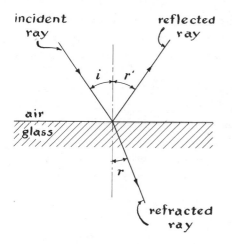

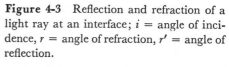

Figure 4-3 Reflection and refraction of a light ray at an interface; i = angle of incidence, r = angle of refraction, r' = angle of reflection.

is the same in all directions, and hence the refractive index is the same for all directions. Such substances are said to be optically isotropic. In all other substances the velocity of light varies according to its direction of vibration in the crystal; such substances are said to be optically anisotropic. A ray of light entering an anisotropic substance is split into two rays vibrating at right angles to each other, traveling generally with different velocities, and thus having differ-

ent refractive indices. The difference in refractive indices, known as the birefringence, is usually quite small (e.g., for quartz it is 0.009) and unobservable
except by instruments, but in calcite the difference is sufficient (0.172) that a
single spot observed through a cleavage fragment appears doubled.

The relationship between refractive indices and crystallography can best be
visualized by drawing, in all directions from the center of a crystal, lines whose
lengths are proportional to the refractive index for the particular vibration
direction. The resulting figure is known as the indicatrix (Fig. 4-4). For noncrystalline and isometric substances the form of the indicatrix is a sphere, since
the refractive index is the same in all directions. For substances crystallizing
in the tetragonal and hexagonal systems the indicatrix has the form of a rotation ellipsoid in which all sections perpendicular to one axis are circular, this
axis coinciding with the *c* axis of the crystal. This form results from the fact
that all rays traveling in the direction of the *c* axis have the same velocity,
since their vibrations are in the plane of the horizontal axes, which are equivalent in these systems. For this reason substances crystallizing in the tetragonal
and hexagonal systems are said to be *uniaxial*. For substances crystallizing in
the orthorhombic, monoclinic, and triclinic systems the indicatrix has lower
symmetry, in agreement with the lower crystallographic symmetry; it is now
a triaxial ellipsoid. A property of such an ellipsoid is that it has only two
circular sections, all others being ellipses; rays traveling at right angles to the
circular sections have the same velocity no matter what the direction of vibration in these sections. The two directions at right angles to the circular sections
are known as the optic axes, and thus orthorhombic, monoclinic, and triclinic
substances are said to be optically *biaxial*.

The orientation of the indicatrix in a crystal is related to the crystallographic
symmetry. In triclinic substances the positions of the three principal axes of the
indicatrix are independent of the positions of the crystallographic axes; in
monoclinic substances one axis of the indicatrix coincides with *b*, the axis of
symmetry, the other two being in the *ac* plane but independent of *a* and *c*; in
orthorhombic substances the axes of the indicatrix coincide in position with
the crystallographic axes.

The optical properties of a substance are therefore closely related to its
crystallographic symmetry, and optical properties have been much used to
determine the crystal system of minerals that do not occur in well-formed crystals. Many other physical properties also show a similar dependence upon
crystallographic direction, among which are thermal and electrical conductivity, compressibility, and thermal expansion.

Optical properties, especially refractive indices, are among the most valuable
determinative properties for mineral identification. For example, the refractive
indices of a tiny grain can readily be determined by immersing the grain in

liquids of known refractive index and examining it with a polarizing microscope. From the refractive indices and other optical properties an unknown mineral can generally be easily identified by referring to tables of these properties.

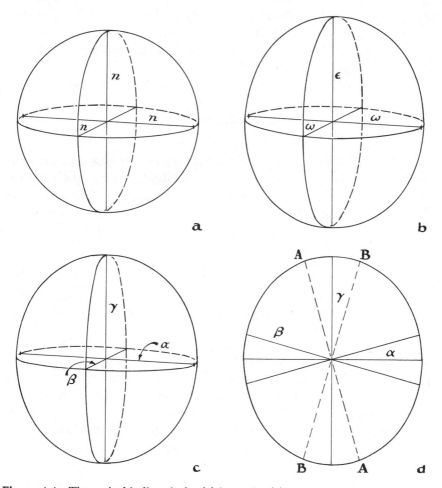

Figure 4-4 The optical indicatrix for (a) isotropic, (b) uniaxial, (c) biaxial substances. In (a) the indicatrix is a sphere, whose radius is proportional to n, the refractive index of the substance. In (b) the indicatrix is a rotation ellipsoid, the horizontal equatorial section being a circle with radius proportional to ω, one of the principal refractive indices, and with the vertical axis proportional to ϵ, the other principal refractive index; ϵ may be greater or less than ω, and its vibration direction is always parallel to the c crystallographic axis. In (c) the indicatrix is a triaxial ellipsoid, the lengths of the principal axes being proportional to α, the smallest index of refraction in the substance, γ, the greatest index, and β, the intermediate index. In (d) is represented the $\alpha\gamma$ section of the ellipsoid; AA and BB are the optic axes, at right angles to the two circular sections of radius β. AOB is the axial angle, designated as $2V$.

Luster

An optical property closely related to reflection and refraction is *luster*. Two main classes of luster are recognized in minerals, the *metallic* and the *nonmetallic*. No sharp division can be made between these two classes, and minerals with a luster intermediate between them are sometimes said to be submetallic.

The impression of luster is produced by the light reflected from the surface of a mineral. The intensity of the luster depends essentially on the quantity of reflected light, and in general is greater the greater the refractive index of the mineral. Luster is largely independent of the color of the mineral.

The relationship between luster and refractive index is approximately as follows.

Metallic luster. Minerals which absorb visible radiation strongly, being opaque or nearly opaque even in very thin fragments (although they may be transparent to infrared radiation), generally have metallic luster. Their refractive indices are 3 or greater. The native metals and most of the sulfides are in this group.

Submetallic luster. Minerals with refractive indices between 2.6 and 3, most of them being semi-opaque to opaque, generally have submetallic luster. Examples are cuprite ($n = 2.85$), cinnabar ($n = 2.9$), and hematite ($n = 3.0$).

Nonmetallic luster. Several varieties of nonmetallic luster are recognized.

1. Vitreous luster is the luster of glass, and is characteristic of minerals with refractive indices between 1.3 and 1.9. This range includes about 70 percent of all minerals, comprising nearly all the silicates, most other oxysalts (carbonates, phosphates, sulfates, etc.), the halides, and oxides and hydroxides of the lighter elements such as Al and Mg.

2. Adamantine luster is the brilliant luster typical of a diamond, and is characteristic of minerals with refractive indices between 1.9 and 2.6. Examples are zircon ($n = 1.92$–1.96), cassiterite ($n = 1.99$–2.09), sulfur ($n = 2.4$), sphalerite ($n = 2.4$), diamond ($n = 2.45$), and rutile ($n = 2.6$). The combination of a yellow or brown color with refractive indices in this range produces a *resinous* luster, a luster like that of resin.

3. Greasy, waxy, silky, and pearly lusters are variants of nonmetallic luster and are caused by the character of the reflecting surface. Diamonds often have a somewhat greasy luster, evidently the result of a microscopically rough surface which scatters the reflected light. Cleavage surfaces of halite have a vitreous luster when fresh, but take on a greasy or waxy appearance after exposure to damp air, which produces a slightly roughened surface. The greasy luster common in nepheline is due to a beginning alteration.

Cryptocrystalline and amorphous minerals, such as chalcedony and opal, commonly have a waxy luster. Porous aggregates of a mineral, such as the clays, scatter incident light so completely that they seem to be without luster and are described as dull or earthy. Silky luster is produced by minerals occurring in parallel-fibrous aggregates, such as asbestos and some varieties of gypsum. Transparent minerals with layer-lattice structures and accompanying perfect lamellar cleavage have a characteristically pearly luster produced by reflection from successive cleavage surfaces; examples are talc, the micas, and coarsely crystallized gypsum.

The luster of minerals has an economic aspect, as evidenced by gemstones. The qualities of beauty in a gemstone include color and transparency as well as luster. Luster is largely responsible for the brilliance of a gemstone, and, other things being equal, the higher the refractive index of a gemstone the greater its brilliance and beauty. The gem varieties of quartz, such as amethyst, have fine transparency and color yet lack the brilliance of diamond or zircon because of the much lower refractive index of quartz.

Color and Streak

In most minerals the impression of color is produced by the absorption of certain wavelengths among those making up white light, the resultant color being in effect white light minus the absorbed wavelengths. Dark-colored substances are those which absorb practically all the wavelengths of white light uniformly. The causes of color in minerals are varied and complex. Sometimes color is a fundamental property, directly related to the chemical composition, as in the blue and green of secondary copper minerals. Sometimes it is unrelated to composition but depends on crystal structure and bond type, as in the contrast between the polymorphs of carbon, diamond being colorless and transparent, graphite black and opaque. Sometimes it is due to impurities, as in the colored varieties of chalcedony. Minerals which have a constant and characteristic color are termed *idiochromatic*; those whose color is variable are called *allochromatic*. Color is one of the most useful of the determinative physical properties, but its utilization as a diagnostic test requires experience and discrimination.

Color directly related to composition is characteristic of substances containing elements belonging to the B subgroups of the periodic table, which have incompletely filled electron shells in their atomic structures. Elements important in this respect in minerals are Ti, V, Cr, Mn, Fe, Ni, Co, and Cu. The production of color is evidently linked with the absorption of part of the radiant energy of light by labile electrons in the atoms of these elements. Color is often intensified by the presence of an element in two valence states. This is well exemplified by iron-bearing compounds: minerals containing iron entirely in the ferrous state or entirely in the ferric state are generally rather pale in color; but most iron minerals contain the element in both valence states and are dark green to black. The mineral vivianite, $Fe_3(PO_4)_2 \cdot 8H_2O$, may be practically colorless when freshly mined, but on exposure to air it turns dark blue or dark green, evidently as a result of partial oxidation of the ferrous iron. A similar characteristic is displayed by synthetic and natural rutile (TiO_2). Synthetic rutile, prepared by melting pure TiO_2, is pale yellow and transparent; rutile as a mineral is always dark red to black and almost opaque. The color of natural rutile may be due to the presence of foreign ions in the structure; for example, Nb^5 can substitute for Ti^4, and electrical neutrality can be maintained by other titanium atoms assuming the Ti^3 state. However, the presence of foreign ions is not essential; synthetic rutile, if it is slightly deficient in oxygen (with an empirical formula about $TiO_{1.97}$) is very dark in color, probably because of the presence of some Ti^3 ions.

Ions or groups of ions which produce characteristic colors are known as *chromophores*. Thus the hydrated Cu^2 ion is the chromophore in the green and blue secondary copper minerals; Cr^3 is the chromophore in the green garnet uvarovite, in green chromian muscovites, and in emerald; the $(UO_2)^2$ ion is the chromophore in the bright yellow and yellow-green secondary uranium minerals.

Sometimes a chromophore may be present in only minute amounts and yet have a strong coloring effect. The amount of chromium in emerald is very small, but it produces an intense green color. The violet color common in lepidolite is due to the presence of a fraction of a percent of Mn^3; it is interesting to note that the chromophoric effect of Mn^3 is dependent upon the absence of iron, since if iron is also present the violet color is not developed.

Many usually colorless minerals, such as quartz, sometimes show strong coloration, which can be attributed to the presence of foreign ions in minute amounts. The colors of amethyst and rose quartz are probably due to traces of titanium or manganese. Sometimes this type of coloration requires not only the presence of foreign ions but also some sort of activation, such as by light or other forms of radiation. The effect is seen in clear bottle glass that has been exposed for a long time to strong sunlight—some of these glasses acquire a deep purple tint. Radiation apparently changes the state of ionization of some of the foreign ions, with consequent distortion of the electronic distribution and the production of color. It has often been notice that quartz associated with radioactive minerals is generally smoky, and this is presumably due to the constant irradiation it has undergone. This phenomenon has been applied commercially to gemstones, since many gemstones are improved in color when irradiated with x-rays or with neutron beams.

Interesting examples of coloration not associated with chromophoric ions are provided by some minerals of the feldspathoid group which contain negative ions other than oxygen in the structure: sodalite is often blue, cancrinite sometimes bright yellow. Coloration in these minerals is probably the result of a disturbance or lack of balance in the electrical field around the ions; the additional negative ions—CO_3^{-2}, Cl^{-1}, S^{-2}, and others—are all very large, and their charge distribution is probably distorted by the unequal attraction of small positive ions at unequal distances.

Many colorations in minerals are due simply to the presence of impurities intimately intermixed with the host mineral. Sometimes these impurities can be seen and identified with a hand lens or a microscope, but often they are very fine-grained and submicroscopic. In certain cases the mineral was homogeneous when crystallized, and the foreign material was produced by exsolution; for instance, the red color of many feldspars is due to the presence of submicroscopic hematite, produced by exsolution of ferric iron which replaced aluminum in the structure when the feldspar crystallized. Fine-grained hematite is a common coloring matter in otherwise colorless minerals, and it produces various shades of red; fine-grained carbon is also a common impurity, producing grays and blacks.

This type of coloring is artificially produced in the staining of agate for use as an ornamental stone. Brightly colored agates do occur naturally, but most such material has been colored by successive soaking in different solutions, which precipitates a fine-grained pigment in the pores of the mineral. Most natural agate is sufficiently porous to permit this.

Some minerals are *pseudochromatic*, that is to say, the color they show is not a true color but rather a play of color produced by certain physical effects. The brilliant colors of precious opal are of this kind, produced by reflection and refraction of light from layers of slightly different refractive index within the mineral; when viewed in transmitted light most precious opal is colorless or pale brown. A similar effect is seen in some feldspars, especially labradorite, and may be due either to reflection and refraction from layers of different refractive index or to reflection from tiny platy inclusions of other minerals (often ilmenite) lying on cleavage surfaces.

Iridescent surface films, similar to the colored films produced by oil on water, are not uncommon on some opaque minerals, especially limonite and hematite. Some minerals tarnish readily and thereby have a surface color quite unlike the true color; thus a freshly broken surface of bornite (Cu_5FeS_4) is bronze-colored, but it rapidly becomes purple-violet as a result of oxidation.

Streak is the color of the finely powdered mineral. It can be obtained by crushing, filing, or scratching the mineral, or by drawing it across a piece of white unglazed porcelain (streak plate). (A streak plate cannot be used with minerals which are harder than the plate.) Because the streak is a more constant, and hence more reliable, diagnostic feature than the color of a mineral, it is a valuable determinative property. Most transparent and translucent minerals have a white streak; dark minerals with a non-metallic luster have a streak usually lighter than the color; minerals with a metallic luster have a streak often darker than the color.

Luminescence

Luminescence is the emission of light resulting from all processes except incandescence. It is usually produced by irradiation, generally with ultraviolet light. *Fluorescence* is the emission of light at the same time as the irradiation; *phosphorescence* is the continued emission of light after irradiation is terminated.

The luminescence of minerals has long been of scientific interest, and has also a number of practical applications. Luminescence can be a valuable aid in prospecting and in mineral dressing, for distinguishing valuable ore minerals with characteristic fluorescence (such as willemite, scheelite, and some uranium minerals). Luminescence is the basis of modern illumination techniques which utilize the fluorescence of inorganic compounds, some of which have long been known as minerals (such as $CaSO_4$, $CaCO_3$, and $ZnSiO_4$).

The fundamental law of luminescence states that the wavelength of the fluorescent light is longer than the wavelength of the exciting radiation. Fluorescence usually involves the absorption of invisible radiation in the ultraviolet part of the spectrum and the simultaneous emission of this energy as visible light with longer wavelength. The absorption of visible light may result in emission in the infrared part of the spectrum—invisible radiation of longer wavelength, undetectable except by special equipment sensitive to infrared radiation. This is known as *infrared luminescence*.

The phenomenon of luminescence is linked with lattice disturbances, which are the result either of a defect lattice or of the presence of foreign ions that function as activators. In many instances it has been shown that these foreign ions are substituting for major elements in the structure—in willemite the activating ion is Mn^2 substituting for Zn^2; in scheelite it is Pb^2 substituting for Ca^2 and Mo^6 for W^6. This explains why willemite from Franklin, New Jersey (which contains Mn), fluoresces, whereas willemite from many other localities (Mn-free) does not. The actual production of luminescence is the result of absorption of energy by the ions and its release in the form of light. Sometimes the absorbed energy is "frozen in" and is released only upon heating the material—this is known as *thermoluminescence*.

Luminescence is often promoted by low temperatures. Nearly all organic compounds and many inorganic compounds fluoresce at the temperature of liquid air. Above 500–600°C most substances which fluoresce at ordinary temperatures no longer do so.

The mineral that always comes to mind in connection with fluorescence is fluorite, from which the term was derived. Fluorite generally fluoresces in blue, and it has been shown that the fluorescence is often due to a small amount of substitution of calcium by rare-earth elements. Calcite is another common mineral which is sometimes fluorescent, generally in red, pink, or yellow; this fluorescence is often caused by the presence of Mn, but sometimes by organic impurities (porphyrins). Scheelite fluoresces in white or bluish white, and the fluorescence becomes yellowish as Mo^6 replaces W^6; the color of the fluorescence is a useful semi-quantitative test for the amount of molydenum replacing tungsten. Some secondary uranium minerals and uranium-bearing opal have a strong green or yellow-green fluorescence, and the bead obtained by fusing uranium-bearing material with sodium fluoride is fluorescent and provides a rapid and sensitive test for this element.

Fluorescence is a rather capricious and unpredictable phenomenon, and few consistent effects are found. The following points should be emphasized:

1. Many localities are characterized by certain typical fluorescent minerals, e.g., green-fluorescing willemite and red-fluorescing calcite at Franklin, New Jersey.
2. Within the same locality the typical fluorescent minerals may also occur in nonfluorescent form.
3. The characteristic features of the fluorescence of certain minerals from one locality do not generally apply to the same minerals from other localities.

An interesting variety of luminescence met with occasionally in minerals is *triboluminescence*, or luminescence induced by crushing, scratching, or rubbing. It is shown by some varieties of sphalerite, fluorite, and lepidolite, and less commonly in feldspar, calcite, magnesite, dolomite, and quartz.

CLEAVAGE AND FRACTURE

If a crystal is strained beyond its elastic and plastic limits it will break; if it breaks irregularly it is said to show *fracture*, and if it breaks along surfaces related to the crystal structure it is said to show *cleavage*. Cleavage is always parallel to a possible crystal face, generally one with simple indices, and its position is described by these indices. The quality of cleavage is described by terms such as "perfect," "good," "distinct," "indistinct." A mineral may show cleavage in more than one crystallographic direction, in which case the quality is generally different in the different directions, allowing them to be distinguished. Perfect cleavage is present when a mineral breaks along the cleavage, giving a smooth lustrous surface, and is difficult to break except along the cleavage (e.g., calcite, muscovite). A mineral has good cleavage when it breaks easily along the cleavage, but such minerals can also be broken transverse to it (e.g., feldspar). Minerals with distinct cleavage break most readily along the cleavage but also fracture easily in other directions; consequently, individual cleavage surfaces are seldom large (e.g., scapolite). In minerals with indistinct cleavage, fracturing takes place as readily as cleavage, and careful inspection is necessary to recognize the cleavage (e.g., beryl). Cleavage is easily recognized in thin sections of minerals, since the cutting and grinding of the section causes the mineral to break along cleavages, which then appear as cracks, more or less straight, depending on the quality.

From the standpoint of technical application, cleavage is an exceedingly important property, and industrial use of many minerals depends upon it. The ability of muscovite to be cleaved into very thin sheets, together with its dielectric properties, is the basis of its use in electrical equipment. A pencil is an efficient writing instrument because the perfect cleavability of graphite causes small cleavage flakes of the mineral to be rubbed off and adhere to the paper. The lubricating qualities of graphite and talc depend upon their softness and the ease with which the minerals part along cleavage surfaces.

Cleavage is a very useful diagnostic property. Careful study of cleavage may make possible the determination of crystal system in fragments of minerals. For example, a mineral with a single direction of cleavage cannot belong to the isometric system, since every form in that system has more than two faces. Similarly, a mineral with three directions of cleavage, all of unequal quality, probably belongs to the orthorhombic, monoclinic, or triclinic systems; if the three directions are at right angles to each other, the system must be orthorhombic. Three equal cleavages correspond to a cube, a rhombohedron, or a hexagonal prism, and the choice can be resolved by the angular relationship (cubic cleavage, all three at 90° to one another; hexagonal prismatic, all three at 60°; otherwise rhombohedral). Four equal cleavages indicate an octahedral, or rarely tetragonal or orthorhombic dipyramidal, cleavage. Six equal cleavages are characteristic of dodecahedral cleavage.

Cleavage is a reflection of internal structure. Minerals possess cleavage because the strength of bonding within the structure is different in different planes. This is particularly prominent in the layer lattices, where the bonding within layers in the structure is very strong, whereas that between the layers is weak. In such structures we find perfect unidirectional cleavage parallel to the layers. This is the cleavage we find in graphite, and in the phyllosilicates (a subclass of the silicates in which the SiO_4 groups form sheet structures, typified by talc and the micas). In graphite the carbon atoms are covalently bonded in sheets, and these sheets are linked together by weak van der Waals' forces, which are easily disrupted. A similar situation is present in talc, which also has strongly bonded sheets linked by weak residual forces. In the micas the sheets are similar to those in talc but are linked by potassium ions; the linkages between the sheets are much weaker than those within the sheets, hence the single perfect cleavage which follows the planes in the structure occupied by the potassium ions.

Cleavage is thus a function of structure, and as such is only indirectly related to chemical composition. Within an isomorphous group the cleavage is the same for all species, although it may vary in quality from one species to another. This is well shown by those isometric substances with the halite structure, all of which have perfect cubic cleavage—i.e., three equal cleavages at right angles to one another. These include not only halite ($NaCl$) itself, sylvite (KCl), and many other univalent halides, but also periclase (MgO) and other bivalent oxides, and galena (PbS). The perfect rhombohedral cleavage of the calcite group is a reflection of a structure similar to that of halite; in the calcite structure the Ca substitutes for Na and the CO_3 group for Cl, but the structure is deformed as though by compression along one trigonal axis, whereby the symmetry changes from isometric to rhombohedral.

Some minerals show *parting*, which resembles cleavage in that it is always parallel to a possible crystal face but differs in that it may or may not be present in a particular specimen. In some minerals parting is a sign of beginning alteration along certain planes; in twinned crystals it often takes place along composition planes.

HARDNESS

Hardness of a mineral is generally defined as its resistance to scratching, and relative hardness has been employed as a useful diagnostic property since the beginning of systematic mineralogy. It was given qualitative precision by the Austrian mineralogist Mohs, who in 1822 proposed the following scale of relative hardness:

1. Talc
2. Gypsum
3. Calcite
4. Fluorite
5. Apatite
6. Orthoclase
7. Quartz
8. Topaz
9. Corundum
10. Diamond

Each of these minerals will scratch those minerals lower in the scale, and will be scratched by those higher in the scale.

It is a tribute to the perspicacity of Mohs that the scale he established is still used today in unaltered form. In his original description Mohs mentions that he endeavored to make the intervals on the scale as nearly equivalent as possible, and at the same time to select common minerals for the individual units. He was aware that the interval between corundum and diamond is greater than that between other units on the scale, but minerals of intermediate hardness were (and still are) unknown. He was also aware that hardness varies somewhat with crystallographic direction, many minerals being softer on cleavage surfaces than in other directions; he therefore recommended finely crystalline specimens of talc and gypsum as standards rather than coarse cleavages.

Many procedures for quantitative determinations of hardness have been devised, and their results show that the intervals on Mohs' scale, except between 9 and 10, are approximately equal. Hence, although it is qualitative, this scale is well suited for comparing relative hardness in minerals and is unlikely to be superseded. Its use requires the simplest of equipment. In the field most mineralogists find the finger nail ($H = 2\frac{1}{2}$) and a pocket knife ($H = 5\frac{1}{2}$) adequate tools. Minerals of hardness 1 have a greasy feel; those of hardness 2 can be scratched with the finger nail; 3, can be cut by a knife; 4, are scratched rather easily; 5, are scratched with some difficulty; those of hardness 6 and over are not scratched by a knife, and, moreover, will scratch glass. A set of small pieces of the standard minerals, or of hardness points (which are conical-shaped pieces of the standard minerals set in brass rods) is useful for testing the harder minerals.

Considered in relationship to crystal structure, hardness is the resistance of the structure to mechanical deformation. The following correlations have been found between hardness and crystal structure. Hardness is greater: (1) the smaller the atoms or ions; (2) the greater their valency or charge; (3) the greater the packing density.

The effect of ionic size can be most clearly seen in an isomorphous group, where the structure is the same in all species. Thus the calcite group comprises carbonates of divalent metals ranging in ionic size from Ca (0.99 Å) to Mg (0.66 Å); hardness increases with decreasing ionic size, from 3 for calcite to $4\frac{1}{2}$ for mangesite. Another good example is furnished by the marked contrast between hematite, Fe_2O_3 ($H = 6$), and corundum, Al_2O_3 ($H = 9$).

The effect of valency or charge can be most clearly seen by comparing compounds with the same structure and similar ionic sizes. Thus soda niter ($NaNO_3$) and calcite have the same structure, and the ionic sizes of Ca and Na are very similar; the hardness of soda niter is 2 and that of calcite is 3. A greater difference exists between the isomorphous minerals niter, KNO_3 ($H = 2$), and argonite ($H = 4$), because here there is not only the difference in charge between potassium and calcium but also a considerable difference in ionic radius (K^1 = 1.33 Å, Ca^2 = 0.99 Å).

The effect of density of packing is well seen in the relationship between density and hardness of different polymorphs. Examples are calcite (G = 2.71 H = 3) and aragonite (G = 2.93, H = 4), quartz (G = 2.65, H = 7) and tridymite (G = 2.26, H = $6\frac{1}{2}$). The same correlation between hardness and density exists between hardness and packing index; the greater the packing index, the greater the hardness. Thus a packing index greater than 6 generally means a hardness of 7 or more.

Since the bonding of a crystal structure is usually different in different directions, the hardness of a mineral may be expected to vary somewhat with crystallographic direction. Even in substances crystallizing in the isometric system the strength of bonding and hence the hardness may not be the same in all directions. Such variation is generally quite small, although occasionally it may be considerable. Thus on the {100} cleavage surface of kyanite the hardness is $4\frac{1}{2}$ in the direction of the c axis and $6\frac{1}{2}$ in the direction of the b axis. A property of diamond that is of great practical importance is the superior hardness of the {111} surfaces. The {111} surface in a diamond crystal is the hardest surface known to man, and this characteristic coupled with the {111} cleavage enables diamond dust to cut all but octahedral planes of other diamonds.

MAGNETIC PROPERTIES

Only a few minerals are ferromagnetic—i.e., strongly attracted by a simple bar or horseshoe magnet. Of these the commonest are magnetite, Fe_3O_4, pyrrhotite, $Fe_{1-n}S$, and a polymorph of Fe_2O_3, maghemite. Sometimes specimens of magnetite and maghemite are themselves natural magnets and will attract iron filings and when suspended will orient themselves with the long axis of the specimen pointing magnetic north and south; such specimens are called *lodestones*, and they were used in the earliest forms of compasses.

All minerals, however, are affected by a magnetic field, although special equipment may be necessary to show this. Minerals which are slightly repelled by a magnet are said to be *diamagnetic*, those which are slightly attracted are said to be *paramagnetic*. Ferromagnetic substances form a subgroup of paramagnetic substances. All iron-bearing minerals are paramagnetic, but also iron-free minerals, such as beryl, can be paramagnetic.

Magnetic separations, which use an electromagnet to produce a high-intensity magnetic field, have considerable application, both in the research laboratory and in ore dressing plants, for the separation of pure concentrates from mixtures of minerals. Sensitive instruments can not only separate para-magnetic minerals from diamagnetic minerals but also separate two paramagnetic minerals from each other (for example, biotite from hornblende).

The magnetic properties of minerals are utilized in geophysical prospecting by the magnetometer, an instrument that allows variations in the earth's magnetic field to be measured and plotted on maps. Such magnetic surveys are valuable for locating ore bodies, in detecting changes in rock type, and in tracing formations with specific magnetic properties. One advantage of magnetometric surveys is that they can be rapidly and easily carried out from an aircraft.

ELECTRICAL PROPERTIES

With regard to electrical properties, minerals can be divided into two groups, the conductors and the nonconductors, although all degrees of conductivity exist. The conductors are those minerals with metallic bonding, and they comprise the native metals and some of the sulfides; they are far less numerous than the nonconductors. Electrical conductivity, like optical and magnetic properties, varies with crystallographic direction in anisotropic substances; for example, the conductivity of hematite is twice as great in directions at right angles to the *c* axis as it is in directions parallel to the *c* axis.

In some nonconducting minerals it is possible to induce electrical charges by changes in temperature (*pyroelectricity*) or directed pressure (*piezoelectricity*).

Pyroelectricity was first observed in gem tourmaline crystals brought from Ceylon by Dutch traders, who noticed that a crystal dropped in warm ashes attracted ash particles at one end only. Pyroelectricity can occur in those substances belonging to crystal classes lacking a center of symmetry.

Pyroelectricity can be demonstrated by dusting a cooling crystal with a mixture of powdered sulfur and red lead (Pb_3O_4), the particles of which have been electrified by blowing them through a silk screen (a rubber bulb with two layers of silk stocking over the mouth makes a simple and satisfactory dusting device). The negatively charged sulfur will be attracted to the positively charged end of the crystal, and the positively charged red lead to the negatively charged end.

The property of piezoelectricity was discovered by the brothers Pierre and Jacques Curie in 1881, when they observed that quartz crystals subjected to properly directed pressure developed positive and negative charges at the ends of the *a* axes. In the following year G. Lippmann suggested that such crystals would become mechanically deformed if they were subjected to an electrical field, and this was confirmed by the Curies. The effect remained a laboratory curiosity until World War I, when experiments were made in transmitting and detecting underwater sound waves by means of quartz plates. The application of piezoelectricity to radio followed soon after; here an alternating electrical field generated by a vacuum tube radio circuit is applied across a quartz

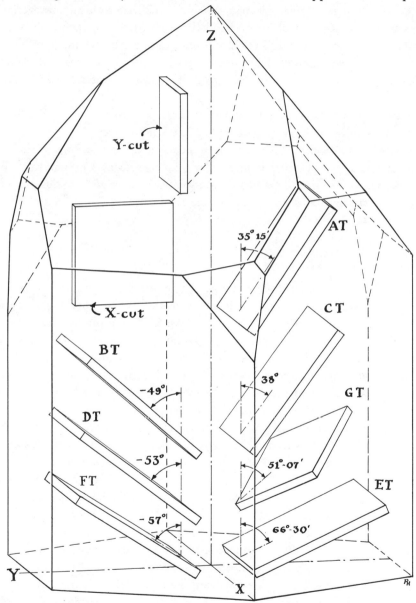

Figure 4-5 Model of a quartz crystal, showing the different directions of cutting to produce oscillator plates of different characteristics.

plate, properly cut and mounted and so dimensioned that one of its natural frequencies of mechanical vibration coincides with the resonant frequency of the circuit, and the frequency of transmission or reception is thereby stabilized and precisely controlled (Fig. 4-5). Theoretically, any substance lacking a center of symmetry is piezoelectric; in many such substances, however, the effect is very weak, and of all materials quartz is the most generally usable because of its chemical and physical stability, high elasticity, and availability. Its availability was greatly strained during World War II, when the annual production of quartz oscillator plates in the United States increased from about 50,000 in 1939 to more than 28,000,000 in 1944! A tremendous search was made for suitable quartz crystal, and much experimentation was carried out on the growing of large quartz crystals in the laboratory.

Static electricity, produced in some nonconductors by rubbing them with silk or fur, has been observed in amber for many centuries. Comparatively few minerals show this effect. It is curious that many gemstones show this property after being cut and polished but not in the rough state.

SURFACE PROPERTIES

Minerals show marked differences in the properties of their surfaces. One such property of great technical significance is called "wettability"—the relative ease with which a surface can be coated with water. According to this property, minerals can be divided into two groups: lyophile minerals are those which are easily wetted, and lyophobe minerals are those which are not easily wetted. Naturally there are all degrees of wettability between extremely lyophile and extremely lyophobe minerals. Minerals with ionic bonding are generally lyophile; those with metallic or covalent bonding are lyophobe.

This difference in surface properties has been applied for many years in the separation of diamonds from accompanying heavy minerals such as garnet. Diamond-bearing rock is crushed, and a concentrate of the diamonds and other heavy minerals is separated by mechanical means. This concentrate is then washed over tables coated with thick grease. The lyophile minerals such as garnet are readily wetted and washed away, but the lyophobe diamonds are not wetted and stick in the grease, from which they are easily recovered.

The principal application of differences in surface properties is, however, in the ore dressing technique known as flotation. Flotation is used primarily for the separation of sulfide minerals from gangue minerals, and individual sulfide minerals from mixtures. In general, the sulfide minerals are lyophobe, the gangue minerals (quartz, calcite, etc.) lyophile. The finely crushed ore is mixed with water, to which are added small amounts of oil (which is attracted to and covers the sulfide grains) and a foam-producing compound. Air is then blown through the mixture, and the foam that is produced carries with it the sulfide minerals, while the gangue sinks. Variations in the type and amount of oil, the foaming agent, and other conditions make possible selective flotation, whereby a complex ore containing several different sulfide minerals can be completely separated into its phases. Flotation techniques are now available that even make possible the separation of lyophile minerals from each other.

RADIOACTIVITY

Radioactivity in minerals is linked with the presence of uranium and thorium. (A few other elements, such as potassium and rubidium, also show weak radioactivity that is detectable with sensitive instruments.) Uranium and thorium atoms disintegrate spontaneously at a constant rate unaffected by the temperature, the pressure, or the nature of the compound in which they occur. Disintegration is accompanied by three types of radiation: alpha radiation, which consists of positively charged helium nuclei (alpha-particles); beta radiation, which consists of negatively charged electrons; and gamma radiation, which is a form of x-rays. Radioactivity can readily be detected by the radiation produced either by its effect on photographic film (Fig. 4-6) or by means of a Geiger counter or scintillometer.

The ultimate product of the disintegration of uranium and thorium is lead, as indicated by the following equations:

$$U^{238} \longrightarrow Pb^{206} + 8He^4$$
$$U^{235} \longrightarrow Pb^{207} + 7He^4$$
$$Th^{232} \longrightarrow Pb^{208} + 6He^4$$

Figure 4-6 Photograph (a) and autoradiograph (b) of uraninite in feldspar, from pegmatite at Grafton, New Hampshire. In the photograph the uraninite is black; in the autoradiograph it is white. [Courtesy the American Museum of Natural History.]

The rate of these reactions is known, and hence the age of a radioactive mineral can be calculated if the amounts of uranium, thorium, and lead are determined and if the mineral contained no primary lead and has not been affected by alteration and leaching. Fresh specimens of radioactive minerals are therefore of great scientific value for the information they can provide on geological age.

The development of atomic energy has resulted in a world-wide search for radioactive minerals (especially uranium minerals, since thorium is not yet used). The ease with which these minerals can be detected by scintillometers and Geiger counters has greatly simplified the search. One result has been the discovery of many new uranium minerals and the thorough investigation of many poorly known species.

Exercises

1. List some important chromophores in minerals, the colors they produce, and give examples of minerals containing them.
2. Calculate packing indices for the following minerals:

	Unit cell dimensions (in Å)	Formula units per unit cell
Spinel, $MgAl_2O_4$	$a = 8.08$	8
Aragonite, $CaCO_3$	$a = 4.96, b = 7.97, c = 5.74$	4
Almandite, $Fe_3Al_2(SiO_4)_3$	$a = 11.53$	8
Enstatite, $MgSiO_3$	$a = 18.22, b = 8.83, c = 5.19$	8
Zircon, $ZrSiO_4$	$a = 6.60, c = 5.98$	4
Cassiterite, SnO_2	$a = 4.74, c = 3.19$	2

3. A specimen of rich gold ore weighs 309.3 g, and its volume is 82.9 cc; quartz ($G = 2.65$) is the only gangue mineral. The gold contains 12.1 percent of silver, the density of this alloy being 17.5. How many grams of pure gold are there in the specimen?
4. Vein material consisting of a mixture of quartz and barite has a density of 3.6. What is the percentage of barite by weight?
5. Calculate the density of the ore whose composition is given in Problem 7, Chapter 3.

Selected Readings

Clark, S. P., *et al.*, 1966, *Handbook of physical constants:* Geol. Soc. America, Memoir 97.

Faul, H., 1954, *Nuclear geology:* New York, Wiley.

Przibram, K., 1956, *Irradiation colours and luminescence:* London, Pergamon Press. Translated by J. E. Caffyn.

Tertsch, H., 1949, *Die Festigkeitserscheinungen der Kristalle:* Vienna, Springer-Verlag.

Wooster, W. A., 1938, *A textbook on crystal physics:* London, Cambridge Univ. Press.

5

The Genesis of Minerals

For many purposes the identification of a mineral is an end in itself, but in the broader view identification is merely an initial step in deciphering the relationship of that mineral to the geological conditions under which it was formed. A mineral is the final product of a complex of natural processes, and its characteristics, its geological environment, and its associated minerals are all clues that when properly evaluated elucidate the conditions under which it formed and its subsequent history.

CHEMICAL COMPOSITION OF THE EARTH'S CRUST

The mineralogy of the earth's crust is ultimately controlled by the abundance and distribution of the chemical elements. Determination of the abundance of the different elements in the earth's crust is a very active field of research, but it is fraught with difficulty because the crust is not homogeneous and because we can sample it only at the surface or at shallow depths in mines or boreholes. But geological processes, such as mountain building and denudation, have exposed at the surface rocks and mineral deposits that must have been formed at depths of several kilometers. Enough is known of the earth's crust to give a fairly reliable estimate of the composition of the outer 10–20 km, at least of the continental areas.

The average composition of this outer 10–20 km is, in effect, that of igneous rocks, since the total amount of sedimentary and metamorphic rocks is insignificant in comparison with the bulk of the igneous rocks. Many thousands of rock analyses have been made, and it is significant that in these analyses only a few elements, and always the same ones, are present as essential constituents. The analysis of a normal rock may be considered reasonably complete if the following constituents are determined: SiO_2, TiO_2, Al_2O_3, Fe_2O_3, FeO, MgO, CaO, Na_2O, K_2O, P_2O_5, H_2O, and CO_2. The rarer elements are not usually

Table 5-I *The average amounts of the elements in crustal rocks in grams per ton or parts per million**

O	464,000	Gd	5
Si	281,500	Dy	3
Al	82,300	Hf	3
Fe	56,300	Cs	3
Ca	41,500	Yb	3
Na	23,600	Er	3
Mg	23,300	Br	3
K	20,900	U	3
Ti	5,700	Be	3
H	1,400	As	2
P	1,050	Sn	2
Mn	950	Ta	2
F	625	W	2
Ba	425	Mo	2
Sr	375	Ge	2
S	260	Ho	1
C	200	Eu	1
Zr	165	Tb	0.9
Y	135	Tl	0.5
Cl	130	Lu	0.5
Cr	100	I	0.5
Rb	90	Tm	0.5
Ni	75	Sb	0.2
Zn	70	Bi	0.2
Ce	60	Cd	0.2
Cu	55	In	0.1
Y	33	Hg	0.08
La	30	Ag	0.07
Nd	28	Se	0.05
Co	25	A	0.04
Sc	22	Pd	0.01
N	20	Pt	0.005
Li	20	Au	0.004
Nb	20	He	0.003
Ga	15	Te	0.002
Pb	13	Rh	0.001
B	10	Re	0.001
Th	10	Ir	0.001
Pr	8	Os	0.001
Sm	6	Ru	0.001

* Omitting those present in amounts of less than 0.001 g/ton: Ne, Kr, Xe, and the short-lived radioactive elements.

determined unless there is reason to expect them (thus ZrO_2 should be determined if the rock is known to contain an appreciable amount of zircon) or unless the research is concerned with the minor and trace elements (those present in amounts of 0.1 percent or less) as well as the major elements. During the last few years many analyses of rocks for minor and trace elements have been made, largely through the availability of sensitive analytical techniques such as spectrographic and colorimetric methods.

The resulting information about the average composition of the earth's crust is summarized in Table 5-I. Since the analyses from which this table has been compiled are largely of rocks from land areas, the average composition is in effect that of the continental crust; the crust under the deep ocean basins is probably more basaltic in composition than this average.

The first feature that stands out in Table 5-I is that only eight elements— O, Si, Al, Fe, Ca, Na, K, Mg—are present in amounts greater than 1 percent, and these eight make up nearly 99 percent of the earth's crust. Four more— Ti, H, P, Mn—are present in amounts between 0.1 percent and 1 percent. The remaining elements together make up less than 0.5 percent of the crust, and these elements include many which are essential to our industrial civilization and to our health and well being! Fortunately, geological processes have led to concentrations of many of these elements, which we call mineral deposits, in particular localities.

Of the major elements, oxygen is absolutely predominant, making up about half the crust by weight. This predominance is even more marked when weight figures are recalculated to atom percent and to volume percent (Table 5-II). In terms of numbers of atoms, oxygen exceeds 60 percent, and in terms of the volume of the different atoms, or rather ions, oxygen makes up more than 90 percent of the total volume occupied by the elements. The crust as a whole is essentially a packing of oxygen anions, bonded by silicon and the ions of the common metals. Oxygen, silicon, and aluminum collectively make up 90

Table 5-II   *The major chemical elements in the earth's crust*

	WEIGHT PERCENT	ATOM PERCENT	IONIC RADIUS (Å)	VOLUME PERCENT
O	46.40	62.19	1.40	94.04
Si	28.15	21.49	0.42	0.88
Al	8.23	6.54	0.51	0.48
Fe	5.63	2.16	0.74	0.49
Mg	2.33	2.05	0.66	0.33
Ca	4.15	2.22	0.99	1.18
Na	2.36	2.20	0.97	1.11
K	2.09	1.15	1.33	1.49

percent of the atoms in the earth's crust. It is obvious, then, that the dominant minerals must be quartz and silicates and aluminosilicates of iron, magnesium, calcium, sodium, and potassium.

As shown by Table 5-I, some of the elements which play an important part in our economic life, and which have long been known to and used by man, are actually quite rare, and, conversely, many unfamiliar elements are relatively abundant. Thus zirconium is more abundant than copper, lead is comparable in abundance to gallium, mercury is rarer than any of the so-called rare earths. The familiarity or otherwise of a particular element is very largely a reflection of its mineralogy, which in turn reflects its crystal chemistry. If an element forms a certain mineral which occurs in quantity in a particular geological envionment, that mineral is likely to attract attention and eventual industrial use. Thus lead, though an element of comparatively low abundance, occurs as the sulfide galena in local concentration in ore deposits, and has been known and used since prehistoric times. Gallium, on the other hand, forms no minerals in which it is present in significant amount; in ionic size and general chemical properties it is so similar to aluminum that most of the gallium in the crust is present in aluminosilicate minerals substituting for aluminum. Gallium is called a *dispersed element*, because although present in the crust in considerable amounts it is dispersed throughout common minerals and never occurs concentrated in any one mineral. Other such elements are rubidium, always dispersed in potassium minerals, and hafnium, always dispersed in zirconium minerals. The longest known and most familiar elements are those forming the major constituents of easily recognized minerals, which occur in local concentrations and are readily converted into useful industrial materials.

GEOCHEMICAL CLASSIFICATION OF THE ELEMENTS

An element is often very specific in regard to the type of minerals it forms. Some elements, such as gold and the platinum metals, nearly always occur in the native state. Others, such as copper, zinc, and lead, are found mainly as sulfides. Many elements, for example, the alkali and alkaline earth metals, generally occur as oxygen compounds, especially silicates or aluminosilicates, and are never found native or as sulfides. The inert gases, as the name implies, form no minerals at all. A useful geochemical and mineralogical classification of the elements is into four groups (Table 5-III): lithophile, those occurring, mainly in oxygen compounds; chalcophile, those occurring mainly as sulfides; siderophile, those occurring mainly as native elements; and atmophile, gaseous elements not readily forming compounds and therefore present mainly in the atmosphere. Some elements may appear in more than one group, since the type of compounds an element forms is dependent not only on the nature of the element but also on the temperature, the pressure, and the other elements present when it formed. For instance, most of the iron in the crust is present as oxides or silicates, but under conditions of low oxygen and high sulfur, iron sulfides are formed, and under highly reducing conditions when little sulfur is present native iron may be produced. Such variations are indicated in Table 5-III by enclosing the element in parentheses under the group of secondary affinity.

Table 5-III *Geochemical classification of the elements*

LITHOPHILE	CHALCOPHILE	SIDEROPHILE	ATMOPHILE
Li, Na, K, Rb, Cs	Cu, Ag	Pt, Ir, Os	Inert gases
Be, Mg, Ca, Sr, Ba	Zn, Cd, Hg	Ru, Rh, Pd	N, (O)
B, Al, Sc, Y	In, Tl, Pb	Au, (Fe)	
Rare earths	As, Sb, Bi		
C, Si, Ti, Zr, Hf	S, Se, Te		
Th, P, V, Nb, Ta	Ni, Co, (Fe)		
O, Cr, W, U	Mo, Re, (Mn)		
H, F, Cl, Br, I	(Ga), (Ge), (Sn)		
Fe, Mn, Ga, Ge, Sn			
(Mo), (Cu), (Zn), (Pb)			
(Tl), (As), (Sb), (Bi)			
(S), (Se), (Te), (Ni), (Co)			

The geochemical character of an element is governed largely by the electronic configuration of its atoms, which, in turn, controls the type of compounds the element forms. Hence geochemical character is closely related to position in the periodic table, as can be seen in Table 5-IV. Lithophile elements are those which ionize readily or form stable oxyanions (such as CO_3^{-2}, SO_4^{-2}, PO_4^{-3}), and in their compounds bonding is largely ionic in character; chalcophile elements ionize less readily, and thus tend to form covalent compounds with sulfur (and selenium and tellurium, when present); the siderophile elements are those for which metallic bonding is the normal condition, and which do not readily form compounds with oxygen or sulfur. A knowledge of the geochemical character of an element enables us to predict the type of minerals it is likely to form and the associations in which it is likely to occur.

MINERALOGICAL COMPOSITION OF THE EARTH'S CRUST

The number of minerals an element forms and the geological environment in which the minerals occur are largely controlled by the abundance of the element and its geochemical character. On first consideration it seems remark-

Table 5-IV *Geochemical Classification of the Elements in Relation to the Periodic Table*

Lithophile: Na (roman)
Chalcophile: *Cu* (italic)
Siderophile: <u>Au</u> (underlined)
Atmophile: **N** (boldface)

H																	**He**
Li	Be											B	C	**N**	O	F	**Ne**
Na	Mg											Al	Si	P	S	Cl	**A**
K	Ca	Sc	Ti	V	Cr	Mn	Fe	*Co*	*Ni*	*Cu*	*Zn*	*Ga*	Ge	*As*	*Se*	Br	**Kr**
Rb	Sr	Y	Zr	Nb	*Mo*		<u>Ru</u>	<u>Rh</u>	<u>Pd</u>	*Ag*	*Cd*	*In*	Sn	*Sb*	*Te*	I	**Xe**
Cs	Ba	La Lu	Hf	Ta	W	*Re*	<u>Os</u>	<u>Ir</u>	<u>Pt</u>	<u>Au</u>	*Hg*	*Tl*	Pb	*Bi*			
			Th		U												

able that the earth's crust, which is made up of more than 80 elements (excluding the short-lived radioactive ones), contains only about 2,000 different compounds (i.e., minerals), and most of these are quite rare. The total number of inorganic compounds is, of course, far greater, but most of them are not found as minerals. Only very stable compounds can occur as minerals; less stable compounds either do not form in nature or soon decompose. Another limitation on the number of minerals is the geochemical association of certain elements. Thus there are no rubidium minerals, even though rubidium is a relatively abundant element; geological processes fail to separate rubidium from the far more abundant potassium, and all the rubidium in the earth's crust is dispersed in potassium minerals. Similarly, the 15 rare-earth elements form very few minerals; indeed, altogether they form fewer than does antimony, an element much lower in abundance than most of them. This is due largely to the very similar ionic radii of the rare earths and their uniformly lithophile character; consequently their crystal chemistry is essentially that of a single element.

Thus the mineralogy of the crust is much simpler than we might expect from its elemental composition. The limitations in mineralogical variation are still more marked when we turn from the earth's crust as a whole to specific geological environments. In general, three major types of environment may be recognized; magmatic, sedimentary, and metamorphic. Each of these can be divided into subsidiary environments, according to the variety of physical and chemical conditions. The mineralogy of each will depend upon the temperatures and pressures of crystallization and the variation in chemical composition of the materials. Thus all sedimentary environments are characterized by a moderate range of temperatures (generally between 0°C and 40°C) and essentially constant (atmospheric) pressure; but the source materials may be igneous rocks, metamorphic rocks, pre-existing sedimentary rocks, ore-bearing veins, or, in fact, any mineral of the crust. Magmatic environments are characterized by high to moderate temperatures, and a wide variation of pressure, but in general a much more limited variation in chemical composition. Metamorphic environments cover a wide range of temperature and pressure, and the materials may be pre-existing rocks of any kind. It is thus convenient to consider the origin and associations of minerals within the framework of these three major groupings.

MINERAL FORMATION AND THE PHASE RULE

In discussing mineral formation we find that three factors control the minerals in a specific environment: the bulk composition, and the temperature and pressure at which crystallization takes place. This qualitative statement has a quantitative expression in a basic law of physical chemistry, known as the phase rule, which states that: *In any system at equilibrium the number of phases (P) plus the number of degrees of freedom (F) is equal to the number of components (C) plus two*, or $P + F = C + 2$. The concepts of system, phases, and components have been discussed in Chapter 3. Degrees of freedom and equilibrium are defined as follows.

Degrees of freedom (also known as variance): This is the number of variables which must be fixed in order to define the condition of the system. The degrees of freedom in geological systems are, in effect, the factors discussed previously, namely, bulk composition, temperature, and pressure.

Equilibrium: Equilibrium exists in any system when the phases exist together in a stable relationship indefinitely if the external conditions are maintained unchanged.

The formula $P + F = C + 2$ may be rewritten in the form $P = C + 2 - F$. This emphasizes that the number of phases (i.e., minerals) increases with the number of components; in other words, the greater the chemical complexity of a rock the larger the number of minerals that may be found in it. In any system the maximum number of phases will occur when $F = 0$, when $P = C + 2$. In any geological environment the condition of absolutely fixed and constant temperature and pressure is extremely improbable. Because these two degrees of freedom normally vary considerably, as during the crystallization of a magma or the metamorphism of a rock, we are usually dealing with bivariant systems, in which $F = 2$. Substituting $F = 2$ into the phase rule we obtain $P = C + 2 - 2$, or $P = C$. This expression is known as the *mineralogical phase rule*, and it can be stated thus: *In a system of* n *components at arbitrary pressure and temperature the maximum number of mutually stable minerals does not exceed* n.

We may now consider a few examples.

The system with n = 1: Any individual element is a one-component system. Sulfur, for example, can exist in two distinct solid phases, orthorhombic and monoclinic; each of these is stable over a particular range of temperature and pressure. At any fixed pressure the two phases can exist together in equilibrium only at a fixed temperature (at atmospheric pressure this temperature is 95°C). Any compound which does not decompose on melting can be considered a one-component system. Thus SiO_2 is a one-component system for which a number of phases (i.e., polymorphs) are known. Each of these polymorphs is the stable form over a considerable range of temperatures and pressures (Fig. 15-5). However, two polymorphs can coexist in equilibrium only under fixed conditions of temperature and pressure; tridymite and quartz, for example, are in equilibrium at 1 atmosphere pressure and 867°C. Hence we normally find only one form of SiO_2 in a particular rock.

The system with n = 2: The mineralogical phase rule predicts that in such a system the maximum number of minerals will be two. This is illustrated by the system $NaAlSiO_4-SiO_2$, in which the following minerals are possible (omitting the high temperature polymorphs of SiO_2): nepheline, quartz, albite ($NaAlSi_3O_8$, i.e., $NaAlSiO_4 + 2SiO_2$), and jadeite ($NaAlSi_2O_6$, i.e., $NaAlSiO_4 + SiO_2$). In igneous rocks we find the associations

nepheline + albite

albite + quartz

In metamorphic rocks we find the associations

albite + quartz

jadeite + quartz

jadeite + albite

Generally speaking, in any rock which has reached equilibrium we find only two of the four possible minerals in this system.

The system with n = 3: Such a system is $MgO-Al_2O_3-SiO_2$, in which some ultrabasic rocks and metamorphosed shales fall. There are no less than fourteen minerals in this system, but in any rock we do not find associations of more than three of them together. Typical associations are forsterite (Mg_2SiO_4) + spinel ($MgAl_2O_4$) + enstatite ($MgSiO_3$) in peridotites, or corundum (Al_2O_3) + spinel + cordierite ($Mg_2Al_4Si_5O_{18}$) in metamorphosed shales.

The mineralogical phase rule is thus a very significant principle in mineral formation, and it expresses the limited mineralogy of most common rocks. Considering an igneous rock, for example, which can be expressed in terms of the components O, Si, Al, Fe, Mg, Ca, Na, and K (i.e., an eight-component system), we might expect a maximum of eight minerals. The actual number of minerals is usually less, because some of these components are not completely independent, being capable of replacing each other, atom for atom, in minerals—for instance, Ca and Na in plagioclase, Fe and Mg in the ferromagnesian minerals.

THE MAGMATIC ENVIRONMENT

Magma is essentially a hot silicate melt, and is the parent material of igneous rocks. We can observe magmas and the formation of igneous rocks in volcanic regions, but much magma solidifies within the crust, and the rocks thereby formed are later exposed at the surface by erosion or by earth movements—hence the classification of igneous rocks into two groups, the volcanic or extrusive rocks, and the plutonic or intrusive rocks.

Igneous rocks are the products of the crystallization of magmas, but they are not the only products. A magma is a hot melt in which the quantitatively important elements are O, Si, Al, Ca, Mg, Fe, Na, and K, but which also contains in minor amounts almost all the remaining elements. The crystallization of minerals from a magma results in a concentration of many of these minor elements in the residual liquid, and a concentration of volatile matter, such as H_2O, CO_2, N_2, compounds of sulfur and boron, and HCl and HF. Thus the crystallization of igneous rocks does not complete the activity of a magma. The residual solutions give rise to pegmatites and hydrothermal veins, sometimes within the already solid igneous rocks but more commonly in fissures in the country rocks; they may even reach the surface in the gaseous state, giving rise to fumaroles, or as hot solutions, forming hot springs. In considering the magmatic environment, four types of mineral occurrence may be distinguished: (1) igneous rocks, (2) pegmatites, (3) hydrothermal veins, and (4) hot spring and fumarole deposits. From the viewpoint of quantity of material, igneous rocks are, of course, vastly more significant than the other three groups. The mineralogy of igneous rocks, however, is comparatively simple, whereas that of pegmatites and hydrothermal veins is very diverse (and very important economically, since their minerals are valuable raw materials of industry). The deposits of hot springs and fumaroles are quantitatively insignificant but mineralogically interesting.

Igneous Rocks

MINERALOGICAL COMPOSITION.   The mineralogy of igneous rocks is comparatively simple. Only seven minerals or mineral groups—quartz, feldspars, feldspathoids, pyroxenes, hornblende, biotite, and olivine—are commonly present in major amounts in igneous rocks. These minerals may be classed as *essential* constituents. A number of other minerals may be present in small amounts; of these the commonest are magnetite, ilmenite, and apatite. These minerals are classed as *accessory* constituents.

The minerals of igneous rocks, whether essential or accessory, are also grouped as leucocratic, or light-colored, and melanocratic, or dark-colored. This grouping is really a chemical one, dividing quartz and the sodium, potassium, and calcium alumino-silicates from the ferromagnesian minerals (the pyroxenes, hornblende, biotite, and olivine). The mineralogy of the igneous rocks can thus be summarized as in Table 5-V.

All of these minerals are never present in a single igneous rocks. In fact, some of them never occur in association; olivine, for example, does not occur in the same rocks as does primary quartz, nor do the feldspathoids. Olivine and the feldspathoids are *incompatible* with quartz. Excess silica reacts with olivine to form pyroxene, thus

$$(Mg,Fe)_2SiO_4 + SiO_2 \longrightarrow 2(Mg,Fe)SiO_3$$

olivine   pyroxene

and reacts with feldspathoids to form feldspars.

The amounts of the essential minerals in different igneous rocks can vary greatly. Some minerals can make up nearly 100 percent of certain igneous rocks—anorthosite is a plagioclase rock, dunite an olivine rock, pyroxenite a pyroxene rock. Most igneous rocks contain two or three essential minerals. A statistical study of about 700 igneous rocks whose mineralogical composition was known gave the following average composition: quartz 12.0 percent, feldspars 59.5 percent, pyroxene and hornblende 16.8 percent, biotite 3.8 percent, ilmenite and sphene 1.5 percent, apatite 0.6 percent, other minerals 5.8 percent.

CHEMICAL COMPOSITION. Igneous rocks, being mixtures of different minerals in various proportions, might be expected to show an extreme variation in chemical composition. This is not so; the range of variation is quite limited. Thus SiO_2 ranges from 30 percent to 80 percent, Al_2O_3 from about 10 percent to 20 percent (but can be very low in rocks with little feldspar or feldspathoid); each of the other major components (CaO, MgO, FeO + Fe_2O_3, K_2O, Na_2O) seldom exceeds 10 percent. Analyses of most igneous rocks generally show some H_2O, but the amount is usually quite low and represents absorbed water except

Table 5-V *Mineralogy of the igneous rocks*

	LEUCOCRATIC	MELANOCRATIC
Essential	*SiO₂* Quartz; tridymite and/or cristobalite in some volcanic rocks	Olivine
	Feldspars K–feldspars: Orthoclase, microcline Perthite: K–feldspar—albite intergrowth Na–Ca feldspars: Plagioclase	*Pyroxenes* Enstatite Hypersthene Augite Aegirine
	Feldspathoids Nepheline Leucite Sodalite Cancrinite	Hornblende Biotite
Accessory	Apatite Muscovite Corundum Sphene Fluorite Zircon	Ilmenite Magnetite Pyrite Pyrrhotite

in rocks containing hornblende and biotite, which have (OH) groups in the structure. Two minor constituents generally present in amounts between 0.1 percent and 1 percent are TiO_2 and P_2O_5.

This restriction in range of chemical composition must reflect a restricted range of the composition of the parent magmas. The restriction is still more marked when the amounts of different igneous rocks are compared. The igneous rocks of the earth's crust belong chiefly to two types: granitic (granites, granodiorites, quartz diorites), and basaltic (basalts and pyroxene andesites). It has been estimated that the granitic rocks comprise over 90 percent of all intrusives and the basaltic rocks over 90 percent of all extrusives—a remarkable fact of fundamental significance in geology.

CLASSIFICATION. Various criteria have been utilized, either singly or in combination, for the classification of igneous rocks. The most significant are mineralogical composition, chemical composition, and geological environment. The first two criteria are not independent, since mineralogical composition and chemical composition are interrelated. The same minerals (if they do not vary in composition) in the same proportion will always give the same bulk chemical composition. The converse, however, is by no means valid; the same chemical composition may give very different rock types, depending upon conditions of crystallization. For example, a hornblendite with over 90 percent hornblende may have practically the same composition as a dolerite consisting of plagioclase, augite, and olivine.

For our purpose a classification based on mineralogical composition is most suitable, and is shown in Table 5-VI. The table is divided into three horizontal sections: rocks with essential quartz, rocks without either quartz or feldspathoids, and rocks with feldspathoids. This division is based on the incompatibility of quartz and feldspathoids. The four vertical divisions of the table are based on the nature of the feldspar: potash feldspar or perthite predominant, potash feldspar and plagioclase in approximately equal amounts, plagioclase feldspars predominant, and little or no feldspar. The rocks with predominant plagioclase feldspars are subdivided into diorite and gabbro families according to the composition of the feldspar; this is by no means a universally accepted distinction, and it is difficult to apply in the field, since plagioclase composition can be determined satisfactorily only by chemical or microscopical examination. However, rocks of the diorite family generally contain larger amounts of leucocratic minerals than do the gabbroic rocks, and hence are usually lighter in color (and lower in density); these qualitative distinctions can often be satisfactorily applied. The factor of geological environment is taken into account by giving coarse-grained and fine-grained equivalents of each mineralogical composition. The grain size reflects the geological environment. The volcanic

Table 5-VI *Mineralogical classification of igneous rocks*

MINERAL CONSTITUENTS	POTASH FELDSPAR OR PERTHITE PREDOMINANT	POTASH FELDSPAR AND PLAGIOCLASE APPROX. EQUAL	PLAGIOCLASE FELDSPARS PREDOMINANT		LITTLE OR NO FELDSPAR
			AN < 50 PERCENT	AN > 50 PERCENT	
Quartz present	A. Granite* B. Rhyolite*	A. Granodiorite B. Quartz latite	A. Quartz diorite (Tonalite) B. Dacite	A. Quartz gabbro B. Quartz basalt	
Neither quartz nor feldspathoids present	A. Syenite B. Trachyte	A. Monzonite B. Latite	A. Diorite B. Andesite	A. Gabbro B. Basalt B. Diabase	A. Pyroxenite A. Peridotite A. Dunite B. Limburgite
Feldspathoids present	A. Nepheline syenite B. Phonolite	A. Nepheline monzonite B. Latite phonolite	A. Essexite and B. Tephrite and	Theralite Basanite	A. Ijolite B. Leucitite B. Nephelinite

* A = plutonic types; B = volcanic types.

types cooled rapidly and so crystallized quickly to give generally fine-grained rocks (often, however, containing larger crystals, or *phenocrysts*, which were floating in the magma when it was erupted). The plutonic types, on the other hand, cooled slowly, and the individual crystals thus had time to grow larger; as a result plutonic rocks are generally medium to coarse-grained.

The composition of the major igneous rocks in terms of essential minerals is illustrated in Fig. 5-1. For each rock type the relative amounts of the essential minerals are given by their intercepts on the vertical scale below the name of the rock type. This diagram illustrates graphically the essential continuity in the sequence of igneous rocks; the different types of igneous rocks grade into one another by the gradual variation in amount of the different essential minerals.

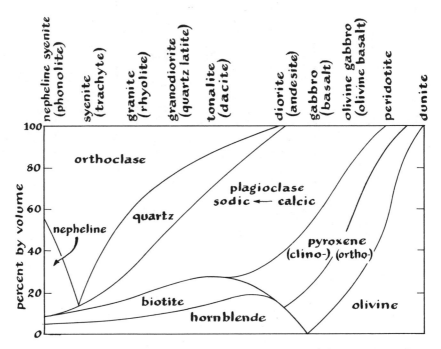

Figure 5-1 The approximate mineralogical composition of the commoner igneous rock types.

Igneous rocks are often described as ultrabasic, basic, intermediate, or acidic, a terminology based on the SiO_2 content, SiO_2 being considered the "acidic" oxide in these rocks. Rocks with less than 45 percent SiO_2, such as the peridotites and dunites, are considered ultrabasic; with 45–53 percent SiO_2, basic (gabbro family); with 53–65 percent SiO_2, intermediate (diorite, syenite, nepheline syenite families); above 65 percent, acidic (granodiorite and granite families).

CHEMICAL FRACTIONATION BY MAGMATIC CRYSTALLIZATION. The silicate minerals crystallizing from a magma remove from it not only the abundant elements but also other lithophile elements whose ionic size and coordination number enables them to enter the structures of these minerals. Thus nickel is largely removed in olivine, in which it substitutes for magnesium; gallium substitutes for aluminum in aluminosilicates, germanium for silicon; titanium, vanadium, and manganese are largely removed in ferromagnesian minerals such as the pyroxenes, hornblende, and biotite.

The lithophile elements which do not readily substitute for the major elements in the essential minerals of igneous rocks remain in solution and hence are enriched in the residual liquid of magmatic crystallization. These elements include B, Be, W, Nb, Ta, Sn, Th, U, Li, Cs, and the rare earths. They eventually crystallize as components of minerals in pegmatites and hydrothermal veins, which are the only economic source for most of them.

Chalcophile elements in a magma combine with sulfur and lesser amounts of arsenic, antimony, bismuth, selenium, and tellurium to form sulfides and related compounds. Pyrrhotite and pyrite may separate from a magma at an early stage of crystallization, but sulfide minerals characteristically segregate at a late stage and are ultimately deposited in hydrothermal veins. This is probably indicative of the low thermal stability of most sulfide minerals. The separation of sulfides must also depend upon a sufficient concentration of sulfide ions in the magma. This may not be attained until a large part of the lithophile elements have been removed by crystallization of the silicate minerals.

Pegmatites

Chemical and geological evidence indicates that the residual melt from the fractional crystallization of a magma will generally be a siliceous liquid rich in alkalies and aluminum, containing water and other volatiles, and with a concentration of those minor elements that are not incorporated in the structures of the common minerals of igneous rocks. Such a residual liquid will probably be unusually fluid for a silicate melt because of the concentration of volatiles. The pressure of volatiles will also provide the driving force to inject the liquid along surfaces of weakness in the surrounding rocks, which may be the already solid part of the same igneous intrusion or the country rock. In this way pegmatites and hydrothermal veins may be formed.

Pegmatites are found in association with many plutonic rocks, but most commonly with granites, as is to be expected if granites themselves are the products of fractional crystallization of a magma. Granite pegmatites consist essentially of quartz and alkali feldspar, generally with some muscovite and biotite also, and are thus similar on composition to granite; the essential differences are in texture, pegmatites being typically very coarse-grained, and in mode of occurrence, pegmatites being usually tabular or pipe-like in form.

Most pegmatites are chemically and mineralogically simple, but the complex ones are spectacular in their content of rare elements and unusual minerals, and hence have been intensively studied. Pegmatites are economically important and have been worked for industrial minerals such as feldspar, muscovite, phlogopite, tourmaline (gem quality), and quartz, and for minerals used as raw materials for rare elements such as beryllium (beryl), niobium and tantalum (columbite-tantalite), lithium (spodumene, lepidolite, and other minerals), and tungsten (wolframite).

Hydrothermal Deposits

Pegmatites often contain minor amounts of sulfides and arsenides. They show many of the characteristics of hydrothermal veins, and examples of pegmatites passing into sulfide-bearing quartz veins are known. Generally, however, hydrothermal deposits represent a later development than pegmatites, having been formed from cooler and more dilute solutions.

The characteristic type of hydrothermal deposit is the sulfide-bearing vein, formed by the filling of fractures or fissures in the country rocks. Many important mineral deposits are of this kind, and they often provide well-crystallized specimens of the ore minerals for collectors and collections. However, not all hydrothermal deposits are veins; many are irregular masses often partly or wholly replacing pre-existing rocks, and in some, such as the famous "porphyry copper" deposits of Utah and Arizona, the ore minerals are widely disseminated in small amounts throughout a large body of rock.

Lindgren, who made a particular study of hydrothermal deposits, pointed out that they could be broadly grouped into three types whose mineralogy and mode of occurrence indicated different conditions of origin. These three types are: *hypothermal* deposits, formed at fairly high temperatures (300–500°C), generally at considerable depths in the crust; *mesothermal* deposits, formed at moderate temperatures (200–300°C); and *epithermal* deposits, formed at comparatively low temperatures (50–200°C). These three types grade into one another, and the boundaries between them are necessarily diffuse, but many mineral deposits can readily be recognized as one or another of these types from their mineralogy and the geological environment. Typical hypothermal deposits are the tin (cassiterite) and tungsten (scheelite and wolframite) veins, and molybdenite deposits such as that at Climax, Colorado. Quartz is the predominant gangue mineral, and it is often accompanied by tourmaline, topaz, and other silicates. Mesothermal deposits typically carry sulfides of iron, lead, zinc, and copper, with a gangue that is generally quartz but which may include carbonates such as calcite, rhodochrosite, or siderite; many important gold-bearing quartz veins are probably mesothermal. The great disseminated copper deposits of the southwestern United States are mesothermal. Epithermal deposits are important producers of antimony (stibnite), mercury (cinnabar), silver (native silver, and silver sulfides) and gold; the lead-zinc deposits of the Tri-State region of Missouri, Kansas, and Oklahoma are probably of this type. Gangue minerals include quartz (often in the form of chalcedony), opal, calcite, aragonite, dolomite, fluorite, and barite.

Alteration of Vein Minerals

Sulfide minerals are especially subject to alteration. They oxidize readily to sulfates, many of which are soluble in water. As a result, the outcrop of a sulfide-bearing vein commonly contains no sulfide, and is a cavernous mass of rusty quartz (*gossan*). The metalliferous material that has been removed in solution is often redeposited at greater depths, giving rise to a zone of secondary enrichment (Fig. 5-2).

Between the ground surface and the top of the water table is a zone of active circulation of air and water. Sulfides oxidize to sulfates, and the metals are carried away in solution, except for iron (oxidation of iron sulfate leads to the precipitation of ferric oxides). The metal-bearing solutions may not travel

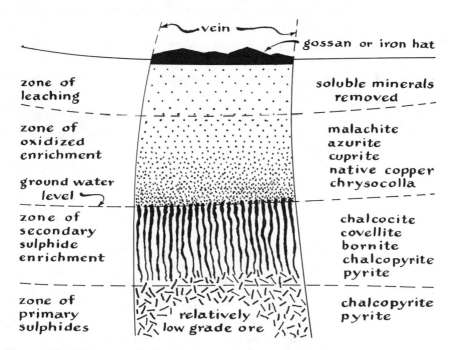

Figure 5-2 Diagrammatic sketch of zones of alteration in a sulfide-bearing vein.

very far, however, before reprecipitation occurs. Carbon dioxide or carbonate rock will precipitate copper as malachite and azurite, zinc as smithsonite, lead as cerussite; other minerals that may be formed are cuprite and native copper, hemimorphite, and anglesite. In this way the metal content may be concentrated and rich ore formed; this zone also produces fine specimens of these secondary minerals.

If the metal-bearing solutions are carried down into the zone of groundwater, conditions alter sharply from oxidizing to reducing, since groundwater is generally deficient in oxygen. A different type of secondary enrichment now takes place, which is controlled by the affinity of the different metals for sulfide. Copper has a strong affinity for sulfur, and copper-bearing solutions react with pre-existing sulfides such as pyrite and chalcopyrite to give *secondary sulfides* richer in copper, such as covellite and chalcocite. In this way a zone of secondary sulfide enrichment is produced, with a marked concentration of copper in comparison with the primary ore, into which the secondary enriched zone grades at greater depth.

Deposits of Hot Springs and Fumaroles

Hydrothermal solutions may eventually reach the surface and appear as hot springs. Generally they have become so diluted with groundwater that their mineral content is quite low, and around most hot springs the only mineral being deposited is opaline silica (siliceous sinter). Sometimes, however, small amounts of sulfur and sulfides are formed, and at Sulphur Bank in California a hot spring deposit has been worked as a mercury mine.

More interesting for a mineralogist are the fumaroles in active volcanic regions, where hot gases are actively depositing minerals. This very restricted (in time and place) environment has been intensively studied, since it is one of the few in which we can actually observe minerals in the process of formation. The common fumarolic minerals are sulfur and chloride minerals, especially ammonium chloride, but numerous common and rare minerals have been found—magnetite, hematite, molybdenite, pyrite, realgar, galena, sphalerite, and many others. The classic locality for such minerals is Mt. Vesuvius in Italy, where over fifty different minerals have been collected from fumaroles.

THE SEDIMENTARY ENVIRONMENT

Sedimentary Processes and Mineral Formation

Sedimentation is, in effect, the result of the interaction of the atmosphere and hydrosphere on the crust of the earth. The original constituents of the crust, the minerals of igneous rocks, are more or less readily attacked by air and natural waters. Having been formed at high temperatures, and sometimes at high pressures as well, they cannot be expected to remain stable under the very different conditions at the earth's surface. Of the common minerals of igneous rocks, only quartz is highly resistant to weathering processes. All other minerals tend to alter; they are attacked by the action of oxygen, carbonic acid, and water, and new minerals are formed which are more stable under the new conditions. The altered rock crumbles under the mechanical effects of erosion, and its constituents are transported by wind, water, or ice and redeposited as sediments or remain in solution.

The primary problem in the mineralogy of sedimentary rocks is the chemical breakdown of some minerals and the formation of others. Since silicates constitute more than 90 percent of the earth's crust (including quartz as a silicate), their stability in the sedimentary environment is a basic factor. Among the silicate minerals, however, there are marked differences in the susceptibility to chemical breakdown by weathering. Of the minerals of igneous rocks the ferromagnesian minerals are more susceptible than the feldspars, and the feldspars are more susceptible than quartz; within the feldspar goup the calcic plagioclases are more readily decomposed than the alkali feldspars. Expressed in structural terms, the resistance to attack by weathering increases as the complexity of silicate linkage increases, the nesosilicates* (olivine) being more readily broken down than the inosilicates* (pyroxene and hornblende), these, in turn, being more readily broken down than the phyllosilicates* (biotite and muscovite) and the tektosilicates* (feldspars and quartz). A corollary of this is that newly formed minerals in sediments are either phyllosilicates or tektosilicates.

The processes by which silicate minerals are broken down chemically during weathering have long been a subject of speculation. It was early found that the univalent and bivalent cations readily go into solution, but the fate of the aluminum and silicon has been less well understood. Older hypotheses assumed hydrolysis of aluminosilicates with the formation of colloidal silicic acid and aluminum hydroxide, which later reacted to give clay minerals. However, recent investigations have shown that during the initial attack a silicate mineral goes into ionic solution, and even the silica and the alumina are, at least for a short time, in true solution. At the surface of a crystal unsatisfied valences exist, which are the loci of reaction with water molecules. Hydration and hydrolysis follow, whereby strong bases such as potassium, calcium, and magnesium are removed and oxygen anions in the crystal lattice may be partly replaced by hydroxyl ions. Aluminum and silicon attract OH ions strongly; aluminum

---

* These terms are explained in chapter 15.

probably groups six OH ions around it to assume its preferred sixfold coordination, whereas silicon remains four-coordinated. When first set free these elements are in ionic solution, but the ions tend to aggregate and form clusters of colloidal size. When first formed these colloidal aggregates are probably amorphous, but, on aging, orientation into definite crystal lattices, such as those of the clay minerals, takes place. Some silicate minerals may not undergo complete breakdown of the lattice during weathering; for example, biotite and muscovite may pass directly to clay minerals by ionic substitution, whereby fragments of the sheet structures are directly incorporated in the new minerals. The ultimate fate of different elements thus depends largely on the relative stability of their ions in water. The most stable are the alkali metal ions, followed by those of the alkaline earths, and they are for the most part carried away in solution. Silicon, aluminum, and iron, on the other hand, are generally soon redeposited as insoluble compounds; new minerals are formed from them at an early stage of weathering.

The behavior of a chemical element during sedimentary processes has been found to depend largely on the relationship between ionic size and ionic charge. This relationship is shown graphically in Fig. 5-3. The diagram shows that the elements fall into three coherent groups: (1) those that form stable soluble

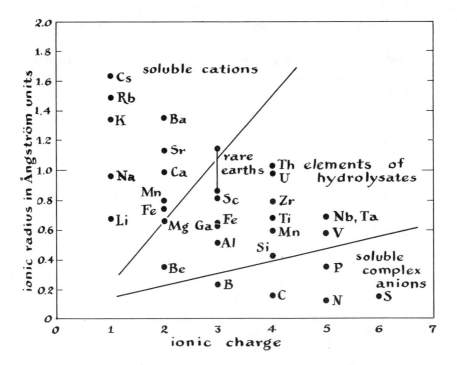

Figure 5-3 Ionic potential and the behavior of elements in sedimentary processes.

cations, and hence are carried away in solution; (2) those whose cations are not stable in natural waters, and which are deposited as oxides, hydroxides, or hydrated aluminosilicates; (3) those that form complex anions with oxygen, and which are more or less soluble in natural waters.

The ratio charge/radius for an ion is known as the *ionic potential*, and is of great significance for many of its properties in the presence of water. Elements with an ionic potential of 3.0 and less belong to the group of soluble cations in Fig. 5-3; those with an ionic potential between 3.0 and 10 are the elements of hydrolysates; and those with an ionic potential greater than 10 form complex anions.

Since the ionic potential is the quotient of charge and radius, it is clear that elements of variable valency have different ionic potentials for each valence stage. This is reflected in their behavior during weathering. For example, iron and manganese in silicate minerals are largely in the bivalent state, and go into solution in this state, since their ionic potentials are lower than 3.0. However, the bivalent ions readily oxidize to the trivalent state in iron and to the quadrivalent state in manganese, and the ionic potentials for these states are much higher and fall within the field of the hydrolysates. Precipitation of iron and manganese minerals in sediments thus involves oxidation followed by hydrolysis.

The other elements which form soluble cations behave in two contrasting ways in sedimentary processes, depending upon the solubility of their carbonates. The carbonates of the alkaline earth elements are insoluble or only slightly soluble in water, and hence these elements are readily precipitated from solution by carbon dioxide in the atmosphere or in natural waters; the carbonates of the alkali metals are soluble, and these elements are not precipitated in this way. The alkali metals thus remain in solution and eventually reach lakes or the ocean, from which their salts may be eventually deposited by evaporation.

Another sedimentary environment that exists in restricted regions is that characterized by reducing conditions. Most sedimentary environments are oxidizing, due to the almost universal presence of oxygen in the atmosphere or in solution. However, in regions cut off from the atmosphere, and in which the natural waters have been exhausted of oxygen, strongly reducing conditions may arise. Under these conditions sulfide minerals may form, such as pyrite or marcasite or, in the absence of sulfide ions, the mineral siderite, $FeCO_3$. Important deposits of native sulfur may result from the reduction of sulfate ions to elemental sulfur, apparently in part at least by bacterial action.

Classification of Sediments

On the basis of these considerations, sediments can be classified into the following broad groups.

1. *Resistates*. These sediments are made up of the primary minerals that resist weathering and are deposited unchanged. The commonest is quartz, but many accessory minerals may be present.

2. *Hydrolysates*. These sediments comprise several distinct types, differing in chemical composition and mineralogy. They include the clays, made up of aluminosilicate minerals; the sedimentary deposits of iron silicates; chemically deposited silica (flint, chert, and opal); and bauxite, which consists essentially of hydrated aluminum oxide.

3. *Oxidates*. The principal types are the sedimentary iron and manganese oxides.

4. *Reduzates*. These sediments include deposits of sedimentary sulfides, sedimentary sulfur, and siderite. Coal and petroleum are reduzate sediments.

5. *Precipitates*. These comprise the limestones and dolomites, made up of calcium and magnesium carbonates (calcite, aragonite, and dolomite); sedimentary apatite (phosphorite) can be included.

6. *Evaporites*. These may be grouped into two genetic types, the marine and the nonmarine. The minerals include calcium sulfates, magnesium chlorides and sulfates, sodium chloride and sulfates (and carbonates, borates, and nitrate in nonmarine deposits), and potassium chloride.

This classification of sediments is summarized in Table 5-VII.

Mineralogy of Resistates

The resistates form the important group of sands and sandstones. The total number of minerals recorded from sands and sandstones is very large, since practically any mineral of igneous or metamorphic origin may have at least a transitory existence in the sedimentary environment. Nevertheless, most minerals can be decomposed and removed by intense weathering. Quartz is the most resistant of the common minerals, and since it is abundant both in igneous and metamorphic rocks it forms the major part of resistate sediments, sometimes to the extent that they contain more than 99 percent of SiO_2, thereby becoming important sources of silica for industrial use. Detrital feldspar is also common in sands and sandstones but is much less resistant than quartz, being decomposed by prolonged weathering. Calcic feldspar is more readily decomposed than alkali feldspar, and is sometimes transformed into calcium zeolites (heulandite, chabazite, and laumontite) in sediments. Other minerals resistant to weathering and which may accumulate in sands and sandstone include zircon, garnet, topaz, columbite, tantalite, andalusite, magnetite, ilmenite, rutile, monazite, cassiterite, gold, and the platinum minerals, and some of them may be extracted for industrial use; such economically valuable deposits are known as *placers*.

Table 5-VII *Classification and composition of sediments*

CLASS	RESISTATES	HYDROLYSATES	OXIDATES	REDUZATES	PRECIPITATES	EVAPORITES
Elements	Si	Al, Si, Fe	Fe, Mn	Fe, S	Ca, Mg	Ca, Mg, Na, K
Minerals	Quartz Accessory minerals	Clay minerals Al hydroxides Glauconite Chamosite	Hematite Goethite Pyrolusite Psilomelane	Pyrite Marcasite Siderite Sulfur	Calcite Aragonite Dolomite	Gypsum Anhydrite Halite Sylvite Carnallite Kernite Soda-niter

Mineralogy of Hydrolysates

The hydrolysate minerals are formed by the chemical decomposition of pre-existing silicate minerals. The most widespread are the clay minerals, which are hydrous aluminosilicates all with the phyllosilicate structure and with very small grain size (generally less than 0.004 mm in diameter and ranging down to colloidal dimensions). The commonest clay minerals are kaolinite, montmorillonite, illite, and chlorite, and they often occur together in a particular sediment.

Much remains to be learned of the mutual relations of the clay minerals and of the conditions favoring the formation of one in preference to another. The primary factors determining the nature of a clay are, firstly, the chemical character of the parent material, secondly, the physicochemical environment in which alteration of this material takes place, and, thirdly, the environment of deposition and diagenesis.* The structure of kaolinite does not accommodate cations other than silicon and aluminum, whereas montmorillonite, illite, and chlorite always contain other elements, especially magnesium and iron. The formation of kaolinite is favored by an acid environment, in which all the bases tend to be removed in solution. Hence kaolinite forms most readily in an environment characterized by a minimum of available components other than silica, alumina, and water, whereas the formation of the other clay minerals requires the presence of additional cations. Alkali feldspars tend to alter to kaolinite, and ferromagnesian minerals, calcic feldspears, and volcanic glass commonly alter to montmorillonite. Montmorillonite is the most reactive of the clay minerals, and generally it does not survive in pre-Mesozoic sediments. It has a close structural relationship to illite and chlorite, and evidently changes readily to these minerals, especially in the marine environment. The comparatively high concentration of potassium and magnesium in sea water promotes this change. Illite is the commonest clay mineral in marine sediments and sedimentary rocks. In general, diagenesis promotes the formation of illite and chlorite and the disappearance of kaolinite and montmorillonite; consequently, shales and argillites consist largely of the former minerals.

Glauconite is essentially a potassium iron aluminosilicate which occurs in small grains in sandstones, limestones, and shales of marine origin. It has been found in material dredged up from the sea floor, where it evidently is forming at the present time. In structure and composition it is related to muscovite; detrital biotite may alter to glauconite in the marine environment, but most glauconite is probably built up from amorphous material. Another iron alumino-

---

* Diagenesis: Those changes of various kinds occurring in sediments between the time of deposition and the time at which complete lithification takes place.

silicate which is an important constituent of some sedimentary rocks, especially some sedimentary iron ores, is chamosite, a member of the chlorite group.

Silica in sedimentary rocks is present mainly as detrital quartz, but important amounts are transported in solution and redeposited, generally as chalcedony in the form of chert and flint. Sometimes opal is formed, and probably much opal has dehydrated and crystallized as chalcedony.

Under tropical conditions, especially those of contrasting wet and dry seasons, weathering sometimes results in complete decomposition of aluminosilicates and removal of silica, leaving a residue largely of hydrated aluminum oxides, gibbsite, $Al(OH)_3$, disapore, $HAlO_2$, or boehmite, $AlO(OH)$. This residue is known as bauxite, and it is important commercially as aluminum ore.

Mineralogy of Oxidates

The commonest oxidate is ferric hydroxide, produced by the oxidation of ferrous compounds in solution. It may be deposited as goethite, $HFeO_2$, or as hematite, Fe_2O_3. Goethite and hematite occur in sedimentary rocks either admixed with sands and clays, which they color brown or red, or as more or less pure bodies, which if large enough are valuable iron ores. Limonite is very fine-grained material which consists essentially of goethite. Manganese is another element that is deposited as an oxidate; it goes into solution in the bivalent form, which is readily oxidized to manganite, $MnO(OH)$, pyrolusite, MnO_2, or psilomelane, a complex mineral consisting largely of MnO_2 with small amounts of other bases, generally barium or potassium.

Mineralogy of Reduzates

Reduzates are relatively uncommon, since reducing conditions exist at the earth's surface only where oxygen is unable to penetrate. Some marine shales were evidently deposited in troughs on the ocean floor where stagnant conditions and decomposing organic matter caused depletion of oxygen and development of H_2S; under these conditions iron sulfide is formed, and it appears in the sediments as pyrite or marcasite. The formation of marcasite is favored by more acid conditions than those giving rise to pyrite. On land, accumulating plant debris which eventually gave rise to coal seams also created highly reducing conditions, which often brought about the deposition of ferrous carbonate, now present as siderite; some formations associated with coal seams have been sufficiently rich in siderite to be worked as iron ores. A special case of mineral formation by reduction is the sulfur associated with salt domes and petroleum in the coastal region of Louisiana and Texas; reducing conditions associated with bacterial activity have produced sulfur from anhydrite, the chemistry of the process being the reduction of sulfate to free sulfur, the calcium being redeposited as calcite.

Mineralogy of Precipitates

The important minerals of precipitate sediments are the carbonates calcite, aragonite, and dolomite. Chert, flint, and siliceous sinter can be considered as precipitate sediments, but they have been discussed previously. The preeminent place of carbonate deposition is in the oceans, especially on the continental shelves. Much of this deposition is through the agency of lime-secreting organisms, but direct precipitation is also significant, especially in warm tropical seas where the water is saturated with calcium carbonate. Calcite is the more stable form of calcium carbonate, and hence deposition is generally as calcite, but aragonite is sometimes formed, particularly in organisms. It generally inverts to calcite within a short time (geologically speaking) but may persist unchanged (when it is probably stabilized by foreign ions in the structure); primary aragonite has been recognized in rocks as old as upper Paleozoic. Considerable amounts of calcite and aragonite are deposited in terrestrial environments, such as in limestone caverns (stalactites and stalagmites), around springs whose water is saturated with $CaCO_3$ (travertine, calc-sinter), from groundwater in semi-arid regions (caliche), and in salt lakes (aragonite oolites are depositing in Great Salt Lake, Utah, at the present time).

The origin of sedimentary dolomites has been the subject of much discussion. Geological evidence indicates that many dolomites have been formed from limestones by the action of magnesium-bearing waters. In many instances sea water acting on calcium carbonate has evidently been responsible; thus, bores in some coral reefs show that the dolomite content of the rock increases with depth, presumably as a result of the continuing action of the magnesium in solution in circulating sea water. Studies on the energy change in the reaction $2CaCO_3 + Mg^2 = CaMg(CO_3)_2 + Ca^2$ show that under the conditions of temperature and magnesium concentration which prevail in the sea, dolomitization will proceed spontaneously, although the rate of the reaction is generally slow. The spontaneous nature of dolomitization is also indicated by the increase in the amount of dolomite relative to limestone in older formations, which have been longer subjected to this process.

One marine precipitate, of unusual occurrence but of great economic importance as a source of phosphate fertilizer, is phosphorite. Phosphorite consists of a variety of apatite and occurs interbedded with other marine sediments sometimes over wide areas. Some especially important deposits are those of North Africa (Cretaceous), Florida (Tertiary), and Idaho and Montana (Permian). These deposits seem to be largely the result of precipitation on a sea floor where little other sedimentation was taking place. Ocean bottom waters are essentially saturated with calcium phosphate, and slight changes in physicochemical conditions will cause the precipitation of phosphorite. If other sedimentation is lacking or in small amount, extensive beds of more or less pure phosphorite will result.

Mineralogy of Evaporites

The evaporites may be divided into two groups, the marine and the nonmarine, according to whether thay have been deposited from bodies of sea water cut off from the ocean or from interior salt lakes. The evaporites are of special interest because of their significance in the interpretation of geological history, as indicators of arid conditions, and because they are a type of deposit whose mode of formation can readily be imitated in the laboratory.

As sea water evaporates under natural conditions calcium carbonate is the first solid to separate. The precipitation of calcium carbonate may be followed by that of dolomite, but there is no evidence that extensive deposits of dolomite have been formed in this way. Indeed, evaporation of sea water in a closed basin cannot give rise to thick carbonate deposits—sea water 1,000 meters deep would give only a few centimeters of limestone unless the waters were continually replenished.

With continued evaporation calcium sulphate is deposited. Depending on temperature and salinity, either gypsum or anhydrite may be formed. In salt solutions of approximately the composition of sea water at 30°C gypsum will begin to separate when the salinity has increased to 3.35 times the normal value; after nearly half the total amount of calcium sulfate has been deposited, anhydrite becomes the stable phase. When the solution has been concentrated to a tenth of the original bulk, halite starts to separate. Anhydrite and halite then precipitate together until the field of stability of polyhalite, $K_2Ca_2Mg(SO_4)_4 \cdot 2H_2O$, is reached.

Most evaporite deposits contain calcium carbonate, calcium sulfate, and sodium chloride; evidently conditions under which other salts could be deposited have seldom been attained. Only when an evaporating body of sea water has been reduced to 1.54 percent of the original volume do potassium and magnesium salts begin to crystallize. Important deposits of these salts of Permian age are worked in Germany, in the Texas–New Mexico area of the USA, and in the province of Perm in the USSR. Very large reserves of potassium salts occur in the Devonian of western Canada, underlying a large section of the great plains in central Saskatchewan.

The nonmarine evaporites are very limited in occurrence and extent, but they are extremely important economically, since they provide most of the world's supply of boron compounds and iodine, all the naturally occurring nitrates, some lithium, much sodium carbonate, sodium sulfate, and sodium chloride, some potassium salts, some bromine, and some gypsum.

The United States produces almost all the boron compounds now used in the world, partly from the brines of Searles Lake, partly from the deposit of sodium borate minerals near Boron, both in the Mojave Desert in southern California. The boron was probably brought to the surface originally by fumarolic and hot spring activity associated with the extensive volcanism of Tertiary and Quaternary time in this region. It was eventually concentrated in lake deposits. Boron compounds are often present in volcanic exhalations, and boric acid is extracted from volcanic steam at Lardarello in Italy.

The only nitrate deposits of commercial importance are those of sodium nitrate in the desert region of northern Chile. In addition to sodium nitrate, they also contain sodium chloride and sodium sulfate and minor amounts of other compounds, including some iodates, from which a large part of the world's supply of iodine is derived as a by-product of nitrate production. No generally accepted explanation of the source of the Chilean nitrate has been developed, but it seems definite that the nitrate was finally collected in saline lakes from which it was deposited by evaporation.

THE METAMORPHIC ENVIRONMENT

Metamorphism may be defined as the sum of the processes that, working below the zone of weathering, cause the recrystallization of rock material. During metamorphism the rocks remain essentially solid; if remelting takes place, a magma is produced, and metamorphism passes into magmatism. Metamorphism is induced in solid rocks as a result of pronounced changes in temperature, pressure, and chemical environment. These changes affect the physical and chemical stability of a mineral assemblage, and metamorphism results from the establishment of a new equilibrium. In this way the constituents of a rock are changed to minerals that are more stable under the new conditions, and these minerals may arrange themselves with the production of textures that are likewise more suited to the new environment. Metamorphism thus results in the partial or complete recrystallization of a rock, with the production of new textures and new minerals.

Heat, pressure, and action of chemically active fluids are the impelling forces in metamorphism. Heat may be provided by the general increase of temperature with depth or by contiguous magmas. Pressure may be resolved into two kinds: hydrostatic or uniform pressure, which leads to change in volume; and directed pressure or shear, which leads to change of shape or distortion. Uniform pressure results in the production of granular, nonoriented structures; directed pressure results in the production of parallel or banded structures. Uniform pressure affects chemical equilibria by promoting a volume decrease, i.e., the formation of minerals of higher density. The action of chemically active fluids is a most important factor in metamorphism, since even when they do not add or subtract material from the rocks they promote reaction by solution and redeposition. When they add or subtract material the process is called *metasomatism*. Probably some degree of metasomatism accompanies most metamorphism. Water is the principal chemically active fluid, and it is aided by carbon dioxide, boric acid, hydrofluoric and hydrochloric acids, and other substances, often of magmatic origin.

Types of Metamorphism and Metamorphic Rocks

Two major types of metamorphism are commonly recognized: *thermal* or *contact metamorphism*, and *regional metamorphism*.

Contact metamorphism is the type of metamorphism developed around bodies of plutonic rocks. Here the temperature of metamorphism has been determined mainly by proximity to the intrusive magma, which may also have given off chemically active fluids that stimulated recrystallization of the country rock. Contact aureoles may vary in width from a few feet to a few thousand feet, measured normal to the igneous contact. The width will vary according to the size of the intrusion—the contact zone around a dike or a sill will be quite narrow, whereas one around a granite pluton may be traceable for hundreds or thousands of feet into the country rock. Broad contact aureoles generally show zones of differing mineralogy outward from the contact, resulting from the decreasing temperatures from the contact into the unaltered country rock and/or differing degrees of metasomatism.

The characteristic rock type of contact metamorphism is *hornfels*, which is a dense, granular rock, sometimes with one or more minerals prominent as larger crystals (porphyroblasts). Different compositional types of hornfels are distinguished by qualifying adjectives such as biotite hornfels, pyroxene hornfels, calc-silicate hornfels.

Sometimes, especially with carbonate rocks (limestone and dolomite) intruded by granitic magmas, a great deal of material has been added to and subtracted from the country rock by volatiles from the magma, and the process can be described as contact metasomatism rather than contact metamorphism. Many ore bodies of sulfides and of iron oxides are the product of contact metasomatism, and they are commonly accompanied by masses of calcium, magnesium, and iron silicates, especially garnets, pyroxenes, and amphiboles; such silicate masses are termed *skarn* or *tactite*.

Regional metamorphism, as the name implies, is metamorphism developed over large regions, often over thousands of square miles in the root regions of fold mountains and in Precambrian terranes. In any large region it is often possible to map zones of increasing degree of metamorphism on the basis of a sequence of mineralogical changes in rocks of uniform composition. This sequence of mineralogical changes evidently reflects a progressive increase in temperature. In metamorphosed argillaceous rocks, for example, successive zones characterized, respectively, by chlorite, biotite, almandite, kyanite, and sillimanite can often be delimited on a map by noting in the field the points at which these minerals first appear. Each of these minerals begins to form when a certain intensity, or *grade*, of metamorphism is reached, and the zonal boundaries, since they are lines of equal grade of metamorphism, are termed *isograds*.

The principal rock types of regional metamorphism are *gneisses* and *schists*. Gneisses are coarsely foliated rocks, the foliae being layers or lenses 1–10 mm thick of contrasting mineralogy, generally alternating light (quartz and feldspar) and dark (ferromagnesian) minerals. Schists are finely foliated rocks with a well-developed lamination along which the rock can be easily broken; the lamination is due partly to the foliation, partly to the parallel or subparallel orientation of flaky minerals such as the micas and chlorites, or prismatic minerals such as the amphiboles. Different kinds of gneisses and schists are generally distinguished by qualifying adjectives indicating prominent minerals—i.e., muscovite biotite gneiss, chlorite schist, hornblende schist—or indicating overall composition, i.e., granitic gneiss.

Low grade metamorphism of fine-grained rocks under high stress produces *slate*, in which only slight recrystallization has taken place, but in which a strongly developed rock cleavage has developed; this cleavage is often at an angle to the original bedding, which can be recognized by color or textural variations on the cleavage surfaces. With somewhat greater metamorphism slates pass into *phyllites*, in which recrystallization has produced a sheen on the cleavage surfaces through the presence of tiny crystals of chlorite and mica. With further metamorphism phyllites pass into schists.

The metamorphism of limestone and dolomite generally results in recrystallization without the production of schistose and gneissose structures, and the resulting product is a *marble*. Similarly, pure quartz sandstones recrystallize to a massive *quartzite*. Ultrabasic rocks are often converted by metasomatism largely into the mineral serpentine, forming serpentine rock or *serpentinite*.

Mineralogy of Metamorphic Rocks

The mineralogy of metamorphic rocks is determined by the bulk composition of the rock and the temperatures and pressures under which metamorphism took place. Metamorphism grades into magmatism when temperature, pressure, and chemical composition are such that part of the rock melts and the material becomes mobile. The onset of metamorphism is difficult to define, since diagenesis generally produces some recrystallization and sometimes the formation of new minerals. Apparently depths of 6–7 km and temperatures of about 150°C are insufficient to initiate metamorphism, since sedimentary rocks brought from these depths and temperatures by oil well drilling show little if any alteration. Such rocks, however, have not been subjected to strong folding and shearing, such as would be prevalent in orogenic belts and which would have a marked effect in promoting metamorphism. Under these conditions metamorphic recrystallization may begin at temperatures of 150–200°C. The temperature at which metamorphism grades into magmatism will depend on the composition of the material; material of granitic composition, under pressure of water vapor, begins to melt at about 650°C, whereas more basic material is more refractory. The average thermal gradient in the crust is about 30°C per kilometer, which indicates that temperatures around 650°C would be reached at depths somewhat greater than 20 km. The average thermal

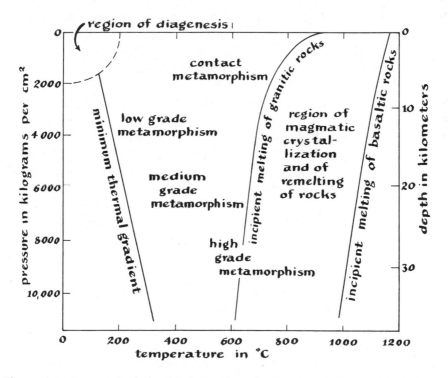

Figure 5-4 Suggested relationship between grade of metamorphism and temperature and pressure (and corresponding depth in the Earth's crust).

gradient varies considerably from place to place, and of course greatly increases in regions of igneous activity.

In considering the mineralogy of metamorphic rocks we may divide the field of regional metamorphism into low, medium, and high grade, as illustrated in Fig. 5-4. These are arbitrary divisions of a continuous phenomenon, and their boundaries are difficult to define. Temperature is the principal controlling factor in metamorphism, but the temperature limits of the different grades of metamorphism are only vaguely known. In practice we estimate grade of metamorphism largely from the mineralogy, since the appearance of some minerals and the disappearance of others, or the change of composition in a solid solution series, is often symptomatic of the intensity of metamorphism. For example, biotite is a common mineral of metamorphic rocks; however, it does not occur in very low grade rocks, its composition then being represented by mixtures of muscovite and chlorite, nor does it occur in very high grade rocks, its place being taken by potash feldspar and garnet. Minerals of the epidote group are characteristic of low grade rocks; at higher grades of metamorphism these minerals are largely converted to anorthite, which enters into plagioclase. Thus parallel with the disappearance of the epidote minerals the anorthite content of the plagioclase usually increases; in lowest grade rocks the plagioclase is nearly pure albite, and it increases in calcium content as the grade of metamorphism increases. Other minerals, such as quartz, can be formed at any grade of metamorphism, and thus appear in all metamorphic rocks of appropriate composition. The relationship between grade of metamorphism and the occurrence of some of the common minerals of metamorphic rocks is shown diagrammatically in Table 5-VIII.

A particularly instructive example of mineral transformations during metamorphism, and one for which the chemical reactions are comparatively simple, is the metamorphism of limestones and dolomites. If calcite is heated in air, it decomposes at about 980°C according to the equation

$$CaCO_3 = CaO + CO_2$$

This reaction seldom takes place under geological conditions for the following reasons: silica is usually present, and calcium silicates, not calcium oxide, are then formed; even if silica is absent, a temperature of 980°C is rarely attained. Calcium oxide has been recorded as having formed in blocks of limestone caught in the basaltic lava at Mt. Vesuvius. In any case CaO is a very reactive material, combining immediately with water to form $Ca(OH)_2$; the latter compound is known as a mineral under the name portlandite, and has been recorded from two localities, where it has probably formed by the hydration of CaO. As a result of the high temperature of decomposition of calcite, metamorphism of a pure limestone usually results in a simple recrystallization into a calcite marble.

Table 5-VIII *Relationship between chemical composition, grade of metamorphism, and minerals formed*

CATIONS PRESENT	GRADE OF METAMORPHISM		
	LOW	MEDIUM	HIGH
Si	————————————Quartz————————————		
Si, Al	———Kyanite———		
	————————Andalusite————————		
		———————Sillimanite———————	
	————————Staurolite————————		
Si, Al, Fe			
Si, Mg	———Serpentine———		
	———Talc———		
	————————Forsterite————————		
Si, Mg, Fe	———Anthophyllite———		
	Cummingtonite		
	———Enstatite———		
	Hypersthene		
Si, Mg, Fe, Al	———Chlorite———		
	————————Cordierite————————		
	———Almandite———		
	———Pyrope———		
Si, Ca		————————Wollastonite————————	
Si, Ca, Mg, Fe	———Tremolite———		
	Actinolite		
	———Diopside———		
	Hedenbergite		
Si, Ca, Mg, Fe, Al	————————Hornblende————————		
	———Augite———		
Si, Ca, Al	———Epidote Group———		
	Ca–zeolites		
	Prehnite		
Si, Ca, Na, Al	———Albite———		
	————————Plagioclase————————		
	————————Scapolite————————		
Si, K, Al	————————Muscovite————————		
	————————Microcline————————		
	———Orthoclase———		
Si, K, Al, Fe, Mg	————————Biotite————————		

The thermal decomposition of dolomite takes place in two stages, the first stage being the decomposition of the magnesian component:

$$CaMg(CO_3)_2 = MgO + CaCO_3 + CO_2$$

This reaction takes place at about 375°C if the CO_2 produced by the reaction can escape freely. The formation of a periclase-calcite rock as a result of the metamorphism of dolomite has been recorded at a number of localities. Periclase hydrates readily to give brucite, and the ultimate product is usually a brucite marble (sometimes known as predazzite or pencatite), often with unaltered cores of periclase enclosed within the brucite. If the temperature is never high enough for the decomposition of the magnesium component, metamorphism of a pure dolomite is a simple recrystallization to give a dolomite marble.

When silica is available, either as admixed quartz or introduced by metasomatizing solutions, the metamorphism of limestones and dolomites follows a different course, with the formation of calcium and magnesium silicates. For a siliceous limestone the first reaction is the formation of wollastonite:

$$CaCO_3 + SiO_2 = CaSiO_3 + CO_2$$

Laboratory investigations have shown that this reaction can take place at about 500–600°C, if the carbon dioxide can escape freely. If the limestone contains only a minor amount of silica, this will be completely converted into wollastonite and the remaining calcite will simply recrystallize, the resulting rock being a wollastonite marble. If the temperature continues to rise, the wollastonite and calcite react to give complex silicate-carbonate minerals and ultimately larnite, Ca_2SiO_4. This reaction may be written

$$CaSiO_3 + CaCO_3 = Ca_2SiO_4 + CO_2$$

The metamorphism of siliceous dolomite is more complicated, in that a greater variety of minerals is possible. However, the reactions reflect the greater affinity of silica for magnesia than for lime, and therefore magnesium-rich silicates form first during the metamorphism of dolomites and continue to form as long as magnesia and silica are available. The net result is the decomposition of dolomite and the liberation of calcite, a process which is sometimes described as *dedolomitization*.

The first reaction between dolomite and quartz results in the formation of talc; it may be represented by the equation

$$3CaMg(CO_3)_2 + 4SiO_2 + H_2O = Mg_3Si_4O_{10}(OH)_2 + CaCO_3 + 3CO_2$$

The next step is the formation of tremolite; it may be produced by the reaction of the previously formed talc with calcium carbonate, or directly from dolomite and quartz:

$$5CaMg(CO_3)_2 + 8SiO_2 + H_2O = Ca_2Mg_5(Si_4O_{11})_2(OH)_2$$
$$+ 3CaCO_3 + 7CO_2$$

The product is a tremolite marble, in which the excess carbonate crystallizes as calcite and dolomite. If sufficient silica is available, all the dolomite will be converted to tremolite and calcite. Under these circumstances a further rise in temperature results in the formation of diopside from tremolite, for which the following equation can be written.

$$Ca_2Mg_5(Si_4O_{11})_2(OH)_2 + 3CaCO_3 + 2SiO_2 = 5CaMg(SiO_3)_2 + 3CO_2 + H_2O$$

However, if the rock is deficient in silica, rising temperatures cause tremolite to react with excess dolomite with the formation of forsterite:

$$Ca_2Mg_5(Si_4O_{11})_2(OH)_2 + 11CaMg(CO_3)_2 = 8Mg_2SiO_4 + 13CaCO_3 + 9CO_2 + H_2O$$

Field observations on the appearance of diopside and forsterite in metamorphosed dolomites, and laboratory experiments, indicate that the formation of forsterite takes place at a higher temperature than that of diopside. All these reactions, however, take place at lower temperatures than that of the reaction of calcite and quartz to give wollastonite. The successive reactions indicate successively higher grades of metamorphism. The minerals formed are therefore indicators of metamorphic grade, and we can recognize the following sequence:

1. Talc   4. Forsterite
2. Tremolite   5. Wollastonite
3. Diopside

Beyond wollastonite we have larnite and a number of uncommon minerals.

METEORITES

A special type of mineral association which deserves mention is that observed in meteorites. Meteorites are rare, but they are of extreme scientific interest since they are the only samples we have of material from outside our planet. Their chemical and mineralogical composition thus provides significant evidence about the composition of other parts of the universe, and since meteorites may be fragments of a disrupted planet the interior of the earth has been inferred to be of similar material. Meteorites consist essentially of an iron-nickel alloy, or of crystalline silicate (mainly olivine and/or pyroxene), or a mixture of these. The silicate meteorites resemble ultrabasic igneous rocks in mineralogy and composition. Many systems of classification have been devised for meteorites, but they may be grouped most simply into three major classes: (1) siderites or irons, (2) siderolites or stony irons, (3) aerolites or stones.

The siderites consist essentially of an iron-nickel alloy (average composition

91 percent Fe, 8.5 percent Ni, 0.5 percent Co), generally with accessory troilite (FeS), schreibersite (Fe,Ni,Co)$_3$P, cohenite (Fe$_3$C), and graphite; diamond has been identified in the Canyon Diablo (Arizona) meteorite. The metal generally shows a definite structure, known as Widmanstätten figures, which is brought out by etching (Fig. 5-5). This structure is due to the intergrowth of lamellae of a nickel-poor alloy (kamacite) in a nickel-rich base (plessite), and is the result of exsolution of an originally homogeneous solid solution.

The siderolites are made up of nickel-iron and silicate in approximately equal amounts. The silicate is generally olivine, and occurs as rather large, rounded grains in a sponge-like network of nickel-iron (Fig. 5-6). Such meteorites are sometimes called pallasites.

The aerolites are divided into two groups, the chondrites and the achondrites, on the basis of texture. The chrondrites are so named because of the presence of chondri, which are small, rounded bodies consisting of olivine or pyroxene. The achondrites are without chondri. In addition to this textural distinction, the two groups of aerolites differ characteristically in composition; the average

Figure 5-5 Section of an iron meteorite showing Widmanstatten structure; black rounded masses are nodules of troilite. (Gibeon, Southwest Africa.) [Courtesy the American Museum of Natural History.]

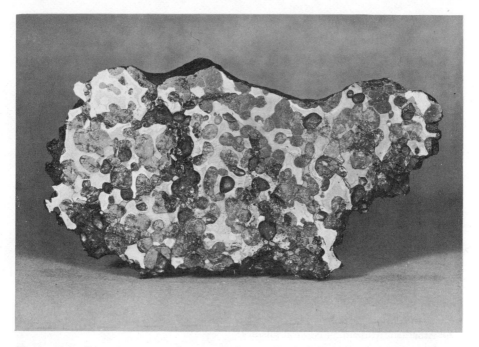

Figure 5-6 Stony-iron meteorite, rounded crystals of olivine in a nickel-iron groundmass. (Brenham, Kansas.) [Courtesy the American Museum of Natural History.]

percentage composition of the chondrites is about 12 nickel-iron, 40 olivine, 30 pyroxene, 6 troilite, and 11 plagioclase, whereas the average percentage composition of the achondrites is about 1 nickel-iron, 9 olivine, 60 pyroxene, and 25 plagioclase. The chondrites are far more common than the achondrites, making up over 90 percent of all aerolites. A typical chondrite is shown in Fig. 5-7.

SUMMARY

The mineralogy of the different associations discussed in this chapter is summarized in the following tabulation.

The Magmatic Environment

1. Igneous Rocks
   a. Silicates: quartz (tridymite, cristobalite), feldspars, feldpathoids, analcime, olivine, enstatite-hypersthene, augite, aegirine, hornblende, biotite, muscovite, zircon, sphene.
   b. Phosphates: apatite, monazite.
   c. Oxides: magnetite, ilmenite, chromite.

Figure 5-7   Stone meteorite, showing black glassy crust with flow lines, and typical fine-grained stony texture where crust has broken off. (Miller, Arkansas.) [Courtesy the American Museum of Natural History.]

      d. Sulfides: pyrite, pyrrhotite. `

      e. Elements: platinum group, diamond (in ultrabasic rocks).

  2. Pegmatites

      a. Silicates: quartz, feldspars (especially microcline and albite), feldspathoids, aegirine, hornblende, biotite, muscovite, phlogopite, tourmaline, spodumene, lepidolite, zircon, thorite, allanite, spessartine, beryl, topaz, scapolite.

      b. Phosphates: apatite, amblygonite, monazite.

      c. Oxides: magnetite, ilmenite, hematite, cassiterite, uraninite, columbite-tantalite, corundum.

      d. Halides: fluorite, cryolite.

      e. Sulfides: arsenopyrite, stibnite, bismuthinite, molybdenite.

      f. Elements: antimony, bismuth.

  3. Hydrothermal Deposits

      a. Silicates: quartz, feldspars, muscovite, chlorite, epidote, hornblende, tourmaline, zeolites, topaz, apophyllite, rhodonite, datolite, axinite, pectolite.

      b. Sulfates: barite.

      c. Carbonates: calcite, dolomite, ankerite, magnesite, rhodochrosite, witherite.

    d. Oxides: cassiterite, magnetite, hematite, ilmenite, uraninite, rutile, anatase, brookite.

    e. Halides: fluorite.

    f. Sulfides: chalcopyrite, bornite, chalcocite, enargite, sphalerite, galena, tetrahedrite, pyrite, marcasite, pyrrhotite, pyrargyrite, proustite, arsenopyrite, stibnite, cinnabar, and many others.

    g. Elements: gold, silver, arsenic, antimony, bismuth.

3A. Secondary Alteration Products of Ore Minerals

    a. Silicates: chrysocolla, hemimorphite.

    b. Sulfates, etc.: anglesite, crocoite, wulfenite.

    c. Phosphates, vanadates: carnotite, tyuyamunite, vanadinite, pyromorphite, mimetite.

    d. Carbonates: malachite, azurite, smithsonite, cerussite.

    e. Oxides: cuprite, hematite, goethite.

    f. Elements: silver, copper.

4. Fumarolic and Hot Springs Deposits

    a. Silicates: quartz (chalcedony), opal, zeolites.

    b. Sulfates: gypsum, alunite, and many others.

    c. Oxides: hematite, magnetite.

    d. Halides: halite, sylvite, salammoniac (NH_4Cl), and many others.

    e. Sulfides: pyrite, cinnabar, stibnite, covellite.

    f. Elements: sulfur.

The Sedimentary Environment

5. Resistates

    a. Silicates: quartz, feldspars, muscovite, biotite, garnet, tourmaline, staurolite, zircon, thorite, topaz, kyanite, andalusite.

    b. Phosphates: monazite.

    c. Oxides: magnetite, ilmenite, corundum, columbite-tantalite, cassiterite, rutile, spinel, chromite.

    d. Elements: gold, platinum metals, diamond.

6. Hydrolysates

    a. Silicates: quartz (chalcedony), opal, clay minerals, glauconite, chamosite.

    b. Oxides: bauxite.

7. Oxidates

    a. Oxides: limonite, hematite, pyrolusite, psilomelane.

8. Reduzates

    a. Carbonates: siderite.

b. Sulfides: pyrite, marcasite.

c. Elements: sulfur.

9. Precipitates

a. Phosphates: apatite (phosphorite).

b. Carbonates: calcite, aragonite, dolomite.

10. Evaporites

a. Sulfates: gypsum, anhydrite, many others.

b. Carbonates: calcite, aragonite, dolomite, sodium carbonates.

c. Borates: kernite, borax.

d. Nitrates: soda-niter.

e. Halides: halite, sylvite, carnallite, many others.

The Metamorphic Environment

11. Low Grade

a. Silicates: quartz, albite, talc, serpentine, chlorite, tremolite-actino-lite, epidote, muscovite, sphene, prehnite, tourmaline, pyrophyllite, spessartine.

b. Carbonates: calcite, dolomite, magnesite, siderite.

c. Oxides: rutile, anatase, brookite, magnetite, hematite, brucite.

d. Sulfides: pyrite, pyrrhotite.

e. Elements: graphite.

12. Medium Grade

a. Silicates: quartz, plagioclase, microcline, orthoclase, kyanite, anda-lusite, staurolite, serpentine, forsterite, anthophyllite, cumming-tonite, cordierite, garnet, hornblende, epidote, muscovite, biotite, tourmaline, scapolite, idocrase.

b. Carbonates: calcite, dolomite.

c. Oxides: rutile, magnetite, hematite, ilmenite, corundum, spinel.

d. Sulfides: pyrite, pyrrhotite.

e. Elements: graphite.

13. High Grade

a. Silicates: quartz, plagioclase, orthoclase, microcline, andalusite, sillimanite, forsterite, pyroxenes, cordierite, garnet, wollastonite, hornblende, scapolite, tourmaline, sphene.

b. Carbonates: calcite.

c. Oxides: magnetite, hematite, ilmenite, corundum, spinel, rutile.

d. Sulfides: pyrite, pyrrhotite.

e. Elements: graphite.

Meteorites

14. Iron Meteorites
    a. Elements: nickel-iron, graphite, diamond.
    b. Sulfides, etc.: troilite (FeS), schreibersite (Fe,Ni,Co)$_3$P, cohenite (Fe$_3$C).

15. Stony Meteorites
    a. Silicates: olivine, enstatite, hypersthene, diopside, plagioclase.
    b. Sulfides: troilite (FeS).
    c. Elements: nickel-iron, graphite.

Exercises

1. Enumerate the minerals you would expect to find in the following materials, showing by equations the possible reactions producing them:
   a) a metamorphosed dolomite which contained some quartz
   b) a metamorphosed limestone containing some kaolinite
   c) a vein containing quartz, chalcopyrite, and pyrite which has been subjected to oxidation and secondary enrichment
   d) a vein containing iron-rich sphalerite, galena, and calcite which has been subjected to oxidation and secondary enrichment

2. How would you extract the minerals with density greater than 2.8 from a friable sandstone? What heavy minerals would you expect to find if the sandstone has been formed by the erosion of an area of (a) high-grade aluminous schists; (b) basaltic rocks; (c) granites and granite pegmatites containing tin, tungsten, and tantalum?

3. Describe the mineralogy of each of the following groups of rocks: andesites, nepheline syenites, marine shales, sedimentary manganese ores, metamorphosed calcareous sandstone.

4. Describe the typical environment in which you would expect to find each of the following minerals: limonite, kyanite, muscovite, cassiterite, quartz, calcite, biotite, epidote, tridymite, glauconite, kaolinite, olivine, labradorite.

5. Give the mode of occurrence of each or the following sets of minerals and add to each group one additional mineral that might be present:
   a. quartz, microcline, lepidolite.
   b. sulfur, cinnabar, pyrite.
   c. limonite, quartz, malachite.
   d. microcline, nepheline, hornblende.
   e. calcite, anhydrite.
   f. wollastonite, diopside.
   g. bauxite.
   h. fluorite, quartz, calcite.
   i. tridymite, sanidine, biotite.
   j. magnetite, ilmenite.

Selected Readings

Barth, T. F. W., 1962, *Theoretical petrology* (2nd ed.): New York, Wiley.

Harker, A., 1939, *Metamorphism* (2nd ed.): London, Methuen.

Mason, B., 1962, *Meteorites:* New York, Wiley.

Mason, B., 1966, *Principles of geochemistry* (3rd ed.): New York, Wiley.

Park, C. F., and R. A. MacDiarmid, 1964, *Ore deposits:* San Francisco, W. H. Freeman and Company.

Pettijohn, F. J., 1957, *Sedimentary rocks* (2nd ed.): New York, Harper.

Shand, S. J., 1947, *Eruptive rocks* (3rd ed.): New York, Wiley.

Turner, F. J., and J. Verhoogen, 1960, *Igneous and metamorphic petrology* (2nd ed.): New York, McGraw-Hill.

6

Determinative Mineralogy

Determinative mineralogy can be defined as the science (and art) of identifying a mineral from its physical and chemical properties. The recognition of an unknown mineral may be instantaneous or may require careful and time-consuming tests, depending upon the identity of the mineral, the quality of the specimen, and the knowledge, experience, and skill of the observer. Many schemes of mineral identification have been devised (one of which is presented in the form of tables in this book), but it should be emphasized that any scheme is valuable only if it is applied with experience and common sense.

Identification of mineral species should be easier than identification of animal and plant species, since the total number of mineral species is comparatively small, about 2,000, and most of these are so rare that the possibility of being faced with them is slight. Nevertheless, to the beginning student even the limited number of minerals encountered in an elementary course may present a formidable problem of identification. To him the facility with which an experienced mineralogist can often put a name to an apparently nondescript specimen merely on sight, aided perhaps by hefting or by scratching with a pocket knife, is an enviable and apparently unattainable one. Here the advantage of familiarity must be emphasized—an experienced mineralogist instinctively sums up the characters of an unknown specimen and compares them with his memory picture of the innumerable specimens he has previously identified. The best analogy is the recognition of a person we have seen many times by the sum total of features, dress, gait, voice, and other characteristics. Thus the best training in mineral identification is the intelligent study of minerals in collections and in the field; careful examination of different specimens of a species will reveal the properties which are suitably diagnostic. Logical schemes of mineral identification are useful guides, but experience and intelligence will often suggest a short cut to the procedure. Even if a rapid examination does not serve to identify the specimen, it should limit the possibilities to comparatively few minerals, and the next step is to select the most suitable diagnostic test, which may be a density measurement, a determination of fusibility, a simple chemical test, or some other procedure. An outline of the various diagnostic properties and methods is given in the following sections.

CRYSTAL FORM

When a mineral occurs in crystals their form may be sufficient to identify the mineral without any further tests. The hexagonal prisms terminated by (generally) unequally developed rhombohedrons characteristic of quartz may be cited as a case in point. Sometimes it is not even necessary to see a free crystal; for example, the rounded triangular cross section of tourmaline crystals serves to identify this mineral immediately.

It is important to remember that crystals may be distorted by the unequal development of different faces, thereby obscuring the true symmetry. For example, it is very unusual to find a cubic crystal that is a perfect symmetrical cube. On the other hand, some minerals may crystallize in forms giving an appearance of false symmetry—the simple crystals of the adularia variety of orthoclase may simulate rhombohedrons. In identifying distorted crystals it should be borne in mind that although the appearance may be unusual, the interfacial angles remain the same, and like faces are physically alike, in type of luster, in striations, and so on. Quartz, for example, is often found in platy crystals resulting from the strong development of one pair of prism faces; however, the angles between each pair of prism faces will be 60°, and each prism face will generally show the characteristic horizontal striations.

Twinning is often a critical diagnostic feature. Some common minerals have highly characteristic twin crystals; e.g., the penetration twins of fluorite, the right-angle and diagonal twins of staurolite, the lattice-like twinned aggregates of cerussite. It is important to remember that repeated twinning may often result in an apparently higher symmetry than that possessed by the mineral. Aragonite, for example, often occurs in crystals which appear to be hexagonal prisms, as a result of repeated twinning on {110}; fortunately there is no reason to misidentify these crystals as calcite, since calcite, although hexagonal, rarely crystallizes in simple hexagonal prisms.

The recognition of polysynthetic twinning by the striations on cleavage planes is particularly important in identifying plagioclase. It is almost the only simple test for distinguishing it from orthoclase. The rare mineral amblygonite shows similar polysynthetic twinning lamellae on cleavage surfaces, and occurs in some pegmatites together with plagioclase; however, it is readily differentiated by its density and by simple chemical tests. Calcite and corundum may show similar striated surfaces, but are easy to distinguish from plagioclase.

CRYSTALLINE AGGREGATES

Most mineral specimens are aggregates of imperfect crystals. Even those specimens which are massive and fine-grained generally show some aspects of crystallinity under the microscope. Type of aggregation can be as useful diagnostically as crystal form. Numerous terms are used to describe aggregation, of which the most important are described below.

Many minerals characteristically occur in aggregates of crystals elongated in a particular direction, which is generally one of the crystallographic axes; the elongation is then described as being parallel to *a*, or *b*, or *c*, as the case may be. When the individual crystals making up the aggregate are fairly large and approximately equidimensional in cross-section, the form is *columnar* (e.g., some tourmaline). If the individual crystals are flattened, the form is *bladed* (e.g., kyanite, stibnite). If the individual crystals are small and needle-like, the form is *acicular* (e.g., pectolite, natrolite). If the crystals are so small they are effectively fibers, the form is *fibrous* (e.g., asbestos, some gypsum). If the crystals seem to be diverging from a common center the form is *radiated* (e.g., pectolite, wavellite).

Some minerals occur in aggregates of flattened plates, a form referred to as *lamellar*. This is particularly characteristic of minerals with a single perfect cleavage, and can be related to the crystallography; e.g., lamellar {001}, or lamellar {010}, and so on. The individual plates or laminae are generally parallel, but can be curved around a common center, giving a *concentric* form. When the laminae are thin and separable, as in the micas, the mineral is said to be *foliated*, and the individual layers are foliae. When a mineral consists of aggregates of small lamellae it is often referred to as *micaceous* (e.g., one variety of hematite).

When the individuals in a crystalline aggregate show no marked elongation, that is, when they are equidimensional, the mineral is *granular;* depending on the size of the individual grains it may be *coarse-granular, medium-granular,* or *fine-granular.* If the grains are too small to be distinguished with the naked eye or a hand lens, the mineral is *compact.* If individual grains can be distinguished with a microscope, the mineral is described as *microcrystalline;* if the individual crystals are even smaller, the mineral is *cryptocrystalline.*

Some distinctive aggregates are not covered by the above terms. Some minerals, deposited in open spaces by dripping waters carrying material in solution, occur in *stalactitic* forms; calcite and aragonite are commonly found in such forms, and sometimes limonite and psilomelane. Sometimes the individual stalactites are curiously curved and twisted, and they are then known as *helictitic* or *coralloid*. A *dendritic* form is branching and tree-like, sometimes seen in crystallized gold, silver, and copper, and in the tree-like staining of pyrolusite and psilomelane occasionally seen on bedding planes of sedimentary rocks. Hematite is sometimes found in rounded or nodular masses, imitative of a bunch of grapes; this form is known as *botryoidal* or *tuberose;* this structure is also seen in some psilomelane, malachite, and other minerals. The internal structure of botryoidal masses is often fibrous, the fibers being at right angles to the surface. Sometimes a mineral occurs as an aggregate of rounded pellets, and the form is *oolitic* if the pellets are small, *pisolitic* if they are large. Calcite and aragonite often occur in oolitic and pisolitic aggregates, and this form is a useful diagnostic property of bauxite.

CLEAVAGE AND FRACTURE

Cleavage has been discussed in Chapter 4, and only those aspects important for mineral identification will be mentioned here. Cleavage is the tendency of a mineral to break in definite directions, which are always parallel to possible crystal faces. The quality of a cleavage may be described as *perfect* when it is difficult to break the mineral in any other direction and the cleavage surfaces are extensive and smooth; as *good* when the mineral breaks readily along cleavage planes but can also be broken in other directions, the cleavage surfaces being smooth but interrupted by other fractures; and as *imperfect* when the ease of fracture along the cleavage planes is only somewhat more pronounced than in other directions and the cleavage surfaces tend to be small and much interrupted.

The direction of cleavage is described by the face to which it is parallel, either by the indices or by the name of the face. In the isometric system {111} cleavage is also known as octahedral cleavage; in the tetragonal and orthorhombic systems it would be pyramidal, in the monoclinic system prismatic, and in the triclinic system pinacoidal or pedial. Similarly, {110} cleavage is dodecahedral cleavage in isometric minerals, prismatic cleavage in other systems (pinacoidal or pedial in the triclinic). The number of directions of cleavage for any crystallographic form will depend on the number of faces in that form; thus {110} cleavage in isometric minerals means six cleavage directions, whereas in tetragonal, orthorhombic, or monoclinic minerals it means two directions, and in triclinic crystals it means one direction. A careful study of cleavage can thus help in determining the crystal system of a mineral without well-developed faces. One useful point is that a mineral with a single direction of cleavage cannot be isometric. (Why not?)

Some minerals often show *parting*, which at first glance is not easily distinguished from cleavage, since it too is fracture along crystallographic planes. However, parting may or may not be present in any individual specimen, whereas cleavage is universal in all specimens of a cleavable mineral. Twinned crystals often part along composition planes.

Minerals without cleavage, and cleavable minerals in fine-grained masses, may nevertheless show a characteristic fracture. The commonest such fracture is *conchoidal*, like that of glass; massive quartz often shows conchoidal fracture, as do many minerals which occur in microcrystalline or cryptocrystalline masses. Other terms applied to fracture are *splintery*, *even*, and *uneven*, which are self-explanatory; *hackly* fracture described a jagged irregular surface like that of broken cast iron.

TENACITY

Tenacity is the resistance that a mineral offers to breaking, crushing, bending, or cutting. Most minerals are *brittle*, that is, they can readily be crushed to a fine powder. However, a few (mainly native metals) are *malleable*, which means that their shape can be changed without breaking; hammering or grinding merely causes the grains to be rolled out into plates. Soft minerals ($H < 3$) which are not brittle are usually *sectile*, that is, they can be cut into shavings with a knife. A mineral is *flexible* if it can be bent easily (e.g., thin plates of gypsum); it is *elastic* if after bending it springs back into its original form (e.g., thin plates of the micas).

HARDNESS

The principles of mineral hardness were outlined in Chapter 4, where it was pointed out that mineralogists use ease of scratching as the criterion of hardness, rating it in terms of an empirical scale devised by Mohs in 1822.

The hardness given for a mineral is that of a smooth clean surface, such as a crystal face or a cleavage plane. Minerals often have a superficial coating of weathered or altered material, and such coatings will give a deceptively low hardness. Similarly, the apparent hardness of a fine-grained friable mass has no relation to that of a well-crystallized specimen; for example, hematite crystals show a hardness of 6, but much red earthy hematite can be scratched with a fingernail. Hardness is an exceedingly valuable diagnostic property, when interpreted intelligently.

In this connection it may be worthwhile to point out the important distinction between hardness and tenacity. Tenacity, the resistance to fracture, is quite a distinct property. Diamond is the hardest substance known, but it has a low tenacity, since it cleaves readily in the {111} direction, as some people who have knocked a diamond against a sharp edge have learned to their dismay. Some minerals, not exceptionally hard, are exceedingly tough, so much so that it is practically impossible to break a massive specimen with a hammer. Nephrite (a compact fibrous variety of actinolite) is a good example.

Some degree of correlation exists between hardness and chemical composition, and is expressed in the following rules:

1. Minerals of the heavy metals, such as silver, copper, mercury, and lead, are soft, their hardness seldom exceeding 3.

2. Most sulfides are relatively soft $(H < 5)$, except those of iron, nickel, and cobalt.

3. Most hydrated minerals are relatively soft $(H < 5)$.

4. Anhydrous oxides and silicates (except those of the heavy metals) are usually hard $(H > 5\frac{1}{2})$.

5. The carbonate, sulfate, and phosphate minerals are relatively soft $(H < 5\frac{1}{2})$.

LUSTER

Luster, discussed in detail in Chapter 4, is a useful diagnostic property. The identifying terms used are largely self-explanatory. Luster can be divided into two types, *metallic* and *nonmetallic*. Different varieties of nonmetallic luster can be recognized, the commonest being *vitreous*, the luster of broken glass; other varieties include *adamantine*, the brilliant luster of diamond; *resinous*, like that of resin; *silky*, often shown by fibrous minerals; *pearly*, commonly seen on perfect cleavages. Compact minerals without luster are described as *dull* or *earthy*.

Minerals with metallic luster are generally opaque; chemically they are the native metals, the sulfides, and the metallic oxides. An imperfect metallic luster is sometimes referred to as *submetallic*, and is observed mainly in black opaque minerals that reflect very little light.

COLOR

In utilizing color for mineral identification it is important to distinguish between a color caused by impurities and one due to the essential elements of the mineral. For many minerals the color is diagnostic, e.g., the yellow of sulfur; whereas some minerals (quartz, for example) may be found with almost any color, and for such minerals color is not diagnostic. When the streak is colored, color is not due to impurities as a rule.

Some elements, especially the heavy metals, have specific coloring effects. Thus secondary minerals containing copper are either green or blue, secondary vanadium minerals are generally red, secondary uranium minerals are yellow, manganese in silicates and carbonates is pink, iron-bearing silicates are often dark green to black.

STREAK

Streak is the color of the powdered mineral, and is much less influenced by impurities than is the color of the mineral itself. Streak is most conveniently observed by rubbing an edge of the mineral on a porcelain streak plate. If a mineral is harder than 6 it will not leave a mark on the streak plate; to observe streak in such minerals they must be powdered with a hammer. For minerals with hardness less than 6 streak can readily be observed by scratching them with a knife.

Minerals with characteristic streak include oxides, sulfides, carbonates, phosphates, arsenates, and sulfates of the heavier metals. Silicates usually have a white streak; for a few the streak is gray or brown.

DENSITY

If a pure fragment of a mineral is available, and is large enough for the density to be determined, such a determination is one of the most useful clues to the identity of a mineral. Density determinations are most readily made by weighing a fragment first in air (W_1) and then in a liquid (W_2); if the density of the liquid is L, the density of the mineral is $(W_1/W_1 - W_2)L$. Usually the liquid is water, in which case L is 1 or very close to 1; for precise work organic liquids such as toluene are more satisfactory, since minute air bubbles do not adhere to the specimen so tenaciously as when water is used.

Practically any moderately precise balance can be used to determine the density of fair-sized fragments. Such fragments are often not obtainable, and special balances have been devised for small pieces. One such is the Jolly balance (Fig. 4-1), in which the weight of the fragment in air and the weight in the liquid is measured by the extension of a spring, thereby obviating the necessity of a set of weights. For very small fragments the Berman balance (Fig. 4-2) is especially suitable.

In mineral identification an exact density determination may not be necessary; in deciding on the identity of a specimen all that may be needed is to know whether its density is greater or less than a certain figure. Here the availability of heavy liquids of known density is very advantageous; a small chip of the unknown can be dropped into a small vial of liquid, and a direct observation as to whether it sinks or floats will put a limit on its density. Suitable liquids are:

Bromoform, $G = 2.9$
Acetylene tetrabromide (tetrabromoethane), $G = 2.96$
Methylene iodide, $G = 3.3$
Clerici solution (a saturated solution of equal amounts of thallous formate and thallous malonate in water), $G = 4.2$ (at room temperature)

REFRACTIVE INDEX

The refractive index is usually characteristic and diagnostic for any substance, and the determination of the refractive index, or indices, of a solid is therefore of great practical importance as a means of identification. Several techniques are available for the determination of the refractive indices of solids, but the method of most general application, and that most widely used, is the immersion method. This is simply a comparative procedure, the solid being immersed in liquids of known refractive index until a match is found. It is an extremely powerful method, since it can be applied to minute grains weighing a millionth of a gram or less, and it is available to anyone who has a microscope and a set of liquids of known refractive index.

The technique of measuring the refractive index of an isotropic transparent solid is extremely simple. The aim is to find which liquid in a set of liquids of known refractive index has the same or nearly the same refractive index as the solid. When the solid particles are immersed in this liquid they become practically invisible; light, in passing from the liquid to the solid and the solid to the liquid, is not refracted, and consequently the edges of the particles cannot be seen; as far as its effect on light is concerned, the whole liquid-solid complex is a homogeneous medium.

The procedure is to immerse particles of the solid [sized grains 0.1–0.2 mm in diameter (70–150 mesh) are most convenient to use] in a drop of liquid of known refractive index on a glass slide, cover the drop with a thin cover glass, and observe the grains, using a low or moderate magnification (10 × objective and 8 × ocular, for instance) and parallel or nearly parallel light. If the grains show up plainly their refractive index differs considerably from that of the liquid; other liquids of different refractive index are then tried, until a liquid is found in which the grains are invisible or very nearly so. The search for this

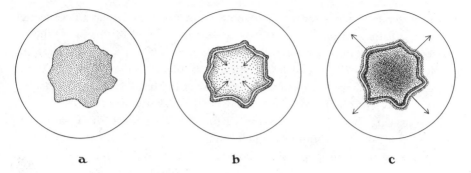

a   b   c

Figure 6-1 The Becke effect. (a) Microscope focused on grain. (b) Microscope tube slightly raised, index of grain higher than that of immersion liquid. (c) Microscope tube slightly raised, index of grain lower than that of immersion liquid.

liquid is not as laborious as one might suppose, because it is possible, by observing the Becke effect (Fig. 6-1), to tell whether the refractive index of the liquid is higher or lower than that of the solid; and with experience one can estimate roughly how much higher or lower. If a grain is focused sharply, a line of light (the "Becke line") can be seen at the edges of the grain; when the objective is raised slightly, the line will either contract within the grain (if the grain has a higher refractive index than the liquid), or expand beyond the grain (if the liquid has a higher refractive index than the grain). The shape of the particles does not matter, since the Becke line always follows the outline of the grain; the determination of the refractive index is just as easy for irregular fragments as for well-formed crystals. The effect is seen most clearly with the substage diaphragm partly closed.

The simplest way of regarding the Becke effect, as well as the best way of remembering which way the line moves, is to think of the grain as a crude lens (Fig. 6-2), which, if it has a refractive index higher than that of the surrounding

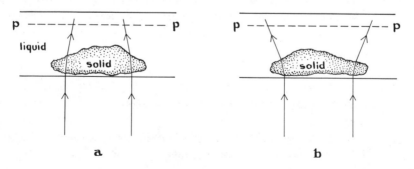

a   b

Figure 6-2 The Becke effect. (a) Refractive index of solid greater than that of liquid. (b) Refractive index of solid less than that of liquid.

medium, tends to focus the light at some point above it; when the objective is raised it is focused on a plane *PP* above the grain, and in this plane the refracted light waves occupy a smaller area than they do in a plane nearer the particle, and thus the Becke line contracts as the objective is raised. If the refractive index of the solid is lower than that of the liquid, the grain acts as a negative lens, and the Becke line expands as the objective is raised.

By observing this effect and trying various liquids in turn, it is possible to find in a few minutes a liquid in which the grains are nearly invisible. In practice, it is convenient to keep a set of liquids with indices differing by a constant interval, generally 0.01 or 0.005. Liquids are available with refractive indices from 1.33 (H_2O) to 2.11; low-melting solids are used for higher indices. Usually, of course, the refractive index of the solid is found to lie between those of two adjoining liquids in the set; its value can be estimated from a comparison of the Becke effects in the two liquids.

The identification of anisotropic substances by determination of refractive indices is somewhat more time-consuming than for isotropic substances, because in anisotropic substances the velocity of light, and hence the refractive index, varies with the direction of vibration of light within the crystal. Grains differing in crystallographic orientation give different refractive indices, and the same grain will give different values for its refractive index according to its orientation relative to the direction of vibration of the light passing through it. In a polarizing microscope the direction of vibration of the light passing through the object is controlled by the polarizer and is either N-S or E-W. Hence by rotating the stage of the microscope the refractive index of a single grain can be tested in various directions in the grain. The difference between the maximum and minimum values for an anisotropic substance is known as the birefringence (Chapter 4).

Anisotropic substances are easily distinguished from isotropic substances when grain mounts are observed in the polarizing microscope with the analyzer inserted. Liquids and isotropic substances are dark and remain so upon rotation of the microscope stage, whereas most anisotropic grains show interference colors and become dark (extinguish) four times at intervals of 90° in a complete rotation of the stage.

The identification of anisotropic substances is usually easiest if the ω index for uniaxial substances or the β index for biaxial substances is determined. Grains suitable for this determination can be recognized in a mount because they show very weak interference color (gray) even at 45° from the extinction position; they behave essentially like isotropic grains in that they show a single refractive index. By observing an interference figure of such a grain it is possible to determine whether it is uniaxial or biaxial. To obtain an interference figure the grain must be observed (with analyzer inserted) in strongly con-

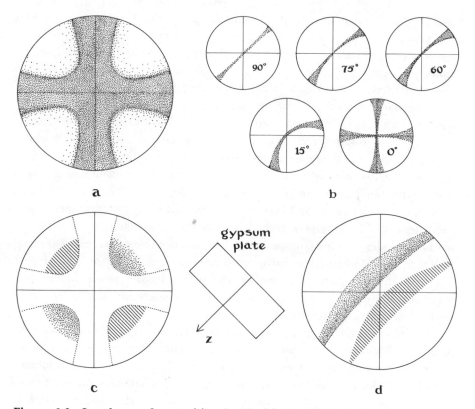

Figure 6-3 Interference figures: (a) uniaxial; (b) biaxial, for different axial angles; (c) interference figure of a uniaxial positive substance after insertion of the gypsum plate; (d) interference figure of a biaxial positive substance after insertion of the gypsum plate. In (c) and (d) the stippling indicates blue, the hatching yellow; the cross in (c) and the bar in (d) are red.

vergent light (a convergent lens is mounted below the microscope stage in a polarizing microscope) with a high-power objective (45 × or 50 ×). The interference figure is seen when the ocular is removed or the accessory Bertrand lens inserted. For a uniaxial substance the interference figure is a black cross (Fig. 6-3). For a biaxial substance the figure may be a cross slightly opened at the center if the axial angle is near 0° (well seen in a cleavage flake of phlogopite), and increasing axial angle results in one part of the cross leaving the field of view and the curvature of the other part decreasing, until for an axial angle near 90° the interference figure is a straight bar across the field (Fig. 6-3).

The optical character, whether positive or negative, can be determined by the use of an accessory plate. If a gypsum plate is inserted, the black cross of a uniaxial figure becomes red, and opposite pairs of quadrants are colored blue and yellow. The positions of the blue and yellow spots indicate the optical sign (Fig. 6-3). It is convenient to establish the relationship between the blue and yellow spots for a particular microscope and gypsum plate by observing the interference figure of a cleavage flake of phlogopite, which is biaxial (pseudo-uniaxial) negative. The interpretation of the effect of a gypsum plate on a biaxial interference figure follows logically (Fig. 6-3).

If necessary, the other refractive indices, ϵ for a uniaxial substance, and α and γ for a biaxial substance, can be determined by examining grains showing maximum birefringence (i.e., highest inference colors). Such grains have come to rest in the plane of the ϵ and ω, or γ and α directions of vibration. A grain of a uniaxial substance showing maximum birefringence will give ω in one extinction position and ϵ in the extinction position at 90°. If the substance is uniaxial positive, ω is the lower index, ϵ the higher; if the substance is uniaxial negative, the reverse is true. For a biaxial substance α is by definition the lowest index, γ the highest index. In biaxial positive substances β is closer to α than to γ; in biaxial negative substances the reverse is true. Thus even if it is not possible to observe an interference figure, it is still possible to determine whether a substance is uniaxial or biaxial and to determine its sign by careful study of the refractive indices of several grains.

Other observations of grains under the microscope can contribute materially to the identification of a mineral. Cleavage directions are revealed by crushed fragments showing straight edges (see, for example, pyroxenes, amphiboles, feldspars). Color is significant; observe whether color changes on rotation of the stage (pleochroism). If pleochroism is present correlate the variation in color with the corresponding refractive indices. If the mineral shows an identifiable crystallographic direction, measure the angle between this direction and the nearest extinction position. This angle is called the extinction angle, and is a useful diagnostic property, characteristic of monoclinic and triclinic minerals. It is especially useful in distinguishing between monoclinic amphiboles and monoclinic pyroxenes, since most amphiboles have maximum extinction angles of less than 20°, whereas the pyroxenes (except aegirine) have maximum extinction angles near 40°.

SPECIAL TESTS

There are certain special tests which are applicable to only a few minerals. Ultraviolet light causes some minerals to fluoresce, but the fluorescence is diagnostic in only a few, since in most it is produced by small amounts of minor and trace elements, and only when these accessory elements are present does the fluorescence appear. Thus calcite from Franklin. New Jersey, fluoresces a bright pink, and it has been shown that it does so because of the presence of manganese: calcites from other localities seldom fluoresce. Scheelite is one mineral for which the fluorescence is diagnostic; it always fluoresces white or yellow in short-wave ultraviolet light. This is an extremely useful property, since scheelite is often difficult to distinguish from the enclosing gangue, which is usually quartz and calcite; prospecting and mining for scheelite has been greatly facilitated by the use of portable ultraviolet lights.

Another property that can be applied to the identification of minerals is radioactivity. All minerals containing uranium and thorium give off radiations that are readily detected by a Geiger counter. With a little practice it is possible to estimate the amount of uranium and thorium in the specimen, which may help to identify the mineral; for example, a specimen of allanite (in which the uranium and thorium content is always quite low) will give a much weaker response on the Geiger counter than a specimen of uraninite approximately the same size held at the same distance from the counter.

Magnetism is readily observable in native iron, magnetite, and pyrrhotite, all of which are attracted by a small hand magnet. Most iron-bearing minerals, if heated strongly in air, become magnetic as a result of the formation of magnetite, and this is a useful test for the presence of iron.

Soluble minerals are few, but most of them have a distinctive taste which is readily observed by touching them to the tongue.

One factor in mineral identification which can hardly be described as a specific property is that of occurrence and association. It is the factor of experience that enables a mineralogist to know which minerals he would expect to find in a certain geological environment and which he would not expect to find. For example, a pale-colored mica in a magnesium-rich rock is probably phlogopite, not muscovite. It is not possible to enumerate all these clues; many are referred to incidentally in other parts of this book. Others will be learned by examining rocks and minerals in collections, and better still, in the field.

CHEMICAL TESTS

The observations and tests outlined previously may be called nondestructive, since a mineral is not decomposed in making them. For many species such observations and tests lead to a certain identification. For others, however, the identity of a specimen may still be in doubt even after all these observations and tests have been made. When this is the case it may be necessary to make some chemical tests to resolve the doubt. Generally these tests are few and simple; careful observation of the physical properties should have narrowed the range of possibilities to a small number of species. A careful consideration of the chemical properties of these species will indicate which tests will provide an unambiguous identification. The possibilities of physical tests should always be exhausted before chemical ones are turned to; physical tests are generally easier, faster, and more readily applied. The field geologist seldom has facilities for chemical tests, and he must usually make his identifications from physical properties alone; hence the student should use chemical tests to confirm a tentative identification based on the physical properties or to discriminate among a small number of possibilities.

Many of the simple chemical tests which have been devised for mineral identification are carried out by means of the blowpipe—a tool for producing a small concentrated flame suitable for dealing with small amounts of material. In the laboratory the blowpipe is generally used in conjunction with a Bunsen burner fitted with a special cap to give a flat flame, but it can give satisfactory results with an alcohol lamp or even a candle—a very important consideration if it is necessary to make difficult mineral identifications in the field.

Blowpiping is a skill that comes with practice and experience. The cheeks should be kept puffed out, while breathing naturally through the mose and allowing a little air to pass into the mouth. After sufficient practice, the use of the blowpipe is neither difficult nor tiring, and it is perhaps best learned by concentrating on the flame itself without giving conscious thought to breathing or blowing.

The simplest procedure, and one that often provides all the information required, is to hold a small chip (about the size of a lead pencil tip) of the unknown mineral with a pair of forceps and heat it in the hottest part of the blowpipe flame (blow hard enough to produce a flame containing no yellow, and hold the chip just beyond the end of the visible flame). The following points should be noted:

1. How easily does the fragment melt?
2. Does it color the flame; if so, is the color characteristic of a certain element?
3. Does it swell up, boil (intumesce) or disintegrate violently (decrepitate)?
4. Is the fragment magnetic after heating? (If so the mineral contains iron.)
5. Is the mineral infusible, and white or pale in color? If so, the fragment should be moistened with cobalt nitrate solution and reheated. If the fragment becomes blue, the mineral contains aluminum (or the mineral is a zinc silicate); if the fragment becomes pink, this indicates the presence of magnesium.

The case of melting is termed *fusibility*, and is rated by an empirical scale, using type minerals, similar in principle to Mohs' hardness scale. The scale of fusibility is as follows:

1. Stibnite   4. Actinolite
2. Chalcopyrite   5. Orthoclase
3. Almandite   6. Enstatite

Practice with fragments of these standard minerals will give familiarity with the fusibility scale. The fusibility scale can also be expressed in the following terms:

1. Fused easily in a closed tube. Dull red heat.
2. Fused with difficulty in a closed tube. Red heat.
3. Easily fused before the blowpipe but not in a closed tube. Orange-red heat.
4. Fused easily on the edges before the blowpipe. Yellow-orange heat.
5. Fused on the edges with difficulty before the blowpipe. Yellow heat.
6. Fused only on the finest points and thinnest edges before the blowpipe. White heat.
7. A fusibility of 7 means that the mineral shows no trace of melting after being heated in the hottest blowpipe flame.

It should be noted that relative fusibility is not necessarily the same as relative melting point. A mineral which conducts heat readily is more difficult to fuse in a blowpipe flame than a poor conductor with the same melting point. A good conductor transfers the heat rapidly away from the hottest point and may be very difficult to raise to the melting point. Copper, for example, has a melting point of 1083°C but is difficult to melt in a blowpipe flame even though the flame is capable of producing a considerably higher temperature than this in a poorly conducting mineral.

The characteristic flame colorations given by different elements are as follows:

Barium: yellow-green
Boron: yellow-green, very brief
Calcium: reddish orange, not very strong
Copper: green
Lithium: red, fairly persistent.
Potassium: violet, fairly persistent
Sodium: yellow, very persistent
Strontium: red, fairly persistent.

Since potassium minerals always contain some sodium, the yellow sodium flame tends to obscure the violet of potassium; this can be overcome by viewing the flame through cobalt blue glass, which absorbs the sodium flame and thereby renders visible the potassium flame. Another useful accessory is the Merwin color screen, devised by H. E. Merwin. It consists of a blue and a violet celluloid sheet, so mounted that they overlap in part, thereby giving three colored strips. It is used for distinguishing potassium, calcium, strontium, and lithium flames in the presence of sodium, whose characteristic flame is absorbed by all three strips.

Flame colorations are more conveniently observed by dipping a clean platinum wire in hydrochloric acid, then in the powdered mineral, and heating in the colorless Bunsen flame.

A useful blowpipe procedure is the reduction (if possible) of a mineral to a metallic bead. This is usually carried out on a charcoal block. Some minerals should be mixed with anhydrous sodium carbonate as a flux, whereas others are readily reduced without flux. The mineral should be finely powdered, and held in a small pit cut in the charcoal block; if the sample tends to blow away it may be moistened to make it stick. The following metals may be obtained by reduction:

Antimony: brittle, gray to white
Bismuth: brittle to malleable, reddish white
Copper: malleable, reddish
Gold: malleable, yellow
Lead: malleable, white, tarnishes
Silver: malleable, white
Tin: malleable, white

Other useful pyrognostic (i.e., heating) tests are those done in closed and open tubes. Closed tubes are made by sealing off one end of 3–inch lengths of $\frac{1}{8}$–inch or $\frac{1}{4}$–inch glass tubing in a Bunsen flame. Mineral powders heated in a closed tube melt if they have a fusibility of 1 or 2, and may give off fumes or sublimates. The following reactions are characteristic:

Arsenic: black, shiny sublimate from arsenides, reddish from sulfides of arsenic.
Mercury: cinnabar gives a black sublimate of mercuric sulfide when heated alone, a sublimate of metallic mercury when heated with Na_2CO_3.
Sulfur: a few sulfides and native sulfur give dark liquid sulfur, which becomes light yellow when cold.
Water: absorbed water (as in clays) is given off at quite low temperatures, water of crystallization is given off at temperatures between about 100°C and 300°C, and water from hydroxyl radicals may not be given off until red heat or greater.

Open-tube tests are made in a 4–inch length of glass tubing which has been bent to an angle of about 120° $1\frac{1}{2}$ inches from one end. A small amount of the finely powdered mineral is placed at the bend, and the tube is then heated with the short part horizontal and the long end toward the vertical. The heat causes a current of air to pass over the sample and oxidize it. Sublimates may form, and the mineral may change color. The following are among the most diagnostic reactions:

Arsenic: white crystalline sublimate
Antimony: dense white fumes and sublimate
Sulfides: fumes of sulfur dioxide
Copper or manganese: mineral turns black by the formation of oxides
Iron: red or brown oxide produced

Borax bead tests are commonly used in mineral identification. A borax bead is prepared by dipping a platinum wire with a small loop on the end into powdered borax and heating it; the borax melts to a colorless liquid and in cooling forms a small glass bead on the loop. While still pasty the bead is touched to the powdered mineral to be tested, and then reheated. The molten borax has great solvent powers, and the resulting glass may show a characteristic color, depending on the elements present in the mineral. The bead should be heated first in the oxidizing flame (the colorless flame produced when oxygen is present in excess) and then in the reducing flame (the yellow flame which contains unburned carbon); the color of the bead may be different in the two flames. Similar tests are made using $NaPO_3$ beads prepared either from $NaPO_3$ or by fusing salt of phosphorus (microcosmic salt), $HNaNH_4PO_4 \cdot 4H_2O$. Diagnostic bead colorations are given in Table 6-I.

Table 6-I   *Bead tests*

| | BORAX BEAD | | NaPO₃ BEAD | |
ELEMENT	OXIDIZING FLAME	REDUCING FLAME	OXIDIZING FLAME	REDUCING FLAME
Cobalt	Blue*	Blue	Blue	Blue
Chromium	Green	Green	Green	Green
Copper	Blue-green	Red, opaque	Blue	Red, opaque
Iron	Yellow	Green	Pale yellow	Colorless
Manganese	Violet	Colorless	Violet	Colorless
Molybdenum	Colorless	Brown	Colorless	Green
Nickel	Brown Violet (hot)	Gray, opaque	Yellow	Yellow
Titanium	Colorless	Violet	Colorless	Violet
Uranium	Yellow	Pale green	Pale green	Green
Vanadium	Colorless	Green	Yellow	Green
Tungsten	Colorless	Yellow	Colorless	Blue

* The colors refer to the cold beads unless otherwise stated.

A generally applicable test is that of solubility in HCl. For this test a small amount of the powdered mineral is placed in a test tube, and about 5 cc of dilute (1:1) hydrochloric acid is added. If no reaction takes place, the test tube is heated carefully over the Bunsen flame. If still no reaction takes place the test is repeated using concentrated HCl. The following reactions are characteristic:

1. Many minerals are completely soluble without effervescence; they include many oxides and hydroxides, some sulfates, and some phosphates and arsenates. If much iron is present, a yellow solution is obtained; copper minerals give a blue or greenish blue solution; cobalt minerals give a pink solution.

2. Solubility with effervescence occurs when the mineral contains a potentially gaseous component, as do the carbonates. All carbonates dissolve in HCl with effervescence; some, such as calcite and aragonite, do so in cold acid, but many do so only on heating. Some sulfides dissolve in HCl and give off H_2S, which is readily recognized by its odor. Oxides of manganese give off chlorine, a greenish poisonous gas, from hot concentrated hydrochloric acid.

3. Decomposition with the separation of an insoluble material is characteristic of many silicates. The insoluble material is generally silica, and it may separate as a fine powder, or as jelly-like masses; in the latter case the mineral is said to gelatinize.

Table 6-II lists the minerals which are soluble or decomposed in HCl.

Table 6-II   *Solubility of minerals in HCl*

METALLIC LUSTER		
Soluble in HCl	Soluble in HCl with	Soluble in HCl with
Goethite (limonite)	evolution of chlorine	evolution of H_2S
	Pyrolusite	Stibnite
Soluble in HCl	Psilomelane	Galena
with difficulty	Manganite	Pyrrhotite
Hematite	Hausmannite	Sphalerite
Ilmenite	Braunite	
Magnetite		

NONMETALLIC LUSTER		
Soluble in HCl	Soluble in HCl with	Decomposed by HCl
Cryolite	formation of silica gel	leaving silica residue
Zincite	Anorthite	Leucite
Brucite	Nepheline	Rhodonite
Colemanite	Sodalite	Wollastonite
Gypsum	Cancrinite	Pectolite
Jarosite	Olivine	Scapolite
Apatite	Willemite	Cordierite
Turquoise	Hemimorphite	Biotite
Carnotite	Datolite	Serpentine
Tyuyamunite	Analcime	Garnierite
Crocoite	Natrolite	Chrysocolla
	Laumontite	Stilbite
Soluble in HCl with		Chabazite
evolution of CO_2		Heulandite
All carbonates.		

TESTS FOR INDIVIDUAL ELEMENTS

For some minerals all that is needed for certain identification is a confirmatory test for a specific element. Simple tests for elements the student may encounter are outlined below:

Aluminum: Light-colored minerals which are infusible can be tested by moistening an ignited fragment with $Co(NO_3)_2$ solution and reheating. Aluminum is present if the mineral turns deep blue. (The zinc silicates hemimorphite and willemite also give this reaction.)

Antimony: Metallic antimony and its compounds with sulfur when heated in the open tube give a dense white smoke of antimony oxide, which deposits on the wall of the tube.

Arsenic: Heated in the open tube, arsenic, arsenides, and sulfides containing arsenic give a very volatile white sublimate of arsenic oxide and a disagreeable odor similar to that of garlic. This odor, which is very characteristic, can often be observed simply on crushing the minerals.

Barium: Barium in minerals gives a yellowish green flame coloration, which can sometimes be intensified by first moistening the mineral with HCl.

Beryllium: There is no simple chemical test for beryllium. Beryl, the only relatively common mineral containing beryllium, can usually be easily recognized by its physical properties.

Bismuth: Bismuth in minerals can be reduced to the elementary state by mixing the powdered mineral with about three parts of anhydrous Na_2CO_3 and heating on charcoal before the blowpipe. The reddish white globules of bismuth thus obtained are readily fusible and brittle (they cannot be hammered out into a thin sheet, as can lead and silver).

Boron: Many boron-containing minerals when heated before the blowpipe give an apple-green color to the flame. The color may be similar to that given by barium, but boron and barium minerals have other distinctive properties.

Calcium: Some calcium minerals give a yellowish red flame coloration. The color is often weak, and in many calcium minerals it does not appear at all. Moistening the powdered mineral with HCl will help to develop the flame coloration.

Carbon: All carbonate-containing minerals effervesce in hydrochloric acid, some in cold acid, others on heating. Silicates containing carbonate, such as cancrinite and scapolite, do not dissolve completely.

Chlorine: Salt minerals containing chlorine can be dissolved in water and the presence of chloride detected by adding $AgNO_3$ solution; white AgCl, soluble in NH_4OH, is precipitated. The presence of chloride in silicates such as sodalite can be detected by dissolving them in nitric acid and applying the $AgNO_3$ test to the solution.

Chromium: Chromium minerals color the borax bead yellow-green in the oxidizing flame, emerald-green in the reducing flame.

Cobalt: Cobalt minerals color the borax bead a fine blue in both the oxidizing and the reducing flames.

Copper: If copper is suspected the powdered mineral is roasted on charcoal, moistened with HCl, and reheated in the blowpipe flame; a strong azure-blue flame coloration, often green on the edges, will be obtained.

Fluorine: Fluorine in minerals can be detected by mixing the mineral with some powdered quartz and heating the mixture with concentrated H_2SO_4 in a test tube. Silicon tetrafluoride is given off, and can be detected by holding a wet glass rod in the test tube; the silicon tetrafluoride reacts with the water and gives a white coating of silica on the rod.

Iron: Iron-bearing minerals give a magnetic residue when heated in the reducing flame (as do nickel and cobalt minerals). Iron-bearing minerals which dissolve partly or wholly in HCl give a yellow solution, from which red-brown $Fe(OH)_3$ is precipitated by NH_4OH.

Lead: Lead minerals are reduced to lead (white malleable pellets, tarnishing gray) by mixing them with four parts of anhydrous Na_2CO_3 and heating on charcoal.

Lithium: Lithium gives an intense red color to the flame, somewhat similar to the strontium flame but deeper.

Magnesium: Light-colored minerals which are infusible can be tested by moistening an ignited fragment with $Co(NO_3)_2$ solution and reheating. If magnesium is present, the mineral becomes pink or flesh-colored.

Manganese: Manganese colors the borax bead violet in the oxidizing flame; in the reducing flame the bead is colorless. A Na_2CO_3 bead is colored bluish green in the oxidizing flame, a very sensitive and specific test.

Mercury: Mercury minerals give a sublimate of metallic mercury when mixed with anhydrous Na_2CO_3 and heated in the closed tube.

Molybdenum: The molybdenum minerals molybdenite and wulfenite are readily recognized from physical properties alone.

Nickel: In the oxidizing flame the borax bead is violet when hot and red-brown on cooling; in the reducing flame the bead becomes gray and turbid from the separation of metallic nickel. An ammoniacal solution containing nickel gives a scarlet precipitate with an alcoholic solution of dimethyl glyoxime.

Niobium: If columbite is suspected, the powdered mineral is fused on charcoal with anhydrous Na_2CO_3 or better, $K_2S_2O_7$, and the melt is dissolved in HCl; on boiling this solution with metallic tin it becomes blue, changing slowly to brown.

Nitrogen (nitrates): When heated on charcoal, nitrates cause the charcoal to burn violently because of the oxygen released.

Phosphorus (phosphates): Most phosphates dissolve in nitric acid; the addition of ammonium molybdate solution produces a yellow precipitate of ammonium phosphomolybdate.

Potassium: The violet flame coloration is characteristic. Since potassium is usually accompanied by sodium the flame must usually be observed through a cobalt blue glass, which absorbs the yellow color due to sodium.

Rare Earths: A rare-earth mineral will generally contain some neodymium, which can be recognized by the spectrographic test described on p. 188.

Silicon: Silicate minerals can be identified by the $NaPO_3$ bead. A $NaPO_3$ bead is prepared in the same way as is a borax bead. When a small amount of a silicate mineral is fused in the $NaPO_3$ bead the mineral is decomposed and the SiO_2 remains as a translucent mass or skeleton, often somewhat hard to detect.

Silver: Silver minerals are readily reduced to silver, a white malleable metal, when mixed with anhydrous Na_2CO_3 and heated on charcoal.

Sodium: Minerals containing sodium give a strong and persistent yellow flame.

Strontium: Strontium in minerals imparts a characteristic crimson-red color to the flame.

Sulfur: Sulfides heated in the open tube give off SO_2, readily recognized by its characteristic odor. Sulfates can be identified by fusing the powdered mineral with a mixture of anhydrous Na_2CO_3 and charcoal dust; the fused mass is placed on a silver coin and moistened, whereupon a black stain of silver sulfide will be formed.

Tantalum: Tantalum minerals always contain some niobium; if a positive niobium test is obtained, tantalum is also present.

Thorium: Thorium (and uranium) minerals can be recognized by their radioactivity. (A specific test for uranium is given below.)

Tin: Minerals containing tin give easily fusible globules of metallic tin when mixed with anhydrous Na_2CO_3 and reduced on charcoal.

Titanium: To test for titanium the mineral should be fused with anhydrous Na_2CO_3 and the melt dissolved in equal amounts of concentrated H_2SO_4 and water; a few drops of H_2O_2 added to the cold solution will give a yellow coloration.

Tungsten: To test for tungsten the mineral should be fused with anhydrous Na_2CO_3 and the melt dissolved in HCl; on heating with metallic zinc, the solution becomes violet.

Uranium: Uranium minerals are radioactive and can be detected with a

Geiger counter. A bead made with a mixture of borax and sodium fluoride and a uranium mineral fluoresces bright yellow in ultraviolet light.

Vanadium: In the oxidizing flame borax beads containing vanadium are yellow when hot and pale yellow-green when cool; in the reducing flame they are dirty green when hot and clear green when cold.

Zinc: Zinc minerals, when heated on charcoal in the reducing flame, give a coating of zinc oxide, which is yellow when hot and white when cold.

Zirconium: Zircon is usually easily recognized by its physical properties. It commonly shows an orange fluorescence in ultraviolet light.

SPECIAL TECHNIQUES

A number of special techniques utilized in mineral identification are mentioned here, because even though they may not be generally available to undergraduate classes the student should be aware of their existence. The most generally applicable of these special techniques are spectrographic analysis, differential thermal analysis, and x-ray diffraction.

Spectrographic Analysis

The principle underlying spectrographic analysis is illustrated by the simple flame tests available for a number of elements. When material containing one of these elements is brought to incandescence, the free atoms emit light of a specific color, or wavelength. When this light is examined in a *spectroscope* these specific wavelengths can be identified and measured. Many elements which do not impart a characteristic color to a flame nevertheless present a diagnostic pattern of lines in the spectroscope. A spectrograph is a development of the spectroscope provided with means for photographing the spectrum and thereby preserving it for reference and detailed examination. Measurement of line intensities makes possible quantitative or semiquantitative determinations of the amounts of the elements present in a mineral. This technique is applicable to the determination of nearly seventy elements, principally the metals. It is particularly valuable in combination with x-ray diffraction; spectrographic analysis indicates the elements present; x-ray diffraction reveals the crystal structure and distinguishes between different compounds of the same elements, and between polymorphs.

An interesting application of spectroscopic examination in routine work has been utilized in the mining and milling of rare-earth minerals. Neodymium, one of the rare-earth elements, has a very strong line in the yellow region of the spectrum. This line can be observed with a pocket spectroscope in light reflected from solids or solutions containing neodymium. The property is used at the deposit in Mountain Pass, California, to recognize rare-earth rich areas during mining and to control the concentrating and extraction of the rare-earth minerals.

Differential Thermal Analysis

A technique of mineral examination which has been greatly developed in recent years is that of differential thermal analysis. In this technique a small sample is heated at a uniform rate and the thermal effects—evolution or absorption of heat—recorded. Loss of combined water or carbon dioxide, change of polymorphic form, recombination of the elements into a different compound, can all be detected by their characteristic thermal effects. In practice, the mineral under examination is packed into a small hole in a metal block with a thermocouple; this thermocouple is in series with another in a similar hole packed with a thermally inert material such as aluminum oxide. The metal block is then heated at a uniform rate to 1000°C or higher. The temperature of the aluminum oxide increases at a uniform rate. The sample under examination also increases uniformly in temperature when no changes are occurring, but during an exothermic reaction (heat evolved) its temperature rises above

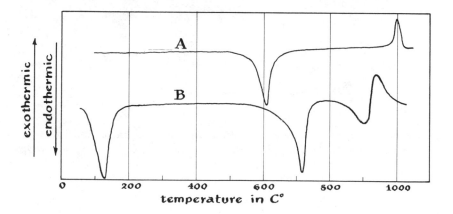

Figure 6-4 Differential thermal analysis curves of kaolinite (a) and montmorillonite (b); upward peaks represent exothermic reactions, downward peaks represent endothermic reactions. In kaolinite the endothermic peak at 600°C results from the loss of OH groups and the breakdown of the crystal structure, and the exothermic peak at 1000°C results from the crystallization of corundum (Al_2O_3). In montmorillonite the endothermic peak just above 100°C results from the loss of weakly held water molecules, the endothermic peak at 700°C results from the loss of OH groups, the small endothermic peak at 900°C represents the complete breakdown of the structure, and the exothermic peak at 950°C results from the crystallization of spinel ($MgAl_2O_4$).

that of the aluminum oxide, whereas during an endothermic reaction (heat absorbed) its temperature lags behind. These differences are plotted on a chart by recording the difference in potential between the two thermocouples. When no change is occurring this difference is close to zero, and the chart shows no inflections. Exothermic reactions cause inflections which are conventionally recorded above the zero line; endothermic reactions cause inflections which

Figure 6-5 Cylindrical x-ray powder cameras, 57.3 mm in diameter at left, 114.6 mm in diameter at right. In each the pinhole is to the left, the exit port trap to the right. On top, the taller knurled knob allows centering of the specimen while it is being observed through the pinhole with a lens (black cap); the other knurled knob clamps the film.

are recorded below the zero line. The position, shape, and magnitude of these inflections are characteristic of the constituents causing them (Fig. 6-4). Differential thermal analysis has proved especially useful in the determination of the mineralogy of clays; clay minerals are not easily identified by classical methods because of their fine grain size and similarity in chemical composition, but they lose their combined water and form new compounds at temperatures which are diagnostic for the individual species.

X-ray diffraction

X-ray diffraction effects form the basis of a generally applicable method of identifying minerals, and indeed all crystalline substances. Qualitative chemical tests will not distinguish between different compounds of the same elements, and even quantitative analyses will not distinguish between polymorphs. Refractive index measurements are applicable only to nonopaque minerals, and the measured refractive indices may not be adequate to characterize the substance, especially if it belongs to a complex solid solution series. X-ray methods are applicable to all crystalline materials, and require only minute amounts of material—a microscopic sliver of a crystal or a few milligrams of powder.

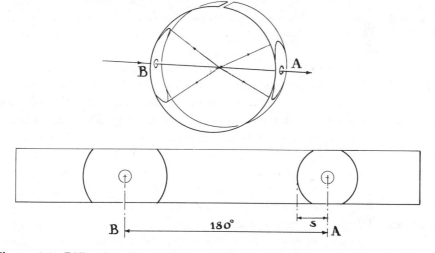

Figure 6-6 Diffractions in relation to a cylindrical powder camera film. At right, a forward reflection (diffraction) forming a nearly circular arc centered at A. At left, a back reflection (diffraction) forming a nearly circular arc centered at B (the entrance pinhole).

The most used method, and the only one that will be discussed here, is the powder method. In its simplest form the method involves mounting a small sample of the finely powdered mineral at the center of a cylindrical camera and irradiating it with a beam of monochromatic (uniform wavelength) x-rays. (Figs. 6-5, 6-6). Cones of diffracted beams are produced by the sample, and each cone of beams is recorded as a pair of lines on the film mounted on the inner circumference of the camera, each pair of lines being symmetrically located relative to the path of the x-ray beam. Each cone consists of a large

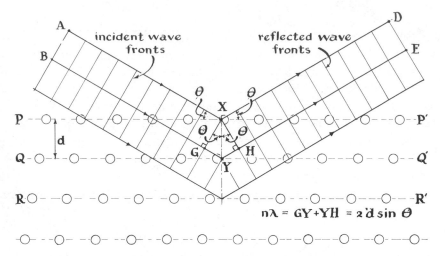

Figure 6-7   Diffraction of x-rays by equally spaced, identical planes of atoms governed by the Bragg law, $n\lambda = 2d \sin \theta$.

number of small diffracted beams produced by a specific plane of atoms in the individual crystalline fragments in the sample. All the diffracted beams in any one cone can be considered as "reflections" from a specific plane of atoms. Such a plane of atoms can reflect monochromatic x-rays only when it is at a definite angle θ to the primary beam; all the crystallites which lie with this plane at the angle θ to the primary beam contribute to this reflection. The angle of reflection is equal to the angle of incidence, hence the reflected beam makes an angle of 2θ with the primary beam. The reflected beams from all the crystallites which have the required orientation therefore form a cone of semivertical angle 2θ having the primary beam as its axis. Different planes of atoms require different angles of incidence, and therefore give reflected beams at different angles to the primary beam. A complete x-ray powder photograph of a mineral thus consists of a pattern of lines, each with a specific position and intensity; it has been aptly described as a "fingerprint" of the mineral, since the x-ray powder pattern is practically unique for each substance. Even minerals of similar composition in a solid solution series can frequently be distinguished by their powder patterns, since both the spacing of lines and their relative intensities usually vary somewhat with composition.

A Geiger counter, proportional counter, or scintillation counter may be used instead of a photographic film to record the diffracted beams. The powdered sample is mounted on a flat surface in the x-ray beam, and the Geiger tube (or other detector) is rotated around the center of the sample. The Geiger counter is connected through electronic circuits to a recorder which continuously records the intensity of radiation received by the counter on a moving-paper chart. The diffracted beams from the sample show up as peaks whose heights are proportional to the intensities of the beams.

The explanation of x-ray diffraction by crystals is illustrated in Fig. 6-7. The horizontal lines PP′, QQ′, etc., represent traces of lattice planes with the spacing d. AX, BY, etc. represent a beam of incident x-rays of wavelength λ. The wavelength λ and the lattice spacing d are of the same order of magnitude, which is why a crystal can serve as a diffraction grating for x-rays. The reflected wave is the resultant of reflections from successive lattice planes, provided that the path difference between reflections from adjacent planes is equal to a whole number of wavelengths of the x-rays used—i.e., $GY + YH = n\lambda$, where n is an integer. If $GY + YH$ is not equal to $n\lambda$ there is no reflection. Since $GY + YH = 2d \sin \theta$, reflection can occur *only* when $2d \sin \theta = n\lambda$. Thus, for a specific lattice plane, reflection takes place only at certain angles θ_1, θ_2, θ_3 for $\lambda = 2d \sin \theta_1$, $2\lambda = 2d \sin \theta_2$, $3\lambda = 2d \sin \theta_3$. . . . For a different lattice plane the value of d will be different, and consequently so will the angle θ. The lattice planes can be identified and referred to by their Miller indices. The different values of n can be designated as the first, second, third, and successive orders of reflection from the specific lattice plane, but it is more usual to multiply the Miller indices of the lattice plane by the order of reflection (n). Thus a powder photograph may have reflections identified as (100), (200), (300), and so on.

Several procedures are available for the identification of a mineral from its powder photograph. If previous examination has suggested a possible identification, and a file of identified powder photographs is available, direct com-

d 4-0784	2.355	2.039	1.230	2.355	Au
I/I₁ 4-0784	100	52	36	100	Gold

Rad. Cu   λ 1.5405   Filter Ni
Dia.   Cut off   Coll.
I/I₁   d corr. abs.?
Ref. Swanson and Tatge, JC Fel. Reports, NBS (1950)

Sys. Cubic (f.c.)   S.G. O_H^5 - Fm3m
a_0 4.0786 b_0 c_0 A   C
α   β   γ   Z 4
Ref. Ibid.

εα   nωβ   εγ   Sign
2V D_x19.302mp   Color
Ref.

Sample purified at NBS laboratory and is about 99.997% Au.
At 26°C
To replace 1-1172, 1-1174, 2-1095

d Å	I/I₁	hkl	d Å	I/I₁	hkl
2.355	100	111			
2.039	52	200			
1.442	32	220			
1.230	36	311			
1.1774	12	222			
1.0196	6	400			
0.9358	23	331			
.9120	22	420			
.8325	23	422			

Figure 6-8 Card 4-0784 for gold, from *X-ray Powder Data File*, published by the American Society for Testing Materials. (By permission)

parison will confirm or deny the identification. Generally, however, it is necessary to determine the *d*-values for the individual reflections, and compare these with published data, usually the ASTM data file (Fig. 6-8). Transparent plastic scales* for specific camera diameters and x-ray wavelength, from which *d*-values can be read directly, are widely used. Otherwise the positions of the lines can be measured by mounting the film on an illuminated screen provided with a scale and cursor. The spacing *s* for each line can be converted to θ from the camera radius, and θ can be converted to *d* by using the relation developed above. Tables are usually available for this. The ASTM data file has an index listing the three strongest lines for most minerals and for many other substances. When an entry in the index is found to correspond to the three strongest lines in the unknown, the whole pattern is then compared with the file card indicated in the index.

Selected Readings

Azaroff, L. V., and M. J. Buerger, 1958, *The powder method in x-ray crystallography:* New York, McGraw-Hill.

Jaffe, H. W., 1949, Visual arc spectroscopic detection of halogens, rare earths and other elements by use of molecular spectra: *Am. Mineralogist,* v. 34, p. 667.

Larsen, E. S., and H. Berman, 1934, *The microscopic determination of the nonopaque minerals* (2nd ed.): Washington, U. S. Government Printing Office.

Nuffield, E. W., 1966, *X-ray diffraction methods:* New York, Wiley.

Paterson, M. J., A. J. Kauffman, and H. W. Jaffe, 1947, The spectroscope in determinative mineralogy: *Am. Mineralogist,* v. 32, p. 322.

Smith, O. C., 1953, *Identification and qualitative chemical analysis of minerals* (2nd ed.): Princeton, Van Nostrand.

Winchell, A. N., and H. Winchell, 1951, *Elements of optical mineralogy; Part II, descriptions of minerals* (4th ed.): New York, Wiley.

* Available from N. P. Nies, 969 Skyline Drive, Laguna Beach, California.

7

The Systematics
of Mineralogy

THE SPECIES CONCEPT IN MINERALOGY

The basic unit in systematic mineralogy is the *species*, but the concept of a mineral species has varied from time to time and even today there are differences of opinion over what constitutes an individual species. In 1923 a committee of the Mineralogical Society of America issued the following definition: "A mineral species is a naturally occurring homogeneous substance of inorganic origin, in chemical composition either definite or ranging between certain limits, possessing characteristic physical properties and usually a crystalline structure."

This definition can be criticized as not being sufficiently precise. For example, with the word *usually* in the last phrase, it could include gases and liquids; air and sea water are homogeneous substances of inorganic origin, with chemical composition ranging between certain limits and possessing characteristic physical properties, although without a crystalline structure. If the qualification "usually" is omitted, gases and liquids are excluded, but so too are the noncrystalline minerals—the amorphous and the metamict species. On the whole it is probably best to include the requirement of a crystalline structure in the definition of a mineral species, and to consider the few solid noncrystalline minerals as exceptions.

Another difficulty in this definition relates to chemical composition. How much variation in composition is to be allowed an individual species? Application of chemical principles would suggest that a mineral species be defined as a naturally occurring solid phase, the term "phase" meaning a homogeneous part of a physicochemical system. For example, any composition in the system

Mg_2SiO_4–Fe_2SiO_4 will crystallize as a homogeneous phase. Naturally occurring phases in this system are called olivine, which, on chemical principles, is therefore an adequate name for any mineral within the system. Nevertheless, mineralogists find it convenient to call the nearly pure Mg_2SiO_4 (which is sometimes found in metamorphosed dolomites) forsterite, and nearly pure Fe_2SiO_4 (which occurs in some granites and pegmatites) fayalite. In addition, natural olivines often contain some substitution of Ca and Mn for Mg and Fe, and thus extend beyond the boundaries of the Mg_2SiO_4–Fe_2SiO_4 system.

Strict application of chemical principles, although leading to logical conclusions, results in practical difficulties. These difficulties are usually overcome by a rather elastic use of the term "mineral species." Thus different varieties of columbite and tantalite are phases of the four-component system $FeNb_2O_6$–$FeTa_2O_6$–$MnNb_2O_6$–$MnTa_2O_6$. Any composition within this system can crystallize as a single phase with the structure of columbite, but has has been found convenient to use two specific names: columbite for phases in which Nb > Ta (in atom percent), tantalite for phases in which Ta > Nb. In contrast to this, any composition between $NaAlSi_3O_8$(Ab) and $CaAl_2Si_2O_8$(An) crystallizes (at least at high temperatures) as a single phase, plagioclase; plagioclase might therefore be considered a single mineral species, but it has been found convenient to distinguish six species: albite, Ab_{100}–Ab_{90}; oligoclase, Ab_{90}–Ab_{70}; andesine, Ab_{70}–Ab_{50}; labradorite, Ab_{50}–Ab_{30}; bytownite, Ab_{30}–Ab_{10}; and anorthite, Ab_{10}–Ab_0 (however, these might better be considered as subspecies of a single species, plagioclase).

Other difficulties arise in the strict application of physicochemical principles. For many minerals we do not know the limits of compositional variation. In some minerals, such as the feldspars, a combination of much work on the minerals and syntheses in the laboratory has led to a fairly complete knowledge; in others, including such an important group as the amphiboles, our knowledge is very imperfect. For example, it is still uncertain whether hornblende and the alkali amphiboles are separate phases or are linked by all intermediate compositions.

CLASSIFICATION OF MINERAL SPECIES

In spite of occasional uncertainty over what constitutes a mineral species, it is the fundamental unit in mineralogy. The exact number of identified species depends on the definition of the term and its interpretation, but today there are about 2,000 known. This number is being increased by the discovery and description of new species, at the rate of about 40 a year.

In order to deal systematically with minerals it is necessary to have some method of classification. The purpose of classification is ultimately to bring like things together and to separate them from unlike things. In the early years of the development of mineralogy as a science, from about 1750 to 1850, many systems of classification were proposed, some based on chemical criteria, others on physical criteria. Of the former, a system originally devised by the Swedish chemist Berzelius gradually became generally accepted. This system groups the mineral species into major divisions, or classes, according to the nature of the anionic group present. As adapted for this book, these classes are as follows:

    I. Native elements
   II. Sulfides (including sulfosalts)
 III. Oxides and hydroxides
 IV. Halides
  V. Carbonates, nitrates, borates, iodates
 VI. Sulfates, chromates, molybdates, tungstates
VII. Phosphates, arsenates, vanadates
VIII. Silicates

It can be seen that this classification, although originally based purely on chemical principles, has a definite significance in terms of crystal structure. The native elements include the metals (with metallic type of bonding), and the semimetals and nonmetals (with covalent bonding); the sulfides include some compounds with metallic bonding, most of those with covalent bonding, and a few with ionic bonding; in the remaining classes practically all of the species have ionic bonding to a greater or lesser degree. In the oxides and hydroxides and in the halides the structures consist of simply packed positive and negative ions; in the remaining classes complex anions are present: XO_3 groupings in the carbonates, nitrates, and borates; SO_4 groupings in the sulfates; PO_4 groupings in the phosphates; and SiO_4 and more complex groupings in the silicates.

The individual classes are divided into subclasses generally on chemical or structural grounds; thus class I, native elements, is divided into two subclasses— the metals, and the semimetals and nonmetals; the silicates are divided into six subclasses according to the structural association of the SiO_4 tetrahedra. The next division is into groups. Most groups include species that are closely related chemically and structurally (i.e., the feldspar group, the amphibole group); sometimes, however, species are grouped on the basis of paragenetic similarity, as are, for example, those of the feldspathoid group and the zeolite group. Groups may be divided into species or series, depending on the variability in chemical composition; the zeolite group comprises a number of individual species of limited variability, whereas the pyroxene group includes the orthorhombic and monoclinic series, each of which is further divided into individual species.

Species themselves may include finer subdivisions, which may be subspecies or varieties. Subspecies are usually arbitrary compositional divisions within a range of composition (labradorite can be considered a subspecies of plagioclase). High- and low-polymorphs (p. 92) of a single mineral are often considered subspecies (e.g., high-quartz and low-quartz are subspecies of quartz). Varieties may have distinctive physical properties (thulite, a pink variety of zoisite), or chemical composition (freibergite, a variety of tetrahedrite which contains silver). The use of separate names for chemical varieties is now being abandoned, such varieties being distinguished by descriptive adjectives. Thus, freibergite is called argentian tetrahedrite, a usage which simplifies nomenclature and provides information about structure and composition. This usage was proposed by Dr. W. T. Schaller in 1930 and is adopted in the seventh edition of *Dana's System of Mineralogy*. The adjective modifiers are formed by adding the suffix *an* (or *ian* or *oan* to represent different valencies) to the name of the substituting element. Thus, magnesite with some Mg replaced by Fe is referred to as ferroan magnesite.

The classification outlined above is not as detailed as that used in the seventh edition of *Dana's System of Mineralogy*. Our classification does not provide specifically for some minerals of very rare occurrence—selenates, selenites, tellurates, tellurites, antimonates, antimonites, arsenites; these minerals may be included with the sulfates. There are also some naturally occurring salts of organic acids (mainly oxalates) and a few solid hydrocarbons. These substances, which fall within our definition of a mineral and are generally accepted as minerals, require two classes in addition to those included here.

THE NAMING OF MINERALS

The names used for some minerals are very old, so old that their origins are lost. In the first century A.D. Pliny listed a number of native or easily reduced elements, common ore minerals, and gem minerals. With the development of mineralogy in the latter part of the eighteenth century, rival systems of nomenclature grew up. Linnaeus, who developed the binomial nomenclature used in botany and zoology, applied it also to minerals, and this system was adopted and extended by J. D. Dana in the first (1837) and second (1844) editions of *The System of Mineralogy*. Thus the genus Baralus included B. ponderosus (barite), B. prismaticus (celestite), B. fusiles (witherite), and B. rubefaciens (strontianite). However, in the third edition of *The System of Mineralogy* (1850)

Dana completely abandoned this nomenclature and adopted the procedure already current in Europe of using a single name for each species.

How are minerals named? However the original describer wishes, with no system other than that the name usually, but not always, ends in *ite*. Many names are derived from Greek or Latin words that give some information about the mineral, such as color (albite, from the Latin *albus*, white), crystal form or habit (sphene, from the Greek *sphen*, wedge), density (barite, Greek *barys*, heavy), or from words that are derived from the chemical composition (calcite, zincite). Two descriptive suffixes are *-clase*, indicating cleavage (Greek *klasis*, fracture), as in ortho*clase*; and *-phyllite*, indicating a flaky mineral (Greek *phyllon*, a leaf), as in pyro*phyllite*. These are useful bases for naming, since they give some indication of the nature of the mineral, but they are limited in their application, since there are too few distinctive properties to go around. Hence the practice has grown of naming minerals after persons, often mineralogists, but the list of persons so honored also includes mineral collectors, mine owners and officials, and public figures. There is a rooseveltite (for Franklin D. Roosevelt) and a mussolinite, but no hitlerite! Many names are also derived from the first locality at which a mineral was found. Examples are anglesite, from the island of Anglesey, off the coast of Wales; aragonite, from Aragon in Spain; strontianite, from Strontian in Scotland; and franklinite, from Franklin, New Jersey.

Sometimes it happens that two or more names are applied to the same species, as is the case in other natural sciences. Under such circumstances the standard procedure is to adopt the name first applied to the species (the rule of priority). Difficulties arise in various ways, especially when different names have become current in different countries. In English-speaking countries the usage in *Dana's System of Mineralogy* is generally accepted as standard, and it is followed in this book.

Part II

DESCRIPTIONS

8

Class I: Native Elements

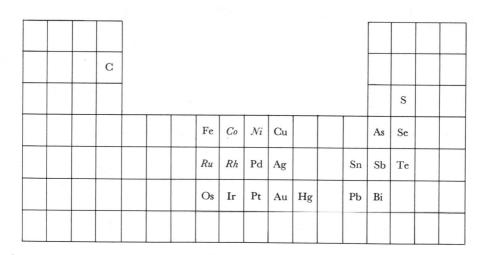

Although a number of chemical elements are found as minerals* in the solid part of the earth's crust, none of them occur in masses large enough to be classed as rocks. However, many elements such as gold, silver, copper, carbon (diamond and graphite), and sulfur are often found in sufficient quantity under certain conditions to have attracted the attention of man from earliest times to the present day. All of these elements are still obtained, in important amounts, from deposits containing the native elements. Deposits of native silver, copper, and, to a lesser extent, sulfur, are becoming less important as economic sources

* Table above shows elements which occur as minerals; those in italics occur as constituents of natural alloys.

of these elements. In addition to the above five elements, platinum and its related metals, which were not recognized until the eighteenth century and later, are also mined in substantial quantities. Other elements—indicated by the table at the opening of this chapter—have been recognized in the native state but are mineralogical curiosities.

The important native elements are classified as follows:

METALS	SEMIMETALS AND NONMETALS
Gold Group	Arsenic Group
Gold, Au	Arsenic, As
Silver, Ag	Antimony, Sb
Copper, Cu	Bismuth, Bi
Platinum Group	Sulfur Group
Platinum, Pt	Sulfur, S
Palladium, Pd	Carbon Group
Platiniridium, (Pt, Ir)	Diamond, C
Iron Group	Graphite, C
Iron, Fe	
Nickel-iron, (Ni, Fe)	

GOLD GROUP

The gold group comprises the well-known metals gold, silver, copper, and lead; the first three are commonly found in small quantities in the earth's crust, whereas lead is a mineralogical curiosity. These native metals are alike in

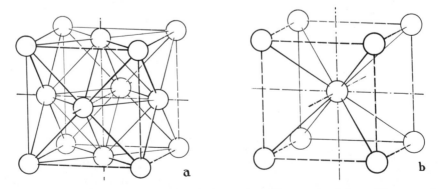

Figure 8-1 (a) Face-centered cubic lattice as found in metals of the gold group; the atoms are actually closer together than shown, and touch along the face diagonals as shown in Fig. 8-4. Each atom has 12 neighboring atoms touching it. (b) Body-centered cubic lattice, illustrating the arrangement in iron.

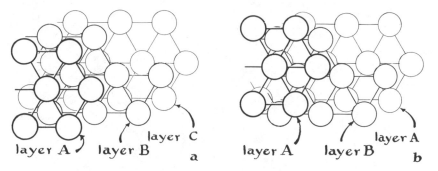

Figure 8-2 (a) Face-centered cubic packing, showing the atoms as spheres about half actual size relative to the space and the layer sequence *ABC*, *ABC*, . . . along a trigonal axis. (b) Hexagonal close packing, showing the atoms as spheres about half actual size and layer sequence *AB*, *AB*, . . . along the hexagonal axis.

structure, with the atoms lying on the points of a face-centered cubic lattice (Figs. 8-1a, 8-2a, 8-3a, 8-4). Each atom is closely coordinated with twelve neighboring atoms, and the unit cube of the structure contains four atoms. The length of the edge of the cubic unit, the *a* length, which depends on the radius of each atom, is given below, together with other physical constants for the pure metals.

Crystals of these metals are of rare occurrence; they all form dendritic and arborescent growths which often show crystal form at their extremities. These metals are also similar in many physical properties (soft, malleable, ductile, good conductors of heat and electricity, opaque to light, high specific gravity).

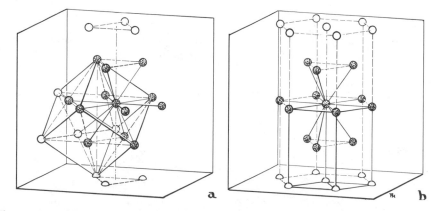

Figure 8-3 (a) Face-centered cubic arrangement, showing the twelvefold coordination and the relation of the layer sequence *ABC*, *ABC*, . . . to a face-centered cubic lattice. (b) Hexagonal cell in the hexagonal close-packed arrangement, showing twelvefold coordination distinct from the face-centered cubic arrangement in Fig. 8-3a.

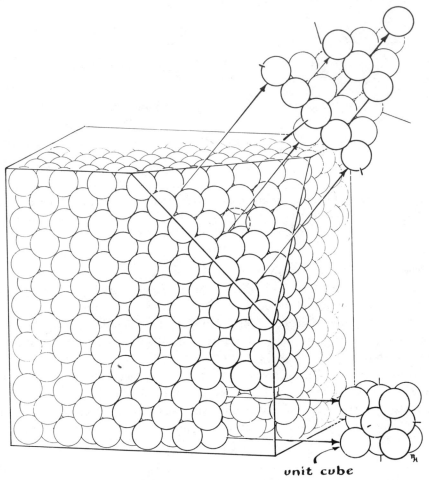

unit cube

Figure 8-4 Face-centered cubic packing, showing the atoms as spheres in their true size relative to the space. The face-centered cubic unit contains four atoms.

Crystal system and class: Isometric; $4/m\ \bar{3}\ 2/m$.

	Au	Ag	Cu	Pb
Cell dimensions:	$a = 4.0786$	$a = 4.0862$	$a = 3.6150$	$a = 4.9505$
Cell content:	$Z = 4$	$Z = 4$	$Z = 4$	$Z = 4$
Hardness:	$2\frac{1}{2}$–3	$2\frac{1}{2}$–3	$2\frac{1}{2}$–3	$1\frac{1}{2}$
Density:	19.3	10.5	8.94	11.37
Color:	Yellow	White	Light rose	Gray-white

As found in nature these metals are rarely pure, but contain various other metals and semimetals in solid solution. As one might expect from the similarity in cell size, a continuous series of solid solutions is found between silver and gold. Gold and copper are also mutually soluble, but silver and copper are almost mutually insoluble. Mercury may be listed with the gold group only because it is often found alloyed with gold or silver. It occurs sparingly in nature as isolated drops in the upper oxidized zone of cinnabar deposits (Calfornia) or with cinnabar in volcanic regions.

Gold, Au

Habit: Crystals rare; usually roughly octahedral, dodecahedral, or cubic; in parallel groups and twinned aggregates. Commonly leafy, dendritic, filiform, or spongy; also massive, in rough, rounded or flattened grains or scales.

Twinning: Common on {111}, usually repeated.

Cleavage and fracture: None; hackly.

Tenacity: Very malleable, ductile and sectile.

Density: 19.3, decreasing with increase in silver, copper, and other alloying elements.

Color and streak: Gold yellow when pure, whiter due to alloyed silver, or orange red due to copper.

Luster and opacity: Metallic; opaque except in very thin foil which transmits blue and green light.

Chemistry: Silver substitutes for gold in the structure, and a complete series extends to pure silver. Native gold may contain as much as 10–15 percent silver. The name *electrum* is applied to natural gold with 20 percent or more of silver. Other metals—palladium, rhodium, copper, bismuth, mercury—are found dissolved in gold to a minor extent.

Diagnostic features: Gold may usually be distinguished by its color and sectility. In small grains it may often be confused with pyrite or chalcopyrite (hence the common designation "fool's gold" for these minerals). Pyrite may be distinguished by its superior hardness, and chalcopyrite by its soft but very brittle nature. Small flakes of biotite partially weathered to a golden color may also be mistaken for gold, especially when wet. Gold may be fused (1062°C) in the blowpipe flame, is unaffected by fluxes, and is insoluble in acids except aqua regia.

Occurrence and production: Gold is present in the earth's crust and in sea water to the extent of about 6 parts in a hundred million. Nearly all gold found in the crust is in the native state. Compounds with tellurium have been recognized at a number of localities, and compounds with bismuth, antimony, and possibly selenium are of very rare occurrence. Gold occurs in the crust in notable amounts in two main types of deposits: hydrothermal veins, and placers, either consolidated or unconsolidated. The latter are derived from the former by weathering and erosional processes; gold, largely unaffected by chemical and mechanical weathering, is concentrated and separated from other minerals because of its high density.

Most gold of hydrothermal origin is found in quartz veins, commonly with pyrite, other sulfides, gold-silver tellurides, rarely lead and mercury tellurides, scheelite, tourmaline, ferroan dolomite (commonly called ankerite), sericite, or the green chromian sericite. Minor amounts of gold are also found in massive hydrothermal deposits with pyrite, chalcopyrite, pyrrhotite, sphalerite; here gold is often recovered as a valuable by-product of base-metal mining.

In placer deposits (both recent unconsolidated and older consolidated placers) the gold occurs as rounded or flattened grains and nuggets associated with other heavy resistant minerals. Gold has been obtained from earliest times largely from unconsolidated placers. The mining of hydrothermal veins and consolidated placers has been of increasing importance in the recovery of the metal since the middle of the nineteenth century.

In recent times gold has been produced in important quantities from conglomerates, placers, and veins. The most important single source is the consolidated quartz-conglomerates of the Witwatersrand, Transvaal, Union of South Africa. Placers are worked along both the eastern and western slopes of the Ural Mountains, and at localities in Siberia, Alaska, and the Yukon. Quartz veins are worked at the Porcupine and Kirkland Lake districts of Ontario, and in many other localities on the Canadian Shield; at the Homestake mine, South Dakota; in the Sierra Nevada, California; at Kalgoorlie, Western Australia; and at Kolar, India. The Union of South Africa, the USSR, Canada, the USA, Australia, and Ghana have produced about 90 percent of the world's estimated annual output of nearly 42 million fine ounces (1960–1964 average). About 60 percent of this total was obtained in the Union of South Africa.

Silver, Ag

Habit: Crystals rare; roughly cubic, octahedral, or dodecahedral; commonly in arborescent and wiry forms; also massive, as scales, leafy forms, or thin plates, filling vein fractures.

Twinning: Common on {111}.

Cleavage and fracture: None; hackly.

Tenacity: Very ductile and malleable.

Density: 10.1–11.1, variable due to dissolved gold, copper, or other metals.

Color and streak: Silver white, often gray to black due to tarnish.

Luster and opacity: Metallic; opaque.

Chemistry: Native silver may contain gold in substitution for silver atoms in the structure (see *gold*). Mercury, arsenic, and antimony are also found dissolved in native silver.

Diagnostic features: Color, tarnish, sectility, and hackly fracture are the most useful identifying features. Silver will fuse (960°C) before the blowpipe to a silver-white globule which will flatten out under pressure when cold. It is soluble in nitric acid. Silver is readily tarnished, with the formation of silver sulfide, by H_2S fumes or solutions of sodium sulfide, providing the well-known silver-coin test for sulfur.

Occurrence and production: Native silver is found in the crust in two principal ways: as small amounts in the oxidized zone of ore deposits, or as deposits from hydrothermal solutions of primary origin. It occurs with sulfides, zeolites, calcite, barite, fluorite, and quartz at Kongsberg, Norway; with arsenides and sulfides of nickel and cobalt, silver minerals, calcite, and barite at Cobalt, Ontario; with uraninite and nickel-cobalt minerals at Joachimstal, Bohemia, and at Great Bear Lake, Canada. Native silver is found associated with native copper at Keweenaw, Michigan.

In recent years much of the silver produced (80 percent of Canadian production) has been obtained as a by-product of gold refining or of lead-zinc mining. Mexico, the USA, Canada, the USSR, Peru, and Australia have produced 75 percent of the annual world output of 240 million ounces during the period 1960–1964.

Copper, Cu

Habit: Crystals rare; cubic or dodecahedral; usually massive filiform or arborescent (Fig. 8-5).

Twinning: Common on {111}.

Cleavage and fracture: None; hackly.

Tenacity: Highly ductile and malleable.

Color: Light rose on fresh surface, quickly changing to copper-red then brown.

Streak: Metallic.

Luster and opacity: Metallic; opaque.

Chemistry: Native copper often contains small to trace amounts of silver, arsenic, iron, bismuth, or antimony.

Diagnostic features: Color, sectility, and hackly fracture are the most significant features. It melts before the blowpipe (1084°C); the globule, which becomes coated with black oxide on cooling in air, is malleable. It dissolves readily in nitric acid.

Occurrence: Copper is most commonly associated with basic extrusive igneous rocks, where it has formed by reaction between copper-bearing solutions and iron minerals. In this type of deposit, which has been extensively mined on

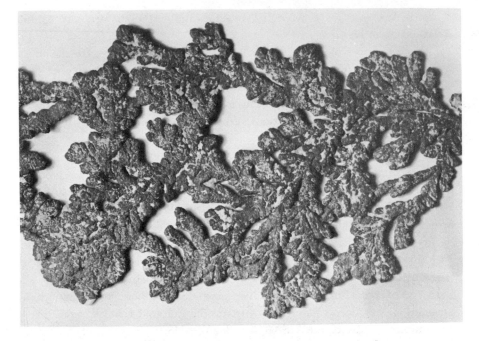

Figure 8-5 Dendritic native copper. [Courtesy Queen's University.]

the Keweenaw Peninsula, Michigan, native copper is associated with chalcocite, bornite, epidote, calcite, prehnite, datolite, chlorite, zeolites, and small amounts of native silver. Copper and other minerals are found filling the cavities in amygdaloidal lavas, or the interstices in interbedded sandstones and conglomerates. Copper is also found replacing the grains and pebbles, or in veins cutting the rocks. Single masses of copper weighing many tons have been found in the Keweenaw district. The series of lavas and sediments containing the copper deposits dip at about 38° to the northwest and outcrop on the surface in a belt about 100 miles long and 2–4 miles wide. Copper was obtained here by prehistoric tribes and by the early settlers. From 1845 to the present, approximately 100 mining companies have produced copper from these deposits; since 1900 the deposits have become less important as a source of copper, and the district is now almost exhausted. Most copper is now recovered from sulfides and from the oxidized minerals (oxides, carbonates, sulfates).

Production and uses: Copper was one of the first metals used by man, for it is believed that in the eastern Mediterranean countries stone implements gave way to those of bronze about 7,000 years ago. Since then the outstanding characteristics of ductility, malleability, resistance to corrosion, and high electrical conductivity have continually increased the demand for copper.

It has been for many years the most important nonferrous metal in quantity and value of world output. During the period 1960–1964 the annual production of new copper for the world from ore, about 5 million tons, came largely from the USA, Chile, Zambia, the USSR, Canada, Japan, and the Congo. The annual world production of copper metal for the same period is in excess of 5 million tons, since reworked scrap copper is included.

PLATINUM GROUP

The platinum group includes the two metals platinum and palladium, and natural alloys of platinum-iridium and iridium-osmium. Native platinum commonly contains iron in solution. Platinum, palladium, and iridium are isometric, with the copper structure. The alloys of iridium and osmium have a hexagonal close-packed structure which is similar to that of magnesium (Figs. 8-2b, 8-3b, 8-6).

Platinum, Pt

Crystal system and class: Isometric; $4/m\ \bar{3}\ 2/m$.
Cell dimensions, and content: $a = 3.9231$; $Z = 4$.

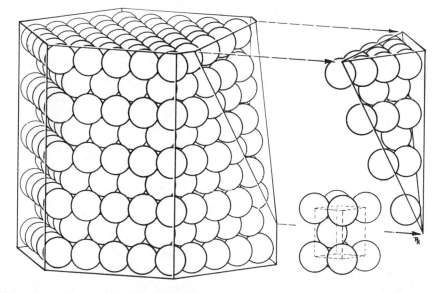

Figure 8-6 Hexagonal close packing, showing the atoms as spheres in their true size relative to the space. The simplest unit cell is a 60° rhombic prism containing two atoms. This structure is typical of magnesium, and is found in some native metals of the platinum group.

Habit: Rarely in cubic crystals; usually in grains, scales or nuggets.

Cleavage and fracture: None; hackly.

Tenacity: Malleable and ductile.

Hardness: 4–4½, increases with iron content.

Density: 14–19 (21.46 for pure Pt).

Color: Whitish steel gray to dark gray.

Luster and opacity: Metallic; opaque.

Chemistry: Native platinum always contains iron, up to about 28 percent; it is distinctly magnetic, often showing polarity when rich in iron. It may also contain palladium, rhodium, iridium, and copper.

Diagnostic tests: Platinum is infusible in the blowpipe flame, and is dissolved only by hot aqua regia.

Occurrence and production: Platinum occurs in basic and ultrabasic igneous rocks, associated with olivine, pyroxene, chromite, and magnetite. It is commonly found as grains or nuggets in river gravels derived from areas containing such ultrabasic rocks. Platinum and a few platinum-bearing minerals are mined from magmatic concentrations in dunite in the Bushveld igneous complex in the Transvaal, Union of South Africa. Large quantities are recovered from river placers along both flanks of the Ural Mountains in the USSR. Minor amounts of platinum are recovered from placers in many other parts of the world.

   At Sudbury, Ontario, where substantial quantities of the platinum metals are recovered from the chalcopyrite-pentlandite-pyrrhotite ores, the metal occurs in the mineral sperrylite, $PtAs_2$. The annual world output of all the platinum metals averages about 1–2 million troy ounces, of which 90 percent is obtained from Canada, the Union of South Africa, and the USSR.

IRON GROUP

This group includes two minerals: iron [body-centered cubic (Fig. 8-1b), containing minor nickel; $a = 2.874$, Disko Island, Greenland] and nickel-iron (face-centered cubic, 77–24 percent nickel; $a = 3.560$, 75 percent Ni, Josephine County, Oregon). Both are very rare as terrestrial material, but they make up the bulk of the metallic constituents of meteorites as kamacite and taenite, respectively. Rarely, iron is found in basalts, and the most notable occurrence is at Disko Island, Greenland, where it is found as large masses and small embedded grains. Nickel-iron has been found as small plates and grains in a number of placer mining districts.

ARSENIC GROUP

The arsenic group includes the native elements arsenic, As; antimony, Sb; and bismuth, Bi. These semimetals crystallize with a rhombohedral lattice. Each atom is bonded to six neighbors but more closely to three of these than to the other three. This is in accordance with its threefold valency and its position in group V of the periodic table. The shorter bonds result in a puckered sheet of atoms, connected to the next sheet by the longer bonds. These sheets lie in the plane (0001), and thus the perfect basal cleavage may occur without breaking of the shorter bonds (Fig. 8-7). In arsenic the short bonds are largely homopolar in character, but in antimony and bismuth they become increasingly metallic in character, accompanied by some decrease in hardness and melting point and by some increase in atomic weight and specific gravity.

Crystal system and class: Trigonal; $\bar{3}\ 2/m$.

	As	Sb	Bi
Axial elements:	$a:c = 2.8053$	$a:c = 2.6174$	$a:c = 2.6089$
Cell dimensions:	$a = 3.760$	$a = 4.307$	$a = 4.546$
	$c = 10.548$	$c = 11.273$	$c = 11.860$
Cell content:	$Z = 6$	$Z = 6$	$Z = 6$
Cleavage:	{0001} perfect	{0001} perfect	{0001} perfect
Hardness:	$3\frac{1}{2}$	$3–3\frac{1}{2}$	$2–2\frac{1}{2}$
Density:	5.7	6.7	9.7–9.8
Color:	Tin white to gray black	Tin white	Silver white, reddish hue
Streak:	Tin white to gray	Gray	Silver white
Luster and opacity:	Nearly metallic; opaque	Metallic; opaque	Metallic; opaque

Arsenic, As

Habit: Natural crystals rare; usually massive, in concentric layers; sometimes reniform or stalactitic.

Chemistry: Arsenic usually contains some antimony, sometimes minor amounts of iron, nickel, silver or sulfur. Although arsenic and antimony form an unbroken series of solid solutions at high temperature in the laboratory, arsenic and antimony are found combined in the mineral *allemontite* in about equal proportions. Allemontite, which may occur in intergrowths with either antimony or arsenic, is probably an intermetallic compound and is treated as a distinct mineral.

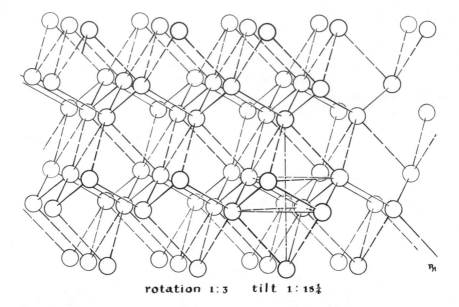

rotation 1:3   tilt 1:18¼

Figure 8-7   Structure of antimony. Solid lines represent short bonds of essentially homopolar character, and longer bonds are dashed; {0001} cleavage horizontal.

Diagnostic features: Although infusible, arsenic is volatile, and gives off an odor like that of garlic, and white fumes of As_2O_3.

Occurrence: Arsenic is commonly found in hydrothermal veins, with silver, cobalt, or nickel ores, and is also associated with barite, cinnabar, realgar, orpiment, stibnite, and galena. Adequate supplies of arsenic for commercial needs are recovered from smelter fumes resulting from the treatment of metallic ores containing arsenic in the form of arsenopyrite, enargite, or tennantite.

Antimony, Sb

Habit: Crystals rare; usually massive; lamellar and distinctly cleavable; also radiated, botryoidal, or reniform.

Twinning: Common on {10$\bar{1}$4}.

Diagnostic test: Antimony melts readily (630°C) before the blowpipe, leaving a brittle metallic globule coated with oxide.

Occurrence and production: Antimony is of hydrothermal origin, occurring in veins with silver ores, often with stibnite, also with sphalerite, pyrite, galena, and quartz. Antimony and the antimony oxides contribute somewhat to commercial sources of antimony, but most is obtained from stibnite and antimonial lead ores. The average annual world output of about 100 million pounds is largely obtained from the Union of South Africa, China, Bolivia, and Mexico.

Bismuth, Bi

Habit: Natural crystals indistinct; usually in reticulated or arborescent shapes; foliated, granular.

Twinning: On {10$\bar{1}$4}, often polysynthetic.

Color and streak: Silver white, with a reddish hue darkening with exposure and developing iridescent tarnish.

Diagnostic tests: Bismuth melts readily at 270°C, forming a brittle globule soluble in nitric acid. It may be distinguished from antimony by its color and low melting point.

Occurrence and production: Bismuth occurs in hydrothermal veins with ores of cobalt, nickel, silver, and tin, and in pegmatites. Most of the bismuth of commerce is obtained from deposits of tin, copper, or silver, or as a refinery by-product from the treatment of lead ores. The average annual world output of about 6 million pounds is produced by Mexico (16 percent), Peru (20 percent), and many other countries.

SULFUR GROUP

This group comprises the mineral sulfur, which is naturally occurring α-sulfur or rhombic sulfur, together with the natural forms of β- and γ-sulfur, which are monoclinic in crystallization. The latter two minerals are of rare occurrence. In orthorhombic sulfur the atoms are bonded into puckered rings of eight atoms (Fig. 8-8d) which may be regarded as S_8 molecules. Each atom is bonded to two neighbors in the ring by homopolar bonds, and the distance between S atoms in different rings is much greater than for atoms within a ring.

Sulfur, S

Crystal system and class: Orthorhombic: $2/m\ 2/m\ 2/m$.

Axial elements: $a:b:c = 0.8133:1:1.9028$.

Cell dimensions, and content: $a = 10.468$, $b = 12.870$, $c = 24.49$; $Z = 128$.

Common forms and angles:

(001) $\wedge$ (011) = 62° 18′	(001) $\wedge$ (113) = 45° 11′
(001) $\wedge$ (101) = 66° 52′	(111) $\wedge$ (11$\bar{1}$) = 36° 39′
(010) $\wedge$ (111) = 53° 13′	(111) $\wedge$ (1$\bar{1}$1) = 73° 34′

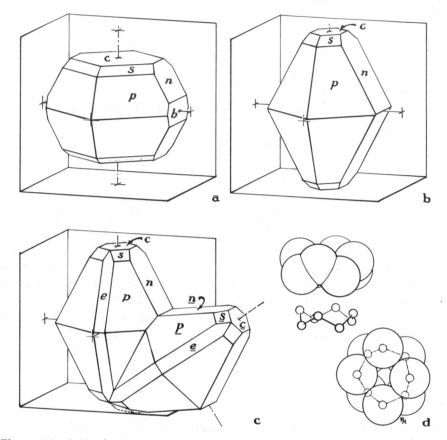

Figure 8-8   Sulfur (rhombic or alpha-sulfur) crystals. Forms: pinacoids $c\{001\}$, $b\{010\}$; rhombic prisms $n\{011\}$, $e\{101\}$; rhombic dipyramids $s\{113\}$, $p\{111\}$. (c) A twin crystal with twin plane $\{011\}$. (d) An S_8 ring (each sulfur atom has two closest neighbors); 16 such rings are contained in the face-centered orthorhombic unit cell.

Habit: Crystals commonly dipyramidal; thick tabular on $\{001\}$ (Figs. 8-8 and 8-9); also massive, in spherical or reniform shapes, incrusting.

Twinning: Rare on $\{011\}$ (Fig. 8-8c).

Cleavage: None.

Fracture and tenacity: Conchoidal to uneven; brittle to slightly sectile.

Hardness: $1\frac{1}{2}$–$2\frac{1}{2}$.

Density: 2.07.

Color: Yellow, to yellowish brown.

Streak: White.

Luster: Resinous to greasy.

Optical properties: $\alpha = 1.958$, $\beta = 2.038$, $\gamma = 2.245$; $\gamma - \alpha = 0.287$; positive, $2V = 70°$.

Diagnostic features: Color, low hardness, brittleness; melts at 113°C, burns at 270°C in air with a blue flame, yielding sulfur dioxide; insoluble in water and unaffected by most acids, soluble in carbon disulfide and some oils.

Figure 8-9 Sulfur. The large crystal is oriented with *b* axis vertical; the large pinacoid *c*{001} is beveled by the rhombic pyramid *p*{111}. (Caltanisetta, Sicily.) [Courtesy the Royal Ontario Museum.]

Occurrence and recovery: Sulfur is frequently found in regions of recent volcanic activity. It may be deposited as a direct sublimation product from volcanic gases. It may result from incomplete oxidation of hydrogen sulfide from volcanic sources, also from the decomposition of hydrogen sulfide in thermal spring waters. The living processes of sulfur-forming bacteria result in the separation of sulfur from sulfates. Sulfur is most commonly found in sedimentary rocks of Tertiary age associated with gypsum and limestone. Very substantial quantities of sulfur occur in the cap-rock of some of the salt domes of the Texas-Louisiana gulf coastal region. Here the sulfur occurs in a cavernous calcite cap-rock and extends into a layer of anhydrite which covers the main body of the salt plug. Although only a few salt domes contain commercial deposits of sulfur, it is estimated that these hold many millions of tons. The low melting point of sulfur permits its recovery from sedimentary beds at depth by the Frasch process: into a well penetrating the sulfur-bearing formation hot water is pumped under pressure, causing molten sulfur to collect at the bottom. It is then forced by hot compressed air up a small inner pipe as red liquid sulfur and piped to a stock pile. Sedimentary series containing sulfur with gypsum, calcite, bituminous limestone, or cellular limestone are found in the USSR, in Sicily, and in other parts of Europe. Large quantities of sulfur are recovered as sulfur dioxide from smelter gases and by-product pyrite. Elemental sulfur is also recovered from pyrite and sour (H_2S–bearing) natural gas. World production in 1964 was nearly 14 million tons, the USA and Mexico producing about half of this.

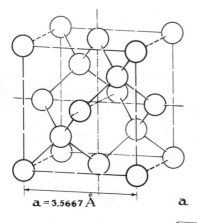

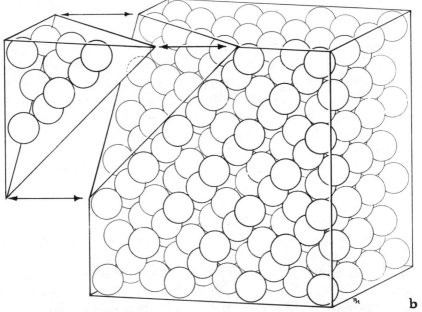

Figure 8-10 Diamond structure. (a) Atoms in the face-centered cubic unit cell, showing the tetrahedral bonds from each atom to the four closest neighbors. (b) Atoms in nearly normal size, showing the relation of the octahedral {111} cleavage to the structure.

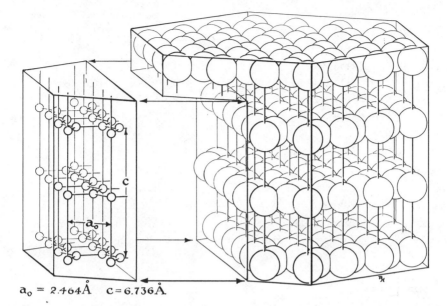

$a_o = 2.464\overset{\circ}{A}$ $c = 6.736\overset{\circ}{A}$

Figure 8-11   Graphite structure. Atoms are closely linked in sheets parallel to {0001} and widely spaced along *c*, resulting in the perfect basal cleavage {0001}.

CARBON GROUP

The carbon group includes the two natural forms of carbon, diamond and graphite. These minerals present the greatest contrast in structure and properties to be found in any pair of polymorphous substances. Diamond is isometric ($a = 3.5670$ Å), with each carbon atom linked tetrahedrally to four neighboring carbon atoms (Fig. 8-10 and Fig. 3-7). The close three-dimensional covalent linking, completing the outer electron shell for each atom by sharing of electrons, is expressed in the great hardness of diamond. Graphite is hexagonal, with the carbon atoms linked together in sheets which are widely spaced along the *c* axis (Fig. 8-11 and Fig. 3-7). This wide spacing of the sheets of carbon atoms explains the perfect and easy cleavage of graphite.

	DIAMOND	GRAPHITE
System:	Isometric	Hexagonal
Crystal class:	$4/m\,\bar{3}\,2/m$	$6/m\,2/m\,2/m$
Axial elements:	—	$a\!:\!c = 1\!:\!2.7338$
Cell dimensions:	$a = 3.5670$	$a = 2.464$
		$c = 6.736$
Cell content:	$Z = 8$	$Z = 4$

	DIAMOND	GRAPHITE
Common forms:	{111}, {110}, {100}	{0001}, {10$\bar{1}$0}, {10$\bar{1}$1}, {10$\bar{1}$2}
Twinning:	{111}	—
Cleavage:	{111} perfect	{0001} perfect, easy
Fracture:	Conchoidal	None
Tenacity:	Brittle	Flexible, not elastic, greasy feel
Hardness:	10	1–2
Density:	3.50	2.09–2.23
Color:	Colorless and varied	Black
Streak:	White	Black
Luster:	Adamantine	Metallic to dull
Opacity:	Transparent to translucent	Opaque

Diamond, C

Optical properties: $n = 2.4175$.

Habit: Crystals commonly octahedral, also dodecahedral, cubic, or tetrahedral; faces often curved; often flattened on {111}.

Twinning: Common on {111}, giving simple and multiple twins.

Color and uses: Pale yellow, brown, white to blue-white, also orange, pink, blue, green, red, black. The color and transparency of diamond vary greatly and have a considerable bearing on its value. The colorless and water-clear material continues to be the most desirable of all substances for cutting as gems. Water-clear diamond in other colors is also used for gem cutting but is not usually valued as highly. Diamond not of gem quality, usually called *bort*, has many specialized industrial uses, most of which depend on its being the hardest known substance. These include dies for drawing very fine wire, cutting tools for special machine work, diamond-drill bits that permit the recovery of a cylindrical sample of rock, abrasives for grinding and cutting wheels or for the fine-grinding or polishing of diamond itself and of many other hard substances. The term bort, in its older and more restricted usage, applies to granular and cryptocrystalline diamond, gray to black in color due to impurities and inclusions. This material shows no distinct cleavage and for this reason is tougher and less brittle than normal diamond. Another variety, which is highly prized for industrial cutting, is called carbonado; it is a black or gray-black bort, massive, with a density less than that of diamond.

Occurrence and production: Diamond is found as scattered crystals in an ultrabasic rock consisting of olivine-rich porphyry or olivine-phlogopite-rich porphyry. This rock, which is called kimberlite, occurs in roughly carrot-shaped intrusive bodies, called "pipes," with diameters up to 2,800 feet and depths of 3,500 feet or more. Many kimberlite pipes are known in South Africa, but only a few contain sufficient diamond to warrant mining. The pipes contain a great variety of inclusions—fragments of rocks from greater depth, and other fragments from above which must have fallen into an open cavity and been caught in the upwelling magma. The magma was not hot enough to alter these fragments appreciably.

The kimberlite rock has been altered by surface weathering to "yellow ground"; at greater depth it is called "blue ground," and here the olivine is largely altered to serpentine. The diamonds may be recovered from the friable yellow ground by passing it, with water, over greased bronze tables, the diamonds adhering to the grease. The harder blue ground, however, must be crushed before tabling. The content of the profitable pipes ranges from about 0.1 to 0.35 carats per ton (1 carat = 0.2 g). The Premier Mine, near Pretoria, South Africa, has produced about $5\frac{1}{2}$ tons of diamonds from 100 million tons of rock (about 0.0000055 percent). This mine also yielded the largest known diamond, the Cullinan, weighing 3,106 carats (1.3 pounds) before cutting.

Diamond is also found in river and beach gravels, and at present over 90 percent of the world's diamond production is obtained from alluvial deposits; however, these deposits yield a lower proportion of good quality gem stones than do the mines. Of the annual production, about half, by weight, is bort suitable only for industrial uses; about 5 percent is suitable for cut stones of 1 carat or larger.

From early times until about 1730 India was the only source of diamonds. Brazil became the chief producer from that time until the discovery of the South African kimberlite pipes in 1867. Diamonds were first discovered in South Africa in the gravels of the Vaal River. This quickly led to the discovery of the pipes and to other river and beach gravels in adjoining parts of South Africa, the Congo, Angola, Ghana, and Tanzania.

Alluvial diamonds have been found in Australia and at scattered points in North America. One occurrence of kimberlite in Arkansas has yielded several thousand small diamonds of good color.

Graphite, C

Habit: Crystals tabular on {0001} with hexagonal prism {1010} and dipyramids {10$\bar{1}$1}, {10$\bar{1}$2}; commonly as embedded foliated masses, isolated tabular grains, and scales; also columnar, radiated, or earthy.

Diagnostic features: Extreme softness, greasy feel, low specific gravity, black color, infusibility, marks paper readily; it is distinguished from molybdenite by its black color and its black streak on glazed porcelain.

Occurrence: Graphite is of common occurrence in metamorphic rocks produced by regional or contact metamorphism. It is found in crystalline limestone, gneiss, schist, quartzite, and metamorphosed coal beds. In thoroughly recrystallized limestone it is present as distinct micaceous flakes. In schists of low-grade metamorphic origin, including some of Precambrian age, the carbonaceous material is massive and apparently amorphous, though commonly called graphite. Graphite is also found in igneous rocks, veins, and pegmatite dikes. Such rocks sometimes are found in contact with metamorphosed graphitic sediments, suggesting that the carbon was probably introduced from the sedimentary rocks. In other deposits the graphite may well have formed from primary constituents of the magma. Graphite in shear zones, contact metamorphic deposits, and hydrothermal veins may have had its origin in magmatic vapors or from adjacent sediments.

The graphite resulting from regional metamorphism of sedimentary rocks probably formed by crystallization of the carbon from organic remains. That found so commonly in rocks of Precambrian age may have originated from the reduction of calcium carbonate by some inorganic process. There is no direct evidence for the nature of such a process.

Production and use: The world's annual graphite production of about 600,000 tons is derived chiefly from the USSR, Korea, Austria, Ceylon, Madagascar, and Mexico. Its high melting temperature (3000°C) and insolubility in acid make it suitable for many uses other than in "lead" pencils. It is utilized in foundry facings, crucibles, lubricants, paints, electrodes, and generator brushes.

9

Class II: Sulfides

														S		
Ca			*V*	*Cr*	*Mn*	Fe	Co	Ni	Cu	Zn	*Ga*	*Ge*	As	Se		
				Mo		*Ru*		Pd	Ag	Cd	*In*	Sn	Sb	Te		
				W				*Pt*	*Au*	Hg	Tl	Pb	Bi			

The sulfides* include a large group of minerals, predominantly metallic in character, with the general formula A_mX_p, in which X, the larger atom, is sulfur, or, to a lesser extent, arsenic, antimony, bismuth, selenium, or tellurium, and the smaller atom is one or more of the metals. In a few, sulfur and either arsenic or antimony are present in about equal amounts. Included also are those minerals known as sulfosalts, which are often placed in a separate class. The sulfosalts have the general formula $A_mB_nX_p$, which may also usually be written as a double sulfide $A_mX_q \cdot B_nX_{(p-q)}$. The common elements

* Table above shows elements which form sulfide minerals; those in italics are of uncommon occurrence.

in the sulfosalts are: as A, Ag, Cu, Pb; as B, As, Sb, Bi, Sn; as X, S. Iron, Ni, and Hg are of rare occurrence; Tl, V, and Ge each occur as major constituents in one or two rare sulfosalt minerals. The elements found in the sulfides and sulfosalts minerals are indicated in the table at the opening of this chapter.

The structural arrangement of most sulfides is now well-known; that of the sulfosalts is more complex and is known only for a few. These few minerals, however, indicate that there is considerable structural similarity between sulfides and sulfosalts. The sulfosalt minerals are of relatively rare occurrence, and only a few of the more common ones are described here.

In atomic bonding these minerals show a wide variation, ranging from typical ionic compounds in some simple sulfides, or homopolar as in pure ZnS, to other sulfides which are practically alloys with distinct metallic characteristics. The sulfur atoms (also As and Sb) are separate in some; in others, such as pyrite, they are linked in pairs by homopolar bonds. Some are sectile; many are opaque or nearly so. Those which are transparent usually have very high indices of refraction, and many transmit only red light. They range in hardness from about 1 in molybdenite to 6, or over, in pyrite and sperrylite.

The sulfides are arranged below according to the decreasing $A:X$ ratio, with the sulfosalts appended and arranged according to the decreasing $(A + B):X$ ratio.

A_2X Type
    Argentite Group: Argentite, Ag_2S
    Chalcocite Group: Chalcocite, Cu_2S
A_3X_2 Type
    Bornite, Cu_5FeS_4
AX Type
    Galena Group: Galena, PbS
    Sphalerite Group: Sphalerite, (Zn,Fe)S
    Chalcopyrite Group: Chalcopyrite, $CuFeS_2$
    Wurtzite Group: Wurtzite, ZnS
    Niccolite Group
        Pyrrhotite, $Fe_{1-x}S$
        Niccolite, NiAs
        Breithauptite, NiSb
    Millerite, NiS
    Pentlandite, $(Fe,Ni)_9S_8$
    Covellite, CuS
    Cinnabar, HgS
    Realgar, AsS
    Orpiment, As_2S_3

Stibnite Group
    Stibnite, Sb_2S_3
    Bismuthinite, Bi_2S_3

AX_2 Type
  Pyrite Group
    Pyrite, FeS_2
    Sperrylite, $PtAs_2$
  Cobaltite Group: Cobaltite, $CoAsS$
  Marcasite Group: Marcasite, FeS_2
  Arsenopyrite Group: Arsenopyrite, $FeAsS$
  Molybdenite, MoS_2
  Krennerite Group
    Krennerite, $(Au,Ag)Te_2$
    Calaverite, $AuTe_2$
    Sylvanite, $(Au,Ag)Te_2$

AX_3 Type
  Skutterudite Series
    Skutterudite, $(Co,Ni)As_3$
    Smaltite, $(Co,Ni)As_{3-x}$
    Chloanthite, $(Ni,Co)As_{3-x}$

A_3BX_3 Type
  Ruby-Silver Group
    Pyrargyrite, Ag_3SbS_3
    Proustite, Ag_3AsS_3
  Tetrahedrite Series
    Tetrahedrite, $(Cu,Fe)_{12}Sb_4S_{13}$
    Tennantite $(Cu,Fe)_{12}As_4S_{13}$

A_3BX_4 Type
  Enargite, Cu_3AsS_4

A_2BX_3 Type
  Bournonite, $PbCuSbS_3$

ABX_2 Type
  Boulangerite, $Pb_5Sb_4S_{11}$

Argentite, Ag_2S

Crystal system and class: Isometric; $4/m\ \bar{3}\ 2/m$ (above 179°C).

Cell dimensions, and content: $a = 4.89$; $Z = 2$. At room temperature Ag_2S (acanthite) is monoclinic, $a = 4.229$, $b = 6.931$, $c = 7.862$, $\beta = 99°\ 37'$, $Z = 8$.

Habit: Crystals cubic, octahedral, rarely dodecahedral; often in groups of parallel indviduals; also arborescent, filiform, massive, or as a coating.

Cleavage: {001}; {011} poor.

Fracture: Subconchoidal.

Tenacity: Very sectile.

Hardness: 2–2½.

Density: 7.2–7.4.

Color: Black

Streak: Black and shining.

Luster and opacity: Metallic; opaque.

Chemistry: The cubic modification of Ag_2S, called argentite, is stable only above 179°C. The cubic crystals are therefore paramorphs and possess a monoclinic internal crystal structure.

Diagnostic features: Color, sectility; in the oxidizing flame on charcoal it fuses with intumescence, emitting sulfurous fumes and yielding a globule of silver.

Occurrence: Argentite is probably the most important primary silver mineral. It occurs widely in hydrothermal sulfide deposits of the low temperature range together with the ruby silvers (pyrargyrite, proustite) and native silver. It may occur as microscopic inclusions in galena; this material has been called argentiferous galena. Fine specimens have been found in Saxony (Freiberg) and Czechoslovakia (Kremnitz, Schemnitz). It occurs in many of the silver producing districts; with native silver at Kongsberg in Norway; in Bolivia (Colquechaca), Peru, Chile (Chanarcillo, Atacama), Mexico (Pachuca, Guanajuato, Zacatecas), Montana (Butte), Nevada (Comstock Lode, Tonopah), and Colorado (Aspen, Leadville).

Chalcocite, Cu_2S

Crystal system and class: Orthorhombic; $2/m\ 2/m\ 2/m$

Axial elements: $a:b:c = 0.5797:1:0.9882$ (for pseudocell).

Cell dimensions and content: $a = 11.881$, $b = 27.323$, $c = 13.491$, $Z = 96$.

Pseudocell: $a/3 = 3.96$, $b/4 = 6.83$, $c/2 = 6.75$, $Z = 4$.

Common forms and angles:

(110) ∧ (010) = 59° 45′	(001) ∧ (021) = 63° 18′
(001) ∧ (111) = 63° 07′	(110) ∧ (1$\bar{1}$0) = 60° 30′

Habit: Crystals prismatic along c or a, or thick tabular on {001} (Fig. 9-1); also compact, massive.

Twinning: Very common on {110}, giving pseudohexagonal forms, on {032} and {112}, and microscopic lamellar twinning.

Cleavage: {110} indistinct.

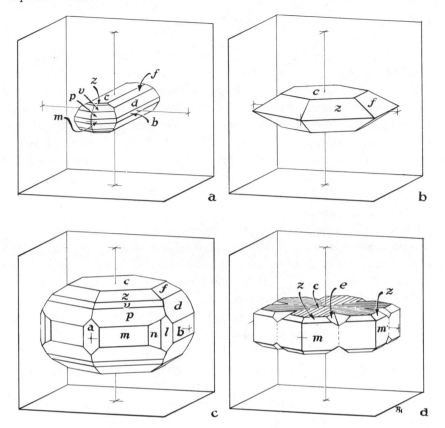

Figure 9-1 Chalcocite crystals. Forms: pinacoids $c\{001\}$, $b\{010\}$, $a\{100\}$; rhombic prisms $l\{130\}$, $n\{230\}$, $m\{110\}$, $f\{012\}$, $d\{021\}$; rhombic dipyramids $z\{113\}$, $v\{112\}$, $p\{111\}$. (d) Twin crystal, twin plane $\{110\}$ or $\{130\}$, composition plane $\{130\}$.

Fracture: Conchoidal.

Tenacity: Brittle to somewhat sectile.

Hardness: $2\frac{1}{2}$–3.

Density: 5.5–5.8.

Color and streak: Black.

Luster and opacity: Metallic; opaque.

Diagnostic features: Chalcocite is distinguished by its black color, slight sectility, and association with other copper minerals, either primary or secondary. It fuses readily at 2–$2\frac{1}{2}$. After roasting on charcoal it gives a globule of copper in the reducing flame. It is soluble in nitric acid. Hexagonal Cu_2S, dimorphous with chalcocite, is stable above 105°C.

Occurrence: Chalcocite occurs most commonly as fine-grained massive material, showing alteration to covellite, malachite, or azurite. It may also be associated with native copper or cuprite. It is one of the most important source minerals for copper.

The principal occurrence of chalcocite is in the supergene enriched zone of sulfide deposits. In arid and semiarid climates oxidizing surface waters attack the primary sulfides, chalcopyrite, bornite, pyrite, enargite, and others, forming soluble sulfates. These soluble sulfates react at greater depth with the primary sulfides, and secondary chalcocite is deposited as a blanket-like deposit at the water table level. This "chalcocite blanket" contains a considerably higher content of copper than does the unaltered "primary ore," and has yielded important amounts of copper at Rio Tinto in Spain, Ely in Nevada, Morenci in Arizona, Bingham Canyon in Utah, and at other localities.

Chalcocite is also found in hydrothermal sulfide veins with bornite, enargite, chalcopyrite, tennantite-tetrahedrite, covellite, pyrite, and quartz, as at Butte, Montana. These veins are concentrated in a faulted and fractured area at the margin of the Boulder batholith of granodiorite. The veins of the central and intermediate zones are considered to be mesothermal and contain the bulk of the copper mineralization. An outer peripheral zone, epithermal in origin, contains sphalerite, rhodochrosite, galena, silver, and argentite with minor chalcopyrite. Thus, single veins which have a strike radial to the zones will change in mineral content from the center to the border of the area. This district, about 2 miles by 4 miles in area, has, since 1879, produced about $2\frac{1}{2}$ billion dollars worth of copper with some silver, gold, zinc, lead, and a minor amount of manganese.

Bornite, Cu_5FeS_4

Crystal system and class: Isometric; $4/m\ \bar{3}\ 2/m$ (?)

Cell dimensions and contents: $a = 5.50$, $Z = 1$ (at 240°C), for the high-temperature, disordered, structure. A metastable form is pseudocubic with $a = 10.94$, $Z = 8$, resulting from twinning of a rhombohedral cell with $Z = \frac{1}{2}$. The low-temperature form is tetragonal, $a = 10.94$, $c = 21.88$, $Z = 16$.

Habit: Crystals cubic, dodecahedral, faces often rough or curved; usually massive.

Twinning: On {111}.

Cleavage: {111} traces.

Fracture: Conchoidal to uneven.

Tenacity: Brittle.

Hardness: 3.

Density: 5.06–5.08.

Color: Copper red to pinchbeck-brown on fresh fracture surface, quickly tarnishing to purplish iridescent.

Streak: Gray black.

Luster and opacity: Metallic; opaque.

Alteration: Bornite may alter to chalcocite, chalcopyrite, covellite, cuprite, chrysocolla, malachite and azurite.

Diagnostic features: The color usually enables identification of bornite and its distinction from chalcocite and chalcopyrite. It fuses at 2 and yields sulfurous fumes in the open tube. On charcoal it fuses in the reducing flame to a brittle magnetic globule. It is soluble in nitric acid with separation of sulfur.

Occurrence: Bornite is a common and widespread copper mineral. Usually of hypogene origin, it occurs in many of the important copper deposits of the world.

Bornite has been found in dikes, in basic intrusives, disseminated in basic rocks, in contact metamorphic deposits, in pegmatites, and in quartz veins. Crystals of bornite are occasionally found in vugs and druses in veins as at Butte, Montana, or at Bristol, Connecticut.

Galena, PbS

Crystal system and class: Isometric; $4/m\ \bar{3}\ 2/m$.

Cell dimensions, and content: $a = 5.936$; $Z = 4$.

Habit: Cubic or cubo-octahedral (Figs. 9-2, 9-3), also octahedral, large crystals often made up of subparallel segments; commonly massive, coarse to very fine granular.

Figure 9-2 Galena cubes on gray chert. (Joplin, Missouri.) [Courtesy Queen's University.]

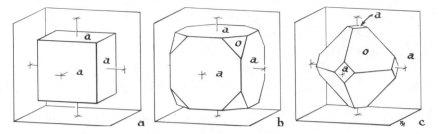

Figure 9-3   Galena crystals. Forms: $a\{100\}$, $o\{111\}$.

Twinning: Twin plane $\{111\}$, penetration or contact twins; twin plane $\{114\}$ with lamellae giving rise to diagonal striations on cleavage faces.

Crystal structure: The atomic arrangement (Fig. 9-4) is the same as that of halite (Fig. 11-1). However, galena and the isostructural minerals PbSe (*clausthalite*) and PbTe (*altaite*), which are of much rarer occurrence, display semimetallic bonding in place of the ionic bonding of halite. In this structure each lead atom is bonded to six sulfur atoms and each sulfur in turn to six lead atoms.

Cleavage: $\{001\}$ highly perfect and easy; the cleavage occurs between atomic planes of greatest spacing in the structure and perpendicular to the Pb–S bonds.

Tenacity: Brittle.

Hardness: $2\frac{1}{2}$.

Density: 7.58.

Color: Lead gray.

Streak: Lead gray.

Luster and opacity: Metallic; opaque.

Chemistry: Commonly, galena is very nearly pure PbS. The silver, arsenic, and antimony reported in chemical analyses are largely due to inclusions of argentite or tetrahedrite, small amounts of which are difficult to detect in a black opaque mineral.

Alteration: In the zone of weathering, galena is readily oxidized to secondary lead minerals including cerussite, anglesite, pyromorphite, or mimetite.

Diagnostic features: Galena is readily distinguishable by its perfect cleavage, metallic luster, color, and hardness. Its cubic cleavage, greater specific gravity, and darker color distinguish it from stibnite, with which it is most easily confused. The perfect cleavage is obscured in the extremely fine-grained material. In the open tube it gives sulfurous fumes; on charcoal it fuses at 2, emits sulfurous fumes, coats the charcoal yellow near assay (PbO) and white with a bluish border at a distance, and yields a globule of metallic lead.

Occurrence: The most important lead mineral in the earth's crust, galena is also one of the most common sulfide minerals. It occurs in many types of deposits: in sedimentary rocks, in hydrothermal veins, and also in pegmatites. Extensive deposits of galena and sphalerite occur as irregular masses in solution cavities and in brecciated zones in limestone. In the typical deposits of this type of the Tri-State district, embracing the adjoining area of Missouri, Kansas, and Oklahoma, and in other sections of Missouri, Wisconsin, Illinois,

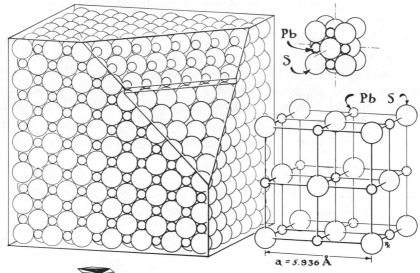

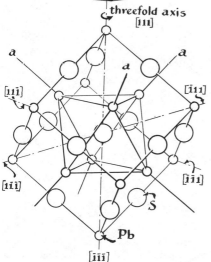

Figure 9-4 Above. Galena structure, showing packing of the atoms in actual relative sizes and in an open view with atoms shown considerably smaller. Left. Galena structure, one unit cube with a cube diagonal [111], a threefold axis, vertical. One coordination octahedron of six Pb around one S is shown.

and Iowa, coarsely crystallized galena and sphalerite occur with chert, marcasite, pyrite, chalcopyrite, dolomite, and calcite. Strikingly well-crystallized specimens were found in these deposits, and examples are displayed in most mineral collections. Similar deposits of low-temperature origin are found in other parts of the world; these deposits usually contain only very low silver values.

Extensive hydrothermal vein deposits containing galena are also important as a source of lead; in addition, these deposits may contain significant silver values. In veins formed at intermediate temperatures, galena, sphalerite, chalcopyrite, pyrite, tetrahedrite, and silver minerals occur with quartz and siderite (Coeur d'Alene, Idaho), with barite and siderite, or with fluorite and barite. In high-temperature veins and replacement deposits the sulfides are found with feldspar, garnet, rhodonite (Broken Hill, New South Wales), or with garnet, diopside, actinolite, biotite (Sullivan mine, Kimberley, British Columbia). In these deposits the galena and sphalerite are usually fine-grained. Large deposits of galena with rich silver values occur as replacements of limestone or dolomite in Mexico. Galena also is found in contact metamorphic deposits.

Production and use: Galena is the most important source mineral for lead, although anglesite and cerussite are important ore minerals in oxidized ore deposits, particularly in Mexico. The average production of lead was about $2\frac{1}{2}$ million tons annually during 1960–1965. Eighty percent of this amount

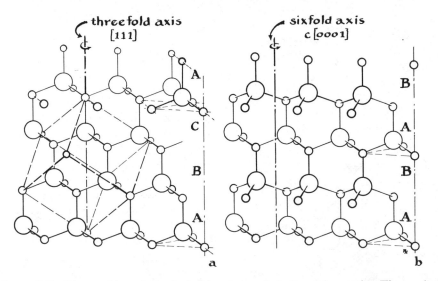

Figure 9-5 The stacking of tetrahedral layers in sphalerite and wurtzite. The vertical axis of wurtzite is sixfold, but in the structure it is in reality a sixfold screw axis involving a translation of one-half *c* for each 60 degrees of rotation.

came from the USA, Australia, Mexico, the USSR, Canada, Peru, Yugo-slavia, Morocco, and China.

SPHALERITE AND WURTZITE GROUPS

These minerals are dimorphous forms of ZnS, sphalerite having face-centered cubic stacking of the ZnS tetrahedral layers, and wurtzite having hexagonal stacking (Fig. 9-5). Polymorphs with more complex stacking sequences have been observed both in nature and in synthetic preparations. Sphalerite changes to wurtzite when heated above 1020°C; this temperature is lower for materials high in iron. Sphalerite is of much commoner occurrence in nature, although many natural minerals apparently contain both kinds of stacking.

	SPHALERITE	WURTZITE
Crystal system:	Isometric	Hexagonal
Crystal class:	$\bar{4}3m$	6 mm
Axial elements:	—	$a:c = 1:1.6387$
Cell dimensions (pure ZnS):	$a = 5.4060$	$a = 3.820$
		$c = 6.260$
Cell content:	$Z = 4$	$Z = 2$
Common forms:	$\{001\}$, $\{111\}$, $\{1\bar{1}1\}$, $\{211\}$	$\{000\bar{1}\}$, $\{10\bar{1}0\}$, $\{10\bar{1}1\}$, $\{10\bar{1}\bar{1}\}$
Habit:	Tetrahedral	Pyramidal
Twinning:	[111] twin axis	—
Cleavage:	$\{110\}$ perfect	$\{11\bar{2}0\}$ easy, $\{0001\}$ difficult
Tenacity:	Brittle	Brittle
Hardness:	$3\frac{1}{2}$ 4	$3\frac{1}{2}$–4
Density (pure ZnS):	4.096	4.089
Color (commonly):	Brown to yellow	Brownish black
Streak:	Brown to light yellow	Brown
Luster:	Resinous to nearly metallic	Resinous
Optical properties:	Isotropic, $n = 2.37$	$\omega = 2.356$, $\epsilon = 2.378$ $\epsilon - \omega = 0.022$, positive

Sphalerite, (Zn,Fe)S

Habit: Crystals tetrahedral (Fig. 9-6) or dodecahedral, often distorted and complex; rough, curved faces common. Often as cleavable masses, coarse to fine granular, fibrous, concretionary, botryoidal.

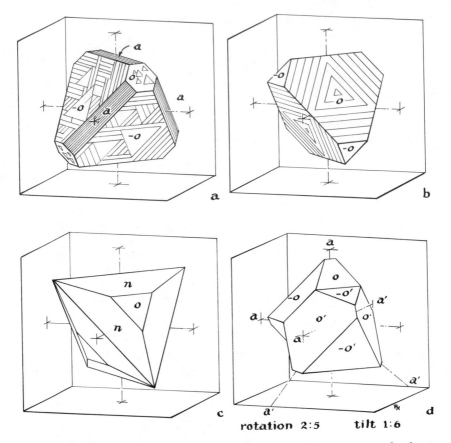

rotation 2:5   tilt 1:6

Figure 9-6 Sphalerite crystals. Forms: cube $a\{100\}$, positive tetrahedron $o\{111\}$, negative tetrahedron $-o\{1\bar{1}1\}$, positive tristetrahedron $n\{211\}$. (d) Contact twin on $\{111\}$.

Twinning: Common on [111], simple or multiple contact twins or lamellar intergrowths, usually on $\{111\}$ as composition plane. Directed pressure causes glide twinning on $\{111\}$.

Crystal structure: The structure of sphalerite (Fig. 9-7, 9-8) is analogous to that of diamond (Fig. 8-10), with four zinc atoms at the points of a face-centered cubic lattice comparable to four of the carbon atoms in diamond, and four sulfur atoms comparable in position to the other four carbon atoms of diamond, at the points of a face-centered cubic lattice displaced one-quarter along the body-diagonal of the first cube. In this structure each zinc atom is coordinated with four sulfur atoms and each sulfur with four zinc. These tetrahedra of SZn_4 are all oriented in the same way, with a triangular face of the tetrahedron parallel to (111), and the ZnS_4 tetrahedra are all oriented in the opposite way, also with a triangular face parallel to

(111); hence the whole structure shows tetrahedral symmetry, as do the crystals. Each {111} face on a crystal is not identical with its parallel and opposite face. This is displayed in Fig. 9-8, which shows the alternate layers of all Zn and all S atoms in the {111} planes. Along the axis [111], normal to (111), the sequence is ZnS–ZnS, etc. to the left, and SZn–SZn, etc. to the right; thus the [111] axes are polar in character and show pyroelectricity. In sphalerite the tetrahedral layers parallel to {111} are stacked in the sequence *ABC ABC ABC.* . . . This results in a cubic structure, as in the face-centered cubic metals, since this sequence results in four identical three-fold axes. The stacking sequence *AB AB* . . . results in a hexagonal structure, as in wurtzite. It has also been noted that the faces of $+o\{111\}$ and $-o\{1\bar{1}1\}$ are affected quite differently by solvents.

The cleavage, parallel to {110}, is seen to break relatively few bonds, and occurs between planes containing equal numbers of Zn and S atoms. The spacing between Zn planes and S planes in {111} is larger but does not result in a cleavage {111}.

The sphalerite structure is characteristic of a large number of *AX* compounds of which those of zinc, cadmium, mercury with sulfur, and mercury

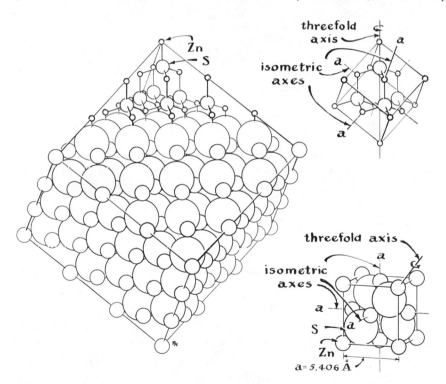

Figure 9-7 Atomic packing in sphalerite.

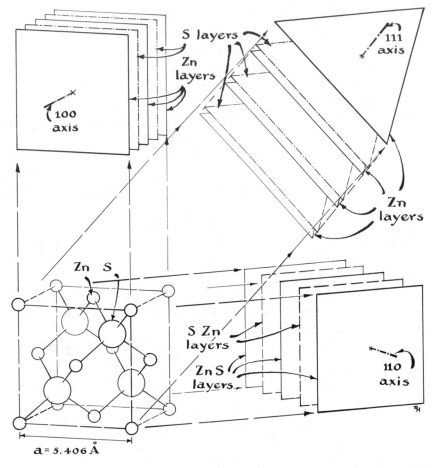

Figure 9-8 Sphalerite structure, showing equal spacing of atomic layers parallel to (100) and (110), and unequal spacing of Zn and S layers parallel to (111) resulting in the nonequivalence of $\{111\}$ and $\{\overline{111}\}$ (the positive and negative tetrahedra).

selenide and telluride are known in nature. Only zinc sulfide is of common occurrence. Solid solution is common in this group, and, whereas there is some substitution of cadmium for zinc in sphalerite, the most common metals substituting for zinc are iron and, to a lesser extent, manganese. Oddly, iron and manganese do not form sulfides isostructural with sphalerite; they normally combine with sulfur in sixfold coordination, with the hexagonal niccolite structure or the cubic halite structure. With increasing substitution of iron, sphalerite is darker in color and has slightly increased cell edge and lower density. The iron content, as recorded in analyses, may be as high as about 26 percent, corresponding to nearly 50 mole percent of FeS. The darker, high-iron varieties are usually of high-temperature origin. The solubility of FeS in ZnS increases with temperature; therefore if sphalerite and an iron sulfide have formed together, the iron content of the sphalerite will give an indication of the temperature of deposition (Fig. 3-5). Cadmium is often present in small amounts replacing zinc in sphalerite, and recorded analyses show a maximum of 1.66 percent.

Color: Commonly brown, black, yellow; less commonly red, green, white, and nearly colorless when free of iron.

Streak: Brown to light yellow or white.

Luster: Resinous to adamantine, almost metallic in high iron varieties.

Opacity: Transparent to translucent. Pyroelectric, polar in direction [111].

Alteration: Sphalerite often weathers, leaving a limonite residue. The zinc sulfate, goslarite, may be found in dry regions. It may also alter to hemimorphite or smithsonite, which are sometimes found as pseudomorphs after sphalerite.

Diagnostic features: The identification of sphalerite is often difficult because of its extremely variable color. Cleavage, hardness, and luster are the most dependable properties in the hand specimens; yellow, yellowish brown, brown, and dark brown are the commonest colors. Sphalerite with other colors is not uncommon and may often deceive even the expert mineralogist. It is fusible with difficulty at 5, and dissolves in hydrochloric acid with evolution of hydrogen sulfide. On charcoal in the reducing flame it gives a coating of zinc oxide which is yellow while hot and white when cold. With cobalt nitrate solution the white coating turns green when heated in the oxidizing flame.

Occurrence: Sphalerite, the most important zinc mineral, is found intimately associated with galena in most of the important deposits of these metals. In some hypothermal replacement bodies it is associated with chalcopyrite, pyrite, pyrrhotite, and magnetite; here galena is usually not present. The Tri-State district referred to under *Galena* has produced tremendous quantities of zinc from sphalerite; galena and other sulfides are less important in quantity. These deposits are considered to be of low-temperature hydrothermal origin, although no related igneous rocks have been found. The masses of sulfide, dominantly sphalerite, occur along solution channels in flat-lying limestone following cherty beds, and around slumped blocks of the cherty beds. The ore bodies occur over a wide area and are found at shallow depths.

Sphalerite occurs widely in hydrothermal veins and replacement bodies with galena and silver minerals at Coeur d'Alene, Idaho; Bingham, Utah; Sullivan Mine, British Columbia; Broken Hill and Mount Isa, Australia; and also at localities in Silesia, Italy, Spain, Yugoslavia, and Burma.

Production and uses: Sphalerite provides, by far, the bulk of the world's zinc,

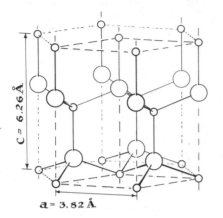

Figure 9-9   Structure of wurtzite.

although in some places smithsonite and hemimorphite are ore minerals of zinc, especially in Mexico; and zincite, willemite, and franklinite are the ore minerals in the unique deposits at Franklin and Sterling Hill, New Jersey. The average annual production of zinc has been about 4 million tons, during 1960–1965, of which 80 percent comes from the USA, Canada, Mexico, the USSR, Australia, Peru, Poland, Italy, Japan, and Germany, in decreasing order of importance. This list shows a close parallel to that for lead. The world's annual cadmium production of about 25 million pounds (1960–1965 average), obtained entirely as a by-product of zinc smelting, being recovered from dust chambers and electrolytic slimes, is provided by the USA (40 percent), Canada, Belgium, and several other countries where zinc smelters are in operation.

Name: The name sphalerite is derived from a Greek word meaning *treacherous*, and the old German miner's name for the mineral, *Blende*, means blind or *deceiving*.

Wurtzite, (Zn,Fe)S

Habit: Crystals hemimorphic pyramidal, short prismatic to tabular on {0001}; usually striated on {10$\bar{1}$0} and {10$\bar{1}$1}; the large basal pedion {000$\bar{1}$} usually dominant, {0001} often absent or small; often fibrous, columnar, or in concentrically banded crusts.

Chemistry: Wurtzite, like sphalerite, usually contains Fe and may also contain Cd and Mn in substitution for Zn; CdS occurs with the wurtzite structure as the mineral greenockite.

Crystal structure: The wurtzite structure may be described as a stacking of tetrahedral sheets of ZnS in a sequence $AB\ AB$ . . . , parallel to (0001) (Figs. 9-5, 9-9). The Zn_4S tetrahedra all point in the same direction, with a triangular face parallel to (0001). The structure is therefore hemimorphic and distinct from sphalerite, in which the ZnS tetrahedral sheets are stacked in a sequence $ABC\ ABC.$ . . .

Occurrence: Wurtzite is found to a minor extent in various sulfidic ores.

CHALCOPYRITE GROUP

Chalcopyrite, CuFeS$_2$

Crystal system and class: Tetragonal; $\bar{4}2m$.

Axial elements: $a\!:\!c = 1\!:\!1.9716.$

Cell dimensions, and content: $a = 5.28$, $c = 10.41$; $Z = 4$

Common forms and angles:

   (001) $\wedge$ (112) $= 54° 21'$
   (100) $\wedge$ (101) $= 26° 54'$
   (001) $\wedge$ (102) $= 44° 36'$
   (112) $\wedge$ (1$\overline{1}$2) $= 109° 50'$

Habit: Crystals commonly tetrahedral in aspect with {112} the dominant form; the disphenoid {112} approaches closely an isometric tetrahedron, since $c/2 = 0.9858$; the faces of {112} are commonly large, dull, striated, whereas those of {1$\overline{1}$2} are small, brilliant, and not striated (Fig. 9-11); commonly massive compact, sometimes botryoidal.

Twinning: On {112}, contact, penetration, or lamellar.

Crystal structure: Though tetragonal in symmetry, the structural arrangement in chalcopyrite is very similar to that in sphalerite. The unit cell is very close in dimension to two cubes. In each pseudo-cubic half-cell the metal atoms occupy face-centered positions, and the sulfur atoms occupy the same positions as those in sphalerite (Fig. 9-10). The arrangement of metals around sulfur is again tetrahedral, the forms {111} and {112} on chalcopyrite are disphenoids, and {112} approximates very closely the isometric tetrahedron. The special distribution of copper and iron atoms, as shown in Fig. 9-10, results in a tetragonal unit cell with c double the length of the pseudocube edge. The unit cell thus contains 4[CuFeS$_2$]. Analyses of chalcopyrite reveal very little divergence from this formula. Selenium may replace sulfur to a minor extent.

Cleavage: {011} sometimes distinct.

Fracture: Uneven.

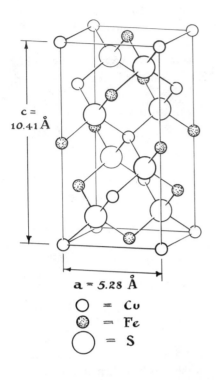

c = 10.41 Å

a = 5.28 Å

○   = Cu

⊛   = Fe

◯   = S

Figure 9-10 Structure of chalcopyrite. Note similarity of a half-cell to the cubic cell of sphalerite (Fig. 9-8).

Tenacity: Brittle.

Hardness: $3\frac{1}{2}$–4.

Density: 4.1–4.3.

Color: Brass yellow, often tarnished slightly iridescent.

Streak: Greenish black.

Luster and opacity: Metallic; opaque.

Diagnostic features: Chalcopyrite is readily distinguished from pyrite by its low hardness, from gold by its brittle character, from pyrrhotite by its color and nonmagnetic character, and from bornite by its color. In the closed tube it often decrepitates and gives a sulfur sublimate. On charcoal it fuses to a magnetic globule. It is soluble in nitric acid, with separation of sulfur.

Alteration: Chalcopyrite alters to chalcocite, covellite, chrysocolla, malachite, and iron oxides. Under the action of weathering, chalcopyrite and other sulfides are oxidized to sulfates; they are carried down into the ore body, leaving much of the iron as an oxide gossan. The sulfate solutions react with primary sulfides to produce supergene sulfides, including chalcocite and covellite, which partially or completely replace chalcopyrite and other primary sulfides in a layer at the level of the water table.

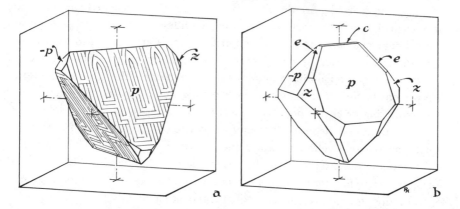

Figure 9-11 Chalcopyrite crystals. Forms: positive tetragonal disphenoid $p\{112\}$, negative tetragonal disphenoid $-p\{1\bar{1}2\}$; tetragonal dipyramids $z\{011\}$, $e\{012\}$, pinacoid $c\{001\}$. Natural etching on p faces is consistent with the tetragonal symmetry.

Occurrence: Chalcopyrite is the most widespread copper mineral, and one of the most important sources of that metal. It has been formed under a great variety of conditions, and most sulfide ore deposits contain some chalcopyrite, but the mesothermal and hypothermal deposits carry important amounts of the mineral. In vuggy and drusy cavities, and in veins, chalcopyrite may be found as small crystals on sphalerite, galena, or dolomite. In veins of moderate to high-temperature origin it usually occurs as irregular blebs and masses. In hypothermal deposits chalcopyrite and pyrite occur with tourmaline (Braden, Chile) or quartz as gangue minerals; cassiterite may be an associated mineral, as in Cornwall, England. In the Rouyn district of Quebec chalcopyrite is found in massive hydrothermal replacement bodies of pyrite and pyrrhotite with more or less sphalerite and magnetite; these ores contain substantial values in gold. It also occurs in sulfide replacement lenses in metamorphic schists at Ducktown, Tennessee, with pyrite, pyrrhotite, and minor amounts of sphalerite, bornite, hematite, magnetite, quartz, chlorite, and other silicates. The surface zone was leached to an iron gossan, and a thin layer of supergene sulfides underlay the gossan.

Chalcopyrite is the important primary copper mineral in most of the "porphyry-copper" deposits. Chalcopyrite, pyrite, and bornite with minor sphalerite and molybdenite are found disseminated and in closely spaced veinlets with quartz in igneous intrusions of monzonite, quartz monzonite, or diorite porphyry. The sulfides are found in the upper fractured zones of the intrusives or in the surrounding intruded rocks. Deposits of this kind, which are important copper producers in Utah, Arizona, Nevada, New

Mexico, and in Chile, are deeply weathered, leaving a leached capping and an enriched zone or blanket of secondary sulfides. In some of these deposits the primary chalcopyrite or other copper minerals are altered to the carbonate, malachite (Ajo, Arizona), or to basic sulfates or chlorides (antlerite, atacamite, etc. at Chuquicamata, Chile). In others, the primary sulfides are present in sufficient quantity to constitute ores. Molybdenite, unaffected by weathering or enrichment, is recovered as an important by-product.

In mesothermal deposits chalcopyrite is associated with large masses of pyrite, as at Mt. Lyell, Tasmania, and at Rio Tinto, Spain. The latter deposits include the largest pyritic bodies in the world. They have been mined for 3,000 years, first for gold, then for copper and sulfur, with minor amounts of lead, zinc, gold, silver, nickel, cobalt, and other metals.

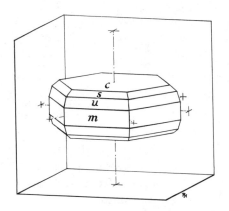

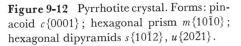

Figure 9-12 Pyrrhotite crystal. Forms: pinacoid $c\{0001\}$; hexagonal prism $m\{10\bar{1}0\}$; hexagonal dipyramids $s\{10\bar{1}2\}$, $u\{20\bar{2}1\}$.

Small amounts of chalcopyrite, pyrite, and other sulfides are often found in sedimentary rocks. Contact metamorphic deposits in limestone, containing chalcopyrite with garnet, tremolite, and other lime silicates, occur at Bisbee, Morenci, and Silver Bell, in Arizona, and at Seven Devils, Idaho. The massive sulfide deposits at Sudbury, Ontario, contain important amounts of chalcopyrite with pyrrhotite and pentlandite.

NICCOLITE GROUP

This group comprises the minerals pyrrhotite, niccolite, and breithauptite; they have the nickel arsenide structure (Fig. 9-13) possessed by many metallic compounds of AX type. Pyrrhotite usually shows a deficiency of iron leading to the general formula $Fe_{1-x}S$. The mineral with composition close to the ideal,

FeS, is known as *troilite*. Both pyrrhotite and troilite possess a superstructure in which the true unit cell is a multiple of the simple cell given below. NiS is known as an artificial material with the nickel arsenide structure, but occurs in nature as millerite with a quite different structure.

Crystal system and class: Hexagonal; $6/m \; 2/m \; 2/m$.

	PYRRHOTITE	NICCOLITE	BREITHAUPTITE
Axial elements:	$a:c = 1:1.6692$	$a:c = 1:1.3907$	$a:c = 1:1.305$
Cell dimensions:	$a = 3.452$	$a = 3.609$	$a = 3.93$
	$c = 5.762$	$c = 5.019$	$c = 5.13$
Cell content:	$Z = 2$	$Z = 2$	$Z = 2$
Cleavage:	None	None	None
Fracture:	Uneven	Uneven	Uneven
Hardness:	$3\frac{1}{2}$–$4\frac{1}{2}$	5–$5\frac{1}{2}$	$5\frac{1}{2}$
Density:	4.58–4.65	7.78	8.23
Color:	Bronze yellow	Pale copper red	Copper red
Streak:	Gray black	Brownish black	Reddish brown
Luster:	Metallic	Metallic	Metallic
Magnetism:	Magnetic, variable	Nonmagnetic	Nonmagnetic

Pyrrhotite, $Fe_{1-x}S$

Habit: Crystals rare; usually tabular to platy on $\{0001\}$ (Fig. 9-12); commonly massive, granular.

Twinning: Rare on $\{10\bar{1}2\}$ as twin plane.

Color: Bronze yellow, tarnishing darker on exposure.

Crystal structure: The structure of pyrrhotite is probably like that of niccolite, with the metal atoms in close-packed hexagonal layers with interleaved close-packed layers of sulfur (or arsenic, in niccolite) atoms with each sulfur nesting between three metals below and three above (Fig. 9-13). Likewise, each metal rests on three sulfurs with three others above. All metal layers have their atoms in rows parallel to c, while the sulfur layers are staggered with repetition along c on the alternate layers. Pyrrhotite analyses show a varying content of sulfur from 50–55 atom percent. It has been shown that specific gravity decreases with increasing proportion of sulfur and the variation agrees with the calculated specific gravity obtained by withdrawing iron (Fig. 3-6). Thus the variation in composition is expressed by the formula $Fe_{1-x}S$, where x may vary from 0 to 0.2. $Fe_{1-x}S$ with 51.9 to 52.5 atom percent sulfur has an hexagonal superstructure and with

53.5 atom percent sulfur the superstructure is monoclinic. Natural pyrrhotites with sulfur 52.5 to 53.5 atom percent are mixtures of hexagonal and monoclinic phases.

Small amounts of nickel, copper, and cobalt are often reported in analyses, but these metals are probably present in minute inclusions of the commonly associated minerals pentlandite and chalcopyrite.

Diagnostic features: Pyrrhotite is rarely found in crystals, and must be distinguished from chalcopyrite by color and magnetism and from pyrite by color and hardness. It can be distinguished from pentlandite only with difficulty. It fuses at about 3 to a magnetic globule. It is decomposed by hydrochloric acid with evolution of hydrogen sulfide.

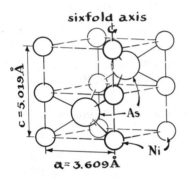

Figure 9-13 Structure of niccolite. Here, as in wurtzite, the vertical axis c, sixfold in crystals, is a sixfold screw axis in the structure.

Occurrence and use: Pyrrhotite occurs principally in basic igneous rocks, either disseminated as minute flecks and blebs or segregated in large masses, as at Sudbury, Ontario. Often chalcopyrite, pentlandite, and other nickel sulfides will be found associated with it. At Sudbury important amounts of platinum and palladium are also found, partly in the rare mineral sperrylite.

Pyrrhotite is also found in pegmatites, in contact metamorphic deposits, and often as early-formed masses with other sulfides in vein and replacement deposits of hypothermal origin. It is found as nodules (troilite) in iron meteorites.

Well-crystallized specimens have been found at several mining districts in Europe; at the Morro Velho gold mine in Minas Geraes, Brazil; at the Potosi mine in Chihauhua, Mexico; and at the Bluebell mine, British Columbia.

Until recently pyrrhotite has not been an economically important mineral. However, since 1955, pyrrhotite concentrates from the Sudbury ores have been converted into a high-grade iron ore, the sulfur being collected for the production of sulfuric acid. Selenium, which probably substitutes for sulfur in pyrrhotite, is also recovered in the smelting processes.

Niccolite, NiAs

Habit: Crystals rare, tabular on {0001} or pyramidal on {10$\bar{1}$1}; usually massive, or reniform with columnar structure.

Diagnostic features: Hardness, color; it fuses at 2, and arsenical fumes are given off.

Crystal structure: As described for pyrrhotite (Fig. 9-13).

Alteration: Niccolite alters readily on the surface to the pale green annabergite. If intergrown with cobalt arsenides, the alteration product is white, becoming pink with increasing cobalt (erythrite).

Occurrence: Niccolite is not a common mineral; it occurs with pyrrhotite, chalcopyrite, and other nickel sulfides in basic igneous rocks and ore deposits derived from them, also in hydrothermal vein deposits with cobalt and silver minerals. It occurs as spheroids with concentric shells in the Natsume nickel mine, Japan, and at several localities in Germany and France. Massive material occurs in veins at Cobalt, Ontario, with cobalt arsenides and silver. It also occurs to a minor extent in the nickel ores at Sudbury, Ontario.

Breithauptite, NiSb.

The rarer mineral, *breithauptite*, NiSb, is isostructural with niccolite (Fig. 9-13), from which it may be distinguished by its darker, distinctly violet-red color. It occurs in veins, associated with silver minerals, niccolite, or cobalt minerals, at Andreasburg in the Harz Mountains, Germany, and at Cobalt, Ontario.

OTHER SULFIDES OF AX TYPE

Millerite, NiS

Crystal system and class: Trigonal; $\bar{3}$ 2/m.

Axial elements: $a:c = 1:0.3273$.

Cell dimensions, and content: $a = 9.620$, $c = 3.149$; $Z = 9$.

Habit: Crystals usually very slender to capillary along c, often in radiating groups; rarely granular.

Cleavage: {10$\bar{1}$1} and {01$\bar{1}$2} perfect.

Tenacity: Brittle, capillary crystals elastic.

Hardness: 3–3$\frac{1}{2}$.

Density: 5.5.

Color: Pale brass yellow.

Streak: Greenish black.

Luster: Metallic.

Diagnostic features: Similar to pyrite in color, millerite is readily distinguished by crystal form, cleavage and hardness. Millerite fuses readily, yielding a magnetic globule.

Occurrence: Normally a mineral of low-temperature origin, millerite occurs frequently as tufts of capillary crystals in cavities in limestone or dolomite, or in carbonate veins, also as an alteration of other nickel minerals. It occurs often as a late mineral in veins containing other sulfides and nickel minerals. Coarse granular material with the color of pyrite, but softer and displaying the perfect cleavage of millerite, occurs at the Panet mines in Quebec. Thin, radiating, fibrous coatings of millerite occur at the Gap mine in Pennsylvania. It is of only minor importance as a source of nickel.

Pentlandite, $(Fe,Ni)_9S_8$

Crystal system and class: Isometric; $4/m \bar{3} 2/m$.

Cell dimensions, and content: $a = 10.04$; $Z = 4$.

Habit: Massive, usually in granular aggregates; large grains often show well-developed octahedral parting planes.

Cleavage: None.

Fracture: Conchoidal.

Tenacity: Brittle.

Hardness: $3\frac{1}{2}$–4.

Density: 4.6–5.0.

Color: Light bronze yellow.

Streak: Light bronze brown.

Luster: Metallic.

Diagnostic features: Pentlandite is nearly always intimately associated with pyrrhotite, from which it can be distinguished in hand specimens only by its marked octahedral parting. It is remarkably similar to pyrrhotite in hardness, density, color, and streak. It is nonmagnetic, but the variable magnetism of pyrrhotite and the prevalence of intergrowths of the two minerals render this feature of little value in identifying hand specimens. It is easily fusible at $1\frac{1}{2}$–2, giving a steel-gray bead. The mineral gives a strong test for nickel.

Occurrence and use: Pentlandite analyses show an iron-to-nickel ratio close to 1:1, but varying somewhat. The variation may in some cases be due to admixed pyrrhotite, which is very difficult to avoid completely in preparing a sample for analysis. Cobalt, to the extent of about 1 percent, is often recorded; this may substitute for iron or nickel and probably accounts for a small cobalt production at Sudbury, Ontario.

Pentlandite is usually found in basic igneous rocks with iron and nickel sulfides and arsenides which probably have accumulated by some process of magmatic segregation. The most important occurrence of the mineral is at Sudbury, where it is the chief source of nickel in the ore. It also occurs in the Bushveld, Transvaal, South Africa; in Norway; at Petsamo in the USSR; and at Lynn Lake and Moak Lake in Manitoba.

The average world production of nickel, amounting to 400,000 tons during 1961–1965, came largely from pentlandite-pyrrhotite ores (Canada, 55 percent; the USSR, 20 percent). About 16 percent comes from residual nickel silicates in New Caledonia and Cuba.

Covellite, CuS

Crystal system and class: Hexagonal; $6/m\ 2/m\ 2/m$.
Axial elements: $a:c = 1:4.310$.
Cell dimensions, and content: $a = 3.792$, $c = 16.344$; $Z = 6$ (Fig. 9-15a).

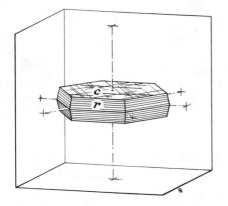

Figure 9-14 Covellite crystal. Typical tabular crystal with large pinacoid $c\{0001\}$ beveled by narrow faces of the hexagonal dipyramid $r\{10\bar{1}1\}$.

Habit: Crystals rare, usually in hexagonal plates with $\{0001\}$ showing hexagonal striations (Fig. 9-14) bounded by steep pyramid faces, $\{10\bar{1}1\}$ and others, horizontally striated; commonly massive; the thin platy habit in a mineral with the lattice dimensions of covellite is consistent with the long c repeat.
Cleavage: $\{0001\}$ highly perfect, flexible in thin leaves.
Hardness: $1\frac{1}{2}$–2.
Density: 4.6–4.76.
Color: Indigo blue or darker, often iridescent in brass yellow and dark red.
Streak: Lead gray to black.
Luster: Metallic.
Opacity: Opaque except in very thin plates.

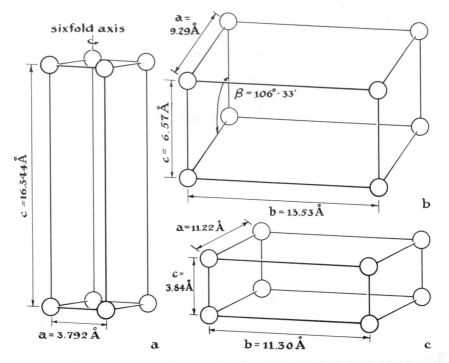

Figure 9-15 (a) Covellite, unit cell. In contrast to the usual tabular habit of covellite crystals, the unit cell has *c* much longer than *a*. (b) Realgar, unit cell of lattice, showing the monoclinic character. (c) Stibnite, unit cell. In contrast to the markedly acicular habit of stibnite, the unit cell has *c* much shorter than *a* or *b*.

Diagnostic features: The color and perfect cleavage distinguish covellite from chalcocite and bornite; covellite fuses at $2\frac{1}{2}$, on charcoal burns with a blue flame and fuses to a globule, yields sulfur in the closed tube.

Occurrence: Covellite approaches very closely the ideal formula CuS: traces of iron are occasionally reported. It is a mineral of much less common occurrence than chalcocite.

Often associated with other copper sulfides, chalcopyrite, chalcocite, and bornite, covellite is usually found in the zone of secondary enrichment where it was formed by alteration of the primary sulfides Primary covellite has also been noted in hydrothermal veins at Butte, Montana, where excellent crystals have been found.

Cinnabar, HgS

Crystal system and class: Trigonal; 32.
Axial elements: $a : c = 1 : 2.2885$.

Cell dimensions, and content: $a = 4.149$, $c = 9.495$; $Z = 3$.

Common forms and angles:

$(0001) \wedge (10\bar{1}1) = 69° 18'$	$(0001) \wedge (10\bar{1}3) = 41° 25'$
$(10\bar{1}0) \wedge (10\bar{1}2) = 37° 04'$	$(10\bar{1}1) \wedge (01\bar{1}\bar{1}) = 71° 47'$

Habit: Crystals rhombohedral or thick tabular on $\{0001\}$ to short prismatic (Fig. 9-16); common in crystalline incrustations; granular, massive, or as an earthy coating.

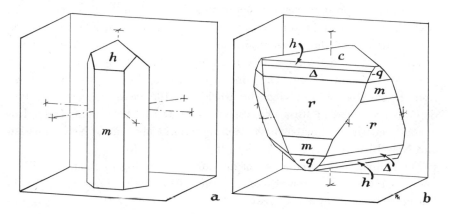

Figure 9-16   Cinnabar crystals. Forms: pinacoid $c\{0001\}$, hexagonal prism $m\{10\bar{1}0\}$; positive rhombohedrons $h\{10\bar{1}3\}$, $\Delta\{10\bar{1}2\}$, $r\{10\bar{1}1\}$; negative rhombohedron $-q\{02\bar{2}1\}$.

Twinning: Common with $\{0001\}$ as twin plane and c axis as twin axis, simple and penetration.

Cleavage: $\{10\bar{1}0\}$ perfect.

Tenacity: Somewhat sect le.

Hardness: $2-2\frac{1}{2}$.

Density: 8.09.

Color: Conchineal red, often inclining to brownish red.

Streak: Scarlet.

Luster: Adamantine, inclining to metallic when dark colored, or dull in friable varieties.

Optical properties: $\omega = 2.905$, $\epsilon = 3.256$; $\epsilon - \omega = 0.351$; positive.

Opacity: Transparent.

Diagnostic features: Color, cleavage; it is volatile before the blowpipe; with sodium carbonate in the closed tube it gives globules of mercury as a sublimate.

Alteration: In the zone of weathering cinnabar may be altered to native mercury, mercury oxide (*montroydite*), and mercurous chloride (*calomel*). Mercury sulfide is also found to a minor extent as *metacinnabar*, which is black and isomorphous with sphalerite.

Occurrence and production: Cinnabar, the most important mercury mineral, is usually found in veins or impregnations formed at low temperatures near recent volcanic rocks or hot-spring areas. It is found in fractures, impregnating sandstone or replacing quartz in sandstone or quartzite. It may often be associated with pyrite, marcasite, and stibnite, with opal, chalcedony, quartz, calcite, or dolomite, also with carbonaceous material in shales and slates.

The most important mercury deposit in the world is at Almaden, Spain, where cinnabar is found impregnating and replacing quartzite interbedded with bituminous shales. These mines have been operating since Roman times, when Pliny recorded that the equivalent of 10,000 pounds of mercury a year was brought to Rome. The mines are still in operation, and have known reserves capable of supplying the world for 100 years. Other important deposits are found at Idria, Yugoslavia, and at Monte Amiata, Italy. Important deposits of cinnabar have been mined in California at New Almaden and New Idria, in Oregon and Nevada, and also in Mexico.

The annual world production of mercury, amounting to about 20 million pounds (1960–1964 average), comes largely from Italy (23 percent), Spain (30 percent), and to a lesser extent Yugoslavia, the USSR, the USA, and Mexico.

Realgar, AsS

Crystal system and class: Monoclinic; $2/m$.

Axial elements: $a:b:c = 0.6878:1:0.4858$, $\beta = 106° 53'$.

Cell dimensions, and content: $a = 9.29$, $b = 13.53$, $c = 6.57$; β,* $Z = 16$ (Fig. 9-15b).

Common forms and angles:

$(010) \wedge (110) = 56° 39'$	$(010) \wedge (021) = 47° 05'$
$(100) \wedge (101) = 43° 50'$	$(001) \wedge (011) = 24° 56'$
$(100) \wedge (001) = 73° 07'$	$(010) \wedge (120) = 37° 14'$

Habit: Crystals usually short prismatic and striated parallel to c (Fig. 9-17); also coarse to fine granular, compact, or as an incrustation.

Cleavage: {010} good; {$\bar{1}$01}, {100}, {120} poor.

Tenacity: Sectile.

Hardness: $1\frac{1}{2}$–2.

Density: 3.56.

Color: Aurora-red to orange-yellow.

---

* In descriptions of monoclinic and triclinic minerals the interaxial angles are given with the axial elements; the same angles are also an essential part of the cell dimensions, but are not repeated in the description.

Streak: Orange-red.

Luster: Resinous to greasy.

Opacity: Transparent when fresh.

Optical properties: $\alpha = 2.538$, $\beta = 2.684$, $\gamma = 2.704$; $\gamma - \alpha = 0.166$; negative, $2V = 46°$.

Alteration: Disintegrates on long exposure to light to a red-yellow powder.

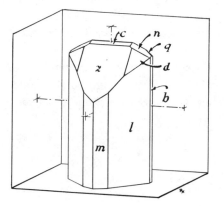

Figure 9-17 Realgar crystal. Forms: pinacoids $c\{001\}$, $b\{010\}$, $z\{101\}$; rhombic prisms $l\{120\}$, $m\{110\}$, $n\{011\}$, $d\{111\}$, $q\{\bar{1}21\}$.

Diagnostic features: Color, hardness; on heating, it is easily fusible at 1, yielding on charcoal a volatile, white sublimate with a garlic-like odor; on heating in the closed tube it gives a red sublimate.

Occurrence: Realgar occurs as a minor constituent in hydrothermal sulfide veins with orpiment and other arsenic minerals, with stibnite, or with lead, silver, or gold ores. It contributes to the arsenic content of such ores, which yield arsenic oxide on smelting. It is found occasionally in limestones or dolomites, or in clay rocks, also as a sublimation product from volcanic emanations or in hot spring deposits.

Orpiment, As_2S_3

Crystal system and class: Monoclinic; $2/m$.

Axial elements: $a:b:c = 1.195:1:0.4421$, $\beta = 90° 27'$.

Cell dimensions, and content: $a = 11.49$, $b = 9.59$, $c = 4.25$; $Z = 4$.

Common forms and angles:

$(010) \wedge (110) = 39° 56'$	$(100) \wedge (101) = 69° 18'$
$(100) \wedge (001) = 89° 33'$	$(010) \wedge (\bar{3}11) = 73° 27'$

Habit: Crystals small, short prismatic on c, with pseudo-orthorhombic appearance or obviously monoclinic with a prominent zone parallel to [103]; commonly in foliated, columnar, or fibrous masses.

Cleavage: $\{010\}$ perfect; lamellae flexible not elastic

Tenacity: Sectile.

Hardness: $1\frac{1}{2}$–2.

Density: 3.49.

Color: Lemon yellow to brownish yellow.

Streak: Pale yellow.

Luster and opacity: Pearly on cleavage surfaces, elsewhere resinous; transparent.

Optical properties: $\alpha = 2.40$, $\beta = 2.81$, $\gamma = 3.02$; $\gamma - \alpha = 0.62$; negative, $2V = 76°$.

Diagnostic features: It is distinguished by its yellow color and excellent cleavage. It reacts to blowpipe tests in the same way as realgar.

Occurrence: Orpiment is typically a very low-temperature hydrothermal mineral; it is found in veins and in hot spring deposits or as a common alteration product of realgar or other arsenic minerals. It occurs with stibnite, realgar, arsenic, calcite, barite, or gypsum.

Stibnite, Sb₂S₃

Crystal system and class: Orthorhombic; $2/m\ 2/m\ 2/m$.

Axial elements: $a:b:c = 0.9928:1:0.3394$.

Cell dimensions, and content: $a = 11.229$, $b = 11.310$, $c = 3.839$; $Z = 4$ (Fig. 9-15c).

Common forms and angles:

(010) $\wedge$ (110) = 45° 12'   (010) $\wedge$ (120) = 26° 44'

(010) $\wedge$ (111) = 72° 12'   (110) $\wedge$ (111) = 64° 17'

Habit: Crystals stout to slender prismatic along c (Fig. 9-18), usually striated or grooved parallel to c; crystals often bent or twisted, commonly in

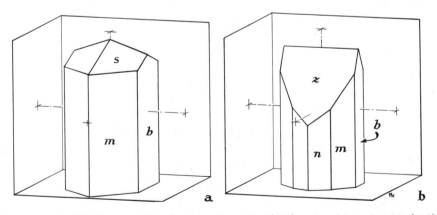

Figure 9-18   Stibnite crystals. Forms: pinacoid $b\{010\}$; rhombic prisms $m\{110\}$, $n\{210\}$, $z\{501\}$; rhombic dipyramid $s\{111\}$. Drawings show only the termination of prismatic crystals.

Figure 9-19 Stibnite, aggregate of acicular crystals. (Manhattan, Nevada.) [Courtesy Queen's University.]

confused aggregates of acicular crystals (Fig. 9-19); also in radiating or columnar masses, or granular.

Cleavage: {010} perfect and easy.

Tenacity: Flexible, not elastic, slightly sectile.

Hardness: 2.

Density: 4.63.

Color and streak: Lead gray; blackish to iridescent tarnish.

Luster and opacity. Metallic, brilliant on cleavage and fresh surfaces; opaque.

Diagnostic features. Distinguished from galena by cleavage, habit, and lower density; stibnite fuses at 1, coloring the flame greenish blue; it is soluble in hydrochloric acid.

Occurrence and use. Stibnite is an important antimony mineral. It probably varies little in composition from the ideal Sb_2S_3, although small amounts of Fe, Pb, Cu, and other metals are reported in analyses. In many localities it is found in close association with lead or iron sulfantimonites, which are closely similar in physical properties. The rarer mineral, *bismuthinite*, Bi_2S_3, is isostructural with stibnite and similar in physical properties.

Stibnite is found typically in hydrothermal vein and replacement deposits of low-temperature origin or in hot-spring deposits. It is the most important ore mineral of antimony, but though it is widely distributed, large deposits are of rare occurrence. It is often associated with realgar, orpiment, galena, lead sulfantimonites, marcasite, pyrite, cinnabar, calcite, ankerite, barite, chalcedony, or quartz. The minerals are usually found filling fissures, joints, and rock pores, or as irregular or lens-like replacement bodies.

Extensive deposits in sandstones have been mined in Hunan province, China, and in limestones at several localities in Mexico. Other deposits occur in Bolivia and Algeria. A large production of antimony metal is obtained as a by-product from the smelting of lead ores in the USA and in British Columbia.

PYRITE AND COBALTITE GROUPS

The AX_2 type of sulfide mineral is classified into several groups; the isometric minerals of the pyrite and cobaltite groups are closely related structurally

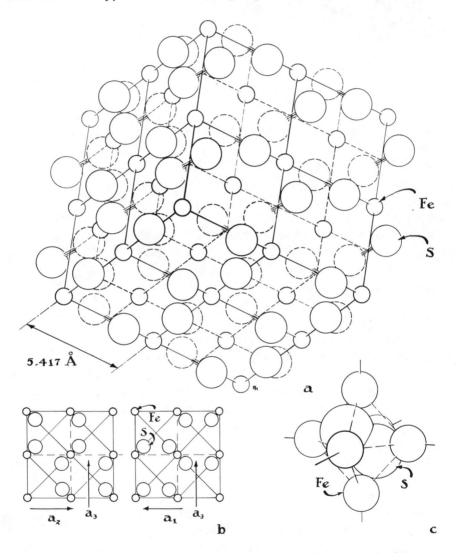

Figure 9-20 Structure of pyrite. The iron atoms are in a face-centered cubic arrangement with sulfur atoms located on the cube diagonals. (b) In the structure projected on (100) and (010), the sulfur pairs are parallel to a_3 on (100) and to a_1 on (010), corresponding to the direction of striations on (100) and (010) on the cube in Fig. 9-21a. (c) The octahedral grouping of iron around one pair of sulfur atoms.

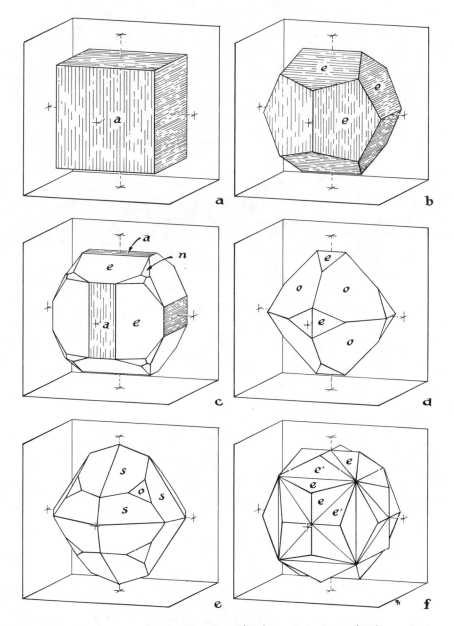

Figure 9-21 Pyrite crystals. Forms: cube $a\{100\}$, pyritohedron $e\{210\}$, octahedron $o\{111\}$, trapezohedron $n\{112\}$, diploid $s\{321\}$. (f) Pyritohedron $e\{210\}$ in twinned position e', twin axis [110], Iron Cross Law. (a), (b), and (c) show the common striations on cube and pyritohedral faces.

and chemically. Although a large number of substances, in which X may be S, Se, Te, As, or Sb, crystallize with the pyrite structure, the naturally occurring compounds are mostly sulfides. Pyrite is by far the commonest of these; sperrylite is a rare but economically important member; MnS_2 (*hauerite*), RuS_2 (*laurite*), $AuSb_2$ (*aurostibite*), and others are of very rare occurrence.

In the pyrite structure (Fig. 9-20) the metal atoms occupy the face-centered cubic lattice positions, and the sulfur atoms lie in pairs along the trigonal axes of the lattice, with each S at about three-eighths the length of the diagonal from an iron atom. The sulfurs are in sixfold coordination about each iron atom. The sulfur atoms in the S_2 pairs are separated by about the same distance as are S atoms in the covalently bonded S_8 rings in native sulfur. The sulfur pairs are in turn coordinated to six iron atoms. The pyrite group includes some of the hardest of the sulfide minerals, with pyrite $6-6\frac{1}{2}$, sperrylite 6-7, and laurite $7\frac{1}{2}$. The compound FeS_2 is also found as a polymorph (marcasite).

Crystal System and class.   Isometric, $2/m\ \bar{3}$.

	PYRITE	SPERRYLITE	COBALTITE
Cell dimension:	$a = 5.417$	$a = 5.967$	$a = 5.57$
Cell content:	$Z = 4$	$Z = 4$	$Z = 4$
Common forms:	{100}, {111}, {210}, {321}	{100}, {111}, {210}	{100}, {210}, {111}
Cleavage:	{100} indistinct	{100} indistinct	{100} perfect
Fracture:	Conchoidal	Conchoidal	Uneven
Tenacity:	Brittle	Brittle	Brittle
Hardness:	$6-6\frac{1}{2}$	6-7	$5\frac{1}{2}$
Density:	5.01	10.58	6.33
Color:	Brass yellow	Tin white	Silver white
Luster:	Metallic	Metallic	Metallic
Magnetism:	Paramagnetic	—	—

Pyrite, FeS_2

Habit: Crystals commonly cubic, also pyritohedral or octahedral (Figs. 9-21, 9-22); the pyritohedral and cubic faces are often striated parallel to the edge (*a*) between them (Fig. 9-21a,b,c); these striae are due to oscillatory growth of these forms which tends to produce rounded faces; crystals often abnormally developed; frequently massive, granular, sometimes radiated, reniform, or globular.

Twinning: Twin axis [110], interpenetrating, known as the Iron Cross law (Figs. 9-21f, 9-23).

Figure 9-22 Pyrite crystals. (a) Striated cube; (b) cube, modified by octahedron (a smaller intergrown cube is protruding from two faces) (Traversella, Italy); (c) pyritohedron $e\{210\}$ with small octahedron faces (Rio Marina, Elba); (d) octahedron $o\{111\}$ (Rio Marina, Elba.). [Courtesy Royal Ontario Museum.]

Color: Pale brass yellow, iridescent when tarnished.

Streak: Greenish black or brownish black.

Diagnostic features: Pyrite is distinguished from most yellow metallic minerals by its superior hardness. It is distinguished from marcasite with certainty only by its crystal form, though marcasite is usually paler in color and more readily altered. Pyrite fuses at $2\frac{1}{2}$–3 and gives abundant sublimate of sulfur in the closed tube; it is insoluble in hydrochloric acid, but the fine powder is completely dissolved in strong nitric acid.

Chemistry: In many occurrences pyrite is very close to the ideal composition FeS_2. Uncommonly, nickel, cobalt, or sometimes both, substitute for iron; if

Figure 9-23 Twinned pyrite crystal, Iron Cross Law, twin axis [110], combining two pyritohedral crystals as in Fig. 9-21f. [Courtesy Queen's University.]

iron predominates these varieties are referred to as nickelian or cobaltian pyrite. The name *bravoite* is used for those rare occurrences of (Ni,Fe)S₂ in which iron is about 50 mole percent or less. Two minerals with the pyrite structure, in which nickel or cobalt is the major metal constituent, have been described from the Congo. It is possible that any composition of (Fe,Ni,Co)S₂ may occur with the pyrite structure. The substitution of Ni for Fe is accompanied by an increase in the edge length of the cubic unit cell (Fig. 9-24), by a decrease in hardness, and by a change in color toward silver-white or gray.

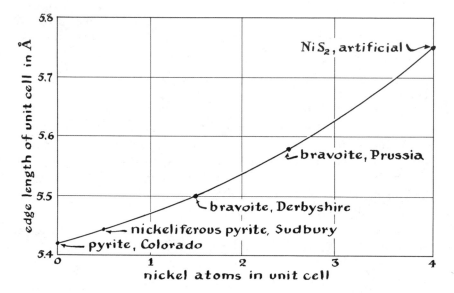

Figure 9-24   Pyrite, variation in edge length of cubic unit cell with nickel content.

Occurrence: Pyrite is by far the most widespread and commonly occurring sulfide mineral. It may be found in almost any type of geological environment: as an accessory mineral in igneous rocks, both acidic and basic types; as a magmatic segregation from igneous rocks; infrequently in pegmatites; in contact metamorphic deposits (Clifton, Arizona); in hydrothermal sulfide veins and replacement deposits over almost the whole temperature range; as a sublimation product (Vesuvius); in sedimentary and metamorphic rocks.

Most massive hydrothermal sulfide bodies contain pyrite, and in some, as at Rio Tinto, Spain, and at Noranda, Quebec, it is the most important sulfide. Pyrite is rarely of economic importance itself, but its presence calls attention to vein or replacement deposits which might, on closer study, be found to contain chalcopyrite, as in many massive sulfide deposits, or gold, as in some of the Precambrian gold deposits, where much of the gold is present as minute grains in and with the pyrite.

Pyrite oxidizes to iron sulfates, which alter to limonite; thus, the presence of extensive pyrite in a rock at some depth may be revealed by a capping of limonite or gossan on the surface. Pyrite yields some of the sulfur of commerce, but most of this pyrite is mined because it is present in ores containing valuable metals. In a few cases local demands for sulfuric acid may result in the mining of pyrite as a source of sulfur.

The oxidation of pyrite to limonite may take place in such a way as to preserve the crystal form of pyrite in a pseudomorph.

Sperrylite, PtAs$_2$

Habit: Crystals usually cubo-octahedral, often with pyritohedron {210}, often with rounded corners and edges.

Occurrence: Sperrylite was first found at the Vermillion mine, Sudbury, Ontario, and was named after the discoverer, Francis L. Sperry, a chemist at Sudbury. Distinct crystals are frequently found in the pyrrhotite-pentlandite-chalcopyrite ores at Sudbury, and the mineral is responsible for the important platinum production from the Sudbury ores. It has also been found in the Bushveld igneous rocks of the Transvaal, South Africa, and as a detrital mineral in certain river gravels.

Cobaltite, CoAsS

Habit: Crystals commonly in cubes {100}, or pyritohedrons {210}, or as combinations of these forms, often with faces striated as in pyrite; also octahedral, or granular to compact.

Color: Silver white inclined to red, also steel gray with a violet tinge.

Streak: Gray-black.

Diagnostic features: Cobaltite is softer than pyrite ($H = 5\frac{1}{2}$), and has perfect {100} cleavage. The reddish tinge helps to distinguish cobaltite from smaltite. Fusible at 2–3; on charcoal it gives a volatile white sublimate of arsenious oxide, which has the characteristic garlic-like odor. The residue gives a deep blue color (cobalt) in the borax bead in the oxidizing flame.

Chemistry and crystal structure: In cobaltite and the rarer mineral *gersdorffite* (NiAsS) the structure is similar to that of pyrite, with As–S taking the place of the S$_2$ in pyrite. In both minerals Fe often substitutes for Co or Ni. In a third member of the group, ullmannite, Sb takes the place of As. If the As or Sb and S take up ordered positions in the structure the symmetry class becomes 2 3.

Occurrence: Cobaltite is found in vein deposits with other cobalt and nickel
sulfides or arsenides (at Cobalt, Ontario, and elsewhere), also disseminated
in metamorphic rocks. Gersdorffite is a minor nickel mineral in the pyrrhotite-
pentlandite-chalcopyrite ores at Sudbury, Ontario.

OTHER SULFIDES OF AX₂ TYPE

Marcasite, FeS₂

Crystal system and class: Orthorhombic; $2/m\ 2/m\ 2/m$.

Axial elements: $a:b:c = 0.8194:1:0.6245$.

Cell dimensions, and content: $a = 4.445$, $b = 5.425$, $c = 3.388$; $Z = 2$.

Common forms and angles:

$(010) \wedge (110) = 50°\ 40'$ $(\bar{1}01) \wedge (101) = 74°\ 38'$

$(110) \wedge (111) = 45°\ 25'$ $(010) \wedge (140) = 16°\ 53'$

Habit: Crystals commonly tabular {010}, also pyramidal (Fig. 9-25a), faces
often curved; often stalactitic, globular, or reniform (Fig. 9-26b) with radiating internal structure, the exterior covered with projecting crystals.

Twinning: Common on {101}; often repeated, resulting in cockscomb and
spear shapes (Fig. 9-26a).

Crystal structure: In the marcasite structure (Fig. 9-25b), which is dimorphous
with pyrite, the metal atoms occupy the lattice points of an orthorhombic
body-centered lattice, and the sulfur atoms surround each iron in sixfold
coordination. The FeS₆ octahedrons are all oriented with two edges parallel
to c and two edges perpendicular to c; the latter edges are shared, resulting

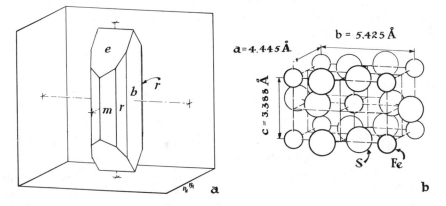

a   b

Figure 9-25   (a) Marcasite crystal. Forms: pinacoid $b\{010\}$, rhombic prisms $m\{110\}$,
$r\{140\}$, $e\{101\}$. (b) Structure of marcasite, showing iron atoms on a body-centered
orthorhombic lattice with pairs of sulfur atoms; each iron atom is octahedrally surrounded by six sulfurs.

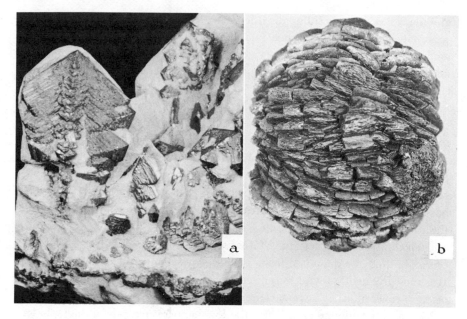

Figure 9-26 Marcasite. (a) Crystals in fine gray chalk. (Folkestone, England.) [Courtesy Royal Ontario Museum.] (b) Radiating nodule of crystals. (Joplin, Missouri.) [Courtesy Queen's University.]

in chains of octahedrons parallel to *c*. The chains centered on the corners of the unit cell have their shared edges inclined to *a* and *b*, whereas the chain at body center of the cell is displaced one-half along *c* and the shared edges are also inclined to *a* and *b* but in a reverse relation to the corner chains. This inclination is such that the sulfurs of the center chain approach those in the corner chains in S_2 pairs, with the same separation as in the S_2 pairs in pyrite. The S_2 pairs have their long axis in the plane (001) and are located between iron atoms along the long axis, *b*. The coordination of S to Fe is identical to that in pyrite. The structure also shows a close similarity to that of rutile, where the octahedra have their shared edges at 45° to the *a* axes, the oxygens do not occur in O_2 pairs, and the symmetry is tetragonal.

Cleavage: {101}, rather distinct.

Fracture and tenacity: Uneven; brittle.

Hardness: 6–6½.

Density: 4.89.

Color: Pale bronze yellow, deepening on exposure, tin white on fresh fracture.

Streak: Grayish or brownish black.

Luster and opacity: Metallic; opaque.

Diagnostic features: Crystal form; hardness; it is fusible at 2½–3 to a magnetic globule; gives odor of sulfur dioxide when heated on charcoal; gives sulfur sublimate in closed tube. When fine powder is treated first with cold nitric acid, then boiled after the initial vigorous reaction has ceased, the mineral is decomposed with separation of sulfur. Under these conditions pyrite is completely dissolved.

Chemistry: Marcasite, dimorphous with pyrite, shows very little variation from the composition FeS_2. It is distinguishable from pyrite only by the crystal form, although marcasite decomposes more readily under atmospheric conditions to ferrous sulfate and sulfuric acid. Marcasite may invert, forming pyrite pseudomorphs, or it may be altered to form limonite pseudomorphs.

Occurrence: Marcasite is found most often in near-surface deposits where it has been formed at low temperatures and from acid solutions, whereas pyrite, the more stable form, is deposited under conditions of higher temperature and alkalinity or low acidity. Marcasite is usually a mineral of supergene origin, but may be deposited from ascending low temperature vein solutions. It is most frequently found in sedimentary rocks, limestone, clays, or lignite, often as concretions or as replacement of fossils.

Marcasite is quite common with the low-temperature galena and sphalerite ores in the Mississippi Valley districts, often as individual crystals or as cockscomb groups. Marcasite is not of any economic value.

Arsenopyrite, FeAsS

Crystal system and class: Orthorhombic; $2/m \ 2/m \ 2/m$.*
Axial elements: $a:b:c = 0.6733:1:0.5958$.
Cell dimensions, and content: $a = 6.43$, $b = 9.55$, $c = 5.69$; $Z = 8$.
Common forms and angles:

$(110) \wedge (1\bar{1}0) = 67° \ 54'$ $(201) \wedge (\bar{2}01) = 121° \ 04'$
$(021) \wedge (0\bar{2}1) = 100° \ 00'$ $(012) \wedge (0\bar{1}2) = 33° \ 10'$

Habit: Crystals common, prismatic with elongation on a or c, commonly as a combination of two prisms, {110} with {021}, {201}, or {012} (Figs. 9-27, 28) faces commonly striated parallel to a; columnar, usually showing distinct rhombic cross sections; granular or compact.

Twinning: Common on {110} as contact or penetration twins (Fig. 9-28), also on {201} as cruciform twins; some arsenopyrite has been found to be monoclinic, but pseudo-orthorhombic crystals result because of twinning on {100} and {001}.

Cleavage: {110} distinct.

---

* Arsenopyrite is sometimes monoclinic (with c ortho. = twofold axis) but pseudo-orthorhombic due to twinning on {100} and {010}.

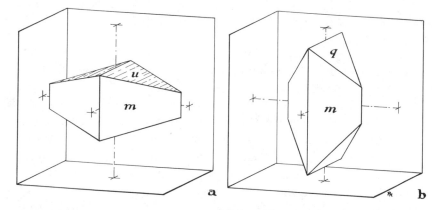

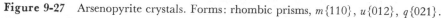

Figure 9-27 Arsenopyrite crystals. Forms: rhombic prisms, $m\{110\}$, $u\{012\}$, $q\{021\}$.

Fracture and tenacity: Uneven; brittle.
Hardness: $5\frac{1}{2}$–6.
Density: 6.07.
Color: Silver white to steel gray.
Streak: Dark grayish black.
Luster and opacity: Metallic; opaque.
Diagnostic features: The crystal form with rhombic cross sections serves to distinguish it from smaltite; fusible at 2, it gives in the closed tube an abun-

Figure 9-28 Arsenopyrite. (a) Crystals in fine-grained quartz, showing striated faces $u\{012\}$ with $m\{110\}$. (Salzburg, Austria.) [Courtesy Royal Ontario Museum.] (b, c) Twinned crystals. Forms $e\{201\}$, $q\{021\}$, $\{032\}$, $\{011\}$ striated, twin plane $\{110\}$. (Deloro, Ontario.) [Courtesy Queen's University.]

dant arsenic sublimate. It is decomposed by nitric acid with separation of sulfur.

Chemistry: Arsenopyrite, essentially FeAsS, often contains some cobalt in substitution for iron, and the composition may grade into *glaucodot*, (Co,Fe)AsS, in which cobalt predominates.

Occurrence: Arsenopyrite, the most abundant arsenic mineral, has usually formed at moderate to high temperatures. It is found most commonly in high-temperature gold-quartz veins, as in South Dakota and Quebec; in high-temperature cassiterite veins, as in Cornwall, England; and with scheelite or in some contact metamorphic deposits with gold and other sulfides. It is less common in low-temperature gold-quartz veins, as in the Mother Lode, California, or in nickel-cobalt-silver veins as in Cobalt, Ontario.

Arsenopyrite is of widespread occurrence in many parts of the world. It is not mined for its arsenic content, but most arsenic of commerce is recovered from smelter fumes resulting from treatment of ores containing the mineral. This source more than satisfies the market for arsenic.

Molybdenite, MoS₂

Crystal system and class: Hexagonal; $6/m\ 2/m\ 2/m$.

Axial elements: $a:c = 1:3.892$.

Cell dimensions, and content: $a = 3.16$, $c = 12.30$; $Z = 2$.

Habit: Crystals hexagonal, thin to thick tabular on $\{0001\}$ with prism $\{10\bar{1}0\}$ and dipyramids $\{10\bar{1}1\}$, $\{10\bar{1}2\}$ as narrow horizontally striated faces; commonly foliated, massive, or in scales.

Cleavage: $\{0001\}$ perfect; laminae very flexible but not elastic.

Crystal structure: MoS₆ octahedrons are linked together by sharing edges to form sheets parallel to $\{0001\}$. Sheets are linked by weak bonds between sulfurs of adjacent sheets, resulting in the perfect basal cleavage (Fig. 9-29).

Tenacity: Sectile.

Hardness: $1-1\frac{1}{2}$.

Density: 4.62–4.73.

Color: Lead gray with bluish tinge.

Streak: Bluish gray on paper, greenish on glazed porcelain.

Luster and opacity: Metallic; opaque.

Feel: Greasy.

Diagnostic features: Molybdenite, together with the rare isostructural mineral *tungstenite*, WS₂, is characterized by a perfect basal cleavage and hardness similar to graphite. Molybdenite has a distinct steel-blue color, as compared with the lead-gray of graphite, and a greenish streak on glazed porcelain.

Molybdenite and graphite may occur in similar environments. Molybdenite is infusible; when heated on charcoal in the oxidizing flame it gives a white coating of molybdic oxide which turns deep blue when touched with the reducing flame. When heated with potassium iodide and sulfur on charcoal or plaster it gives a deep blue sublimate.

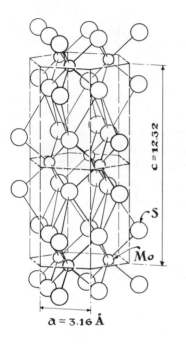

$c = 12.32$

$a = 3.16 \text{ Å}$

Figure 9-29 Structure of molybdenite, MoS_2. Octahedral layers parallel to {0001} are loosely bonded to adjacent layers, resulting in basal cleavage similar to that in graphite.

Occurrence and production: It is the most common molybdenum mineral; it is found as an accessory in some granites, often in pegmatites, sometimes in commercially important quantities. Molybdenite also occurs in deep-seated veins with scheelite, wolframite, topaz, and fluorite, also in contact metamorphic deposits with lime silicates, scheelite, or chalcopyrite; in this environment graphite might also occur.

Molybdenite is the principal source of molybdenum, which has found many applications in special steel alloys since 1913. In recent years the major sources of molybdenite have been quartz veinlets in granite at Climax, Colorado, and the copper mines of Utah, New Mexico, and Arizona. Minor amounts are recovered in Mexico, Chile, Canada, and Norway. Molybdenite occurs at numerous scattered pegmatite localities in the southern part of the Canadian Precambrian shield, but only one of these is now being mined. The annual world production of about 45 thousand tons of molybdenum (1961–1965 average) is obtained from the USA (70 percent) and the USSR (15 percent), with small amounts from Canada, Chile, and China.

KRENNERITE GROUP

This group comprises three minerals: *krennerite*, $(Au,Ag)Te_2$; *calaverite*, $AuTe_2$; *sylvanite*, $(Au,Ag)Te_2$.

These three minerals are very nearly unique in that they are the most important naturally occurring compounds of gold and are also three of a somewhat larger group of minerals which contain tellurium as a major constituent. They are usually found in low-temperature hydrothermal veins, but are found also in some hydrothermal veins of higher temperature origin. All three minerals have been identified in gold mines of Kalgoorlie, Western Australia; Nagyag, Transylvania; Cripple Creek district, Colorado; and the Rouyn district, Quebec. Calaverite and sylvanite, more widespread in occurrence than krennerite, also occur in the Mother Lode, California (including Calaveras county, for which locality calaverite is named), and in the Kirkland Lake gold district of Ontario. In these localities the gold tellurides are associated with quartz, pyrite, native gold, and very rare tellurides of bismuth, lead, mercury, and nickel.

	KRENNERITE	CALAVERITE	SYLVANITE
Crystal system:	Orthorhombic	Monoclinic	Monoclinic
Crystal class:	$mm2$	$2/m$	$2/m$
Axial elements:	$a:b:c$	$a:b:c$	$a:b:c$
	$= 0.533:1:0.270$	$= 1.632:1:1.152$	$= 1.995:1:3.257$
		$\beta = 90° \pm 30'$	$\beta = 145° 26'$
Cell dimensions:	$a = 16.54$	$a = 7.19$	$a = 8.96$
	$b = 8.82$	$b = 4.41$	$b = 4.49$
	$c = 4.46$	$c = 5.08$	$c = 14.62$
Cell content:	$8[(Au,Ag)Te_2]$	$2[AuTe_2]$	$4[(Au,Ag)Te_2]$
Cleavage:	{001} perfect	None	{010} perfect
Hardness:	2–3	$2\frac{1}{2}$–3	$1\frac{1}{2}$–2
Density:	8.62	9.24	8.16

Calaverite, $AuTe_2$

Habit: Crystals bladed or lath-like, short prisms usually elongated and striated parallel to b; often massive, granular.

Twinning: Common on {101}, {310}, or {111}.

Color: Brass yellow to silver white.

Streak: Yellowish, to greenish gray.

Luster and opacity: Metallic; opaque.

Blowpipe: It fuses readily at 1, and on charcoal gives a bluish-green flame and metallic globules of gold.

Sylvanite, (Au,Ag)Te₂

Habit: Crystals short prismatic along c or b, also tabular on {100} or {010}; often skeletal or bladed, also granular.

Twinning: Common on {100} as simple contact, lamellar, or penetration twins.

Color and streak: Steel gray to silver white.

Luster: Brilliant metallic.

Blowpipe tests: It fuses at 1 on charcoal to a dark-gray globule and gives reactions for tellurium; on continued heating it gives a malleable metallic bead, but less readily than calaverite.

SKUTTERUDITE SERIES

This series comprises three minerals: *skutterudite*, $(Co,Ni)As_3$; *smaltite*, $(Co,Ni)As_{3-x}$; *chloanthite*, $(Ni,Co)As_{3-x}$.

These mineral names are applied to portions of a continuous solid solution series. The name smaltite, the oldest and best known, is generally preferred and applies to the most common occurrences. The minerals of the series are not sharply separated nor readily distinguishable. The high cobalt members yield pink erythrite on weathering, and the rarer high nickel members yield green annabergite. In addition to the varying cobalt-nickel content, iron may substitute to about 12 percent; this series with $Fe:(Co + Ni)$ greater than 1:1 is not represented among minerals. The arsenic to metal ratio varies from a maximum of 3:1 in skutterudite to about 2:1 in some occurrences of smaltite or chloanthite. The last two minerals have the same structure as skutterudite but are deficient in arsenic.

Figure 9-30 Smaltite. Fine-grained colloform smaltite with calcite. (Cobalt, Ontario.) [Courtesy Queen's University.]

Crystal system and class: Isometric; $2/m\,\bar{3}$.

Cell dimensions and content: $a = 8.19$–8.29; $Z = 8$.

Habit: Crystals usually cubic, cubo-octahedral, or octahedral, rarely modified by dodecahedron or pyritohedron; often massive, dense fine granular, also colloform (Fig. 9-30).

Cleavage: {100}, {111}, distinct, but variable and not characteristic.

Fracture and tenacity: Conchoidal to uneven; brittle.

Hardness: $5\frac{1}{2}$–6.

Density: 6.5.

Color: Tin white to silver gray, sometimes tarnished iridescent.

Luster and opacity: Metallic, often brilliant; opaque.

Diagnostic features: Distinguished from arsenopyrite by crystal form; in the open tube these minerals give a white sublimate of arsenious oxide, and in the closed tube a mirror sublimate of metallic arsenic; on charcoal they give a strong odor of arsenic and magnetic globule.

Occurrence: The smaltite minerals are usually found in moderate temperature veins with other cobalt and nickel minerals, especially cobaltite and niccolite, also arsenopyrite, silver, and bismuth. Typical occurrences are at Skutterud, Norway, the Co–Ni–Ag veins of Saxony, and similar deposits at Cobalt and South Lorrain, Ontario; they are found also with silver and pitchblende at Great Bear Lake, Northwest Territories, Canada. The mineral has yielded important amounts of cobalt (and the by-product arsenic) from the ores of Cobalt, Ontario, where it has been a by-product of silver mining since 1904.

RUBY-SILVER GROUP

Pyrargyrite, Ag₃SbS₃; Proustite, Ag₃AsS₃

Crystal system and class: Trigonal; $3m$.

	PYRARGYRITE	PROUSTITE
Axial elements:	$a:c = 1:0.7892$	$a:c = 1:0.8039$
Cell dimensions:	$a = 11.047$	$a = 10.816$
	$c = 8.719$	$c = 8.695$
Cell content:	$Z = 6$	$Z = 6$
Hardness:	$2\frac{1}{2}$	2–$2\frac{1}{2}$
Density:	5.85	5.57
Color:	Deep red	Scarlet vermilion
Streak:	Red	Vermilion
Luster:	Adamantine	Adamantine

	PYRARGYRITE	PROUSTITE
Opacity:	Translucent	Translucent
Optical properties:	$\epsilon = 2.881$,	$\epsilon = 2.792$,
	$\omega = 3.084$	$\omega = 3.088$
	$\omega - \epsilon = 0.203$	$\omega - \epsilon = 0.296$
	negative	negative

Habit: Crystals commonly prismatic with hexagonal form due to $\{11\bar{2}0\}$, commonly with hemimorphic development with trigonal pyramids $\{01\bar{1}2\}$, $\{10\bar{1}1\}$, or $\{10\bar{1}\bar{1}\}$, also massive compact.

Twinning: Common on $\{10\bar{1}4\}$ and $\{10\bar{1}1\}$.

Cleavage: $\{10\bar{1}1\}$ distinct.

Chemistry: The two minerals are isostructural, but analyses record only minor amounts of As substituting for Sb in pyragyrite, and the same is true of Sb in proustite. There is no evidence for a very extended solid solution series.

Occurrence: These minerals are commonly called ruby-silver ores. They are found in the low-temperature silver veins usually as late-forming primary minerals, but they may form during the processes of secondary enrichment.

Pyrargyrite is more common than proustite and often constitutes an important ore of silver. It is found with proustite, argentite, tetrahedrite, silver, calcite, dolomite, or quartz. Fine crystals have been found in the silver veins of Saxony and Bohemia, and important quantities are found in many silver mines in South America, Mexico, Nevada, Idaho, and Colorado; it is also found with silver at Cobalt, Ontario. Proustite is often found in the same vein with pyrargyrite and silver.

TETRAHEDRITE SERIES

Tetrahedrite, $Cu_{12}Sb_4S_{13}$; Tennantite, $Cu_{12}As_4S_{13}$

Crystal system and class: Isometric; $\bar{4}3m$.

Cell dimensions and content: $a = 10.21$ (tennantite); $a = 10.37{-}10.48$ (tetrahedrite), increasing with silver content: $Z = 2$.

Habit: Crystals commonly tetrahedral (Figs. 9-31a,b,c, 9-32); often massive, granular to compact.

Twinning: Twin axis [111] as contact or penetration twins.

Cleavage: None.

Fracture and tenacity: Subconchoidal to uneven; brittle.

Hardness: $3{-}4\frac{1}{2}$ (tennantite harder).

Density: 4.6–5.1 (increasing with antimony and silver content).

Color: Flint gray to iron black.

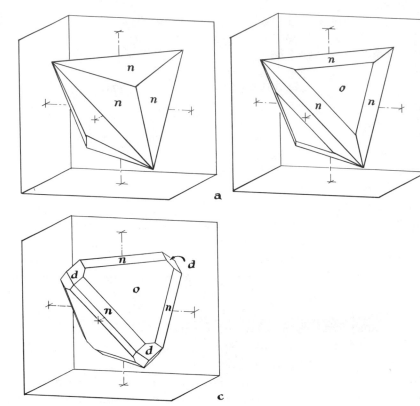

Figure 9-31 Tetrahedrite crystals. Forms: dodecahedron $d\{110\}$, positive tetrahedron $o\{111\}$, positive tristetrahedron $n\{211\}$.

Streak: Black to brown.

Luster and opacity: Metallic; opaque.

Chemistry: These minerals are the end-members of a continuous solid solution series; the name tetrahedrite is used when Sb predominates, and tennantite is used when As predominates. Copper is always the most important metal, but iron is always present in substitution from about 1 percent to possibly 13 percent. Zinc is reported in many analyses to 8 or 9 percent, but may be due in part to admixed sphalerite. Silver is often present and has been reported up to 18 percent, but is usually less than 5 percent; this silver-bearing variety has been called *freibergite*. The presence of mercury in tetrahedrite has also been recorded.

Alteration: Usually alters to malachite, azurite, and antimony oxides.

Occurrence: Tetrahedrite is probably the most widespread and economically important sulfosalt mineral. It is often an important copper ore mineral, and the argentian varieties add appreciably to the silver values of sulfide ores in which they occur. It is also an important antimony-bearing mineral in these ores. The minerals are commonly found in low- to moderate-temperature hydrothermal veins of copper, lead, zinc, and silver minerals; they are of rare occurrence in high-temperature veins (Cornwall, England; and Broken Hill, New South Wales) or contact metamorphic deposits. Tennantite is of less common occurrence than tetrahedrite.

Figure 9-32 Tetrahedrite crystals, showing $o\{111\}$, with negative tetrahedron $-o\{\bar{1}11\}$, and $d\{110\}$, with calcite. (Locality unknown.) [Courtesy Queen's University.]

Well-crystallized specimens have come from several mining regions of Europe. The minerals are of considerable commercial importance in the mining districts of western North America, particularly in Idaho, Montana, Utah, Colorado, New Mexico, Arizona, in British Columbia, and at Tsumeb, Southwest Africa.

OTHER SULFOSALTS

Enargite, Cu_3AsS_4

Crystal system and class: Orthorhombic; $2mm$.
Axial elements: $a:b:c = 0.864:1:0.829$.
Cell dimensions and content: $a = 6.41$, $b = 7.42$, $c = 6.15$; $Z = 2$.
Common forms and angles:
  $(010) \wedge (110) = 49° 10'$ $(001) \wedge (101) = 43° 49'$
  $(100) \wedge (120) = 59° 57'$ $(110) \wedge (111) = 53° 05'$
Habit: Crystals tabular on $\{001\}$ or prismatic along c; often massive, granular, or prismatic.
Twinning: Common on $\{320\}$, often as star-shaped cyclic trillings.
Cleavage: $\{110\}$ perfect; $\{100\}$, $\{010\}$ distinct.
Fracture and tenacity: Uneven; brittle.
Hardness: 3.

Density: 4.45.

Color: Grayish black to iron black.

Streak: Black.

Luster and opacity: Metallic, tarnishing dull; opaque.

Chemistry: Enargite is copper arsenic sulfide (Cu_3AsS_4) in which Sb may substitute for As to about 6 percent by weight, and minor iron substitutes for copper.

Alteration: Enargite may alter to tennantite, or, on oxidation, to a variety of copper arsenates, as at Tintic, Utah.

Occurrence: The mineral is found in vein and replacement deposits formed at moderate temperatures, at times in sufficient quantity to be important as an ore of copper. It may also occur as a late mineral in some low-temperature deposits. Enargite is usually associated with pyrite, sphalerite, bornite, galena, tetrahedrite, covellite, chalcocite, barite, and quartz.

　　The principal localities for enargite are in Chile, Peru, and Mexico; at Butte, Montana, and at Bingham and Tintic, Utah.

Bournonite, $PbCuSbS_3$

Crystal system and class: Orthorhombic; $2/m\ 2/m\ 2/m$.

Axial elements: $a:b:c = 0.937:1:0897$.

Cell dimensions and content: $a = 8.16$, $b = 8.7$, $c = 7.8$; $Z = 4$.

Common forms and angles:

$(100) \wedge (110) = 43° 09'$	$(010) \wedge (011) = 48° 08'$
$(001) \wedge (101) = 43° 45'$	$(100) \wedge (210) = 25° 06'$

Habit: Crystals usually short prismatic along c to tabular on $\{001\}$; ($hk0$) faces striated parallel to c, and ($h0l$) parallel to b; often subparallel aggregates; massive, granular.

Twinning: Very common on $\{110\}$, often repeated forming cruciform or wheel-like aggregates.

Cleavage: $\{010\}$ imperfect.

Fracture and tenacity: Subconchoidal; brittle.

Hardness: $2\frac{1}{2}$–3.

Density: 5.83.

Color and streak: Steel gray to iron black.

Luster and opacity: Metallic, often brilliant; opaque.

Chemistry: Bournonite may contain up to about 3 percent As; there is no evidence for extended solid solution to the rarer mineral $PbCuAsS_3$ (*seligmannite*).

Alteration: Bournonite alters to antimony oxides, cerussite, malachite, or azurite.

Occurrence: Bournonite is one of the more common sulfosalts, occurring typically in moderate temperature hydrothermal veins, associated with galena, tetrahedrite, sphalerite, chalcopyrite, pyrite, siderite, quartz and more rarely with stibnite, the antimony sulfosalts, rhodochrosite, dolomite, and barite. It contributes to the lead, copper, and antimony recovery from some complex sulfide vein deposits. Well-crystallized material has been found in certain sulfide veins of the old mining districts in Romania, Bohemia, Saxony, and Cornwall, England. It occurs also in sulfide veins in Bolivia and Peru, at Broken Hill in Australia, and in Sonora, Mexico. Large crystals were found with siderite and sphalerite at Park City, Utah, also in Arizona (Boggs Mine), in California (Cerro Gordo Mine), and at Austin, Nevada.

Boulangerite, $Pb_5Sb_4S_{11}$

Crystal system and class: Monoclinic; $2/m$.

Axial elements: $a:b:c = 0.916:1:0.344$; $\beta = 100° 48'$.

Cell dimensions and content: $a = 21.56$, $b = 23.51$, $c = 8.09$, $Z = 8$.

Habit: Crystals long prismatic to acicular and deeply striated parallel to c, rarely terminated by well-developed faces; plumose, fibrous, compact fibrous masses.

Cleavage and tenacity: {100} good; brittle; thin fibers flexible.

Hardness: $2\frac{1}{2}$–3.

Density: 6.23.

Color: Bluish lead gray.

Streak: Brownish gray.

Luster and opacity: Metallic; opaque.

Diagnostic features: Gives blowpipe tests for lead and antimony; boulangerite is similar in physical properties to stibnite and a number of other lead-antimony sulfosalts. It often occurs as the fibrous material called feather-ore, and it is impossible to distinguish boulangerite, with certainty, from several other lead-antimony sulfosalts without quantitative chemical analysis or x-ray study.

Occurrence: It is of frequent occurrence in vein deposits formed at low or moderate temperature, associated with galena, stibnite, sphalerite, pyrite, quartz, siderite, and other lead sulfosalts. It has been found in lead-zinc veins at several localities in Europe, in Idaho, Montana, and Colorado, and in British Columbia.

10

Class III:
Oxides and Hydroxides

H																	
	Be	B													*O*		
	Mg	Al	Si														
	Ca		Ti	V	Cr	Mn	Fe	Co	Ni	Cu	Zn	Ga	*Ge*	As	Se		
		Y	Zr	*Nb*	Mo						Cd	In	Sn	Sb	Te		
		La Lu	*Hf*	*Ta*	W						Hg	Tl	Pb	Bi			
			Th		U												

This large class of minerals includes those compounds in which atoms or cations, usually of one or more metals, are combined with oxygen.* In some of these hydrogen is one of the cations and is present as hydroxyl or as water of hydration. The oxides include primarily those compounds for which detailed studies have shown the presence of ionic bonding of the isodesmic type. Thus, the oxide class does not include those compounds which have discrete anionic

* Table above shows elements which form oxide and hydroxide minerals; those in italics occur only as constituents of complex oxides.

radicals in the structure—the ionic compounds of anisodesmic type, such as carbonates and sulfates. Included here are multiple oxides such as chromite, $FeCr_2O_4$, and columbite, $(Fe,Mn)Nb_2O_6$, which on chemical grounds alone are sometimes referred to as iron chromate and iron-manganese niobate. Since modern crystallochemical studies indicate that these substances are of the isodesmic type, they are here considered as oxides.

The oxides are classified conveniently on the basis of the $A:X$ ratio, and are followed by the hydroxides.

OXIDES

A_2X Type

Cuprite, Cu_2O

AX Type

Periclase Group: Periclase, MgO

Zincite Group: Zincite, ZnO

AB_2X_4 Type

Spinel Group

Spinel, $MgAl_2O_4$

Magnetite, Fe_3O_4

Franklinite, $(Zn,Mn,Fe)(Fe,Mn)_2O_4$

Chromite, $(Mg,Fe)Cr_2O_4$

Hausmannite, $MnMn_2O_4$

Chrysoberyl, $BeAl_2O_4$

A_2X_3 Type

Hematite Group

Corundum, Al_2O_3

Hematite, Fe_2O_3

Ilmenite, $FeTiO_3$

Braunite, $(Mn,Si)_2O_3$

Pyrochlore-Microlite Series:

$NaCaNb_2O_6F$–$(Na,Ca)_2Ta_2O_6(O,OH,F)$

Psilomelane, $(Ba,H_2O)_2Mn_5O_{10}$

AX_2 Type

Rutile Group

Rutile, TiO_2

Cassiterite, SnO_2

Pyrolusite, MnO_2

Wad

Plattnerite, PbO_2

Anatase, TiO_2

Brookite, TiO_2

Columbite-Tantalite, $(Fe,Mn)(Nb,Ta)_2O_6$
Uraninite Group
    Uraninite, UO_2
    Thorianite, ThO_2

HYDROXIDES
    Brucite, $Mg(OH)_2$
Lepidocrocite Group
    Lepidocrocite, $FeO(OH)$
    Boehmite, $AlO(OH)$
      Bauxite
    Manganite, $MnO(OH)$
Goethite Group
    Diaspore, $HAlO_2$
    Goethite, $HFeO_2$
      Limonite
Gibbsite, $Al(OH)_3$

OXIDES

Cuprite, Cu_2O

Crystal system and class: Isometric; 432.

Cell dimensions and content: $a = 4.2696$; $Z = 2$.

Habit: Crystals usually octahedral, less often dodecahedral or cubic; sometimes greatly elongated along a into capillary fibers, known as *chalcotrichite;* also massive, granular, or earthy.

Cleavage: {111} interrupted.

Fracture and tenacity: Conchoidal; brittle.

Hardness: $3\frac{1}{2}$–4.

Density: 6.10.

Color: Red, sometimes nearly black.

Streak: Brownish red.

Luster: Adamantine to submetallic or earthy.

Opacity: Red by transmitted light, yellow in thinner splinters.

Optical properties: $n = 2.849$ (red light).

Diagnostic features: Cuprite is softer than hematite but harder than cinnabar and proustite; the latter two also have a lighter streak. It remains unaltered when heated in the closed tube; on charcoal it blackens, then fuses, and is reduced to a metallic globule of copper. It is soluble in concentrated hydrochloric acid, and on cooling and diluting with cold water yields a heavy, white precipitate of cuprous chloride.

The symmetry of cuprite presents an unusual problem. Crystals frequently show faces of the general forms {*hkl*} which display gyroidal symmetry, and this has long been accepted as the correct crystal class. The structure, one of the first analyzed by Bragg (1915),* apparently possesses the full isometric symmetry $4/m\ \bar{3}\ 2/m$. Etch figures also indicate the higher symmetry.

Alteration: Crystals of cuprite are often altered to malachite pseudomorphs. It may also alter to copper or other secondary copper minerals.

Occurrence: Cuprite is often found in the oxidized zone of copper deposits. It is commonly associated with copper, malachite, azurite, chalcocite, iron oxides, clays, and a black copper oxide (*tenorite*, CuO). Crystals of cuprite are often closely associated with native copper. Cuprite is widespread in its occurrence, especially in those parts of the world where the results of supergene alteration on primary copper deposits have not been destroyed by glaciation. It is sometimes an important ore of copper, as at Bisbee, Arizona. It occurs commonly in the copper deposits of the southwestern United States (Ely, Nevada) and in Chile (Chuquicamata), and also in Australia (Mount Isa) and Africa (Katanga, Congo). Fine specimens have been found in the tin and copper mines of Cornwall, England, and at Chessy, near Lyon, France.

Periclase, MgO

Crystal system and class: Isometric; $4/m\ \bar{3}\ 2/m$.

Cell dimensions and content: $a = 4.213$; $Z = 4$.

Habit: Crystals octahedral; usually in irregular or rounded grains.

Cleavage: {100} perfect.

Hardness: $5\frac{1}{2}$.

Density: 3.58.

Color: Colorless to grayish white, also yellow to brown or black due to iron or foreign inclusions.

Streak: White.

Luster and opacity: Vitreous; transparent.

Optical properties: $n = 1.736$, increasing with iron content.

Diagnostic features: Infusible in the blowpipe flame, and easily soluble in dilute hydrochloric or nitric acid.

Chemistry: Periclase has been found containing as much as about 8 percent Fe substituting for Mg. Artificial MgO forms a complete solid-solution series with FeO. Periclase is a rare mineral, occurring as disseminated grains and clusters of grains in marbles. It is formed by dissociation of dolomitic lime-

* See Selected Readings at the end of Chapter 15

stones in high-temperature metamorphism; it alters readily to brucite or hydromagnesite (a hydrated magnesium carbonate). Periclase has the rock salt (halite) structure. *Lime*, CaO, and *manganosite*, MnO, have the same structure.

Occurrence: Periclase has been recognized in limestone blocks ejected from Monte Somma, Vesuvius, Italy, and in metamorphosed limestones in Sweden and Moravia, and at Crestmore, California, where it is associated with forsterite, magnesite, brucite, chondrodite, or spinel.

Zincite, ZnO

Crystal system and class: Hexagonal; 6mm.
Axial elements: $a:c = 1.6020$.
Cell dimensions and content: $a = 3.249$, $c = 5.205$; $Z = 2$.
Habit: Natural crystals rare, hemimorphic (Fig. 10-1); usually massive, foliated, also compact, in rounded masses.

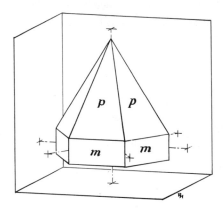

Figure 10-1   Zincite crystal. Forms: hexagonal prism $m\{10\bar{1}0\}$; hexagonal pyramid $p\{10\bar{1}1\}$; pedion $-c\{000\bar{1}\}$ (not visible).

Crystal structure: Zincite has a hemimorphic hexagonal structure which is closely similar to that of the hexagonal zinc sulfide, wurtzite.
Cleavage: $\{10\bar{1}0\}$ perfect, but difficult.
Fracture and tenacity: Conchoidal; brittle.
Harness: 4.
Density: 5.68 (pure).
Color: Orange yellow to deep red; the color of natural ZnO is probably due to manganese (pure ZnO is white).
Streak: Orange yellow.
Luster: Subadamantine.
Optical properties: $\omega = 2.013$, $\epsilon = 2.029$; $\epsilon - \omega = 0.016$; positive.
Diagnostic features: Zincite is infusible before the blowpipe; on charcoal in

the reducing flame it gives a coating of zinc oxide which is yellow while hot and white when cool. The white zinc oxide sublimate turns green when moistened with cobalt nitrate and heated in the oxidizing flame.

Occurrence: Zincite is a rare mineral, apart from its exceptional occurrence in the zinc deposits of Franklin and Sterling Hill, New Jersey, where it amounts to about 1 percent of the ore. Here it occurs with willemite and franklinite in calcite, usually in granular form, rarely as larger masses.

SPINEL GROUP

The term spinel is used generally for a large number of oxides found in nature, in artificial laboratory preparations, and in slags resulting from metallurgical operations. These structures are usually described as double oxides AB_2X_4 in which A is one or more divalent metals (Mg,Fe,Zn,Mn,Ni), B is one or more trivalent metals (Al,Fe,Cr,Mn or Ti^4) and X is oxygen. Most of the natural spinels fall into three series—spinel, in which B is mainly Al; magnetite, in which B is mainly Fe^3; and chromite, in which B is mainly Cr^3. Extensive solid solution is present in the minerals of each series, but solid solution occurs to a lesser extent between minerals that are members of different series. Much broader solid solution variations in composition are found in artificial preparations.

The face-centered cubic unit cell of spinel, with $a = 8.0$–8.5, contains $8[AB_2X_4]$. In structure (Fig. 10-2) the oxygen atoms lie approximately in cubic closest packing. The eight A atoms, which have the same arrangement as the carbon atoms in diamond (Fig. 8-10), are in fourfold coordination between a tetrahedral group of oxygen atoms, and the sixteen B atoms are in sixfold coordination between an octahedral group of oxygen. In turn, each oxygen is linked to one A and three B atoms. This ideal spinel structure presents an anomaly, since in most mineral compositions the B metal ions are smaller than the A and occur in higher coordination with oxygen; this is not in conformity with the rules determining coordination numbers. In some species— $MgFe_2O_4$ (magnesioferrite), $Ti(Fe,Mg)_2O_4$ (ulvöspinel), and some artificial spinels—the fourfold coordinated A positions are occupied by 8 of the smaller trivalent atoms, leaving, in $MgFe_2O_4$, 8 Mg and 8 Fe^3 distributed at random over the 16 sixfold coordinated B positions. In ulvöspinel, Mg and Fe^2 fill the A positions, and Ti^4 with Fe and some Al probably fill the B positions. Many spinels have not been studied in sufficient detail to indicate the type to which they belong.

The occurrence of natural intergrowths of two minerals suggests that solid solution in the spinels is limited in many instances at the temperature of forma-

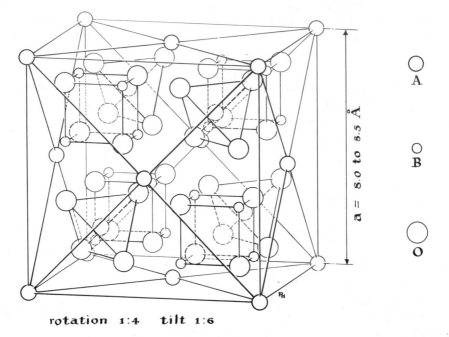

rotation 1:4   tilt 1:6

Figure 10-2 Structural arrangement in spinel. Of the medium circles representing the eight *A* atoms, four are shown in tetrahedral coordination with oxygen (larger circles) as four tetrahedra within the cube. The small circles representing the sixteen *B* atoms are in octahedral coordination with oxygen. The *B* atoms are shown at the alternate corners of four small cubes with edge length $a/4$.

tion, although the solid solution may be more extended or complete at higher temperature. Intergrowths of magnetite and ilmenite indicate that Ti does not readily enter the spinel structure at normal temperatures of formation.

Spinels may also vary in the $A:B$ ratio, as in some synthetic spinels with $MgO \cdot nAl_2O_3$, where n may be greater than unity. This is in agreement with the fact that Al_2O_3 and Fe_2O_3 in their gamma modifications possess the spinel structure. The structure still includes 32 oxygen atoms per unit cell, but some of the *A* and *B* metal sites are vacant, giving an overall composition of B_2O_3; this is an example of a defect structure. The substance γ-Fe_2O_3 is recognized in nature as the mineral *maghemite*.

	SPINEL	MAGNETITE	CHROMITE	FRANKLINITE
Crystal class:	$4/m\ \bar{3}\ 2/m$	$4/m\ \bar{3}\ 2/m$	$4/m\ \bar{3}\ 2/m$	$4/m\ \bar{3}\ 2/m$
Cell dimension:	$a = 8.080$ (pure)	$a = 8.391$	$a = 8.36$	$a = 8.42$
Content:	$8[MgAl_2O_4]$	$8[Fe_3O_4]$	$8[(Mg,Fe)Cr_2O_4]$	$8[(Zn,Mn,Fe)(Fe,Mn)_2O_4]$

	SPINEL	MAGNETITE	CHROMITE	FRANKLINITE
Hardness:	$7\frac{1}{2}$–8	$5\frac{1}{2}$–$6\frac{1}{2}$	$5\frac{1}{2}$	$5\frac{1}{2}$–$6\frac{1}{2}$
Density:	3.581 (pure)	5.20 (pure)	5.09 (pure)	5.07–5.22

Spinel, $MgAl_2O_4$

Habit: Crystals usually octahedral, sometimes modified by the cube or dode-cahedron; also massive, coarse granular to compact, as irregular or rounded grains.

Twinning: Common on {111} (Spinel Law, Fig. 2-40) often flattened on the composition plane (111).

Cleavage: None; indistinct parting on {111}.

Fracture and tenacity: Conchoidal; brittle

Density: 3.581 (pure) increasing with substitution of iron, zinc, or manganese for magnesium.

Color: Variable, red (ruby spinel) to blue, green, brown, nearly colorless; dark green to brown or black for impure varieties.

Streak: White to gray, green or brown.

Luster: Vitreous.

Optical properties: $n = 1.719$, increasing with substitution of Mg and Al by other elements.

Diagnostic features: Spinel is transparent, becoming less so with substitution of Fe^2, Zn, Mn, or Fe^3 in the composition; it is infusible, and insoluble in acids except concentrated sulfuric; fine powder dissolves completely in salt of phosphorus bead.

Chemistry: In spinel, Fe^2, Zn, and less often Mn^2, may be found substituting for Mg in all proportions. Al is the dominant trivalent oxide, but Fe^3, Cr, Mn^3, V, and Ti may substitute for Al to a considerable extent.

Occurrence: The spinels, usually high-temperature minerals, are found as accessory minerals in basic igneous rocks, in highly aluminous metamorp..ic rocks, in contact-metamorphic limestones, rarely in ore veins and pegmatites, in gravels and sands.

Spinels of gem quality, including the rose-red balas ruby, spinel-ruby, and some other colors, are found in contact-metamorphic limestone and the alluvium derived from it mainly in Ceylon, Burma, India, and Afghanistan.

Magnetite, Fe_3O_4

Habit: Crystals octahedral, sometimes dodecahedral; often massive, granular, coarse, or fine.

Twinning: Common on {111} (Spinel Law).

Cleavage: None, commonly shows distinct parting on {111}.

Color: Black.

Streak: Black.

Luster and opacity: Splendent to dull metallic; opaque.

Magnetism: Magnetic; some specimens show polarity (lodestone).

Diagnostic features: Black color and streak and strong magnetic character distinguish magnetite from other minerals high in iron; infusible; dissolves slowly in hydrochloric acid.

Chemistry and structure: Magnetite has the general formula AB_2O_4, in which A is chiefly Fe^2, but Mg, Zn, Mn, and less often Ti and Ni, may substitute for it. B is essentially Fe^3 with small amounts of Al, Cr, Mn^3, and V substituting for it. Whereas in most natural spinels the $A:B$ ratio is close to $1:2$, in magnetite the ratio of $Fe^2:Fe^3$ may decrease toward $\gamma\text{-}Fe_2O_3$ (maghemite). This mineral is isostructural with magnetite, and has the same number of oxygen atoms but some iron positions vacant; it may form as an alteration product of magnetite.

Occurrence: Magnetite is one of the most widespread oxide minerals. It is commonly found (1) as minor accessory mineral in igneous rocks; (2) as a magmatic segregation deposit with apatite and pyroxene, of which important deposits are mined at Kiruna in Sweden and in the Adirondacks, New York; (3) in contact-metamorphic deposits, such as limestones, with garnet, diopside, olivine, pyrite, hematite, and chalcopyrite as at Iron Springs (Utah), at Cornwall (Pennsylvania), and in the Ural Mountains (the USSR); as crystals in chlorite schists; (4) as replacement deposits associated with biotite, amphiboles, epidote, and feldspars; (5) in some high temperature sulfide veins; (6) and as a detrital mineral in beach or river sands.

Magnetite is often found in deposits large enough for commercial exploitation. Such ores often contain P as apatite, Ti as ilmenite, and S in sulfides, and require special treatment. Important magnetite ores such as those at Kiruna in Sweden rank that country sixth in world iron ore production. Large deposits of magnetite are also mined in the Urals, and at other localities as mentioned above.

Chromite, $(Mg,Fe)Cr_2O_4$

Habit: Crystals rare, octahedral; commonly massive, fine granular to compact.

Cleavage: None.

Fracture and tenacity: Uneven, brittle.

Color: Black.

Streak: Brown.

Luster and opacity: Metallic, almost opaque.

Diagnostic features: Weak magnetism and brown streak distinguish chromite from magnetite: infusible; gives green borax bead; commonly associated with serpentine.

Chemistry: Chromites approximate the composition $(Mg,Fe)Cr_2O_4$, with 8 formula units per unit cell (Fig. 10-3). In the majority of nearly 200 analyses assembled by Stevens (1944) Mg ranges from 4 to 6 atoms, Fe^2 from 2 to 4, Cr from 7 to 14, Al from two to nine, and Fe^3 from zero to two atoms per unit cell. A very few analyzed samples had from 3 to 11 atoms of Fe^3 per unit cell

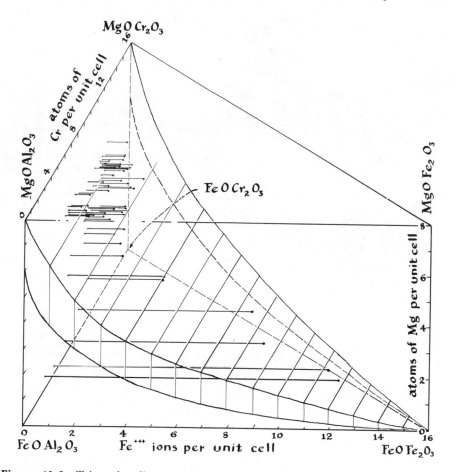

Figure 10-3 Triangular diagram, showing the compositional range of natural chromites in terms of six end-member components. Observed compositions are represented as points at the extremities of each horizontal line. These points lie in a blanket-like space extending from the $MgO \cdot Al_2O_3$—$MgO \cdot Cr_2O_3$ join on the upper left to the $FeO \cdot Fe_2O_3$ corner at the lower right. Most points fall toward the upper rear corner of the blanket, corresponding with the range of composition indicated above. [After Stevens, R. E., *Am. Mineralogist*, v. 29, p. 1, 1944.]

(with low Mg content), which provides some evidence for extensive solid solution between chromite and magnetite. The Cr_2O_3 content of most analyzed chromites ranges from 30 percent to 61 percent (maximum 67.91 percent in pure ferrochromite, and 79.04 percent in magnesiochromite). Metallurgical grade chromite must have a minimum ratio of Cr:Fe of 2.5:1, which corresponds roughly to 42 percent Cr_2O_3. Chromite ore with 33–48 percent Cr_2O_3 and 12–30 percent Al_2O_3 is suitable for chrome refractory brick.

Figure 10-4 Franklinite, crystals with large octahedron faces $o\{111\}$ and minor dodecahedron faces $d\{110\}$, embedded in calcite. (Franklin, New Jersey.) [Courtesy Royal Ontario Museum.]

Occurrence: Chromite is usually found as an accessory mineral in ultrabasic igneous rocks, such as peridotites, and in serpentinites derived from them. The mineral may occur as segregated masses, lenses, or as disseminated grains in sufficient quantity to serve as an ore. In a few localities it has been concentrated in detrital sands. Olivine, serpentine, pyroxene, chromian spinel, chromium garnet (uvarovite), chromian chlorites, magnetite, and pyrrhotite are often associated with chromite.

Production and use: Chromite is the only ore mineral of chromium, which is of great value as a ferro-alloy metal, especially in stainless steels. The mineral chromite is also used for making refractories.

The major producing countries for chromite are Turkey, the Union of South Africa, the USSR, the Philippines, Rhodesia, Yugoslavia, Albania, and Iran (aggregating about 90 percent of the annual world production of about $4\frac{1}{2}$ million tons). In Rhodesia layered bands of chromite occur with ultrabasic igneous rocks, now largely altered to serpentine, in the Great Dike. Platinum is also found in some layers of this rock. In South Africa chromite is found in similar layered bands within the Bushveld Igneous Complex.

Small amounts of chromite are mined in California, Oregon, and other states where it occurs in serpentines. During wartime minor amounts of chromite were mined in the asbestos-serpentine area of Quebec.

Franklinite, $(Zn,Mn,Fe)(Fe,Mn)_2O_4$

Habit: Crystals octahedral (Fig. 10-4); also massive.
Cleavage: None; parting on {111}.
Color: Black to brownish black.
Streak: Reddish brown.
Luster: Metallic to semimetallic.
Magnetism: Weak.
Chemistry: Franklinite is dominantly $ZnFe_2O_4$, but is always found with Mn^2 and Fe^2 substituting for Zn, and Mn^3 substituting for Fe^3.
Occurrence: It is the dominant ore mineral of the zinc deposits of Franklin and Sterling Hill, New Jersey, where it forms thick beds in crystalline lime stone. It is associated with zincite, willemite, and rhodonite.

OTHER OXIDES OF AB_2X_4 TYPE

Hausmannite, $MnMn_2O_4$

Crystal system and class: Tetragonal; $4/m\ 2/m\ 2/m$.
Axial elements: $a:c = 1:1.639$.
Cell dimensions, and contents: $a = 5.76$, $c = 9.44$; $Z = 4$.
Common forms and angles:
  (101) $\wedge$ (10$\bar{1}$) $= 62°\ 46'$ (001) $\wedge$ (103) $= 28°\ 39'$
Habit: Crystals often pseudo-octahedral with dipyramid {101}; also massive granular.
Twinning: Common on {112}, repeated.
Crystal structure: Hausmannite has a distorted spinel structure; the degree of distortion decreases on heating, and above about 1160° Mn_3O_4 is isometric.
Cleavage: {001} perfect.
Fracture and tenacity: Uneven; brittle.
Hardness: $5\frac{1}{2}$.
Density: 4.84.
Color: Brownish black.

Streak: Chestnut brown.

Luster and opacity: Submetallic; almost opaque.

Diagnostic tests: Infusible; soluble in hot hydrochloric acid, with evolution of chlorine.

Occurrence: Hausmannite is found typically in high temperature hydrothermal veins, in contact-metamorphic deposits, and as a product of recrystallization of sedimentary or residual manganese deposits. Probably it is an important constituent of manganese deposits formed by circulating meteoric waters and in residual deposits.

    Hausmannite is found at numerous manganese-mineral localities in Europe, particularly in Germany (Ilmenau, Ilfeld), and Sweden (Långban), also in the USA in the Batesville district of Arkansas, with pyrolusite and psilomelane.

Chrysoberyl, $BeAl_2O_4$

Crystal system and class: Orthorhombic; $2/m\ 2/m\ 2/m$.

Axial elements: $a:b:c = 0.5823:1:0.4707$.

Cell dimensions and content: $a = 5.476$, $b = 9.404$, $c = 4.427$; $Z = 4$.

Common forms and angles:

    $(001) \wedge (111) = 43° 05'$ $(010) \wedge (110) = 59° 47'$

    $(001) \wedge (121) = 51° 08'$ $(010) \wedge (130) = 29° 48'$

Habit: Crystals tabular on $\{001\}$.

Twinning: Common on $\{130\}$ as twin plane; contact or penetration; often repeated, forming pseudo-hexagonal crystals (Fig. 10-5).

Cleavage: $\{110\}$ distinct; $\{010\}$ imperfect.

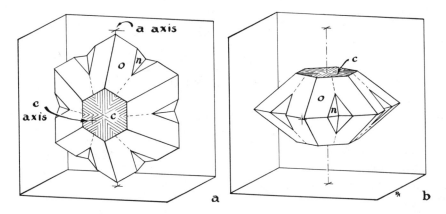

Figure 10-5 Chrysoberyl, twinned crystal. Forms: pinacoid $c\{001\}$; rhombic dipyramids $o\{111\}$, $n\{121\}$; cyclic twin on $\{130\}$ as twin plane and composition plane.

Crystal structure: The formula of chrysoberyl is similar to the spinel formula, but the symmetry is orthorhombic; although the structure is similar to spinel, in that Be is in fourfold coordination with O and Al is sixfold coordination with O, it is isostructural with olivine—Be and Si, Al and Mg occupy corresponding positions in the two structures; the hexagonal close-packed oxygen arrangement results in a pseudo-hexagonal lattice with $a:b$ close to $1:\sqrt{3}$ and common twinning on $\{130\}$.

Hardness: $8\frac{1}{2}$.

Density: 3.75.

Color: Green, greenish white, and yellowish green.

Streak: White.

Luster and opacity: Vitreous; transparent.

Optical properties: $\alpha = 1.745$, $\beta = 1.748$, $\gamma = 1.754$; $\gamma - \alpha = 0.009$; positive, $2V = 45°$.

Diagnostic features: Color, hardness, twinning; infusible and insoluble; fine powder dissolves in salt of phosphorus bead.

Occurrence and use: Chrysoberyl is usually found in granite pegmatite and aplite, also in mica schists, and often as a detrital mineral with diamond, corundum, garnet, cassiterite. When transparent and of good color it is used as a gem. The emerald-green variety *alexandrite*, which is red by transmitted light and usually green by artificial light, is prized as a gem. Another gem variety exhibits a chatoyant effect due probably to oriented, needle-like inclusions.

Chrysoberyl occurs in pegmatites in Maine (Newry) and New Hampshire (Orange Summit), in Madagascar, in Germany, and in Italy. It is found in mica schists in the Ural Mountains, and in placer gravels in Ceylon, Rhodesia, and Japan.

HEMATITE GROUP

This group includes corundum, hematite, and the ilmenite series. All are rhombohedral and virtually isostructural; the first two have symmetry $\bar{3}\ 2/m$ and the members of the ilmenite series, $FeTiO_3$, (ilmenite), $MgTiO_3$, $MnTiO_3$, have the lower symmetry $\bar{3}$.

The rhombohedral unit cell of these structures has $a:c = 1:2.72–1:2.79$, in which c is double the value used in the past for describing crystals of these minerals. The structure contains $2[Al_2O_3]$ in the rhombohedral unit or $6[Al_2O_3]$ in the hexagonal unit. The oxygen atoms are arranged approximately in hexagonal closest packing with the close-packed layers parallel to $\{0001\}$. Cations may lie between these layers, each coordinated to six oxygen atoms. If all such spaces were filled, a composition AX would result, but in Al_2O_3 only two-thirds of these positions are occupied by cations, each in octahedral co-ordination with three oxygen atoms above (along c) and three below. Groups of three oxygen atoms form a triangular face parallel to $\{0001\}$ and common to two octahedra. Thus three oxygen atoms with an aluminum above and below constitute Al_2O_3 groups with the form of a trigonal dipyramid. There are two orientations of these groups in the structure, which are related by the vertical symmetry planes (glide planes in this case).

In the ilmenite minerals, where the cation positions in corundum and hematite are occupied in an ordered fashion by two cations, Fe and Ti, Mg and Ti, or Mn and Ti, the symmetry plane parallel to $\{11\bar{2}0\}$ is not present, and the symmetry is $\bar{3}$. Both cations are in octahedral coordination with oxygen. Corundum and hematite show very little solid solution but ilmenite forms a series in which Mg or Mn may substitute for Fe in all proportions. The high magnesium and manganese minerals of the ilmenite series are of rare occurrence and will not be discussed further here.

	CORUNDUM	HEMATITE	ILMENITE
Crystal system:	Trigonal	Trigonal	Trigonal
Symmetry:	$\bar{3}\,2/m$	$\bar{3}\,2/m$	$\bar{3}$
Axial elements:	$a:c = 1:2.7303$	$a:c = 1:2.7307$	$a:c = 1:2.7606$
Cell dimensions:	$a = 4.758$	$a = 5.039$	$a = 5.093$
	$c = 12.991$	$c = 13.76$	$c = 14.06$
Content:	$6[Al_2O_3]$	$6[Fe_2O_3]$	$6[FeTiO_3]$
Cleavage	None	None	None
Parting:	$\{0001\}$ $\{01\bar{1}2\}$	$\{0001\}$ $\{01\bar{1}2\}$	$\{0001\}$ $\{01\bar{1}2\}$
Hardness:	9	5–6	5–6
Density:	4.0–4.1; 3.98 (pure)	5.26; 5.256 (pure)	4.72; 4.79 (pure)

Corundum, Al_2O_3

Common angles and forms:

(0001) $\wedge$ (11$\bar{2}$3) = 61° 11'   (0001) $\wedge$ (01$\bar{1}$2) = 57° 35'

(0001) $\wedge$ (11$\bar{2}$1) = 79° 37'   (0001) $\wedge$ (10$\bar{1}$1) = 72° 23'

Habit: Crystals commonly tabular on $\{0001\}$ to short prismatic along c with $\{11\bar{2}0\}$ (Fig. 10-6), also steep pyramidal with $\{11\bar{2}3\}$, $\{11\bar{2}1\}$, and others; commonly rough, rounded, barrel-shaped crystals (Figs. 10-6c, 10-7a); rarely rhombohedral; prism and pyramid faces horizontally striated due to oscillatory combination of faces with different slope; massive granular (emery), in rounded grains.

Twinning: Common on $\{01\bar{1}2\}$, and $\{0001\}$, often lamellar producing a lamellar structure and striations on $\{0001\}$ and $\{01\bar{1}2\}$ (Fig. 10-7b).

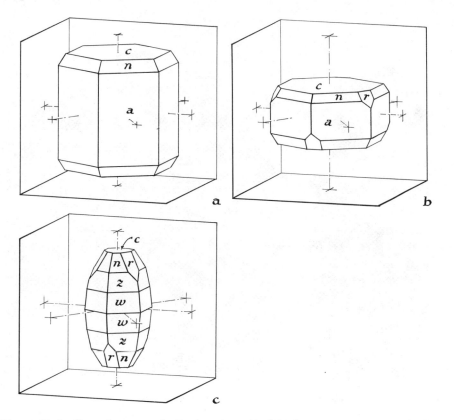

Figure 10-6 Corundum crystals. Forms: pinacoid $c\{0001\}$; hexagonal prisms $a\{11\bar{2}0\}$; negative rhombohedron $r\{01\bar{1}2\}$, hexagonal dipyramids $n\{11\bar{2}3\}$, $z\{11\bar{2}1\}$, $\omega\{7.7.\bar{14}.3\}$.

Parting: Common on $\{0001\}$ and $\{01\bar{1}2\}$, with marked striations on the parting planes.

Tenacity: Brittle, very tough when compact.

Color: Blue (sapphire), pink to blood red (ruby) also yellow, yellow brown, green, purple to violet; some crystals show color zones (Fig. 10-7b); pure Al_2O_3 is white.

Streak: White.

Luster and opacity: Adamantine to vitreous; transparent to translucent.

Optical properties: $\epsilon = 1.760$, $\omega = 1.768$; $\omega - \epsilon = 0.008$; negative.

Diagnostic features: Hardness; crystal form (barrel shape is characteristic), parting; infusible; insoluble; fine powder moistened with cobalt nitrate and ignited intensely gives a blue color due to aluminum.

Occurrence and use: Corundum has long been recognized as an important mineral because of its value both as an abrasive and as a gemstone. Among natural minerals corundum is next in hardness to diamond; it has also

Figure 10-7 Corundum. (a) Rough, greenish brown, barrel-shaped crystal, showing a hexagonal dipyramid close to $\{44\bar{8}3\}$. (Locality unknown.) (b) Basal parting plane (0001), showing striations due to the trace of parting planes $\{01\bar{1}2\}$ as horizontal lines, and of parting planes $\{1\bar{1}02\}$ at 30° to right of vertical; the parting surface $\{01\bar{1}2\}$ shows at the bottom of the picture; vertical lines and those inclined 60° to left of vertical are lines of differing growth color due to zones parallel to the prism $\{11\bar{2}0\}$ or a hexagonal dipyramid $\{h \cdot h \cdot \overline{2h} \cdot l\}$. Color bronze-brown. (Craigmont, Raglan Township, Ontario.) [Courtesy Queen's University.]

found wide applications as an abrasive in the form of emery. Emery is more properly the name for a black or gray-black rock which is a granular mixture of corundum with magnetite or hematite and spinel.

Corundum of gem quality is found in a wide variety of colors: yellow (oriental topaz), green (oriental emerald), purple (oriental amethyst), blue (sapphire), red (ruby). The color of ruby is apparently due to a small Cr content, and the color of sapphire to Fe or Ti. The varieties star sapphire and star ruby are highly prized for the six-rayed star formed by internal reflection in a strong light. Most varieties prized for gem purposes are transparent and should be free of the parting so common to some corundum. Good quality corundum which is not up to the standards of color and transparency required for gems is widely used for jewel bearings in watches and other instruments.

Corundum is most often found in rocks which have a lower silica and higher alumina content than that required for feldspar. These rocks are syenites containing feldspar with nepheline, sodalite, and often corundum, but no quartz. Corundum is a common constituent of nepheline syenites and nepheline-feldspar pegmatites, at Bancroft and Craigmont, Ontario. It may result from recrystallization or metamorphism of highly aluminous rocks in contact with igneous rock, as in the metamorphism of bauxite. The finest gem corundum is found in Burma in recrystallized limestone and in the deeply weathered surface zone; placer deposits in Ceylon yield most of the near-gem material used in industry.

Emery has long been obtained commercially from Naxos, Samos, and other Greek islands; it has also been mined at Chester, Massachusetts.

Large amounts of artificial corundum (alundum) for abrasive purposes are produced by melting bauxite in an electric furnace. This material and artificial silicon carbide have displaced natural corundum for many abrasive uses. Emery is still preferred by many for the fine-grinding of glass in the manufacture of precision optical components, and natural corundum is preferred for some abrasive-wheel applications. Artificial rubies and sapphires can also be made by melting and recrystallizing alumina in an oxy-hydrogen flame. Gems cut from carefully prepared artificial "boules" colored by chromium, cobalt, or titanium salts are not easily distinguished from natural stones.

Hematite, Fe_2O_3

Common forms and angles:

$(0001) \wedge (01\bar{1}2) = 57° 36'$	$(0001) \wedge (01\bar{1}8) = 21° 30'$
$(0001) \wedge (11\bar{2}3) = 61° 13'$	$(0001) \wedge (11\bar{2}1) = 79° 37'$

Habit: Crystals thick to thin tabular on $\{0001\}$, often as subparallel growths on $\{0001\}$, also rhombohedral (Fig. 10-8); micaceous to platy, compact columnar, fibrous and radiating, in reniform masses with smooth fracture and in botryoidal or stalactitic shapes, commonly earthy and frequently admixed with clay and other impurities, also granular friable to compact, concretionary, or oolitic.

Twinning: On $\{0001\}$ as penetration twins and on $\{01\bar{1}2\}$ usually lamellar.

Parting: On $\{0001\}$ and $\{01\bar{1}2\}$ due to twinning.

Fracture and tenacity: Uneven; crystals very brittle; thin laminae elastic.

Hardness: 5–6, quite soft in earthy varieties.

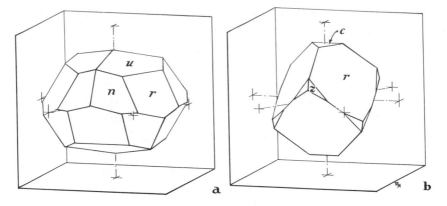

Figure 10-8   Hematite crystals. Forms: pinacoid $c\{0001\}$; negative rhombohedrons $r\{01\bar{1}2\}$, $u\{01\bar{1}8\}$; hexagonal dipyramids $n\{11\bar{2}3\}$, $z\{11\bar{2}1\}$.

Color: Steel-gray (crystals and specularite), sometimes iridescent; earthy and compact material dull to bright red.

Streak: Red or red brown.

Luster and opacity: Metallic (crystals and specularite) to submetallic in dull earthy varieties; thin splinters deep blood red by transmitted light.

Chemistry: Hematite varies but little from the ideal composition, with 69.94 percent Fe. Some Ti can be present, and up to several percent of water may be present in the fibrous or ocherous varieties.

Diagnostic features: Hematite characteristically shows a red-brown streak; the color in massive and earthy varieties is also red to red brown, whereas in the variety called specular hematite or specularite the color is black and the luster metallic and splendent. The specular material may be foliated or micaceous and very brittle; in this form the red-brown color is difficult to obtain on the streak plate. It is infusible in the blowpipe flame, and becomes magnetic in the reducing flame, then nonmagnetic again when heated in the oxidizing flame. It is soluble in concentrated hydrochloric acid.

Occurrence and use: Hematite is the most important and widely used source mineral for iron. It is exceeded in iron content by magnetite, but it is much more abundant and provides over 90 percent of the iron produced in North America. As a primary mineral, hematite is frequently found as an accessory in igneous rocks, and as a minor constituent in high-temperature hydrothermal veins and in contact metamorphic deposits associated with magnetite.

The huge deposits of hematite worked as iron ore are mainly of sedimentary origin, with subsequent concentration or enrichment by meteoric waters or possibly by hydrothermal solutions. The extensive Clinton iron beds of Upper Silurian age outcrop over a length of about 700 miles in New York,

Pennsylvania, and Alabama. These deposits are sedimentary beds of oolitic hematite, associated with clay and limonite. They are probably of marine shallow-water deposition. The iron, resulting from weathering of surface rocks, was carried by sluggish streams as bicarbonate, or in colloidal form, and deposited in shallow, quiet seas. Hematite was quickly precipitated when the fresh stream water mixed with salt water. Some of the hematite is found coating shell fragments or as hematite muds. Similar beds of Ordovician age outcrop in the Wabana basin on the east coast of Newfoundland and they extend, with a low angle dip, beneath the ocean bottom for several miles.

To the south and west of Lake Superior, extensive deposits of hematite have been mined; the principal districts are: Marquette and Menominee in Michigan; Penokee-Gogebic in Wisconsin; and Mesabi, Vermilion, and Cuyuna in Minnesota. The richest portions of these districts, consisting of hematite varying from hard specular to earthy types, with minor limonite and magnetite, have resulted from enrichment of hard ferruginous siliceous sediments of Precambrian age. These original iron-bearing sediments have been regionally metamorphosed and are generally known as taconite or banded iron-formation. These rocks are now composed of silica as chert, and iron as hematite, magnetite, siderite, or greenalite. The taconites average about 25 percent iron and are now being developed as a commercial source of iron. The enrichment resulted from removal of silica and carbonates by meteoric or hydrothermal waters, leaving a high-grade iron ore as a residue. The Lake Superior districts have produced about 2,500 million tons of high-grade hematite, of which the annual output constitutes nearly 80 percent of the iron ore obtained in the USA.

High-grade hematite deposits are also found in England, Brazil, Venezuela, Australia, Liberia, China, and Canada.

Extensive hematite deposits have resulted from surface weathering and leaching of limestones containing siderite lenses in North Africa, and from leaching of serpentine in Cuba.

The average annual world production of iron ore during the years 1961–1965 was about 500 million tons. Large producers are the USA (15 percent), the USSR (25 per cent), France (11 percent), China (6 percent), and Canada (6 percent). The importance of iron and steel in modern life is evident to everyone, since these substances are essential in nearly all forms of construction, transport, communications, and manufacturing. A continued supply of iron ore, available at a reasonable cost, is essential for the maintenance of present-day living and commerce. The tremendous tonnages of ore in reserve in the newly developed Quebec-Labrador iron belt assure North America of adequate supplies for many years to come.

Ilmenite, FeTiO₃

Common forms and angles:

$(0001) \wedge (01\bar{1}2) = 57° 54'$ $(0001) \wedge (11\bar{2}3) = 61° 29'$

$(0001) \wedge (10\bar{1}4) = 38° 33'$

Habit: Crystals commonly thick tabular on $\{0001\}$ (Fig. 10-9), with hexagonal prisms $\{10\bar{1}0\}$, $\{11\bar{2}0\}$, and one or more rhombohedrons; commonly compact massive, as embedded grains or loose in sand.

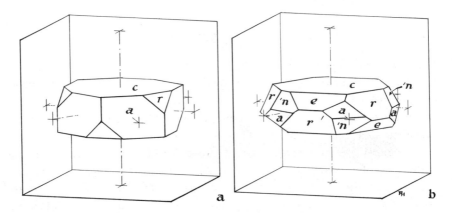

Figure 10-9 Ilmenite crystals. Forms: pinacoid $c\{0001\}$; hexagonal prism $a\{11\bar{2}0\}$; negative rhombohedron $r\{01\bar{1}2\}$; positive rhombohedron $e\{10\bar{1}4\}$; negative rhombohedron $n'\{11\bar{2}3\}$.

Twinning: Common on $\{0001\}$, also lamellar on $\{01\bar{1}2\}$.

Parting: On $\{0001\}$ and $\{01\bar{1}2\}$.

Fracture: Conchoidal.

Color: Iron black.

Streak: Black.

Luster: Metallic to submetallic.

Chemistry: Ilmenite is essentially FeTiO₃ with magnesium substituting for Fe and grading into *geikielite* (Mg,Fe)TiO₃. Manganese also substitutes for iron, although less commonly.

Diagnostic features: Ilmenite is readily distinguished from hematite by its black streak and from magnetite by its nonmagnetic character. Ilmenite often appears slightly magnetic due to intergrown magnetite. Like hematite, ilmenite is infusible in the blowpipe flame; it will yield a test for titanium by the method given on p. 187.

Occurrence and use: It occurs principally in close association with gabbros, diorites, and anorthosite rocks as veins, disseminated deposits, or large masses.

It is also found as an accessory mineral in igneous rocks, in pegmatites, and occasionally in quartz veins with chalcopyrite and hematite. Some heavy black beach sands in India and Florida contain important amounts of ilmenite.

The important sources are: the ilmenite-hematite masses in anorthosite near Allard Lake, Quebec (the largest body contains 125 million tons of ore with 35 percent TiO_2 and 40 percent iron); a body of ilmenite-magnetite in Norway (Blaafjeldite, Sogndal); the titaniferous magnetite deposits with anorthosite and gabbro in the Adirondacks of New York (Elizabethtown); and the beach placers of Australia. In Virginia (Roseland) ilmenite is found in massive dikes with rutile and apatite, and in the adjacent syenite.

The annual world production of ilmenite, a little over 2 million tons (1961–1965 average), is obtained from the USA (40 percent), Australia (13 percent), Norway (12 percent), Canada (21 percent), and other countries.

OTHER OXIDES OF A₂X₃ TYPE

Braunite, $(Mn,Si)_2O_3$

Crystal system and class: Tetragonal; $4/m\ 2/m\ 2/m$.
Axial elements: $a:c = 1:1.9904$.
Cell dimensions, and content: $a = 9.38$, $c = 18.67$; $Z = 8$, with $Mn:Si = 7:1$.
Common forms and angles:
  (001) ∧ (112) = 54° 36′   (001) ∧ (213) = 56° 01′
  (213) ∧ (123) = 30° 24′   (001) ∧ (102) = 44° 52′
Habit: Crystals pyramidal, with {101} and {311}; also granular massive.
Twinning: On {112}, contact twins.
Cleavage: {112}, perfect.
Fracture and tenacity: Uneven; brittle.
Hardness: $6–6\frac{1}{2}$.
Density: 4.72–4.83.
Color and streak: Dark brownish black to steel gray.
Luster and opacity: Submetallic; opaque.
Diagnostic features: Cleavage form distinct from other manganese minerals hausmannite and pyrolusite; infusible; soluble in HCl with evolution of Cl, leaving a residue of gelatinous silica.
Occurrence: Braunite is commonly associated with other manganese oxide minerals and barite; in veins and lenses resulting from the metamorphism of manganese oxides and silicates; as a secondary mineral formed under weathering conditions with pyrolusite and psilomelane.

Braunite has been noted at numerous localities, and it often becomes an ore mineral for manganese. Good crystals have been found at Ohrenstock near Ilmenau, Thuringia, at Ilfeld in the Harz Mountains (Germany), at St. Marcel, Piedmont (Italy), at Jacobsberg and Långban (Sweden), at Nagpur (India). Other localities are Miguel Burnier near Ouro Preto, Minas Geraes (Brazil); Nombre de Dios (Panama), where it occurs with pyrolusite and psilomelane; Spiller Manganese mine, Mason County, Texas, where it occurs as lenses in quartzite; and the Batesville district, Arkansas.

Pyrochlore-Microlite Series: $NaCaNb_2O_6F-(Na,Ca)_2Ta_2O_6(O,OH,F)$

Crystal system and class: Isometric; $4/m\ \bar{3}\ 2/m$.

Cell dimensions and content: $a = 10.37–10.41$; $Z = 8$.

Habit: Crystals usually octahedral, also in irregular masses or embedded grains.

Cleavage or parting: {111} sometimes distinct.

Fracture and tenacity: Subconchoidal to uneven; brittle.

Hardness: $5–5\frac{1}{2}$.

Density: 4.2–6.4, increasing with tantalum content.

Color: Brown with yellowish or reddish shades to black (pyrochlore); pale yellow to brown, sometimes red, olive buff or green (microlite).

Luster and opacity: Vitreous to resinous; dark varieties nearly opaque.

Optical properties: $n = 1.9–2.2$, usually isotropic.

Diagnostic features: Crystal form; often radioactive; usually infusible.

Chemistry: Most specimens of these minerals are metamict, but an isometric structure like that of the nonmetamict material is restored on heating. A large number of elements, including rare earths, uranium, and thorium, are found in substitution for Na and Ca. Nb predominates in pyrochlore and Ta in microlite; Ti is present in rather minor amounts.

Occurrence: Pyrochlore is usually found associated with zircon, apatite, and other rare-earth minerals in pegmatites derived from alkalic rocks, particularly in Norway, and in Hastings County, Ontario. It also occurs as an accessory mineral in nepheline syenite (and other alkalic dike rocks) and in carbonates associated with alkalic intrusives, as in the Alnö region of Sweden, in East Africa, at Oka in Quebec, and at Fen in Norway. Pyrochlore has been mined as a source of niobium.

Microlite is often found with tantalite or columbite in the albitized parts of some granite pegmatites, particularly in Norway, at Varuträsk (Sweden), and in the northeastern USA. It was mined as a source of tantalum from the Harding pegmatite in New Mexico during World War II. It is also found in placers in Western Australia.

Psilomelane, $(Ba,H_2O)_2Mn_5O_{10}$

Crystal system and class: Monoclinic; $2/m$.

Axial elements: $a:b:c = 3.319:1:4.809$; $\beta = 92° 30'$.

Cell dimensions and contents: $a = 9.56$, $b = 2.88$, $c = 13.85$.

Habit: Massive as botryoidal, reniform, or mamillary crusts; stalactitic, also earthy.

Hardness: 5–6.

Density: 4.7.

Color: Iron black to dark steel gray.

Streak: Brownish black to black.

Luster and opacity: Submetallic; opaque.

Diagnostic tests: Hardness, botryoidal form, infusible; gives manganese test with the sodium carbonate bead; flame test for barium.

Chemistry: Psilomelane is one of a number of hard, hydrous manganese oxides occurring as fine-grained, massive, or botryoidal material. The name was formerly applied to material of this sort from many localities, but modern methods of investigation have segregated other distinct mineral species in these hard ores. Psilomelane now is used for the mineral in which Ba predominates, *cryptomelane* for that with K, *coronadite* for that with Pb.

Occurrence: Psilomelane is typically a secondary mineral formed under surface conditions of temperature and pressure, often associated with pyrolusite, goethite, and hausmannite. It may be found as large residual deposits resulting from the weathering of manganous carbonates or silicates, also as concretionary masses in clays and in lake and swamp deposits.

RUTILE GROUP

The rutile group includes rutile, TiO_2, pyrolusite, MnO_2, cassiterite, SnO_2, and the rare mineral plattnerite, PbO_2. The members of the group are isostructural but show almost no tendency to form series between the species, probably because of the rather large differences in the radii of the metal ions. Rutile and cassiterite occur in granite rocks, formed probably at high temperature, whereas pyrolusite forms at much lower temperatures. Anatase and brookite are polymorphous forms of TiO_2, but these structures are not found in any other minerals, though the structure of columbite is related to that of brookite.

The structure of cassiterite as shown here (Fig. 10-10) is typical of the structure of this group. This structure is tetragonal with $2[SnO_2]$ per unit cell. The metal atoms are located at the points of a body-centered tetragonal lattice. Each metal is coordinated to six oxygens, two along [110] in line with the

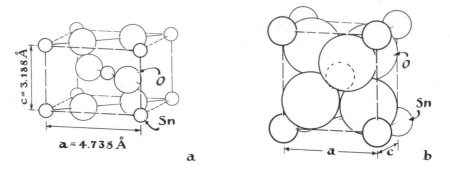

Figure 10-10 Structure of cassiterite. Each Sn atom is octahedrally coordinated to six oxygens: (a) atoms shown as small circles, (b) atoms shown in their true relative sizes.

body-centered metal atom, two above and two below along [1$\bar{1}$0] in line with the corner metal atoms. The octahedrons lie with two edges parallel to *c* and two parallel to [110] or [1$\bar{1}$0]; they form chains parallel to *c* linked by shared edges. The chains containing the corner metal atoms have the shared edges parallel to [110], and those containing the body-centered metal atoms have their shared edges parallel to [1$\bar{1}$0]. The chains are linked by sharing corner oxygen atoms. This structure is similar to, but more symmetrical than, that of marcasite.

Analyses of rutile occasionally indicate a minor tin content, but more commonly they show substantial amounts of iron (to about 11 percent Fe_2O_3 or 15 percent FeO), niobium (to 32 percent Nb_2O_5), or tantalum (to 35 percent Ta_2O_5). Fe^2 occurs with Nb or Ta. If this content of Fe, Nb, and Ta is present in solid solution in all cases, the composition leads to a general formula $Fe_x(Nb,Ta)_{2x}Ti_{1-3x}O_2$, where *x* reaches a maximum of 0.2.

Recent studies on niobian rutile (ilmenorutile) have shown that several specimens are in fact intergrowths of rutile and columbite; therefore some chemical analyses recorded in the literature undoubtedly represent a mixture of minerals. In contrast, an iron tantalum mineral from Ross Lake, Northwest Territories, has the rutile structure.

Artificial $FeTa_2O_6$ and the minerals *mossite-tapiolite*, $Fe(Nb,Ta)_2O_6$, have the trirutile structure in which the *c* length is three times the *c* length of rutile. This structure is dimorphous with columbite-tantalite $Fe,Mn(Nb,Ta)_2O_6$. Tapiolite, in which Ta > Nb, has been identified in pegmatites in Mecejana, Ceará, Brazil; and near Tazenakht, Morocco; Ross Lake, Northwest Territories; Custer County, South Dakota; Kimito, Finland. Members of this series with Nb > Ta (mossite) are of rare or doubtful occurrence. In rutile with appreciable iron and niobium or tantalum the structure is a disordered form of the trirutile structure.

Artificial $FeNb_2O_6$, $MnNb_2O_6$, $MnTa_2O_6$, and many natural occurrences of columbite-tantalite have an orthorhombic structure. Minerals with $Nb > Ta$ greatly predominate; some supposed occurrences of tantalite have proved to be tapiolite, while rare occurrences of minerals high in manganese and tantalum are tantalite.

	RUTILE	PYROLUSITE	CASSITERITE	TAPIOLITE
Crystal system:	Tetragonal	—	—	—
Crystal class:	$4/m\ 2/m\ 2/m$	—	—	—
Axial elements:	$a:c$ $= 1:0.6439$	$a:c$ $= 1:0.651$	$a:c$ $= 1:0.6728$	$a:c$ $= 1:1.9411$
Cell dimensions:	$a = 4.594$ $c = 2.958$	$a = 4.39$ $c = 2.86$	$a = 4.738$ $c = 3.188$	$a = 4.754$ $c = 9.228$
Cell content:	$2[TiO_2]$	$2[MnO_2]$	$2[SnO_2]$	$2[FeTa_2O_6]$
Cleavage:	$\{110\}$ distinct	$\{110\}$ perfect	$\{100\}$ imperfect	None
Hardness:	$6-6\frac{1}{2}$	$6-6\frac{1}{2}$	$6-7$	$6-6\frac{1}{2}$
Density:	4.25 (calc.)	5.06	6.99 (calc.)	8.17 (calc.)
Common color:	Brown	Black	Brown	Black
Optical properties:	$\omega = 2.621,$ $\epsilon = 2.908$ $\epsilon - \omega$ $= 0.287$ positive	Opaque	$\omega = 2.006,$ $\epsilon = 2.097$ $\epsilon - \omega$ $= 0.091$ positive	Nearly opaque

Rutile, TiO_2

Common forms and angles:
  $(110) \wedge (111) = 47° 41'$ $(100) \wedge (101) = 57° 13'$
  $(011) \wedge (0\bar{1}1) = 65° 34'$ $(010) \wedge (130) = 38° 26'$

Habit: Crystals commonly prismatic to slender or acicular; prism faces vertically striated, usually terminated by dipyramids $\{101\}$ or $\{111\}$ (Figs. 10-11, 10-12); also granular massive.

Twinning: Very common on $\{101\}$; simple contact (Figs. 10-11c, 10-12b); repeated contact twins giving geniculate forms (Figs. 10-11b, 10-12d), or cyclic forms with 6 or 8 individuals (Figs. 10-11d, 10-12c, 2-41); also polysynthetic.

Cleavage: {110} distinct, {100} less distinct.

Fracture and tenacity: Conchoidal to uneven; brittle.

Color: Reddish brown, rarely yellowish, bluish violet or black; pure TiO_2 is white; artificial rutile crystals used as gemstones under the name "titania" are usually very pale yellow.

Streak: Pale brown or grayish black; white for pure TiO_2.

Luster and opacity: Adamantine to metallic; transparent in thin pieces.

Diagnostic tests: Rutile is infusible in the blowpipe flame and insoluble in acids; it is decomposed on fusion with sodium carbonate and yields a test for titanium. Most material yields a test for iron and, less commonly, tantalum, niobium, chromium, or tin.

Occurrence and use: Rutile is often found as an alteration product of other titanium minerals, especially sphene and ilmenite. Titanium dioxide occurs in three polymorphous forms, of which rutile is by far the most common in nature.

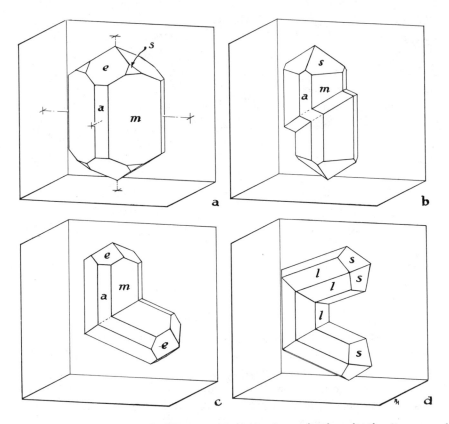

Figure 10-11   Rutile crystals. Forms: tetragonal prisms $a\{100\}$, $m\{110\}$; ditetragonal prism $l\{310\}$; tetragonal dipyramids $e\{101\}$, $s\{111\}$.

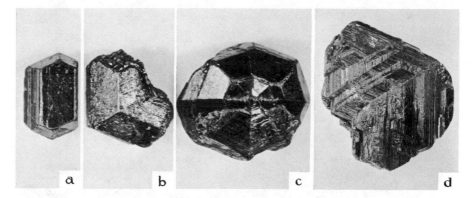

Figure 10-12 Rutile. (a) Single crystal with tetragonal prisms {110}, {100}, striated, and tetragonal dipyramid {101}. (b) Simple twin on (101) as twin plane. (c) Cyclic twin with eight individuals with {101} as twin plane. (d) Geniculate twinning on {101}; striations parallel to [001] are vertical in the largest individual. (Locality unknown.) [Courtesy Queen's University.]

The mineral is typically formed at high temperature. It is widespread as an accessory mineral in igneous rocks, gneisses, and schists. It is also found in pegmatites, in metamorphosed limestones, in high-temperature veins with apatite or quartz, and in quartzite. Minute needles of rutile are found in quartz and micas. It is also a common constituent of beach sands.

Large amounts of rutile occur in apatite veins in Norway. It occurs as fine crystals in druses in crevices and veins in metamorphic rocks in the Alpine regions of Switzerland, France, Italy, and Germany. It is found with ilmenite at Roseland, Virginia, and in minor occurrences at many other localities in the USA.

Production and use: Most of the world's rutile production (about 210,000 tons in 1964) comes from beach sands in Australia, with minor production from sands in Florida and South Carolina. The bulk of rutile and ilmenite produced is used as a source of titania for paint pigments, but metal production is steadily increasing. Although rutile is preferred for metal production, supplies probably will not be adequate and ilmenite will also be used as a source of titanium metal.

Pyrolusite, MnO_2

Common forms and angles:

$(110) \wedge (111) = 47° 22'$ $(010) \wedge (011) = 56° 56'$

Habit: Rarely as well-developed prismatic crystals; usually massive, columnar, or fibrous and divergent; in reniform coatings and in concretionary forms; granular to powdery; as dendritic growths on fracture surfaces of

Figure 10-13   Pyrolusite. Dendritic growth on fracture surface of sandstone. (Locality unknown.) [Courtesy Royal Ontario Museum.]

rocks (Fig. 10-13); common as pseudomorphs with orthorhombic form after manganite.

Cleavage: {110} perfect.

Fracture and tenacity: Uneven; very brittle.

Hardness: 6–6½ (crystals), 2–6 (massive material); fibrous or granular material often soils fingers easily and marks paper.

Density: 5.06 (crystals), ranging down to 4.4 (massive material).

Color: Light to dark steel gray to iron gray, sometimes bluish.

Streak: Black.

Luster and opacity: Metallic; opaque.

Diagnostic tests: Pyrolusite is found well-crystallized in tetragonal forms (formerly known as *polianite*) or as pseudomorphs after orthorhombic manganite. The massive or columnar forms may show a water content to several percent, and small amounts of alkali metals or Ba due to admixed psilomelane or other manganese minerals. Pyrolusite is usually distinguished from other manganese minerals by its black streak, low hardness, and low water content. Pyrolusite is infusible before the blowpipe, and readily yields tests for manganese in the borax or sodium carbonate beads. It commonly yields a little water in the closed tube.

Occurrence and use: Pyrolusite, one of the commonest manganese minerals, is always formed under highly oxidizing conditions. It is most often found as bog, lake, or shallow marine deposits; in the oxidized zone of ore deposits or rocks containing manganese; and as deposits formed by circulating meteoric waters. It is usually associated with other manganese and iron oxides and hydroxides. It may replace other manganese minerals, such as rhodonite, alabandite, hausmannite, and rhodochrosite. It is one of the important ore minerals of manganese.

Substantial manganese deposits have not been found in North America. This is a serious shortage for highly industrialized countries, since about 95 percent of the manganese produced is used in steel alloys and the remainder in dry batteries and chemicals. Minor production of manganese results from the mining of manganiferous iron ores in Minnesota and in Labrador. The bulk of the world production of 15 million tons (1960–1964 average) is obtained from the USSR, India, the Union of South Africa, Brazil, Gabon, and China, in decreasing order of importance.

The term *wad* is commonly used as field term for substances consisting mainly of manganese oxides, in much the same sense as is limonite for hydrous iron oxide and bauxite for hydrous aluminum oxides. Wad is usually a fine-grained mixture, chiefly pyrolusite and psilomelane or other complex hydrous manganese oxides identifiable only by very detailed study. Some varieties of wad contain other metallic oxides, such as BaO, CuO, K_2O, Na_2O, CoO, Fe_2O_3, Al_2O_3, PbO_2, Li_2O. Some of these oxides (such as BaO, in psilomelane) are present as constituents of the component manganese minerals.

Habit: Earthy or compact masses, reniform, concretionary, incrusting, or as stains; ordinarily without internal structure, but occasionally fine, fibrous, banded, or scaly.

Hardness: Usually very soft, soiling the fingers.

Density: 2.8–4.4, often loose and porous.

Color: Black, bluish or brownish black.

Streak: Brown to black.

Luster: Dull.

Occurrence: Wad is usually found as a bog or lake deposit, in clays and in shallow marine sediments, in the oxidized zone of ore deposits, and as a residual product of weathering in areas of manganese-bearing rocks. It is formed under oxidizing conditions at normal temperatures and pressures as a direct precipitate due to chemical or biogenic action. It is of widespread occurrence in many parts of the world, especially in the manganese ore-producing areas.

Cassiterite, SnO_2

Common forms and angles:

(110) $\wedge$ (111) = 46° 25'   (100) $\wedge$ (101) = 56° 04'

(321) $\wedge$ (231) = 20° 57'   (101) $\wedge$ (011) = 46° 30'

Habit: Crystals usually short prismatic, less commonly pyramidal, with prisms {100}, {100}, {210}, dipyramids {111}, {101}, {321} (Fig. 10-14); also massive, in radially fibrous botyroidal crusts or concretionary masses (wood-tin); as brown rounded pebbles with a concretionary structure, or fine sand-like grains (stream tin).

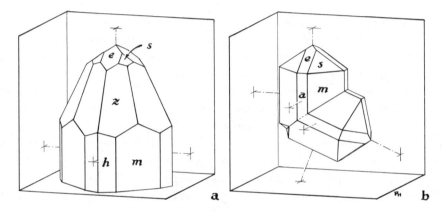

Figure 10-14 Cassiterite crystals. Forms: tetragonal prisms $a\{100\}$, $m\{110\}$; ditetragonal prism $h\{210\}$; tetragonal dipyramids $e\{101\}$, $s\{111\}$; ditetragonal dipyramid $z\{321\}$. (b) Twinned on (011) as twin plane and composition plane.

Twinning: Very common on {101}; contact (Fig. 10-14b) or penetration twins, often repeated.

Cleavage: {100} perfect, {110} imperfect.

Tenacity: Brittle.

Color: Usually yellow or red-brown to brown-black; color may be unevenly distributed within crystals in growth bands; white (pure SnO_2).

Streak: White, grayish or brownish.

Luster: Adamantine and splendant.

Opacity: Light colored material transparent.

Chemistry: Well-crystallized cassiterite is close to SnO_2 in composition, sometimes with as much as 3 percent Fe^3 substituting for Sn, and less often with minor niobium and tantalum. The material commonly called wood-tin, showing botryoidal and reniform shapes, concentric and radially fibrous internally, is usually brown in color, with lower specific gravity, and has inclusions of hematite and silica.

Diagnostic features: Cassiterite, often a difficult mineral to identify, is best distinguished by its color, specific gravity, hardness, and crystal form. It has the highest specific gravity among the common light-colored, nonmetallic minerals; specimens of wolframite have similar color and specific gravity but are readily scratched by steel ($H = 4-4\frac{1}{2}$). It is infusible before the blowpipe but is reduced to metallic tin when heated with sodium carbonate on charcoal. It is only slowly attacked by acids, but a fragment, placed in dilute hydrochloric acid with a little metallic zinc, becomes coated with a dull gray deposit of metallic tin which turns bright when rubbed.

Occurrence: Cassiterite is the most important ore of tin and one of the few tin minerals. It occurs typically in high-temperature hydrothermal veins or metasomatic deposits which are genetically associated with siliceous igneous rocks. Wolframite, tourmaline, topaz, quartz, fluorite, arsenopyrite, muscovite, Li-micas, bismuthinite, bismuth, and molybdenite are often associated with cassiterite. The coarse-grained rock containing these minerals, often called greisen, is formed by hydrothermal alteration along fissures. It may also occur in pegmatite dikes. Important amounts of cassiterite occur in the unusual moderate-temperature Bi–Pb–Ag veins in Bolivia with rare tin sulfide minerals. Cassiterite is a common alluvial mineral in areas of granitic rocks.

Production and use: Tin was known to the ancients and was used in making bronze at least as far back as 3200 B.C. The Phoenicians brought tin from Spain and Cornwall for trade in Mediterranean countries as early as 1100 B.C. Vein deposits containing cassiterite were mined in Cornwall and other areas in Europe from the early fifteenth century. The Cornish deposits became the world's most important source of tin during the eighteenth and nineteenth centuries. These ancient sources of tin have dwindled, and today Malaya, Indonesia, Bolivia, Thailand, Nigeria, the USSR, and China produce 90 percent of the world's requirements of tin ore (185,000 tons, 1961–1965 average). Except for the vein deposits of Bolivia, which constitute the only source in the Western Hemisphere, nearly all of this production is from alluvial deposits.

OTHER OXIDES OF AX₂ TYPE

Anatase, TiO_2; Brookite, TiO_2

These two minerals are crystalline modifications of TiO_2 polymorphous with rutile.

	ANATASE	BROOKITE
Crystal system:	Tetragonal	Orthorhombic
Crystal class:	$4/m\ 2/m\ 2/m$	$2/m\ 2/m\ 2/m$
Axial elements:	$a:c = 1:2.514$	$a:b:c = 0.5942:1:0.5601$
Cell dimensions:	$a = 3.793$	$a = 5.456$
	$c = 9.51$	$b = 9.182$
		$c = 5.143$
Cell content:	$4[TiO_2]$	$8[TiO_2]$
Cleavage:	{001}, {101} perfect	{120} indistinct
Hardness:	$5\frac{1}{2}$–6	$5\frac{1}{2}$–6
Density:	3.90	4.14
Habit:	Crystals pyramidal {101}, or tabular {001}	Crystals tabular {010}, prismatic along c with {120}
Color:	Yellowish and reddish brown to black; also colorless, greenish, lilac, gray	Brown, yellowish to reddish brown; dark brown to black
Streak:	Colorless to pale yellow	Colorless to grayish
Luster:	Adamantine	Metallic adamantine
Optical properties:	$\epsilon = 2.488,\ \omega = 2.561$ $\omega - \epsilon = 0.073$ negative	$\alpha = 2.583,\ \beta = 2.584,$ $\gamma = 2.700$ $\gamma - \alpha = 0.117$ positive $2V = 10°$

Occurrence: Anatase and brookite are of rarer occurrence than rutile; they occur typically in vein or crevice deposits formed from leaching of gneisses or schists by hydrothermal solutions. Anatase is found occasionally in pegmatites, brookite rarely; both minerals are often found as detrital grains. Neither is of economic value as a source of titanium. In some industrial uses the anatase form of titanium dioxide is preferred to the rutile form. Anatase is converted to rutile on heating to about 915°C, and brookite is converted at about 750°C.

Columbite-Tantalite Series: $(Fe,Mn)Nb_2O_6$–$(Fe,Mn)Ta_2O_6$

Crystal system and class: Orthorhombic; $2/m\ 2/m\ 2/m$.

Axial elements: $a:b:c = 0.402:1:0.357$.

Cell dimensions and content: $a = 5.74$, $b = 14.27$, $c = 5.09$ (17 percent Ta_2O_5); ($a = 5.77$, $b = 14.46$, $c = 5.06$ for red manganotantalite); $Z = 4$.

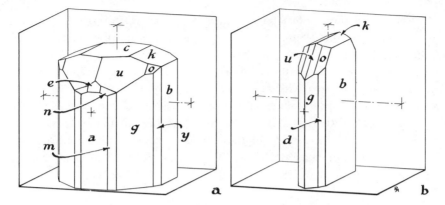

Figure 10-15 Columbite-tantalite crystals. Forms: pinacoids $c\{001\}$, $b\{010\}$, $a\{100\}$; rhombic prisms $m\{110\}$, $g\{130\}$, $y\{160\}$, $d\{170\}$, $k\{011\}$; rhombic dipyramids $u\{111\}$, $o\{131\}$.

Common forms and angles:

$(010) \wedge (110) = 68° 06'$ $(010) \wedge (011) = 70° 21'$

$(100) \wedge (130) = 50° 20'$ $(001) \wedge (111) = 43° 45'$

Habit: Crystals short prismatic along c, thin tabular on $\{010\}$ (Fig. 10-15) or thick tabular on $\{100\}$; often in large groups of parallel or subparallel crystals; massive.

Twinning: Common on $\{201\}$, as contact twins, also as penetration or repeated twins.

Cleavage: $\{010\}$ distinct, $\{100\}$ less distinct.

Fracture and tenacity: Subconchoidal; brittle.

Hardness: 6 (columbite) to $6\text{-}6\frac{1}{2}$ (tantalite).

Density: 5.2 (columbite) to 7.95 (tantalite).

Color: Iron black to brownish black (manganotantalite may be dark red); often tarnished iridescent.

Streak: Dark red to black.

Luster: Submetallic, often brilliant, subresinous.

Opacity: Transparent in thin splinters, increasing in high manganese varieties.

Fusibility: $5\text{-}5\frac{1}{2}$, with difficulty.

These minerals form an almost continuous series of solid solutions within the range shown in the formula. The name columbite is used for those with Nb > Ta and tantalite for those with Ta > Nb. Fe or Mn may predominate in either mineral. Recent studies indicate that many minerals in the ferrotantalite composition range are in fact tapiolite. The specific gravity increases linearly with increase in Ta, and is useful in estimating Ta content. The Fe:Mn ratio has only a slight affect on the specific gravity. This series includes the most abundant niobium-tantalum minerals and constitutes the most important source of these metals. These minerals are usually found in granite pegmatites, particularly those with albite, Li silicates, and Li–Mn–Fe phosphate phases. They may be associated with albite, microcline, beryl, lepidolite, muscovite, tourmaline, spodumene, amblygonite, apatite, microlite, and cassiterite; they may also be found detrital. These minerals have been found in many of the pegmatite areas of the world: in the northeastern USA; in South Dakota (Keystone), Colorado (Pikes Peak), and California; in Madagascar; in Norway and Sweden; and in the USSR.

Uraninite, UO_2

Crystal system and class: Isometric; $4/m\ \bar{3}\ 2/m$.

Cell dimensions, and content: $a = 5.4682$ (pure); $Z = 4$. The cell edge decreases (to about 5.4) with oxidation of U^4 to U^6; it increases due to substitution of Th for U ($a = 5.5997$, pure ThO_2).

Habit: Crystals usually cubic, octahedral, or cubo-octahedral; also massive (pitchblende); dense botryoidal, or reniform, with a banded structure; in dendritic aggregates of small crystals.

Fracture and tenacity: Uneven to conchoidal; brittle.

Hardness: 5–6.

Density: 10.95, decreasing with oxidation of U^4 to U^6, also with substitution of thorium or cerium for uranium; crystals 8–10; massive 6.5–8.5.

Color: Steely black, brownish black to black.

Streak: Brownish black or grayish.

Luster and opacity: Submetallic to pitch-like, also dull; nearly opaque.

Diagnostic features: Density, color, pitchy luster, and radioactivity; it has the highest density of any oxide mineral when pure; almost infusible before the blowpipe; dissolves in fluxes, gives distinctive bead tests for uranium.

Chemistry: Although uraninite is essentially UO_2 with the fluorite structure, the natural material is always more or less oxidized, and grades toward U_3O_8 in actual composition. The crystallized material, called "uraninite proper" by some writers, contains 1 percent or more of thorium and rare-earth metals. The substitution of Th for U, to the extent of over 40 percent ThO_2 in nature, may be complete in artificial preparations, extending to pure ThO_2, which occurs in nature as the mineral thorianite. Cerium or yttrium earths substitute for U to at least 10 percent. Recently the isostructural substance CeO_2 (cerianite) has been found as a mineral. Pegmatite occurrences are almost all of this type, and the specific gravity usually lies in the range 8–10.

The massive or colloform variety of uraninite, commonly called pitchblende, usually contains less than 1 percent of Th and rare-earth metals substituting for U. The specific gravity is usually in the range 6.5–8.5. Lead is always present, since it is the stable end-product of the radioactive disintegration of uranium and thorium; helium is also present for the same reason.

Occurrence: The principal occurrences of uraninite can be grouped under several types:

1. Uraninite, usually containing more than 1 percent Th and rare-earths, found in granite and syenite pegmatites. It is associated with zircon, tourmaline, monazite, mica, and feldspar, also with complex oxides of the rare-earths with Nb–Ta–Ti. The mineral occurs either with distinct crystal form or massive, without colloform structure. Typical occurrences are numerous in the southern parts of the Canadian Precambrian shield, as at Bancroft, Ontario, and at Villeneuve, Quebec, also in some of the pegmatites in New Hampshire, in southern Norway, and in East Africa.

2. Colloform crusts (pitchblende) in high temperature hydrothermal tin veins with cassiterite, pyrite, chalcopyrite, and arsenopyrite, particularly in Cornwall, England.

3. Colloform masses in moderate-temperature hydrothermal veins with pyrite, chalcopyrite, galena, carbonates, barite, fluorite, bismuth, silver, and Co–Ni–As minerals. This type of association is important in the Czechoslovakian deposits and those of Great Bear Lake, Canada.

4. In veins similar to those mentioned in Number 3, but lacking the Co–Ni–Ag minerals, as at Gilpin County, Colorado, and South Alligator River, Northern Territory, Australia.

5. In quartz-pebble conglomerates as minute grains, as in the Transvaal gold-bearing conglomerates, and in the Blind River district of Ontario. These low-grade ores contain large reserves of uranium dioxide. Although their origin is a subject of considerable speculation, the distribution of the uraninite indicates an origin approximately contemporaneous with the formation of the conglomerate. Other minerals containing U, Th, and rare-earths, and commonly found as detrital grains, also occur in the conglomerates.

HYDROXIDES

Brucite, $Mg(OH)_2$

Crystal system and class: Trigonal; $\bar{3}\ 2/m$.
Axial elements: $a:c = 1:1.5154$.
Cell dimensions and content: $a = 3.147$, $c = 4.769$; $Z = 1$.

Habit: Crystals usually broad tabular on {0001}, also as subparallel aggregates of plates; commonly foliated massive; fibrous with fibers separable and elastic.

Cleavage: {0001} perfect, foliae flexible, sectile.

Crystal structure: Brucite has a layer structure in which each layer, parallel to {0001}, consists of two sheets of OH in hexagonal close packing, with a sheet of Mg atoms between them (Fig. 10-16). Each Mg atom is in sixfold coordination between OH, and each OH fits into three OH of the next layer. The layers are held together by weak secondary forces between adjacent OH sheets.

Hardness: $2\frac{1}{2}$.

Density: 2.39 (2.368, calculated).

Color: White to pale green, gray, or blue.

Streak: White.

Luster and opacity: Pearly on cleavage surfaces, elsewhere waxy to vitreous; transparent.

Optical properties: $\omega = 1.560$, $\epsilon = 1.580$; $\epsilon - \omega = 0.020$; positive.

Chemistry: Manganese and iron commonly substitute for Mg in brucite to the extent of a few percent. $Mn(OH)_2$ (pyrochroite) and $Ca(OH)_2$ (portlandite) are isostructural with brucite.

Diagnostic features: The mineral is found in three very distinct forms: the platy or foliate type; the fibrous material usually elongated on an *a* axis; and massive material with a soapy appearance. Brucite, in foliate form, is harder than talc and gypsum, is not greasy as is talc, is not elastic like muscovite, is not fusible, and is more readily soluble in hydrochloric acid than gypsum. The fibers are not as silky or fine as those of chrysotile. Brucite is infusible and glows before blowpipe; it is easily soluble in hydrochloric acid, giving standard chemical test for magnesium.

Occurrence: Brucite, usually associated with calcite, aragonite, talc, magnesite, and other rarer minerals, is typically a low-temperature hydrothermal vein mineral. It occurs in serpentine, in chlorite or dolomitic schists, or in crystalline limestones as an alteration product of periclase.

   Fibrous brucite is found in serpentine in New Jersey, and commonly in the asbestos mines of Asbestos and Black Lake in Quebec. Granular brucite, pseudomorphous after periclase, occurs in crystalline limestone at Crestmore, California. It is recovered commercially from crystalline limestone at Wakefield, Quebec, and from deposits in the Paradise Range, Nye County, Nevada.

   Brucite is one of several minerals used as a source for magnesium metal.

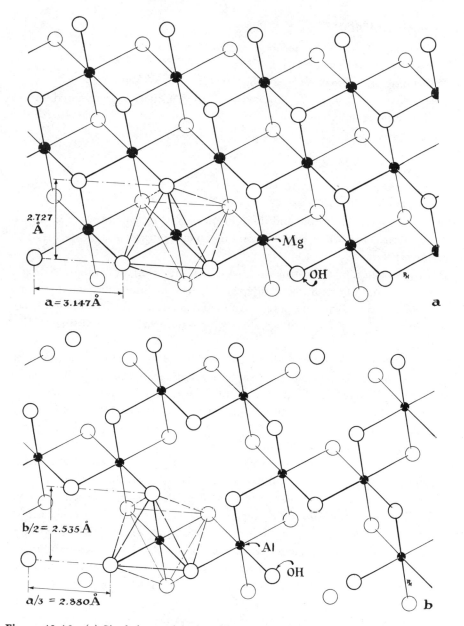

Figure 10-16 (a) Single layer of the brucite structure with Mg in octahedral coordination with (OH); the hexagonal *a* axis is horizontal. (b) Single layer of the gibbsite structure with Al in octahedral coordination with (OH); only two-thirds of the possible octahedral sites are occupied. The *a* axis of the monoclinic lattice is horizontal.

Boehmite, AlO(OH)

Crystal system and class: Orthorhombic; $2/m\ 2/m\ 2/m$.

Axial elements: $a:b:c = 0.2346:1:0.3026$.

Cell dimensions and content: $a = 2.868$, $b = 12.227$, $c = 3.700$; $Z = 4$.

Habit: Microscopic crystals tabular on $\{001\}$; usually disseminated or in pisolitic aggregates.

Cleavage: $\{010\}$.

Density: 3.0.

Color: White.

Optical properties: $\alpha = 1.64$, $\beta = 1.65$, $\gamma = 1.66$; $\gamma - \alpha = 0.02$; positive, $2V = 80°$.

Diagnostic features: Infusible and insoluble; yields water when heated in closed tube; turns blue when moistened with cobalt nitrate and heated in the oxidizing flame.

Chemistry: Boehmite is isostructural with the rare iron mineral *lepidocrocite*, FeO(OH), but not with manganite, MnO(OH). Boehmite and lepidocrocite are polymorphous with diaspore, $HAlO_2$, and goethite, $HFeO_2$. Boehmite and lepidocrocite possess a sheet structure which gives rise to the distinct $\{010\}$ cleavage.

Occurrence and use: The mineral, widely distributed as a major constituent of most *bauxite*, is impossible to identify megascopically because of its very fine-grained character. As an important constituent of bauxite it constitutes an important ore mineral of aluminum. It and the other minerals of bauxite, diaspore, gibbsite, and iron oxides, result from weathering, under tropical conditions, of aluminum silicate rocks low in free quartz.

The important world producers of bauxite—Surinam, Guyana, the USA, France, Jamaica, Hungary, the USSR, Greece, Australia, and Guinea— provide 85 percent of the average annual production of about 30 million tons. The recovery of aluminum metal requires great quantities of electrical power. Three countries, the USA, Canada, and the USSR, supply 75 percent of the world production of aluminum metal. The availability of cheap hydroelectric power in Canada makes it economically feasible to transport ore great distances by sea from South and Central America and from West Africa to smelters in Canada. Thus, Canada has become an important metal producer although it produces no aluminum ore.

Manganite, MnO(OH)

Crystal system and class: Monoclinic; $2/m$ (pseudo-orthorhombic).

Axial elements: $a:b:c = 1.693:1:1.087$; $\beta = 90° 00'$.

Cell dimensions and content: $a = 8.94$, $b = 5.28$, $c = 5.74$; $Z = 8$.

Common forms and angles:

(010) $\wedge$ (210) = 49° 45'   (010) $\wedge$ (111) = 47° 33'

(8$\bar{1}$0) $\wedge$ (810) = 23° 54'   (101) $\wedge$ ($\bar{1}$01) = 65° 24'

Habit: Crystals striated and short, or long prismatic parallel to c (Fig. 10-17); columnar to fibrous.

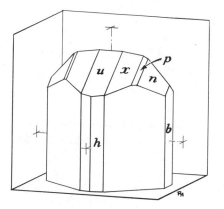

Figure 10-17 Manganite crystal. Forms: monoclinic pinacoids $b\{010\}$, $u\{101\}$; prisms $m\{210\}$, $h\{810\}$, $x\{818\}$, $p\{212\}$, $n\{111\}$. The crystals usually develop with pseudo-orthorhombic form.

Twinning: Twin plane $\{011\}$ as contact or penetration twins, often repeated.

Cleavage: $\{010\}$ very perfect; $\{110\}$ and $\{001\}$ less perfect.

Fracture and tenacity: Uneven; brittle.

Hardness: 4.

Density: 4.33.

Color: Dark steel gray to iron black.

Streak: Reddish brown to black.

Luster and opacity: Submetallic; almost opaque.

Diagnostic features: Infusible; soluble in concentrated hydrochloric acid with evolution of chlorine; yields much water on heating in closed tube; gives bead test for manganese.

Occurrence: Manganite is found in low-temperature hydrothermal veins with barite, calcite, siderite, hausmannite; in deposits formed by meteoric waters associated with pyrolusite, goethite, psilomelane; or with other manganese minerals in residual clays. It is often altered to pyrolusite, or less often to other manganese oxides. Manganite is seldom an important constituent of manganese ores. It is found at Ilfeld, and Ilmenau, Germany; and at the Jackson mine, Negaunee, Michigan; and at many other localities.

Diaspore, $HAlO_2$

Crystal system and class: Orthorhombic; $2/m$ $2/m$ $2/m$.

Axial elements: $a:b:c = 0.4664:1:0.3017$.

Cell dimensions and content: $a = 4.396$, $b = 9.426$, $c = 2.844$; $Z = 4$.

Common forms and angles:

(010) ∧ (110) = 65° 00'   (110) ∧ (111) = 54° 29'

Habit: Crystals thin platy or tabular on {010}, elongated to acicular parallel to c; foliated massive and in thin scales, sometimes stalactitic; disseminated.

Cleavage: {010} perfect, {110} less perfect.

Fracture and tenacity: Conchoidal; very brittle.

Hardness: $6\frac{1}{2}$–7.

Density: 3.3–3.5 (3.380, calculated).

Color: White, grayish white, colorless, rarely greenish brown, lilac, or pink.

Luster: Brilliant vitreous, pearly on cleavage faces.

Optical properties: $\alpha = 1.68$–1.70, $\beta = 1.70$–1.72, $\gamma = 1.73$–1.75; $\gamma - \alpha = 0.05$; positive, $2V = 85°$.

Diagnostic features: Diaspore tests are similar to those for boehmite, with which it is polymorphous; good cleavage and hardness are characteristic for coarsely crystalline material.

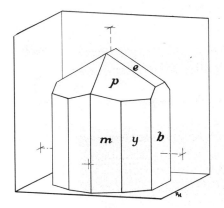

Figure 10-18 Goethite crystal. Forms: pinacoid $b\{010\}$; rhombic prisms $y\{120\}$, $m\{110\}$, $e\{021\}$; rhombic dipyramid $p\{121\}$.

Occurrence: Diaspore is of widespread occurrence as massive fine-grained material with boehmite and gibbsite in bauxites, especially in those of Hungary, France, and the USA (Arkansas, Missouri). It is found with corundum in emery rock and in crystalline limestone; as a late hydrothermal mineral in some alkalic pegmatites; or as a hydrothermal alteration of other aluminous minerals.

Goethite, HFeO₂

Crystal system and class: Orthorhombic; $2/m\ 2/m\ 2/m$.

Axial elements: $a:b:c = 0.4616:1:0.3034$.

Cell dimensions and content: $a = 4.596$, $b = 9.957$, $c = 3.021$; $Z = 4$.

Common forms and angles:

(010) ∧ (120) = 47° 17'	(010) ∧ (021) = 58° 45'
(110) ∧ (1$\bar{1}$0) = 49° 34'	(121) ∧ (1$\bar{2}$1) = 53° 48'

Habit: Crystals prismatic and striated parallel to *c* (Fig. 10-18); also thin tabular and scaly parallel to {010}; aggregates of capillary to acicular crystals; usually massive; reniform, botryoidal, or stalactitic masses (Fig. 10-19) with internal radiating fibrous or concentric structure; also bladed or columnar; compact; earthy; sometimes pisolitic or oolitic, also loose and porous.

Cleavage: {010} perfect, {100} less perfect.

Fracture: Uneven.

Hardness: 5–5½.

Figure 10-19 Goethite, stalactitic mass, showing internal radial fibrous structure and lustrous black surface. (Horhausen, Germany.) [Courtesy Royal Ontario Museum.]

Density: 3.3 to 4.3.

Color: Blackish brown to yellow or reddish brown to brownish yellow in massive and earthy forms.

Streak: Brownish yellow, orange yellow.

Luster and opacity: Crystals adamantine—metallic to dull; fibrous material silky; transparent in thin splinters.

Diagnostic features: Streak and color distinguish goethite from hematite; becomes magnetic in reducing flame; gives water in closed tube.

Occurrence: The greater part of the common yellow-brown and brown ferric oxides usually known as *limonite* properly belong to this species. Goethite usually results from weathering of iron minerals, siderite, pyrite, magnetite, or glauconite under oxidizing conditions at ordinary temperatures; it also forms by direct precipitation from marine or meteoric waters in bogs or lagoons. Goethite, as the main constituent of limonite, is found as a gossan or weathered capping on veins or replacement deposits rich in iron-bearing sulfides. Residual limonite cappings are also found over rocks containing ferrous-iron minerals where rock decay has progressed for a long time; these are called laterites. Large deposits resulting from the weathering of serpentine are found in Cuba.

Oolitic limonite is mainly responsible for the iron content of the extensive "Minette" ores of eastern France. They are sedimentary deposits, outcropping in Lorraine, along the French-German border, and dipping west. After many years, mining is extending deeper and to the west. Large tonnages of residual limonite are now being mined in the "Labrador Trough," located along the Quebec-Labrador boundary.

Goethite is often found as crystals, or botryoidal or stalactitic forms, in near-surface fissures and pockets in many kinds of rocks.

Gibbsite, Al(OH)$_3$

Crystal system and class: Monoclinic; $2/m$.

Axial elements: $a:b:c = 1.7043:1:1.9170$; $\beta = 94° 34'$.

Cell dimensions and content: $a = 8.641$, $b = 5.070$, $c = 9.719$; $Z = 8$.

Habit: Crystals tabular on {001} with {100} and {110}; radiating spheroidal concretions; also stalactitic and encrusting with a smooth surface; compact earthy.

Twinning: Common with [130] as twin axis or {001} as twin plane.

Cleavage and tenacity: {001} perfect; tough.

Crystal structure: Gibbsite has a layer structure (Fig. 10-16b) which is similar in some respects to that of brucite. Each layer, one half of the c-axis period in thickness, is composed of two sheets of OH ions approximately in the hexagonal closest-packing arrangement with Al lying between the two sheets. Each Al is in octahedral coordination with six OH, but only two-thirds of the possible octahedral sites are occupied, in contrast with brucite, where all are occupied. In gibbsite the layers are packed together such that the OH in adjoining layers are in line parallel to c instead of nesting in the trigonal depressions, as in brucite. This arrangement requires a c length slightly more than double that of brucite. The relatively weak attraction between OH in adjoining layers results in the perfect cleavage {001}.

Hardness: $2\frac{1}{2}$–$3\frac{1}{2}$.

Density: 2.4.

Color: White, grayish, greenish or reddish white.

Luster and opacity: Pearly on cleavage surfaces, vitreous on others; transparent.

Optical properties: $\alpha = 1.56$, $\beta = 1.56$, $\gamma = 1.58$; $\gamma - \alpha = 0.02$; positive, $2V = 0$–$40°$.

Diagnostic features: Gives strong clay odor when breathed upon; when crystalline it is softer than diaspore; not distinguishable from boehmite when fine-grained.

Occurrence: Gibbsite is another of the important constituent minerals of bauxite. It occurs with boehmite and diaspore in earthy, oolitic, or pisolitic bauxite, constituting the chief mineral in some occurrences. It may also be found as a low-temperature hydrothermal mineral in veins or cavities in alkalic and aluminous igneous rocks.

11

Class IV: Halides

H																	
Li		*B*												N	O	F	
Na	Mg	Al	*Si*													Cl	
K	Ca				Mn	Fe			Cu							Br	
	Y								Ag							I	
	La Lu									Hg	Pb	Bi					

The halides comprise those minerals which are primarily compounds of the halogen elements—F, Cl, Br, I—in which a halogen is the sole or principal anion.* They include simple isodesmic structures such as halite, as well as anisodesmic structures such as cryolite, in which anion groups, $(AlF_6)^{-3}$, are found. Some halides contain water of crystallization, and others are oxyhalides or hydroxyhalides. In addition, some minerals classified as silicates, phosphates, etc. contain minor amounts of halogen elements, particularly fluorine and chlorine.

* Table above shows elements which form halide or oxyhalide minerals; those in italics occur in complex fluorides or in fluorite (Y).

Chlorine and fluorine are far more abundant (130 and 625 parts per million, respectively) in the zone of observation of the earth's crust than bromine and iodine (2.5 and 0.5 parts per million). From a mineralogical point of view this is enhanced by the fact that bromine is more concentrated in the ocean waters (65 parts per million) and is consequently of very rare occurrence in minerals.

Fluorine is of strikingly different geological occurrence from the other halogens. It is found almost exclusively in minerals of igneous origin—in minerals of igneous rocks, pegmatites, hydrothermal veins, or rocks affected or altered by pyrometasomatic action. Chlorine, on the other hand, is found as a minor element in a few minerals of igneous or metamorphic origin—for example, apatite, sodalite, scapolite—but the great bulk of the chlorine of the earth's crust is found dissolved in the sea waters or in solid stratiform deposits of soluble salts which resulted from the evaporation of sea water at various periods in geological time.

The important halide minerals are:

Halite	NaCl	Cryolite	Na_3AlF_6
Sylvite	KCl	Carnallite	$KMgCl_3 \cdot 6H_2O$
Fluorite	CaF_2		

In addition, chlorides, bromides, and iodides (in decreasing order of importance) occur in the oxidized zone of many ore deposits, especially in the arid regions of the earth. The basic copper chloride, atacamite, is an important constituent of the ore in the Chilean copper deposits. The insoluble halides of silver—cerargyrite (AgCl), bromyrite (AgBr), and iodyrite (AgI)—have been recognized at numerous localities and have constituted an important part of the ore, especially in the early days of development, in the silver mines of Saxony, Mexico, New South Wales, and the southwestern USA. Halides of lead, mercury, and bismuth have also been noted.

HALITE GROUP

The crystal structure of halite (Fig. 11-1) or rock salt, was the first to be analyzed by x-ray diffraction methods. The halite structure is found in a large number of AX compounds with a radius ratio between 0.41 and 0.73, and includes minerals of the halite group, galena group, and periclase group. The important members of the halite group are halite and sylvite. At ordinary temperatures there is virtually no substitution of Na by K in the halite structure. In the halite structure one kind of ion occurs at the points of a face-centered cubic lattice and the other kind of ion occurs midway between each pair of the first kind along the [100] directions. Each ion is in octahedral coordination with six ions of the opposite kind, and not to one in particular. The analysis

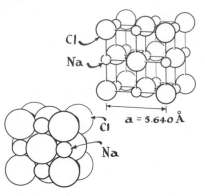

Figure 11-1 Structure of halite. Na and Cl are each octahedrally surrounded by six atoms of Cl or Na.

of this structure first established an important point in the crystal chemistry of typical inorganic compounds and minerals: there is no grouping of atoms into discrete molecules.

	HALITE	SYLVITE
Crystal system:	Isometric	Isometric
Crystal class:	$4/m\ \bar{3}\ 2/m$	$4/m\ \bar{3}\ 2/m$
Cell edge:	$a = 5.6402$	$a = 6.2931$
Cell content:	$Z = 4$	$Z = 4$
Cleavage:	{100} perfect	{100} perfect
Hardness:	$2\frac{1}{2}$	$2\frac{1}{2}$
Density:	2.164 (calc.)	1.9865 (calc.)
Refractive index:	1.544	1.490

Halite, NaCl

Habit: Usually as cubes, often with cavernous and stepped faces, hopper crystals (Fig. 11-2); also massive, coarsely granular to compact; rarely columnar.

Color: Colorless, white, gray, yellow, red, rarely blue or purple.

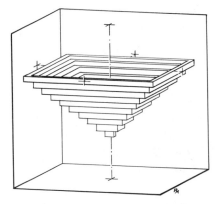

Figure 11-2 Halite, hopper crystal with cube faces. This form often results from growth of crystals floating on the surface of natural brines.

Streak: White.

Luster and opacity: Vitreous; transparent to translucent.

Taste: Saline.

Diagnostic features: Halite may be distinguished by its salty taste, low hardness, and cubic cleavage; it is sometimes fluorescent.

Occurrence: Halite is by far the commonest water-soluble mineral, occurring as extensive beds formed by evaporation of enclosed bodies of sea water. These beds are found interstratified with and covered by beds of shale, limestone, dolomite, and gypsum or anhydrite. Sodium chloride occurs in solution in springs that arise from salt beds, and in the waters of the ocean, seas, and salt lakes. Halite is also found as a sublimation in volcanic regions and as an efflorescence in arid areas. Halite, along with other salts, occurs in playa deposits in dried inland lake basins such as the Great Salt Lake desert in Utah.

Stratiform deposits of halite are of widespread occurrence throughout the world in sedimentary basins of various geological ages from Palaeozoic to Recent. Beds as much as several hundred feet thick occur under a large area that includes southwestern Ontario and Michigan; under the plains of Saskatchewan; in Nova Scotia; in New Mexico, New York, Kansas, and Texas; and in many countries of the world.

Deformation of thick salt beds, usually at considerable depth, may result in the local extrusion of plug-like or chimney-like masses, from a few hundred feet to several miles in width, into overlying sediments, resulting in the so-called salt domes or salt plugs. Salt domes are found in Germany, Spain, Romania, Iran, and in the Gulf Coast region of the USA, where approximately 250 have been discovered in the search for oil and gas concentrations. The tops of salt plugs may be near-surface, as at Avery Island, Louisiana, or at depths as much as 14,000 feet, as at Lafitte, Louisiana. Drills have penetrated salt plugs for depths of 5,000 feet or more in solid salt. It is considered likely that the salt was squeezed upward from salt beds at depths of from 15 to 35 thousand feet along some line of weakness in overlying strata. Vast reserve deposits of salt, sufficient to supply world demand for 30 thousand years, are available in such salt domes. The USA, the USSR, England, China, India, Germany, and France are the chief producers, but almost all countries have some supplies of salt available. World production in 1964 was estimated to be 109 million tons.

Sylvite, KCl

Habit: Usually as cubes; massive; coarsely granular to compact.

Tenacity: Not as brittle as halite, distinctly sectile as compared with halite when scratched with a knife point.

Color: Colorless or white, also grayish, bluish, yellowish red, or red, the red tints usually due to included particles of hematite.

Streak: White.

Luster and opacity: Vitreous, transparent to translucent.

Taste: Saline, more bitter than halite.

Diagnostic features: Sylvite may be very similar in appearance to halite, and they may occur together in the evaporite salt beds. Sylvite may best be distinguished from halite by its bitter taste and its tendency to sectility. On scratching a smooth surface, a knife point raises a distinct powder with halite, but very little with sylvite. The latter test is dependable and easier to apply than tasting, especially if one is examining many feet of drill core. In a certain locality sylvite may have slightly different transparency, and often fills in around euhedral cubes of halite, clearly indicating the later formation of sylvite.

Occurrence: Sylvite occurs principally in some of the bedded basin-like salt deposits of halite and gypsum, but is much less common than halite. The most famous locality for sylvite and other potash salts is at Stassfurt, in Germany, where about 30 saline minerals have been recognized. Important quantities of sylvite are now being mined from the bedded salt deposits of New Mexico and Texas. Well drilling has outlined a very large area in Saskatchewan, underlain by salt beds which contain important reserves of sylvite in the upper zone of the Prairie Evaporites, of Devonian age, at depths of 3,000–5,000 feet.

Use: Until the discovery of potassium salts in Germany in 1861 most potash was obtained from wood ash. With the development of salt mining, potash minerals became widely used for fertilizers, and Germany supplied most of the world demand until 1925, when other countries began developing potash-salt deposits. The USA, the USSR, France, Canada, and Spain now produce important amounts of potassium salts. The USSR and Canada estimate reserves at about 50 billion tons each. About 90 percent of the world production is used in fertilizers.

OTHER HALIDES

Cerargyrite, AgCl

Crystal system and class: Isometric; $4/m\ \bar{3}\ 2/m$.

Cell dimensions and content: $a = 5.5491$; $Z = 4$.

Habit: Crystals uncommon, usually as cubes; ordinarily massive, in crusts and waxy coatings or as wax or horn-like masses.

Tenacity: Tough, sectile and ductile, plastic.

Hardness: $2\frac{1}{2}$.

Density: 5.56, increasing with bromine substituting for chlorine.

Color: Colorless when pure and fresh, usually gray, also yellowish, greenish brown; on exposure to light becomes violet brown or purple.

Luster and opacity: Resinous to adamantine; horn-like, usually translucent.

Optical properties: $n = 2.07$.

Diagnostic features: Color, sectility; easily fusible at 1 before blowpipe; gives a silver globule when heated on charcoal.

Chemistry: A complete solid-solution series extends to *bromyrite* (AgBr) by substitution of Br for Cl. The name *embolite*, commonly used in the past for intermediate members of this series, is not considered necessary now. Cerargyrite is adequate when Cl > Br, and bromyrite when Br > Cl. These names may be modified by the adjectives "bromian" or "chlorian" to indicate the presence of substantial substitution. The substitution of I for Cl or Br, usually only in small amount, leads to the names iodian cerargyrite or bromyrite. Cerargyrite is isostructural with halite, whereas natural silver iodide is hexagonal.

Occurrence: These minerals are secondary in origin and are usually found in the upper, oxidized zone of silver deposits, especially in arid regions. They may be associated with silver, jarosite, limonite, also cerussite, atacamite, malacite, pyromorphite.

Cerargyrite was obtained in great quantity in the early mining at several localities in Saxony. It is found in mines in Chile, Bolivia, Peru, Mexico, and at Broken Hill, New South Wales; it was also very common in mining districts of the western USA.

Fluorite, CaF$_2$

Crystal system and class: Isometric; $4/m\ \bar{3}\ 2/m$.

Cell dimensions and content: $a = 5.4626$; $Z = 4$.

Habit: Crystals usually cubes (Figs. 11-3, 11-4), less often octahedrons, dodec-

Figure 11-3 Fluorite. Cubes with minor octahedron faces, on calcite with nodules of dark sphalerite. (Durham, England.) [Courtesy Royal Ontario Museum.]

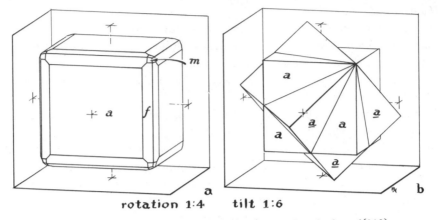

rotation 1:4   tilt 1:6

Figure 11-4 Fluorite crystals. Forms: cube $a\{100\}$; tetrahexahedron $f\{310\}$; trapezo-hedron $m\{311\}$. (b) Interpenetrating cubes twinned on [111] as twin axis.

ahedrons; cubes or octahedrons modified by $\{110\}$, $\{310\}$, $\{311\}$, or $\{421\}$; cube faces smooth and lustrous, octahedron faces rough and dull; often composite with many small crystals in parallel aggregation; massive, coarse- to fine-grained or compact.

Twinning: [111] as twin axis, interpenetrating (Fig. 11-4b).

Cleavage: $\{111\}$ perfect.

Crystal structure: The structure of fluorite (Fig. 11-5) is unique among natural halides, although it is also found in uraninite, UO_2, and the related oxides ThO_2 and CeO_2. The Ca atoms occupy face-centered cubic positions, the F atoms occupy the body-center of each cubelet with half the cube-edge length of the unit cell. The unit cell contains $4[CaF_2]$. Each Ca atom is in eightfold coordination with eight F atoms at the corners of a cube with $a/2$,

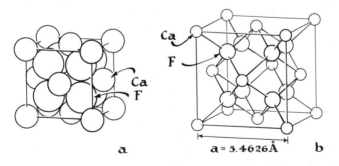

a $a = 5.4626Å$ b

Figure 11-5 Structure of fluorite. Each fluorine is coordinated to four calcium atoms, and each calcium to eight fluorine atoms.

and each F atom is in fourfold coordination with four Ca atoms at the corners of a tetrahedron.

Hardness: 4.

Density: 3.18.

Color: Colorless and transparent when pure, commonly wine yellow, green, greenish blue, violet blue, also white, gray, sky blue, purple, bluish black, or brown; the color is often distributed in zones parallel to the crystal faces; massive material may also exhibit zones of different color; usually fluorescent under ultraviolet light, or cathode rays.

Luster: Vitreous.

Optical properties: $n = 1.433$.

Diagnostic features: Distinguished by octahedral cleavage; softer than feldspars; harder than calcite and lacks effervescence with acid; fusible at 3; gives a red flame, indicating calcium; when heated with potassium disulfate in a closed tube, hydrofluoric acid is generated which etches the wall of the glass tube, and a white deposit of silica is formed on the upper, cooler wall of the tube.

Chemistry: Although most fluorite is fairly pure, Y, Ce, and traces of other rare-earth elements may substitute for Ca.

Occurrence: Fluorite is found in mineral deposits of widely different character but mostly of primary origin. It occurs most commonly as a vein mineral, in some veins as the principal constituent, in others as a gangue mineral with lead and silver ores, usually associated with quartz, calcite, dolomite, or barite. It may also be found in cavities in dolomite or limestone, associated with celestite, anhydrite, gypsum, dolomite, sulfur, or millerite.

Fluorite is also characteristic: (1) of some pneumatolytic deposits such as greisens; (2) of high-temperature cassiterite veins associated with topaz, tourmaline, lepidolite, apatite, or quartz; (3) less commonly as a late hydrothermal mineral in cavities and joints in granite; (4) in pegmatites. A dark purple variety occurs with calcite, minor apatite, hornblende, biotite, and uraninite, in vein-like bodies cutting syenite at Wilberforce, Ontario. The occurrences of fluorite are very widespread.

Production and use: The principal producing countries are the USA, France, Germany, Mexico, the USSR, Canada, England, China, and Italy. Although the USA (Hardin and Pope Counties, Illinois) and Canada are among the larger producers both countries must import additional amounts for their metallurgical requirements. A small amount of colorless, transparent material is used for optical purposes; large quantities are used in metallurgical operations as a flux (especially in the manufacture of open-hearth steel and in the smelting of aluminum) and in the manufacture of chemicals.

Atacamite, $Cu_2(OH)_3Cl$

Crystal system and class: Orthorhombic; $2/m\ 2/m\ 2/m$.

Axial elements: $a:b:c = 0.658:1:0.749$.

Cell dimensions and content: $a = 6.02$, $b = 9.15$, $c = 6.85$; $Z = 4$.

Common forms and angles:

   $(010) \wedge (120) = 37° 14'$ $(110) \wedge (1\bar{1}0) = 66° 42'$

   $(010) \wedge (011) = 53° 10'$ $(010) \wedge (111) = 63° 41'$

Habit: Commonly in slender prismatic crystals elongate and striated parallel to *c*, also tabular on {010} (Fig. 11-6); in crystalline aggregates, massive, fibrous or granular.

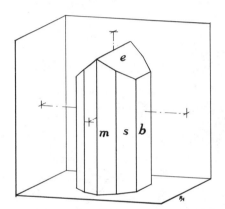

Figure 11-6 Atacamite crystal. Forms: pinacoid $b\{010\}$; rhombic prisms $m\{110\}$, $s\{120\}$, $e\{011\}$.

Cleavage and fracture: {010} perfect; brittle.

Hardness: $3-3\frac{1}{2}$.

Density: 3.76.

Color: Bright green to dark green.

Streak: Apple green.

Luster: Vitreous.

Optical properties: $\alpha = 1.831$, $\beta = 1.861$, $\gamma = 1.880$; $\gamma - \alpha = 0.049$; negative, $2V = 75°$.

Diagnostic features: Atacamite is characteristically green in color and may be distinguished from malachite by the lack of effervescence in acids, and from brochantite and antlerite by the copper chloride flame test; fusible, giving azure-blue flame of copper chloride; gives globule of copper when fused with sodium carbonate on charcoal.

Occurrence: Atacamite is found as a secondary mineral in the oxidized zone of copper deposits, especially under saline desert conditions. It may be associated with malachite, cuprite, brochantite, chrysocolla, gypsum, and limonite.

Named from the Atacama desert of Northern Chile, where it was first found, it occurs at many localities in the arid coastal belt of Chile (Paposa, Copiapo, Chuquicamata) particularly in the oxide-ore of the copper mines. It is found sparingly in some copper mines of Utah (Tintic district) and Arizona (Bisbee, United Verde mine, Mammoth mine).

Carnallite, $KMgCl_3 \cdot 6H_2O$

Crystal system and class: Orthorhombic; $2/m\ 2/m\ 2/m$.
Axial elements: $a:b:c = 0.596:1:1.406$.
Cell dimensions and content: $a = 9.56$, $b = 16.05$, $c = 22.56$; $Z = 12$.
Habit: Crystals not common, pseudohexagonal due to nearly equal development of $\{hhl\}$ and $\{0kl\}$ forms, tabular on $\{001\}$; usually massive, granular.
Cleavage: None.
Fracture: Conchoidal.
Hardness: $2\frac{1}{2}$.
Density: 1.60.
Color: Colorless to milk-white, often reddish due to enclosed oriented scales of hematite.
Luster: Greasy, dull.
Taste: Bitter.
Optical properties: $\alpha = 1.467$, $\beta = 1.475$, $\gamma = 1.494$; $\gamma - \alpha = 0.027$; positive, $2V = 70°$.
Diagnostic features: It is distinguished from other salts by the lack of cleavage, conchoidal fracture, and deliquescent nature; easily soluble in water, giving standard tests for chlorine and magnesium; easily fusible $(1-1\frac{1}{2})$ with violet flame.
Occurrence: Carnallite is found in the upper layers of saline evaporite deposits associated with sylvite, halite, and *kieserite* ($MgSO_4 \cdot H_2O$). These minerals contribute to the magnesium content of natural brines from which some magnesium is produced, and carnallite also provides some potash. The principal occurrence is in the salt deposits of Stassfurt, Germany. It is also found with sylvite in the potash-bearing salt beds of New Mexico, Texas, and Saskatchewan. Specimens of this mineral must be kept in sealed bottles, for in a humid atmosphere they will deliquesce.

Cryolite, Na_3AlF_6

Crystal system and class: Monoclinic; $2/m$.
Axial elements: $a:b:c = 0.964:1:1.389$; $\beta = 90° 11'$.
Cell dimensions and content: $a = 5.40$, $b = 5.60$, $c = 7.78$; $Z = 2$.

Common forms and angles:
   (110) ∧ (1Ī0) = 87° 54'   (001) ∧ (101) = 55° 18'
   (001) ∧ (011) = 54° 15'   (001) ∧ (Ī01) = 55° 11'

Habit: Crystals rare, cuboidal (pseudocubic) with {001} and {110}; {101}, {Ī01}, and {011} together simulate an octahedron; massive, coarsely granular.

Twinning: Very common with several twin laws occurring together; [110], [021], [Ī11], [100] as twin axes as well as others, repeated or polysynthetic, reflecting the pseudo-isometric character of the cell bounded by {001} and {110}.

Cleavage: None; parting on {001} and {110} giving cuboidal forms; brittle.

Hardness: $2\frac{1}{2}$.

Density: 2.97.

Color: Colorless to white; also brownish or reddish.

Streak: White.

Luster: Vitreous to greasy.

Optical properties: $\alpha = 1.339$, $\beta = 1.339$, $\gamma = 1.340$; $\gamma - \alpha = 0.001$; positive, $2V = 43°$.

Diagnostic features: It is characterized by the cuboidal form of the parting fragments, the white color, and the greasy luster; easily fusible, giving a strong yellow flame due to sodium; a test for fluorine is easily obtained, as for fluorite.

Occurrence: Cryolite is not a common mineral, and is found in important quantities at only two localities—Ivigtut in West Greenland, and Miask, in the USSR. At Ivigtut it occurs in a pegmatitic body associated with a small granite stock. Microcline, quartz, siderite, fluorite, and minor amounts of sulfides occur with the cryolite. It is often altered to other rare fluorides. The importance of this unusual mineral derives from its use, in molten form, as the electrolyte in the Hall-Héroult process for the electrolytic reduction of aluminum ores to metallic aluminum. Much of the cryolite used in this operation is now manufactured from fluorite by a process developed when supplies of cryolite were liable to interruption during World War II.

12

Class V:
Carbonates, Nitrates, Borates

H																	
			C											*N*	*O*	*F*	
Na	Mg	*Al*												*P*	*S*		
K	Ca				Mn	Fe	Co	Ni	Cu	Zn							
	Sr									Cd							
	Ba	La Lu										Pb	Bi				
			U														

The carbonates include some very common and widespread minerals.* In those whose structures are known, the fundamental anionic unit is $(CO_3)^{-2}$, which is a planar group with C at the center of an equilateral triangle of three O atoms. The structures of the anhydrous carbonates are anisodesmic oxysalts, but the structures of most of the hydrated and basic carbonates are not known. The nitrates, which are of rare occurrence as minerals, include some which are isostructural with carbonates—for example, calcite and soda niter.

* Table above shows elements which form carbonate minerals; those in italics are constituents of a few complex carbonates.

In the borates the anionic group may be simple $(BO_3)^{-3}$, as in orthoboric acid, or in the form of anion complexes which include both BO_3 triangles and BO_4 tetrahedra (borax and colemanite). In these complexes some of the oxygen positions are occupied by (OH). Boron may also occur in fourfold coordination with oxygen in some borosilicates.

Some carbonate minerals form extensive rock masses in sedimentary and metamorphosed sedimentary rocks. Thus calcite is the principal constituent of limestone and marble, and dolomite is the principal constituent of the rock of the same name and of dolomitic marbles. Other carbonates occur in evaporite deposits. Carbonates are also commonly found in hydrothermal vein deposits, and often accompany sulfide minerals in hydrothermal replacement zones.

Nitrates, being very soluble minerals, are found only in very arid, virtually rainless, sections of the earth's crust. Borates are found for the most part in dry lake basins in mountainous regions or in lake bed formations of Tertiary age.

Carbonates

The most common carbonate minerals are members of the calcite, dolomite, and aragonite groups:

CALCITE GROUP	DOLOMITE GROUP	ARAGONITE GROUP
Calcite, $CaCO_3$	Dolomite, $CaMg(CO_3)_2$	Aragonite, $CaCO_3$
Magnesite, $MgCO_3$	Ankerite, $CaFe(CO_3)_2$	Witherite, $BaCO_3$
Siderite, $FeCO_3$	Kutnahorite,	Strontianite, $SrCO_3$
Rhodochrosite,	$CaMn(CO_3)_2$	Cerussite, $PbCO_3$
$MnCO_3$		
Smithsonite, $ZnCO_3$		
Symmetry: $\bar{3}\,2/m$	Symmetry: $\bar{3}$	Symmetry: $2/m\,2/m\,2/m$

Each group includes three or more isostructural substances. $CaCO_3$ is polymorphous as calcite and aragonite. The other members of the calcite group with metal ions of smaller radius than Ca^2 do not crystallize with the aragonite structure, and the members of the aragonite group with larger metal ions than Ca^2 do not crystallize with the calcite structure.

Dolomite is closely related to calcite; in fact they are almost isostructural, the rhombohedral lattices in both having similar dimensions and the minerals of both groups having similar rhombohedral cleavage form with interfacial angles of 73–75° and 105–107°. In calcite the Ca ions are all structurally equivalent and occur in rows parallel to the threefold axis (Fig. 12-1), whereas in dolomite these positions are occupied alternately by Ca and Mg ions. This structure differs from the calcite structure in not possessing twofold axes parallel to the hexagonal *a* axes or the symmetry planes perpendicular to them. An exactly parallel relationship exists between hematite, Fe_2O_3, and ilmenite, $FeTiO_3$.

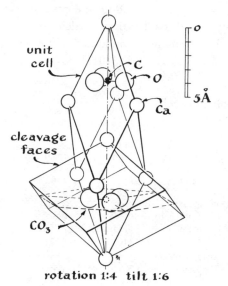

unit
cell

cleavage
faces

CO_3

rotation 1:4 tilt 1:6

Figure 12-1 Structure of calcite. The axial ratio given in the text corresponds to the cleavage rhombohedron as the unit cell which is a face-centered rhombohedron, whereas the simpler structural (rhombohedral) unit cell containing $2[CaCO_3]$ is shown in this figure.

The many chemical analyses of minerals of these groups show that they have a wide range of atomic substitution. However, since the conditions of temperature, pressure, and availability of ions at the time of formation are not known for most mineral specimens, it is not possible to outline in detail the limits of substitution of one ion for another in any of these structures or the temperatures and pressures favorable to such substitution. Laboratory investigations of artificial materials are slowly providing information of this sort. The extent of substitution occurring in the minerals is indicated under each species.

CALCITE GROUP

The structure of calcite (Fig. 12-1) is analogous to that of halite (Fig. 11-1) if we consider the unit cube of halite to be shortened along one trigonal axis to give interaxial angles of 101° 55′ (the interedge angles in the calcite cleavage rhombohedron) instead of 90° in halite. The Ca atoms are situated at the points of a face-centered rhombohedron (corresponding to face-centered cubic positions in halite, Cl in Fig. 11-1). The CO_3 ions are situated midway between Ca atoms along rows parallel to the rhombohedral cleavage edges (corresponding to the Na positions in Fig. 11-1); thus one CO_3 group is situated at the body center of the rhombohedron. The CO_3 ions are trigonal in form, planar parallel to $\{0001\}$, and each group is turned $60°$ with respect to the next one (note orientation of CO_3 groups along c in Fig. 12-1). The simplest unit cell of the structure is a rhombohedron with an interedge angle of $46° 4'$, corresponding to hexagonal dimensions $a = 4.989$, $c = 17.062$. This unit cell corresponds to the rhombohedron $\{40\bar{4}1\}$ in the conventional axial ratio for calcite based on the cleavage rhombohedron as $\{10\bar{1}1\}$. In the following mineral descriptions all indices apply to the axial ratio in which the cleavage is $\{10\bar{1}1\}$.

Crystal system and class: Trigonal; $\bar{3}\ 2/m$.

Cell content: $Z = 2$ (for simplest rhombohedral cell), $Z = 6$ (hexagonal unit cell).

	CALCITE	MAGNESITE	SIDERITE	RHODO-CHROSITE	SMITHSONITE
Axial elements:					
Morphology $(a{:}c)$	$1{:}0.8550$	$1{:}0.8102$	$1{:}0.819$	$1{:}0.820$	$1{:}0.804$
Structure $\quad(a{:}c)$	$1{:}3.4199$	$1{:}3.2409$	$1{:}3.275$	$1{:}3.280$	$1{:}3.215$
Cell dimensions:	$a = 4.989$	$a = 4.633$	$a = 4.72$	$a = 4.777$	$a = 4.653$
	$c = 17.062$	$c = 15.015$	$c = 15.46$	$c = 15.67$	$c = 15.028$
Cell content:	$6[CaCO_3]$	$6[MgCO_3]$	$6[FeCO_3]$	$6[MnCO_3]$	$6[ZnCO_3]$
Cleavage:	$\{10\bar{1}1\}$	$\{10\bar{1}1\}$	$\{10\bar{1}1\}$	$\{10\bar{1}1\}$	$\{10\bar{1}1\}$
Hardness:	3	4	4	4	$4{-}4\frac{1}{2}$
Density:	2.711 (calc.)	3.009 (calc.)	3.96 (pure)	3.69 (calc.)	4.43 (pure)
Optical properties:	$\epsilon = 1.486$	$\epsilon = 1.509$	$\epsilon = 1.635$	$\epsilon = 1.597$	$\epsilon = 1.625$
	$\omega = 1.658$	$\omega = 1.700$	$\omega = 1.875$	$\omega = 1.816$	$\omega = 1.850$
	negative	negative	negative	negative	negative

Calcite, $CaCO_3$

Common forms and angles:

$(0001) \wedge (10\bar{1}1) = 44° 38'$ $(01\bar{1}0) \wedge (01\bar{1}2) = 63° 44'$

$(10\bar{1}1) \wedge (\bar{1}101) = 74° 57'$ $(21\bar{3}1) \wedge (3\bar{1}\bar{2}1) = 35° 35'$

$(40\bar{4}1) \wedge (\bar{4}401) = 114° 10'$ $(0001) \wedge (02\bar{2}1) = 63° 08'$

Habit: Crystals common and extremely varied in development, prismatic along c, tabular on $\{0001\}$, rhombohedral $\{01\bar{1}2\}$, $\{02\bar{2}1\}$, $\{40\bar{4}1\}$, scalenohedral $\{21\bar{3}1\}$, (Fig. 12-2, 12-3; Fig. 2-33) parallel or subparallel aggregates; massive, coarse to very fine granular; stalactitic, nodular, tuberose, and coralloidal shapes; oolitic or pisolitic (Fig. 12-4), usually concentrically banded and internally radiating.

Twinning: Two twin laws are commonly recognized on calcite: (1) Twin plane and composition plane $\{0001\}$ (see Fig. 12-2a; Fig. 2-42a); re-entrant angles are often present about the equator of the twinned crystal except when bounded by $\{10\bar{1}0\}$. (2) Twin plane and composition plane $\{01\bar{1}2\}$, as simple contact twins (Fig. 2-42b); also lamellar (Fig. 12-5), and produced by pressure, in metamorphism, or artificially, often resulting in visible

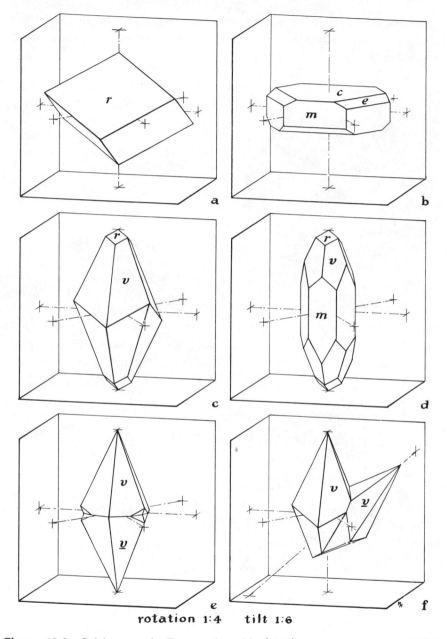

rotation 1:4   tilt 1:6

Figure 12-2 Calcite crystals. Forms: pinacoid $c\{0001\}$; hexagonal prism $m\{10\bar{1}0\}$; positive rhombohedron $r\{10\bar{1}1\}$; negative rhombohedron $e\{01\bar{1}2\}$; hexagonal scaleno-hedron $v\{21\bar{3}1\}$. (e) Twinned on (0001) as twin plane and composition plane. (f) Twinned on $(\bar{2}20\bar{1})$ as twin plane and composition plane.

striations on the $\{10\bar{1}1\}$ cleavage planes in crystalline limestone (Fig. 12-6). Less commonly, twin crystals are found with twin plane $\{10\bar{1}1\}$ (Fig. 12-7) or $\{02\bar{2}1\}$ (Fig. 12-2f).

Figure 12-3 Two crystals of honey-colored calcite in random intergrowth, with forms $r\{10\bar{1}1\}$ and $v\{21\bar{3}1\}$. (Joplin, Missouri.) [Courtesy Queen's University.]

Chemistry: Calcite is the stable form of $CaCO_3$ at most temperatures and pressures, but recent work shows that it can change to aragonite through high-pressure metamorphism; it is also less stable than aragonite below $-60°C$. Calcite may be pure $CaCO_3$, but often contains other metals in substitution for Ca (Fig. 12-8). Minor substitution of Mg occurs in calcite, but the tend-

Figure 12-4 Pisolitic limestone, broken surface showing concentric layering in individual spherules (4 mm maximum diameter). (Carlsbad, Germany.) [Courtesy Queen's University.]

ency to form the ordered phase dolomite is very great, in keeping with the large difference (26.5 percent) in ionic radii between Ca and Mg. Fe^2 substitutes for calcium to a somewhat greater extent (21.7 percent difference in ionic radii), and the ordered phase ankerite is less common. Mn^2 substitutes extensively for Ca (14.3 percent difference in radius), and the disordered phase shows extensive solubility; however, the ordered phase kutnahorite has been identified at several localities. At Franklin, New Jersey, kutna-

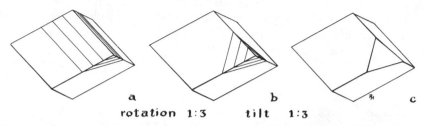

Figure 12-5 (a) Lamellar twinning in calcite, with (01$\bar{1}$2) as twin plane with striations parallel to long diagonal of the rhombic cleavage plane. (b) Lamellar twinning in dolomite, with (02$\bar{2}$1) as twin plane with striations parallel to short diagonal of rhombic cleavage plane. (c) Edge between (10$\bar{1}$1) and (02$\bar{2}$1) as short diagonal of rhombic (10$\bar{1}$1) face.

horite, calcian rhodochrosite, and manganoan calcite have been identified. At this locality solid solution extends from pure $CaCO_3$ to 40 weight percent $MnCO_3$, and from pure $MnCO_3$ to about 25 weight percent $CaCO_3$. Zn and Co have been found substituting for Ca to a minor extent.

Density: 2.71 (pure), increases with substitution of Fe, Mn, or Zn.

Color: Colorless and transparent or white when pure; varying widely with substitution or with mechanically included material: yellow, brown, pink, blue, lavender, greenish, gray, black, greenish due to chlorite, reddish brown due to hematite; sometimes fluorescent and phosphorescent under ultraviolet, x-rays, cathode rays, or sunlight.

Figure 12-6 Lamellar twinning on {01$\bar{1}$2} in salmon-pink calcite, showing striations (horizontal light lines on cleavage plane) parallel to long diagonal of cleavage rhomb. Black crystals are tourmaline with typical triangular cross sections and prismatic striations. (Frontenac County, Ontario.) [Courtesy Queen's University.]

Streak: White to grayish.

Luster: Vitreous.

Diagnostic features: Calcite is infusible. It effervesces freely in cold dilute hydrochloric acid. It is characterized by its hardness (3), perfect rhombohedral cleavage with angles between adjacent planes of 75° or 105°, and vitreous luster. It is distinguished from dolomite by the ready effervescence of fragments (as well as powder) in acid, as noted above, and by the direction of twin striations (Fig. 12-5); it is distinguished from aragonite by its lower specific gravity and cleavage.

Figure 12-7   Calcite. Twin crystal on $\{10\bar{1}1\}$ as twin plane, with prism $m\{10\bar{1}0\}$ and negative rhombohedron $e\{01\bar{1}2\}$. Alternate faces of $\{10\bar{1}0\}$ are horizontally striated, due to narrow $\{h0\bar{h}l\}$ faces. (Sudbury, Ontario.) [Courtesy Queen's University.]

Occurrence: Calcite is a very common and widely distributed mineral in the earth's crust. It is an important rock-forming mineral in sedimentary and metamorphosed sedimentary rocks. It occurs in nearly pure form in great thicknesses of chalk and limestone, and as a cementing material in other sedimentary rocks. Crystalline limestone, known as "marble," may vary from almost pure calcite, as in the pure white statuary marble (Carrara, Italy), to a varied assemblage of minerals including calcite, diopside, tremolite, and many others, depending on the composition of the original limestone, the conditions of metamorphism, and the materials added from adjacent rock masses or igneous sources. Calcite is a common constituent of altered basic igneous rocks, where it develops by alteration of calcium silicates; it also develops as a late hydrothermal deposit in veins, both of meteoric and hydrothermal origin, in which it is typically associated with sulfides, quartz, barite, fluorite, dolomite, and siderite. Crystallized material is often found lining vugs and fractures in limestone, also in larger caves and solution channels as stalactites and stalagmites. From springs and streams it is often deposited as travertine, tufa, or calc-sinter.

Calcite of optical quality has been obtained from a cavernous zone of basalt in Iceland (hence the common name Iceland spar for optical material), as well as from New Mexico and elsewhere. In the United States exceptional crystals have been found in the Michigan copper deposits and in the low-temperature lead-zinc veins of the central states.

Production and uses: Calcite, in its wide variety of occurrences, has many uses. The clear transparent material has long been used for optical purposes, particularly in the construction of polarizing prisms. From very early times

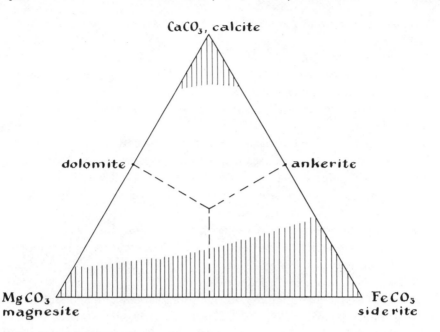

Figure 12-8 Triangular diagram showing the compositional variations of calcite, magnesite, and siderite. Shaded areas indicate the limits of solid solution observed in nature.

calcite or limestone has been burned to quicklime (CaO), slaked to hydrated lime $Ca(OH)_2$, and mixed with sand to make mortar. It also finds wide use in chemical industries and as a fertilizer. Portland cement, now widely used in concrete for building purposes, is made by burning a finely ground mixture of 75 percent $CaCO_3$ and 25 percent clayey material. Limestone is the raw material for some rock-wool insulating material. Limestone and lime find varied uses in the metallurgical processes for smelting both iron and nonferrous metals, where they are usually used as a flux to help remove impurities in the slag. Finally, limestone of many types is widely used in building, both as dimension stone for actual construction or as a facing or veneer on concrete walls, as interior ornamental finish stone, or for carvings and statuary. Crushed limestone is also widely used as the coarse aggregate in concrete and as railroad ballast, and is mixed with asphalt for road surfacing.

Magnesite, $MgCO_3$

Common forms and angles:

$(10\bar{1}1) \wedge (\bar{1}101) = 72° 33'$ $(0001) \wedge (10\bar{1}1) = 43° 06'$

Habit: Crystals rare, rhombohedral $\{10\bar{1}1\}$ or prismatic along c; commonly massive, coarse- to fine-grained, or very compact, earthy to chalky, lamellar or coarsely fibrous.

Chemistry: In $MgCO_3$, Fe^2 may substitute for Mg, and a complete series extends to siderite. Mn and Ca substitute for Mg to a small extent (Fig. 12-8).

Density: 3.00 (pure), increasing with substitution of Fe for Mg.

Color: Colorless, white, grayish white, yellowish to brown.

Streak: White.

Luster and opacity: Vitreous; transparent to translucent.

Diagnostic features: Infusible; scarcely affected by cold hydrochloric acid, but dissolves with effervescence in hot hydrochloric acid. The addition of excess ammonia and sodium phosphate to the solution will give a white precipitate of ammonium magnesium phosphate.

Occurrence: The compact massive forms of magnesite may contain SiO_2 or magnesium silicates as impurities, and they resemble chert in appearance but are inferior in hardness. Magnesite is much less common than calcite and only rarely occurs as a sedimentary rock. It may be formed in a variety of ways: through alteration of rocks consisting largely of magnesium silicates (serpentine, olivine, pyroxene) by carbonated waters, usually producing cryptocrystalline material; as crystalline stratiform beds of metamorphic origin with talc-chlorite- or mica-schists; through replacement of calcite rocks by magnesium-bearing solutions (dolomite is found as an intermediate product).

The replacement deposits are usually distinctly crystalline and are found associated with, or in, dolomite. Extensive deposits of this type are mined in Washington, and in Austria, Manchuria, Czechoslovakia, and Quebec. Veins, and masses resulting from alteration of magnesian silicates, are mined extensively in California, and in Greece, India, the USSR, and Yugoslavia. Magnesite is calcined for the manufacture of refractory bricks, cements, and flooring, and it is used for making magnesium metal.

Siderite, $FeCO_3$

Common forms and angles:

(10$\bar{1}$1) $\wedge$ ($\bar{1}$101) = 73° 00′   (0001) $\wedge$ (05$\bar{5}$1) = 78° 03′

Habit: Crystals commonly rhombohedral {10$\bar{1}$1}, less often {01$\bar{1}$2}, {02$\bar{2}$1}, {40$\bar{4}$1}; also thin to thick tabular on {0001}; crystal faces often curved or composite; massive, coarse- to fine-grained, botryoidal or globular, oolitic, earthy, or stony.

Chemistry: Mg may substitute for Fe^2 to form a complete series to magnesite (Fig. 12-8), and Mn^2 may also substitute for Fe^2. Ca may substitute for Fe^2

to a minor extent, but the difference in radius between these ions precludes extended solid solution and the ordered phase ankerite often occurs.

Density: 3.96, decreasing with substitution of Mn^2, Mg, or Ca for Fe^2.

Color: Yellowish brown, and grayish brown to brown and reddish brown, also gray, yellowish gray, greenish gray; the dark colors are due to partial oxidation of Fe^2.

Streak: White.

Luster and opacity: Vitreous; translucent.

Diagnostic features: Becomes magnetic on heating; soluble in hot hydrochloric acid with effervescence. Siderite may be distinguished from other carbonates by its color, specific gravity, and alteration to iron oxides, and from sphalerite by its rhombohedral cleavage.

Occurrence: Siderite is widespread as a bedded deposit in sedimentary rocks, associated with clay, shale, or coal seams. It is usually massive and fine-grained, sometimes concretionary; impure varieties include clay-ironstones and bituminous or black-band ores. These deposits have been used as iron ores in Poland (Radom), in England (Somerset, Durham, Yorkshire), and in Pennsylvania. They are also found in a metamorphosed condition, as in Austria (Erzberg, Huttenberg) and in Ontario (Michipicoten).

Siderite is often present in hydrothermal veins as a gangue mineral of primary origin, as in the silver-lead veins of the Coeur d'Alene district, Idaho, the Slocan district, British Columbia, and at Freiberg, Germany. It occasionally is the predominant mineral in some veins. It may also occur in some quantity as a hydrothermal replacement of limestone by iron-bearing solutions. Deposits of this type constitute important iron-ores at Bilbao, Spain, and in Algeria. It is found rarely in pegmatites, as it is in Greenland with cryolite. The mineral readily alters to hematite or limonite, and important residual bodies of these minerals have been formed in this way, particularly in Algeria.

Rhodochrosite, $MnCO_3$

Common forms and angles:

$(10\bar{1}1) \wedge (\bar{1}101) = 73°\ 04'$ $\qquad$ $(1\bar{1}02) \wedge (01\bar{1}2) = 43°\ 30'$

Habit: Distinct crystals uncommon, rhombohedral $\{10\bar{1}1\}$; often rounded or composite; massive coarsely granular to compact; columnar, incrusting, also globular and botryoidal.

Chemistry: Rhodochrosite often contains small amounts of Fe^2 substituting for Mn^2. A complete series may extend to siderite. Ca commonly substitutes for Mn^2, and a series extends to about 25 weight percent $CaCO_3$. With larger amounts of $CaCO_3$ the ordered phase kutnahorite with the dolomite structure is the stable mineral.

Color: Various shades of pink, rose, and rose red, also yellowish gray to brown.

Streak: White.

Luster and opacity: Vitreous; translucent.

Diagnostic features: Infusible; soluble in hot hydrochloric acid with effervescence; turns black but remains nonmagnetic when heated on charcoal. It is distinguished by its pink color, rhombohedral cleavage, and hardness of 4. The pink manganese silicate, rhodonite, has a hardness of 6.

Occurrence: Rhodochrosite is often found as a primary gangue mineral in low- or moderate-temperature hydrothermal veins. It occurs with silver, lead, zinc, and copper sulfide ores at Butte, Montana, and at other localities in the western USA, and in Europe; it may be associated with calcite, siderite, dolomite, fluorite, barite, quartz, manganite, alabandite (MnS), tetrahedrite, and sphalerite. It also occurs with rhodonite, garnet, hausmannite, and other minerals in high-temperature metasomatic deposits, or as a secondary mineral in residual or sedimentary manganese oxide deposits. A substantial proportion of the high-grade manganese ore produced in the USA is rhodochrosite, obtained from veins in the Butte district of Montana.

Smithsonite, $ZnCO_3$

Common forms and angles:

  $(10\bar{1}1) \wedge (\bar{1}101) = 72° 12'$ $(01\bar{1}2) \wedge (1\bar{1}02) = 42° 46'$

Habit: Distinct crystals rare, rhombohedral $\{10\bar{1}1\}$ or $\{02\bar{2}1\}$; faces curved and rough, or composite; usually botryoidal reniform or stalactitic; as crystalline incrustations; coarse-grained to compact; earthy, or as porous or cavernous masses.

Chemistry: Fe^2, Ca, Co, Cu, Mn, Cd, and Mg may substitute for Zn but are rarely found in excess of a few percent. Smithsonite is isostructural with calcite, in which the metal ion is in octahedral coordination, an unusual coordination for zinc. Zinc is normally found in tetrahedral coordination with oxygen (hemimorphite, willemite, zincite) and with sulfur (sphalerite, wurtzite).

Color: Grayish white to dark gray, greenish or brownish white, green to apple green, bluish green, blue, yellow, brown, white.

Streak: White.

Luster and opacity: Vitreous; translucent.

Diagnostic features: Infusible; soluble in warm hydrochloric acid with effervescence; when heated in the reducing flame on charcoal, it gives a coating of zinc oxide, yellow when hot, white when cold. Smithsonite is distinguished by its effervescence in acids, tests for zinc, hardness of 4, and high specific gravity.

Occurrence: Smithsonite is a secondary mineral found in the oxidized zone of ore deposits and derived from the alteration of primary zinc minerals. It may also be found replacing calcareous rocks adjacent to the ore deposits. It is usually associated with hemimorphite and secondary lead and copper minerals such as cerussite, malachite, anglesite, and pyromorphite.

Although smithsonite is often found in the surficial oxidized zone of lead-zinc deposits, and has provided some of the zinc produced from those deposits, it is rarely reported as an important ore of zinc. It is a major ore mineral in the mining districts of northern Mexico. Smithsonite also occurs at Laurium, Greece; Beuthen, Silesia; Bytom, Poland; Moresnet, Belgium; Tsumeb, South West Africa; Monarch and Leadville, Colorado; Tintic, Utah; and other localities.

DOLOMITE GROUP

The dolomite structure results from the regular substitution of alternate atoms of Ca in calcite by another divalent metal, usually Mg, Fe, or Mn. This is an ordered structure which results because of the difference in radii between Ca and the other metals. Of the minerals in this group, dolomite is the most common, consistent with the large difference in radii between Ca and Mg and the very limited solid solution between calcite and magnesite. Ankerite is less common, and kutnahorite, wherein the radius difference is less marked, is of rare occurrence. The inequivalence of Ca with the second metal results in a lower symmetry for this ordered structure as compared with calcite.

Crystal system and class: Trigonal, $\bar{3}$.
Cell content: $Z = 1$ for simplest rhombohedral cell, $Z = 3$ for hexagonal unit cell.

	DOLOMITE	ANKERITE	KUTNAHORITE
Axial elements:			
Morphology			
($a:c$)	1:0.8235	1:0.835	1:0.842
Structure ($a:c$)	1:3.2941	1:3.340	1:3.369
Cell dimensions:	$a = 4.842$	$a = 4.832*$	$a = 4.85$
	$c = 15.95$	$c = 16.14$	$c = 16.34$
Cell content:	$3[CaMg(CO_3)_2]$	$3[Ca(Mg,Fe)(CO_3)_2]$	$3[CaMn(CO_3)_2]$
Cleavage:	$\{10\bar{1}1\}$	$\{10\bar{1}1\}$	$\{10\bar{1}1\}$

* Dimensions for mineral with Fe:Mg = 1:1.1.

	DOLOMITE	ANKERITE	KUTNAHORITE
Hardness:	$3\frac{1}{2}$–4	$3\frac{1}{2}$–4	$3\frac{1}{2}$–4
Density:	2.85	3.01	3.12
Optical properties:	$\epsilon = 1.500$	$\epsilon = 1.51 - 1.55$	$\epsilon = 1.535$
	$\omega = 1.679$	$\omega = 1.69 - 1.75$	$\omega = 1.727$
	negative	negative	negative

Dolomite, $CaMg(CO_3)_2$

Common forms and angles:

$(10\bar{1}1) \wedge (\bar{1}101) = 73° 16'$ $(0001) \wedge (40\bar{4}1) = 75° 16'$

Habit: Crystals commonly rhombohedral $\{10\bar{1}1\}$ (Fig. 12-9) or $\{40\bar{4}1\}$ also prismatic with $\{11\bar{2}0\}$ or tabular on $\{0001\}$; $\{10\bar{1}1\}$ is often striated horizontally, also commonly curved or made up of subparallel individuals passing into saddle-shaped forms; massive, coarse to fine granular, columnar; as a rock-forming mineral usually compact, massive, like ordinary limestone, or granular.

Figure 12-9 Dolomite rhombohedra, white translucent, with honey-colored, etched calcite scalenohedron, from a vug in dolomite. (Penfield, New York.) [Courtesy Queen's University.]

Twinning: Common on $\{0001\}$, also on $\{10\bar{1}0\}$ and $\{11\bar{2}0\}$; lamellar twinning on $\{02\bar{2}1\}$ (Fig. 12-5b) common in grains of dolomite in marble.

Chemistry: Fe^2 is commonly found substituting for Mg in dolomite, and a complete series probably extends to ankerite. This series is now arbitrarily divided at Mg:Fe = 1:1, although in the past much material now referred to as ferroan dolomite was called ankerite. Mn is present only to the extent of a few percent, then usually with iron; manganese may also substitute for iron in ankerite. Co and Zn may also substitute for Mg to a minor extent.

Density: 2.85 for pure dolomite, increasing to about 3.02 for a mineral with Mg:Fe = 1:1.

Color: Colorless and transparent or white, gray or greenish, becoming yellowish brown or brown with increasing Fe^2 content; also pink or rose. Ferroan dolomite and ankerite turn dark brown or reddish on weathering, as does siderite.

Streak: White.

Luster: Vitreous to pearly.

Diagnostic features: Dolomite is only slightly attacked by cold dilute acids, in contrast to calcite, but dissolves readily in warm acids; it is infusible before the blowpipe and glows brightly; ankerite and ferroan dolomite darken in color and become magnetic when heated in the blowpipe flame. In coarsely crystalline granular material the direction of twinning lamellae distinguish calcite and dolomite (Fig. 12-5).

Occurrence: Dolomite is of widespread occurrence in sedimentary strata. In this form it is believed to have originated through transformation of limestone or coral (dolomitization) by magnesium-bearing solutions. Most dolomite rocks are mixtures of dolomite and calcite. Dolomites may be recrystallized by metamorphism, and siliceous dolomite rocks may develop diopside, tremolite, and other silicates in this process. The mineral also occurs in hydrothermal veins with fluorite, barite, calcite, siderite, quartz, and metallic ore minerals; in cavities and veins in limestone or dolomite rocks; or as veins or crystals in serpentine, talcose rocks, and altered basic igneous rocks. It occurs commonly as a gangue mineral with the lead-zinc deposits of Missouri and adjoining areas; in sulfide veins of many mining districts of the western USA; also in the metalliferous veins of Cornwall, England.

The rock-forming dolomite is usually low in iron, but ferroan dolomite is frequently found with iron ores. It is also found as a gangue mineral in gold-quartz veins and as a replacement mineral in the adjoining rocks in the Mother Lode district of California, and in the Porcupine and Larder Lake gold camps of Ontario and Quebec. In these occurrences it is often found with bright green chromian muscovite (or sericite). Dolomite is used as a source of magnesium or calcium metals, and of magnesia for refractory bricks or blast furnace fluxes.

ARAGONITE GROUP

Aragonite is polymorphous with calcite. The structure of aragonite is stable for the larger metal ions, whereas the calcite structure is stable for calcium and smaller ions. A parallel is found in the nitrates, wherein soda niter ($NaNO_3$) is isostructural with calcite, and niter (KNO_3), with a larger cation, is isostructural with aragonite.

The aragonite structure includes planar trigonal CO_3 groups parallel to {001}. The calcium atoms lie approximately in the positions of hexagonal close packing, which gives aragonite a pseudo-hexagonal character that is responsible for its great tendency to form twin crystals. Each CO_3 group lies between six

Ca atoms, but arranged such that each O is linked to three Ca atoms, whereas in calcite each O is linked to two Ca atoms. Calcium is again in octahedral coordination with oxygen.

Crystal system and class: Orthorhombic; $2/m\ 2/m\ 2/m$.

	ARAGONITE	WITHERITE	STRONTIANITE	CERUSSITE
Axial elements:				
$(a:b:c)$	0.6224:1:0.7205	0.5967:1:0.7221	0.6070:1:0.7165	0.6158:1:0.7292
Cell dimensions:				
	$a = 4.959$	$a = 5.313$	$a = 5.107$	$a = 5.195$
	$b = 7.968$	$b = 8.904$	$b = 8.414$	$b = 8.436$
	$c = 5.741$	$c = 6.430$	$c = 6.029$	$c = 6.152$
Cell content:	$4[CaCO_3]$	$4[BaCO_3]$	$4[SrCO_3]$	$4[PbCO_3]$
Hardness:	$3\frac{1}{2}$–4	3–$3\frac{1}{2}$	$3\frac{1}{2}$	3–$3\frac{1}{2}$
Density:	2.930 (calc.)	4.308 (calc.)	3.785 (calc.)	6.582 (calc.)
Optical properties:				
α	1.530	1.529	1.518	1.803
β	1.680	1.676	1.665	2.074
γ	1.685	1.677	1.667	2.076
$2V(-)$	18°	16°	8°	8°

Aragonite, $CaCO_3$

Common forms and angles:

   $(010) \wedge (110) = 58°\ 06'$ $(011) \wedge (0\bar{1}1) = 71°\ 33'$

   $(001) \wedge (111) = 53°\ 45'$ $(010) \wedge (091) = \ \ 8°\ 46'$

Habit: Untwinned crystals are very rare; short to long prismatic along c, also acicular or chisel-shaped with a steep prism $\{091\}$ and pyramids $\{991\}$

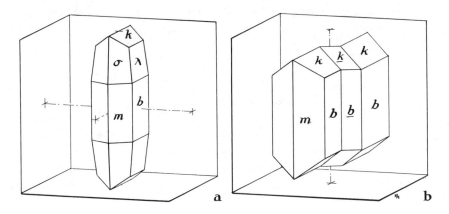

Figure 12-10   Aragonite crystals. Forms: pinacoid $b\{010\}$; rhombic prisms $m\{110\}$, $k\{011\}$, $\lambda\{091\}$; rhombic dipyramid $\sigma\{991\}$. (b) Twinned on $\{110\}$ as twin plane and composition plane.

(Fig. 12-10a); pseudohexagonal symmetry is often marked; sometimes in columnar aggregates or crusts, as radiating groups of acicular crystals; in coralloid, reniform, or globular shapes, pisolitic with a radially fibrous and concentrically zoned structure; stalactitic.

Twinning: Twin plane {110}, very common, usually repeated (Fig. 12-10b), resulting in pseudohexagonal aggregates of both contact and penetration types (Fig. 12-11), also as thin polysynthetic lamellae producing fine striations on {001} or parallel to *c*.

Figure 12-11 Aragonite. Pseudohexagonal twinned crystal due to cyclic interpenetration twinning on {110} as twin plane. [Courtesy Queen's University.]

Cleavage: {010} distinct.

Fracture and tenacity: Subconchoidal; brittle.

Chemistry: Aragonite is polymorphous with calcite, which is the stable form under most natural conditions, and it may change spontaneously into calcite. Aragonite will begin to invert to calcite upon heating to 400°C in dry air or at lower temperatures in contact with water. When it is heated to dull red in the closed tube aragonite whitens and falls to powder as calcite. Sr and Pb are found to substitute for Ca to the extent of a few percent. Aragonite is formed under a much narrower range of conditions than calcite and is much less widespread, being found only in low-temperature near-surface deposits, and in rocks metamorphosed at high pressures and comparatively low temperatures.

Color: Colorless to white, also gray, yellowish, blue, green, pale to deep violet, rose red.

Streak: White.

Luster: Vitreous, transparent to translucent.

Diagnostic features: Infusible; soluble in cold dilute hydrochloric acid with effervescence; may be distinguished from calcite by cleavage and higher specific gravity. Columnar material has cleavage parallel to elongation, whereas calcite shows rhombohedral cleavage cross-cutting the columnar crystals.

Occurrence: The principal modes of occurrence are: (1) as crystals, pisolites, or sinter deposits from hot springs and geysers or as stalactites in caves; (2) as disseminated crystals or masses in gypsum or clay; (3) with limonite, calcite, malachite, smithsonite, and other secondary minerals in the oxidized zone of ore deposits; (4) in veins and cavities with calcite, dolomite, and magnesium minerals in serpentine, altered basic igneous rocks, and basalt; (5) with celestite and sulfur, as in Sicily; (6) in fossils or in the hard parts of living organisms, such as oysters, and in pearls.

Witherite, $BaCO_3$

Common forms and angles:

$(010) \wedge (110) = 59° 11'$ $(010) \wedge (021) = 34° 42'$

Habit: Crystals always twinned; repeatedly twinned on {110} giving pseudo-hexagonal dipyramids, also short prismatic parallel to c; faces usually rough and horizontally striated; also in globular, tuberose, and botryoidal forms; columnar, granular, or coarse fibrous.

Twinning: Universal on {110}.

Cleavage: {010} distinct.

Color: Colorless to milky, white, or grayish, also weakly tinted yellow, brown, or green.

Streak: White.

Luster and opacity: Vitreous; transparent to translucent.

Chemistry: In nature witherite varies little in composition, although a complete series with $SrCO_3$ does exist in artificial materials.

Diagnostic features: Fuses at 3; soluble with effervescence in dilute hydrochloric acid; gives a yellowish green flame (barium). Witherite is characterized by its high specific gravity, easy fusibility, green flame color, and solubility in HCl, the solution even when very dilute giving a precipitate on adding sulfuric acid.

Occurrence: It is not a common mineral, though as a barium mineral it is next in importance to barite ($BaSO_4$). It is found usually in low-temperature hydrothermal veins with barite and galena. Witherite occurs in large amount with barite in Mariposa County, California; also at Alston Moor, Cumberland, and at the Morrison mine, Durham, England; and with fluorite at Rosiclare, Illinois.

Strontianite, SrCO₃

Common forms and angles:

$(010) \wedge (110) = 58° 45'$	$(010) \wedge (021) = 34° 54'$
$(001) \wedge (111) = 54° 06'$	$(011) \wedge (0\bar{1}1) = 71° 14'$

Habit: Crystals short or long prismatic parallel to c, often pseudo-hexagonal due to equal development of $\{110\}$ and $\{010\}$, with $\{111\}$ and $\{021\}$; massive, columnar to fibrous, granular; in rounded masses.

Twinning: Very common on $\{110\}$ as contact twins, often repeated.

Cleavage: $\{110\}$ nearly perfect.

Density: 3.785 (calc. for pure SrCO₃), 3.72 (3.3 percent CaO), and less with increased Ca content.

Color: Gray; yellowish, or greenish.

Luster and opacity: Vitreous; translucent.

Chemistry: Strontianite almost always includes some Ca in substitution for Sr.

Diagnostic features: Infusible; on intense heating swells up, throws out fine sprouts, and colors the flame crimson (strontium). It effervesces in dilute hydrochloric acid, and a moderately dilute solution gives a precipitate of strontium sulfate on addition of sulfuric acid. It is characterized by its high specific gravity and its effervescence in acid; it is distinguished from witherite and aragonite by its flame color, and from celestite by its poor cleavage and solubility in acid.

Occurrence: It is typically a low-temperature hydrothermal mineral, usually associated with barite, celestite, and calcite in veins in limestone, less frequently with metallic minerals. It also occurs as concretionary masses in limestone or clay. Strontianite is less common than celestite but is preferred as an ore of strontium. Commercial quantities are obtained in the vicinity of Münster, Germany, as small veins in marl deposits. Large deposits occur in limestone in the Strontium Hills, San Bernadino County, California; it occurs with celestite in the cap rock of some salt domes in Texas and Louisiana.

Cerussite, PbCO₃

Common forms and angles:

$(010) \wedge (110) = 58° 22'$	$(010) \wedge (021) = 34° 31'$
$(001) \wedge (111) = 54° 16'$	

Habit: Crystals tabular on $\{010\}$, or dipyramidal and pseudohexagonal (Fig. 12-12); stellate and reticular aggregates due to twinning (Fig. 12-13); massive, granular to dense and compact.

Twinning: Almost universal; most common on $\{110\}$, repeated contact types

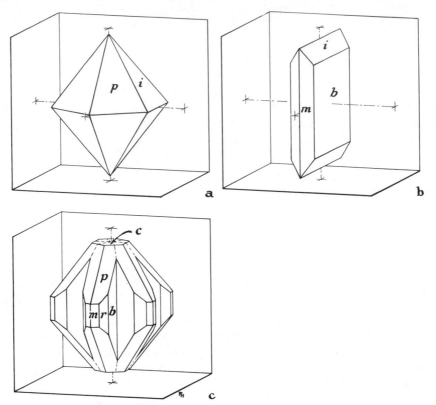

Figure 12-12 Cerussite crystals. Forms: pinacoids $c\{001\}$, $b\{010\}$; rhombic prisms $m\{110\}$, $r\{130\}$, $i\{021\}$; rhombic dipyramid $p\{111\}$. (c) Cyclic twin on $\{110\}$ as twin plane.

producing stellate aggregates (Fig. 12-12c) or as twin lamellae; also on $\{130\}$.

Cleavage: $\{110\}$, $\{021\}$ distinct.

Fracture and tenacity: Conchoidal; very brittle.

Color: Colorless to white and gray or smoky, sometimes darker colors due to impurities; white.

Streak: White.

Luster and opacity: Adamantine; translucent.

Diagnostic features: Fuses very easily; soluble in warm dilute nitric acid with effervescence; when heated with sodium carbonate on charcoal it gives a globule of lead and a yellow to white coating of lead oxide. It may be distinguished by its high specific gravity, white color, and adamantine luster; the effervescence in acid distinguishes it from anglesite.

Occurrence: Cerussite is a common mineral in the upper oxidized zone of ore deposits containing galena. It occurs as crystalline crusts or dense masses, concentrically banded and often with a core of unaltered galena. Anglesite, $PbSO_4$, may have formed as a first step in the oxidation and later altered to cerussite. It is usually associated with anglesite, limonite, malachite, smith-

Figure 12-13 Cerussite. Reticulated twin aggregate, crystals elongated on the *a*-axis and twinned on {110} as twin plane. (Broken Hill, Australia.) [Courtesy Royal Ontario Museum.]

sonite, and other secondary minerals. Fine crystalline material, usually twinned or in reticulate aggregates, has been found in many lead mining districts, particularly at Broken Hill, in New South Wales and in Zambia; at Leadhills, Scotland; at Tsumeb, South West Africa; and in the southwestern USA.

OTHER CARBONATES

Malachite, $Cu_2(CO_3)(OH)_2$

Crystal system and class: Monoclinic; $2/m$.
Axial elements: $a:b:c = 0.7935:1:2705$; $\beta = 98° 45'$.
Cell dimensions and content: $a = 9.502$, $b = 11.974$, $c = 3.240$; $Z = 4$.
Habit: Distinct crystals very rare; usually prismatic to fine acicular parallel to *c*, and grouped in tufts or rosettes; commonly massive or incrusting, with surface mammillary, botryoidal (Fig. 12-14), or tuberose forms; internally divergent and fibrous, and banded in color.
Twinning: Twin plane {100} very common.
Cleavage: {$\bar{2}01$} perfect, {010} fair.
Fracture: Uneven on massive material.
Hardness: $3\frac{1}{2}$–4.
Density: 4.05 (down to 3.6 for massive material).

Color: Bright green.

Streak: Pale green.

Luster and opacity: Silky, velvety, or dull; translucent.

Optical properties: $\alpha = 1.655, \beta = 1.875, \gamma = 1.909; \gamma - \alpha = 0.254$; negative, $2V = 43°$.

Diagnostic features: Fusible at 3, giving a green flame; upon heating with fluxes on charcoal it gives a copper globule. It may be distinguished by its green color, effervescence in acids, and common botryoidal forms.

Figure 12-14   Malachite. Section through botryoidal material; light green to very dark green. [Courtesy Queen's University.]

Occurrence: Malachite is widespread as a secondary copper mineral, often associated with the less common mineral azurite. It may be found as pseudomorphs after azurite or cuprite. It occurs commonly in the upper oxidized zone of copper deposits, especially in regions where limestone is present; it is also associated with limonite, calcite, chalcedony, chrysocolla, and less commonly with other secondary copper, lead, or zinc minerals.

At one famous locality in Siberia—Mednorudiansk, near the town of Nizhni Tagil—it occurred in such large banded masses that it was cut up and used as an ornamental stone in the form of table tops and other decorative objects. Important amounts are also found in copper mining districts of Africa, Australia, and the southern USA.

Azurite, $Cu_3(CO_3)_2(OH)_2$

Crystal system and class: Monoclinic; $2/m$.

Axial elements: $a:b:c = 0.8569:1:1.7686$; $\beta = 92° 27'$.

Cell dimensions and content: $a = 5.008$, $b = 5.844$, $c = 10.336$; $Z = 2$.

Common forms and angles:

    $(001) \wedge (100) = 87° 35'$ $(001) \wedge (\bar{1}02) = 47° 10'$

    $(010) \wedge (110) = 49° 38'$ $(001) \wedge (012) = 41° 22'$

Habit: Crystals varied in habit, tabular on $\{001\}$ or $\{102\}$, short prismatic along *c* or *b* (Fig. 12-15); also massive, or stalactitic, with a columnar or radial structure; earthy.

Cleavage: {011} perfect, {100} fair.

Fracture and tenacity: Conchoidal; brittle.

Hardness: $3\frac{1}{2}$–4.

Density: 3.77.

Color: Azure blue to very dark blue in crystals, lighter in massive or earthy types.

Streak: Light blue.

Luster and opacity: Vitreous; transparent.

Optical properties: $\alpha = 1.730$, $\beta = 1.758$, $\gamma = 1.838$; $\gamma - \alpha = 0.108$; positive, $2V = 68°$.

Diagnostic features: Chemical tests give the same results as for malachite; readily distinguished by its blue color.

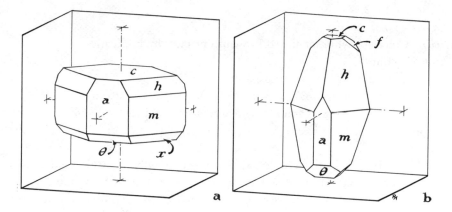

Figure 12-15 Azurite crystals. Forms: pinacoids $c\{001\}$, $a\{100\}$, $\theta\{\bar{1}02\}$; prisms $m\{110\}$, $f\{012\}$, $h\{111\}$, $x\{\bar{1}12\}$.

Occurrence: Azurite forms in the upper oxidized zone of copper deposits by reaction between carbonated waters and copper minerals, or between copper sulfate solutions and limestone. It is almost always associated with malachite, although of less common occurrence, and also commonly associated with limonite, calcite, chalcocite, chrysocolla, copper oxides, and other secondary copper minerals. Although azurite is often found interbanded with malachite in botryoidal material it also is often found as distinct crystals implanted on malachite or other secondary minerals, the crystals often altered to malachite pseudomorphs. Notable localities have been Chessy, France; Tsumeb, South West Africa; Broken Hill, New South Wales; and Bisbee, Arizona.

Soda niter, NaNO₃

Crystal system and class: Trigonal; $\bar{3}\ 2/m$.

Axial elements: $a{:}c = 1{:}3.3196$ (structure cell); $a{:}c = 1{:}0.8299$ (cleavage rhombohedron $\{10\bar{1}1\}$.

Cell dimensions and content: $a = 5.0696$, $c = 16.829$; $Z = 6$.

Habit: Crystals rhombohedral $\{10\bar{1}1\}$, usually massive and granular or as an incrustation.

Twinning: Common on $\{01\bar{1}2\}$, $\{0001\}$ or $\{02\bar{2}1\}$.

Cleavage: $\{10\bar{1}1\}$ perfect.

Crystal structure: Soda niter is isostructural with calcite, but no substitution of Ca or CO_3 occurs.

Tenacity: Somewhat sectile.

Hardness: $1\frac{1}{2}$–2.

Density: 2.25.

Color: Colorless, white or tinted brownish or gray by impurities.

Streak: White.

Luster and opacity: Vitreous; transparent.

Optical properties: $\epsilon = 1.337$, $\omega = 1.585$; $\omega - \epsilon = 0.248$; negative.

Diagnostic features: Water soluble with a cooling taste; easily fusible; gives red vapors of nitrogen dioxide when heated with potassium disulfate in the closed tube.

Occurrence: It is a water-soluble salt found commonly in arid regions as a surface impregnation or efflorescence, often associated with gypsum, halite, and other nitrates and sulfates. The only significant deposits of soda niter occur in a narrow belt about 450 miles long in the deserts of northern Chile, along the virtually rainless eastern slopes of the coast range. These deposits, which consist of a near-surface layer from a few inches to a few feet in thickness, and which contain numerous other salts, have supplied important amounts of nitrate for the world's fertilizer demands. In recent years artificial nitrates prepared by fixation of atmospheric nitrogen have cut seriously into the market for Chilean nitrate. The nitrate rock also contains iodates, which have provided the only mineral source of iodine.

Kernite, Na₂B₄O₆(OH)₂·3H₂O

Crystal system and class: Monoclinic; $2/m$.

Axial elements: $a{:}b{:}c = 1.6976{:}1{:}0.7609$; $\beta = 108°\ 52'$.

Cell dimensions and content: $a = 15.55$, $b = 9.16$, $c = 6.97$; $Z = 4$.

Habit: Crystals usually equant with common forms $\{001\}$, $\{\bar{2}01\}$, $\{100\}$, $\{110\}$; as cleavable masses.

Cleavage: $\{100\}$, $\{001\}$ perfect, $\{\bar{2}01\}$ fair.

Hardness: $2\frac{1}{2}$.

Density: 1.91.

Color: White.

Streak: White.

Luster and opacity: Vitreous to satiny; transparent.

Optical properties: $\alpha = 1.454$, $\beta = 1.472$, $\gamma = 1.488$; $\gamma - \alpha = 0.034$; negative, $2V = 80°$.

Diagnostic features: Before the blowpipe it swells and eventually fuses to a clear borax glass; slowly soluble in cold water. It is characterized by long splintery cleavage fragments, parallel to b, and low specific gravity.

Occurrence: Kernite occurs with borax and other borates as a large mass, and as veins in clay-shales of the Boron borax deposit, Kern County, California. The deposit, which includes crystals of kernite several feet thick, is apparently a somewhat altered buried lake deposit. A large proportion of the boron production of the United States is obtained from this deposit.

Borax, $Na_2B_4O_5(OH)_4 \cdot 8H_2O$

Crystal system and class: Monoclinic; $2/m$.

Axial elements: $a:b:c = 1.1109:1:1.1427$; $\beta = 106° 35'$.

Cell dimensions and content: $a = 11.858$, $b = 10.674$, $c = 12.197$; $Z = 4$.

Common forms and angles:

(001) $\wedge$ (100) $= 73° 25'$ (11$\bar{1}$) $\wedge$ (110) $= 37° 49'$

(010) $\wedge$ (110) $= 43° 12'$ (010) $\wedge$ ($\bar{1}$12) $= 60° 56'$

Habit: Crystals usually short prismatic parallel to c, somewhat tabular on $\{100\}$ (Fig. 12-16).

Cleavage: $\{100\}$, $\{110\}$ perfect.

Crystal structure: The crystal structure contains anionic complex groups $B_4O_5(OH)_4^{-2}$ made up of two triangular and two tetrahedral groups; one corner of each tetrahedron and each triangle is occupied by (OH). All the oxygen atoms are shared by two tetrahedrons or by a tetrahedron and a triangle.

Fracture and tenacity: Conchoidal; brittle.

Hardness: $2-2\frac{1}{2}$.

Density: 1.71.

Color: Colorless to white.

Streak: White.

Luster and opacity: Vitreous; translucent to opaque.

Optical properties: $\alpha = 1.447, \beta = 1.470, \gamma = 1.472; \gamma - \alpha = 0.025$; negative, $2V = 40°$.

Diagnostic features: Easily fusible with much swelling; characterized by crystal form and tests for boron.

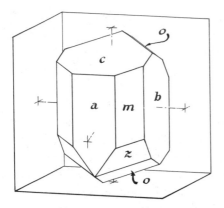

Figure 12-16   Borax crystal. Forms: pinacoids $c\{001\}$, $b\{010\}$, $a\{100\}$; prisms $m\{110\}$, $o\{\bar{1}12\}$, $z\{\bar{1}11\}$.

Occurrence: Borax is an evaporite mineral associated with halite, sulfates, carbonates, other borates, and muds in dried lakes and playas. Large deposits of borax and borate brines are worked as a source of borates in California. The most important deposits are at Boron, and also at Searles Lake, a pan-shaped desert basin, about 5 miles by 10 miles in area, that contains many substances besides borates. The Boron deposit of kernite, borax, and other borate minerals is the largest reserve of boron compounds in the world.

Colemanite, $CaB_3O_4(OH)_3 \cdot H_2O$

Crystal system and class: Monoclinic, $2/m$.

Axial elements: $a:b:c = 0.7762:1:0.5417; \beta = 110° 07'$.

Cell dimensions and content: $a = 8.743$, $b = 11.264$, $c = 6.102$; $Z = 4$.

Common forms and angles:

$(001) \wedge (110) = 73° 52'$	$(001) \wedge (100) = 69° 53'$
$(010) \wedge (110) = 53° 55'$	$(010) \wedge (\bar{1}21) = 44° 37'$

Habit: Crystals usually equant, rather varied (Fig. 12-17); common forms $\{110\}$, $\{001\}$, $\{100\}$, $\{\bar{2}01\}$, $\{\bar{1}21\}$, $\{011\}$, $\{021\}$, $\{111\}$, $\{\bar{1}11\}$, $\{\bar{2}21\}$; massive, cleavable to granular or compact.

Cleavage: $\{010\}$ perfect, $\{001\}$ distinct.

Crystal structure: In colemanite the BO_3 and BO_4 groups are linked by sharing oxygens to form endless chains in which the repeat unit, $B_3O_4(OH)_3$, includes one triangle and two tetrahedral groups.

Fracture and tenacity: Uneven; brittle.

Hardness: $4\frac{1}{2}$.

Density: 2.42.

Color: Colorless, white, yellowish, or gray.

Luster and opacity: Vitreous, often brilliant; transparent to translucent.

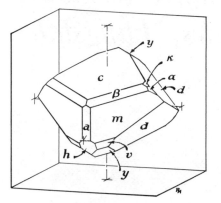

Figure 12-17 Colemanite crystal. Forms: pinacoids $c\{001\}$, $a\{100\}$, $h\{\bar{2}01\}$; prisms $m\{110\}$, $k\{011\}$, $\alpha\{021\}$, $\beta\{111\}$, $y\{\bar{1}11\}$, $v\{\bar{2}21\}$, $d\{\bar{1}21\}$.

Optical properties: $\alpha = 1.586$, $\beta = 1.592$, $\gamma = 1.614$; $\gamma - \alpha = 0.028$; positive, $2V = 56°$.

Diagnostic features: Before the blowpipe it exfoliates, crumbles, fuses imperfectly, and gives a green flame (boron); it gives water in the closed tube; it is characterized by its crystal form, one perfect cleavage, and the exfoliation on heating.

Occurrence: Colemanite is found in many of the California borax deposits, where it occurs principally in geodes in the sedimentary rocks. It has apparently been formed by the action of meteoric waters on borax and other borates, precipitated originally in a playa deposit and later buried by sedimentation. Colemanite also occurs in borate deposits at Faraskoy, western Turkey, and in the Inder region, Kazakhstan, the USSR.

13

Class VI: Sulfates, Chromates, Molybdates, Tungstates

H																	
			C												N	O	F
Na	Mg	Al														S	Cl
K	Ca			V	*Cr*	Mn	Fe	Co	Ni	Cu	Zn	*Ga*	*Ge*				
	Sr									*Ag*					Sb		
	Ba										Hg		Pb				
				U													

This group includes those anisodesmic salts with anionic groups of the general type $(XO_4)^{-n}$, in which X is a hexavalent ion in tetrahedral coordination with oxygen, yielding $(XO_4)^{-2}$. In the sulfates,* chromates, and a few very rare selenates and tellurates the anion groups are symmetrical tetrahedrons $(XO_4)^{-2}$.

The sulfates comprise a large number of minerals, some of primary origin in veins, others of secondary origin in the oxidized zone of ore deposits or in

* Table above shows elements which form sulfate minerals; those in italics are constituents of complex sulfate minerals (Ag in argentojarosite).

evaporite deposits. The sulfates are subdivided into anhydrous, hydrated, and those containing hydroxyl or a halogen. Chromates are very scarce as minerals, but include one well-known species, crocoite ($PbCrO_4$), and usually occur as secondary minerals in ore deposits. Chromates of potassium are of very rare occurrence with iodates in nitrate deposits. Selenates have been recognized as very rare secondary minerals in the oxidized zone of ore deposits, particularly those containing the rare copper selenide minerals.

The molybdates and tungstates are $A_m(XO_4)_n$ compounds in which the $(XO_4)^{-2}$ anionic group is a distorted tetrahedron, and X is Mo or W. The X atom in these structures is appreciably larger in radius than S, and there is no substantial substitution of S for Mo or W. Partial or complete series exist between some molybdates and tungstates.

Anhydrous Sulfates

Type AXO_4
   Barite, $BaSO_4$
   Celestite, $SrSO_4$
   Anglesite, $PbSO_4$
   Anhydrite, $CaSO_4$

Hydrated Sulfates

Type $AXO_4 \cdot xH_2O$
   Gypsum, $CaSO_4 \cdot 2H_2O$
   Chalcanthite, $CuSO_4 \cdot 5H_2O$
   Melanterite, $FeSO_4 \cdot 7H_2O$
   Epsomite, $MgSO_4 \cdot 7H_2O$

Anhydrous Sulfates Containing Hydroxyl

Type $A_m(XO_4)_p Z_q$
   Brochantite, $Cu_4(SO_4)(OH)_6$.
   Antlerite, $Cu_3(SO_4)(OH)_4$

Type $A_2(XO_4) Z_q$
   Alunite Group
      Alunite, $KAl_3(SO_4)_2(OH)_6$
      Jarosite, $KFe_3(SO_4)_2(OH)_6$

Anhydrous Chromates

   Crocoite, $PbCrO_4$

Molybdates, Tungstates

    Type AXO_4
        Wolframite, $(Fe,Mn)WO_4$
        Scheelite, $CaWO_4$
        Wulfenite, $PbMoO_4$

ANHYDROUS SULFATES

The principal anhydrous sulfates of natural occurrence are the members of the barite group—barite, celestite, and anglesite—and the mineral anhydrite, not isostructural with barite. In the barite structure S is in tetrahedral coordination with oxygen, and the cations Ba, Sr, or Pb are in twelvefold coordination with oxygen. Anhydrite, with the calcium ion of smaller radius, has a different structure in which calcium is in eightfold coordination with oxygen.

Crystal system and class: Orthorhombic; $2/m\ 2/m\ 2/m$.

	BARITE	CELESTITE	ANGLESITE	ANHYDRITE
Axial elements:				
(a:b:c)	1.6290:1:1.3123	1.5618:1:1.2829	1.5709:1:1.2890	0.9993:1:0.8916
Cell dimensions:	$a = 8.878$	$a = 8.359$	$a = 8.480$	$a = 6.991$
	$b = 5.450$	$b = 5.352$	$b = 5.398$	$b = 6.996$
	$c = 7.152$	$c = 6.866$	$c = 6.958$	$c = 6.238$
Cell content:	$4[BaSO_4]$	$4[SrSO_4]$	$4[PbSO_4]$	$4[CaSO_4]$
Cleavage:	{001}, {210} perfect	{001} perfect, {210} good	{001} good, {210} distinct	{010} perfect, {100} nearly perfect, {001} good
Hardness:	$3–3\frac{1}{2}$	$3–3\frac{1}{2}$	$2\frac{1}{2}–3$	$3\frac{1}{2}$
Density:	4.48 (calc.)	3.971 (calc.)	6.323 (calc.)	2.963 (calc.)
Optical properties:				
α	1.636	1.621	1.878	1.569
β	1.638	1.623	1.883	1.574
γ	1.648	1.630	1.895	1.609
$2V(+)$	37°	50°	70°	43°

Barite, $BaSO_4$

Common forms and angles:
    $(001) \wedge (101) = 38° 51'$ $(210) \wedge (2\bar{1}0) = 78° 20'$
    $(001) \wedge (011) = 52° 42'$ $(001) \wedge (211) = 64° 18'$

Habit: Crystals common, tabular on {001}, also prismatic and elongated parallel to c, a, or b (Fig. 13-1a,b,c,d, Fig. 13-2).

Color: Colorless to white, also yellow, brown, reddish, gray, greenish or blue; color may be distributed in growth zones.

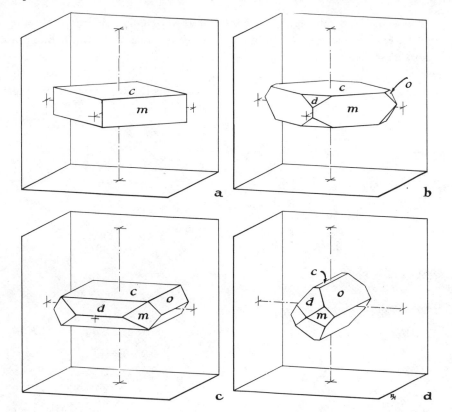

Figure 13-1 Barite crystals. Forms: pinacoids $c\{001\}$; rhombic prisms $m\{210\}$, $o\{011\}$, $d\{101\}$.

Streak: White.

Luster and opacity: Vitreous; often transparent.

Diagnostic features: Before the blowpipe it decrepitates and fuses with difficulty, giving a yellowish-green barium flame; gives a silver coin test for sulfur after heating with sodium carbonate in a reducing flame. It is distinguished by its cleavage with angles of 90°, 78° 20′, and 101° 40′, its crystal form, and its specific gravity.

Chemistry: Barite, $BaSO_4$, is isostructural with celestite, $SrSO_4$, and anglesite, $PbSO_4$, but not with anhydrite, $CaSO_4$, which has a different structure due to the smaller size of Ca^2. Sr substitutes for Ba, and a complete solid solution series probably exists. In most mineral occurrences, however, the composition is close to one of the end-member compositions $BaSO_4$ or $SrSO_4$.

Occurrence: Barite, the most common barium mineral, occurs principally as a gangue mineral in hydrothermal metalliferous veins which have formed at moderate or low temperatures, in rare cases forming the major constituent of the vein. It is often associated with fluorite, calcite, siderite, dolomite, quartz, galena, manganite, stibnite, and celestite. It is widely distributed as veins, lenses, cavity fillings, or replacement deposits in limestones where it was formed by either hypogene or groundwater solutions. Because of its insolubility it often occurs in residual clay deposits resulting from the weathering of limestones.

Figure 13-2 Barite crystals. (a) Tabular on {001} with {210} and {101}, with intergrown stibnite needles; some crystals show growth zones. (Felsöbanya, Romania.) [Courtesy Royal Ontario Museum.] (b) Elongate on *b* with {001}, {210}, {101}, with dolomite. (Cumberland, England.) [Courtesy Queen's University.]

Barite occurs at numerous localities in Germany (Freiberg, Saxony; Claustal, Harz; and at Meggen in Westphalia); in Romania (Felsöbanya); in England at the Dufton lead mines in Westmorland, at Alston Moor, Cumberland. It occurs in the USA at Cheshire, Connecticut; in the lead ores of Missouri and Wisconsin; as "desert roses" and concretions at Norman and elsewhere in central Oklahoma; near El Portal and Jerseydale, Mariposa County, California. It occurs in Canada near Walton, Hants County, Nova Scotia.

Celestite, SrSO₄

Common forms and angles:

(001) ∧ (101) = 39° 24′   (210) ∧ (2̄10) = 75° 58′
(001) ∧ (011) = 52° 04′   (001) ∧ (211) = 64° 22′

Habit: Crystals usually tabular on {001}, also elongated along *a*, *b*, or *c*. (Fig. 13-3a,b,c).

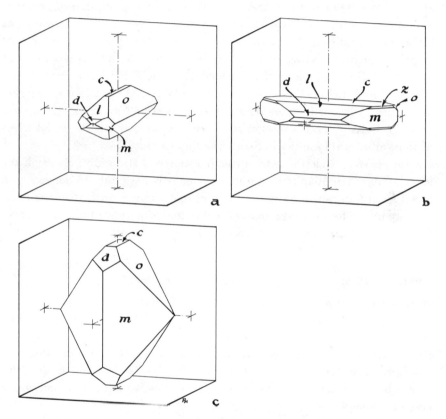

Figure 13-3   Celestite crystals. Forms: pinacoid *c*{001}; rhombic prisms *m*{210}, *o*{011}, *l*{102}, *d*{101}; rhombic dipyramid *z*{211}.

Color: Colorless to pale blue, also white, reddish, or greenish.

Streak: White.

Luster and opacity: Vitreous; transparent to translucent.

Diagnostic features: Before the blowpipe it decrepitates and fuses with difficulty, giving a crimson strontium flame; gives silver coin test for sulfur. Distinguished from barite by its lower specific gravity, by the fine fibrous character of some material, and, with certainty, by flame test for strontium.

Chemistry: Celestite is isostructural with barite and often contains small amounts of Ba substituting for Sr. Extensive substitution has sometimes been noted.

Occurrence: It occurs chiefly in sedimentary rocks, in stratiform deposits of gypsum, anhydrite, or halite, often associated with sulfur; and in cavities, veins, and disseminated in limestone or dolomite with fluorite or gypsum. It may have been deposited directly from sea water or precipitated in favorable horizons by meteoric waters carrying strontium in solution. Celestite is also found as a primary mineral in hydrothermal veins. Celestite occurs at numerous localities in Germany: at Giershagen, Westphalia, in sediments; at Scharfenberg, Saxony, in veins. It occurs with sulfur, aragonite, and gypsum at Girgenti, Sicily. In England large deposits occur near Bristol and other localities in Gloucestershire. In the USA and Canada it is common in cavities in Paleozoic limestones and dolomites in New York, Ohio, and Ontario. Celestite occurs in California with gypsum and halite in lake bed deposits at several localities in San Bernadino County.

Production and use: Celestite is the principal source of strontium, although the less abundant strontianite is also used, since it is less costly to convert to a useful strontium salt. Strontium salts find many minor uses, as in the beet-sugar industry, for fireworks, in electrical batteries, paints, rubber, and glass. England and Germany are the chief producers of celestite.

Anglesite, $PbSO_4$

Common forms and angles:

  (001) $\wedge$ (101) = 39° 22'   (210) $\wedge$ ($2\bar{1}0$) = 76° 18'
  (010) $\wedge$ (011) = 37° 48'   (001) $\wedge$ (111) = 56° 48'

Habit: Crystals often well-developed, tabular on {001}, or prismatic and elongate parallel to *c, a* or *b*; (Fig. 13-4a,b) commonly massive, granular to compact, nodular, also massive with concentric banding and enclosing a core of galena.

Fracture and tenacity: Conchoidal; brittle.

Color: Colorless to white, often tinged gray, yellow, green.

Streak: White.

Luster and opacity: Adamantine; transparent to opaque.

Diagnostic features: Fuses readily before the blowpipe; heating on charcoal with sodium carbonate results in a lead globule with a yellow to white coating of lead oxide; gives silver coin test for sulfur. The high specific gravity and adamantine luster are typical of anglesite.

Occurrence: Anglesite is a secondary mineral usually found in the oxidized zone of ore deposits containing galena. Usually formed by oxidation of galena, it is associated with cerussite and other secondary lead minerals, gypsum, and silver halides.

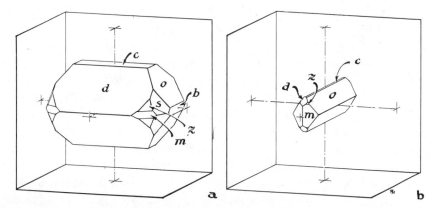

Figure 13-4 Anglesite crystals. Forms: pinacoids $c\{001\}$, $b\{010\}$; rhombic prisms $m\{210\}$, $o\{011\}$, $d\{101\}$; rhombic dipyramids $z\{211\}$, $s\{232\}$.

Anglesite occurs at Matlock and Cromford in Derbyshire, England; on the island of Anglesey, Wales; at Leadhills, Scotland. It occurs in Africa at Sidi-Amor-ben-Salem, Tunisia; at Tsumeb (South West Africa); at Broken Hill, New South Wales; and Dundas, Tasmania. It occurs in the USA at Tintic district, Utah; Coeur d'Alene, Idaho; at Goodsprings and Eureka, Nevada.

Anhydrite, CaSO₄

Habit: Crystals rare; usually massive, fine granular, fibrous, and radiated or plumose.

Cleavage: The cleavages $\{010\}$, $\{100\}$, and $\{001\}$ though of distinctly different quality simulate cubic cleavage in massive material.

Color: Colorless to bluish or violet, often gray to dark gray.

Streak: White to grayish white.

Luster and opacity: Vitreous to pearly; pearly on $\{010\}$, vitreous on $\{001\}$ and $\{100\}$; transparent to translucent.

Diagnostic features: In the blowpipe flame it is fusible at 3 to a white enamel; gives an orange-red calcium flame when moistened with hydrochloric acid. Soluble in hot hydrochloric acid, and a dilute solution added to barium chloride solution gives a white precipitate of barium sulfate. The three cleavages at right angles are a distinctive feature of anhydrite; it is harder than gypsum and has a higher specific gravity than calcite. The axial ratio, lattice, and cleavage differ markedly from those of barite. It may include only minor amounts of Sr in substitution for Ca.

Occurrence: It is an important rock-forming mineral, found as extensive beds interstratified with gypsum, limestone, dolomite, and salt. Anhydrite is deposited directly from the evaporation of sea water at temperatures of 42°C or higher, or at lower temperatures with increased salinity. At lower temperatures, or lower salinity, gypsum is deposited. Sedimentary beds of anhydrite may, in part, have been formed by dehydration of earlier gypsum beds, or anhydrite may also be converted to gypsum by the action of meteoric waters. It is also found as an accessory mineral in sedimentary rocks, to a minor extent as a gangue mineral in metalliferous veins. Anhydrite is an important mineral in the cap rock of many salt domes in Texas and Louisiana. It occurs in the salt deposits in Germany at Stassfurt; in Poland at Wieliczka, Krakow; in the Pyrenees; in the Salt Range, Punjab, India; in Canada as bedded deposits in Nova Scotia. It is extensively mined at Billingham, England, as a source of sulfate for making ammonium sulfate.

HYDRATED SULFATES

Gypsum, $CaSO_4 \cdot 2H_2O$

Crystal system and class: Monoclinic; $2/m$.
Axial elements: $a:b:c = 0.374:1:0.414$; $\beta = 113° 50'$.
Cell dimensions and content: $a = 5.68$, $b = 15.18$, $c = 6.29$; $Z = 4$.
Common forms and angles:

(010) ∧ (120) = 55° 37'   (010) ∧ (011) = 69° 16'
(010) ∧ ($\bar{1}$11) = 71° 50'   (100) ∧ (001) = 66° 10'

Habit: Crystals commonly simple in habit, tabular on {010} (Figs. 13-5a,b, 13-6); crystals frequently have warped or curved surfaces, and long prismatic crystals are often bent or curled; also granular, massive, coarse to very fine, foliated, or fibrous veinlets.

Twinning: Contact twinning on {100} or {$\bar{1}$01} very common, giving swallowtail twins, also as cruciform penetration twins (Fig. 2-45b, Fig. 13-7a,b).

Cleavage: {010} very good, yielding thin, polished foliae, also distinct {100}, {011} giving rhombic fragments with angles 66° and 114°; fragments flexible, not elastic.

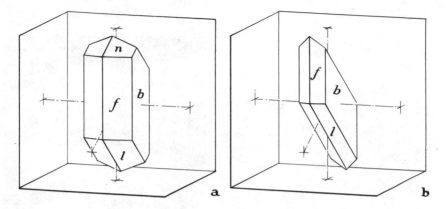

Figure 13-5 Gypsum crystals. Forms: pinacoid $b\{010\}$; prisms $f\{120\}$, $n\{011\}$, $l\{\bar{1}11\}$.

Hardness: 2.
Density: 2.32.
Color: Colorless and transparent, also white, gray, yellowish when massive.
Streak: White.

Figure 13-6 Gypsum. Crystal with $\{010\}$, $\{120\}$, $\{\bar{1}11\}$, $[101]$ vertical.

Figure 13-7 Gypsum (a) Simple twin with {$\bar{1}01$} as twin plane, [101] vertical; rough faces on top are {120}; trace of {001} cleavage visible at upper left. (b) Cleavage fragment from a swallowtail twin, twin plane {100} vertical; inclined lines are fibrous fractures parallel to *a*. In each photograph {010} is parallel to page. (Localities unknown.) [Courtesy Queen's University.]

Luster and opacity: Vitreous, often pearly on {010} cleavage; transparent to translucent.

Optical properties: $\alpha = 1.519$, $\beta = 1.523$, $\gamma = 1.529$; $\gamma - \alpha = 0.010$; positive, $2V = 58°$.

Diagnostic features: Fuses in the blowpipe flame at 3; gives water in closed tube and turns white; soluble in hot dilute hydrochloric acid, the solution giving, with barium chloride, a white precipitate of barium sulfate; gives a silver coin test for sulfur, as do other sulfates. The low hardness (2), and the perfect cleavage with two poorer cleavages, are diagnostic for gypsum, and the presence of much water distinguishes it from anhydrite.

Chemistry: Gypsum shows almost no variation in composition, but some varieties are colored by the presence of iron or clay impurities. It converts slowly to the hemi-hydrate in air at about 70°C, and more rapidly above 90°; commercially, the hemi-hydrate, known as plaster of Paris, is made by heating gypsum at 190–200°C; at higher temperatures anhydrite is produced.

Varieties: Several varieties of gypsum are recognized: the coarsely crystallized material, often colorless and transparent, is called selenite; aggregate material with a parallel fibrous structure is called satin spar; and the fine-grained massive material is commonly called alabaster.

Occurrence: Gypsum is found as extensive sedimentary deposits interbedded with limestone, red shales, sandstone, clay, and rock salt. It is normally the first salt deposited in the evaporation of sea water, followed by anhydrite and halite as the salinity increases, which are followed, in rare cases, by the more soluble sulfates and other salts of Mg and K. Gypsum may also occur in many other ways: as saline lake deposits; with native sulfur (Girgenti, Sicily; Krakow, Poland); around volcanic fumaroles (Vesuvius); as an efflorescence on soils or in limestone caves; in the cap rock of salt domes; and in the gossan over pyritic mineral deposits in a limestone area.

Gypsum is common in Tertiary sediments of the Paris basin in France; in the salt and potash deposits of Northern Germany; as very large crystals at Naica, Chihuahua, Mexico; in commercial deposits in Paleozoic rocks of New York, Michigan, and Ontario; in extensive deposits at Hillsborough, New Brunswick, and in the Windsor district, Hants County, Nova Scotia.

Production and use: Gypsum is extensively mined in many parts of the world for use in the construction industry, especially for plasters, gypsum wallboard, roof tiles, in cements, and as a filler in paper and paints. It is also often used as a soil conditioner (landplaster) and fertilizer.

In the five-year period 1960–1964 the annual world production of about 50 million tons came from the USA (21 percent), Canada (14 percent), England (10 percent), France (10 percent), and many other countries.

Chalcanthite, $CuSO_4 \cdot 5H_2O$

Crystal system and class: Triclinic; $\bar{1}$.

Axial elements: $a:b:c = 0.5725:1:0.5575$; $\alpha = 97° 35'$, $\beta = 107° 10'$, $\gamma = 77° 33'$.

Cell dimensions and content: $a = 6.122$, $b = 10.695$, $c = 5.96$; $Z = 2$.

Common forms and angles:

(100) $\wedge$ (1$\bar{1}$0) =   26° 04'   (010) $\wedge$ (110) = 69° 49'

(100) $\wedge$ ($\bar{1}\bar{1}$1) = 120° 27'   (010) $\wedge$ (130) = 33° 54'

Habit: Crystals commonly short prismatic parallel to [001] (Fig. 13-8), also massive and granular, stalactitic, or reniform.

Cleavage: {1Ī0} imperfect.
Fracture: Conchoidal.
Hardness: 2½.
Density: 2.28.
Color: Sky blue of different shades.
Streak: White.
Luster and opacity: Vitreous; subtransparent.
Optical properties: $\alpha = 1.515, \beta = 1.539, \gamma = 1.546; \gamma - \alpha = 0.030$; negative, $2V = 55°$.
Diagnostic features: Water soluble; infusible and loses water on heating; gives copper globule when fused with sodium carbonate on charcoal.

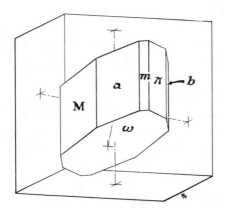

Figure 13-8   Chalcanthite crystal. Forms: pinacoids $b\{010\}$, $a\{100\}$, $m\{110\}$, $\pi\{130\}$, $M\{1\bar{1}0\}$, $\omega\{\bar{1}\bar{1}1\}$.

Occurrence: This soluble mineral, often known as bluestone, is found with other hydrated sulfates of copper and iron in the oxidized near-surface zone of copper sulfide ore deposits. Copper and iron sulfates result from the oxidation of chalcopyrite and other sulfides by meteoric waters; often in arid regions these salts are deposited in the oxide zone as substantial deposits. They are of commercial importance particularly in the Chilean copper deposits (Chuquicamata, Quetena, and Copaquire). Copper and sulfate ions are commonly present in mine waters, from which chalcanthite may crystallize as stalactites or crusts in mines; the copper is often recovered from such waters by precipitating it with metallic iron.

Melanterite, $FeSO_4 \cdot 7H_2O$

Crystal system and class: Monoclinic; $2/m$.
Axial elements: $a:b:c = 2.168:1:1.694$; $\beta = 105° 15'$.
Cell dimensions and content: $a = 14.11$, $b = 6.51$, $c = 11.02$; $Z = 4$.

Habit: Crystals equant to short prismatic along *c* with {110} and {001}; usually in stalactitic or concretionary forms; as fibrous to capillary aggregates and crusts; massive and pulverulent.

Cleavage: {001} perfect, {110} distinct.

Fracture and tenacity: Conchoidal; brittle.

Hardness: 2.

Density: 1.898 (pure).

Color: Green of various shades, passing into greenish blue and blue with increasing substitution of Cu for Fe; greenish white.

Streak: White.

Luster: Vitreous.

Optical properties: $\alpha = 1.470$, $\beta = 1.479$, $\gamma = 1.486$; $\gamma - \alpha = 0.016$; positive, $2V = 80°$.

Taste: Sweetish, astringent, and metallic.

Chemistry: Cu commonly substitutes for Fe^2, and a series extends to about Fe:Cu = 1:1.9. Mg also substitutes for Fe, and a partial series extends to artificial $MgSO_4 \cdot 7H_2O$, the monoclinic polymorph of epsomite. Less commonly, Zn, Co, Ni, and Mn are also found in substitution for Fe. In many occurrences several metal ions substitute for Fe simultaneously. Melanterite dehydrates to the pentahydrate or to lower hydrates at room temperature, depending on the relative humidity.

Diagnostic features: Soluble in water; yields water in the closed tube and SO_2 on strong heating; on charcoal turns brown to black, becoming magnetic.

Occurrence: Melanterite and its compositional varieties are secondary minerals formed by the oxidation of pyrite, marcasite, and copper-bearing pyritic ores. They often occur as an efflorescence on the walls and timbers of mine workings in the oxidized zone of pyritic ore bodies, especially in arid regions. Melanterite may be associated with epsomite, chalcanthite, gypsum, and other hydrous or basic sulfates. Important localities at which melanterite has been identified include: Rammelsberg in the Harz, Germany; Rio Tinto in Spain; the Alma mine, Alameda County, California; Ducktown, Tennessee; Butte, Montana; Bingham Canyon, Utah; Falun, Sweden.

Epsomite, $MgSO_4 \cdot 7H_2O$

Crystal system and class: Orthorhombic; 222.

Axial elements: $a:b:c = 0.9891:1:0.5720$.

Cell dimensions and content: $a = 11.86$, $b = 11.99$, $c = 6.858$; $Z = 4$.

Common forms and angles:

$(110) \wedge (1\bar{1}0) = 89° 22'$ $\qquad$ $(111) \wedge (\bar{1}\bar{1}1) = 77° 38'$

Habit: Artificial crystals short prismatic along *c*; natural crystals rare; usually fibrous to hair-like or acicular crusts, fiber axis c; as efflorescences, botryoidal or reniform masses.

Cleavage: {010} perfect, {101} distinct.

Hardness: 2–2$\frac{1}{2}$.

Density: 1.678 (calc).

Color: Colorless and transparent in crystals, massive material white and translucent.

Luster: Vitreous, silky to earthy on fibrous material.

Optical properties: $\alpha = 1.433, \beta = 1.455, \gamma = 1.461; \gamma - \alpha = 0.028$; negative, $2V = 50°$.

Chemistry: Minor substitution of Zn, Fe, Co, Ni, Mn for Mg occurs in epsomite. The substitution may be more extensive in artificial crystals, and complete series extend to $NiSO_4 \cdot 7H_2O$ and $ZnSO_4 \cdot 7H_2O$. In dry air at ordinary temperatures epsomite loses as much as $1H_2O$, changing to *hexahydrite*.

Diagnostic features: Very soluble in water; on heating in the closed tube it dissolves in its own water of crystallization, acid water being driven off at higher temperatures; fuses before the blowpipe to alkaline $MgSO_4$.

Occurrence: Epsomite occurs as crusts or fibrous efflorescences in mine workings in coal or metal deposits; in limestone caves; in sheltered places on outcrops of any magnesian rock particularly in the presence of pyrite or pyrrhotite; in the oxidized zone of pyritic deposits in arid regions; in mineral spring deposits, salt lake deposits, and oceanic salt deposits. It has long been known as a deposit from mineral waters at Epsom, Surrey, England; and at Sedlitz, Bohemia. Epsomite is found in salt lake deposits in Albany County, Wyoming; in Nevada; near Oroville, Washington; near Ashcroft, British Columbia; in salt deposits of marine origin at Carlsbad, New Mexico, and in Stassfurt, Germany.

ANHYDROUS SULFATES CONTAINING HYDROXYL

Brochantite, $Cu_4SO_4(OH)_6$

Crystal system and class: Monoclinic; $2/m$.

Axial elements: $a:b:c = 1.328:1:0.611; \beta = 103° 22'$.

Cell dimensions and content: $a = 13.08, b = 9.85, c = 6.02; Z = 4$.

Common forms and angles:

(100) ∧ (210) = 32° 52'   (110) ∧ (1̄10) = 75° 28'

(001) ∧ (100) = 76° 38'   (001) ∧ (1̄11) = 39° 54'

Habit: Crystals usually prismatic to acicular parallel to c (Fig. 13-9); commonly as loosely coherent aggregates of acicular crystals, also in groups or drusy crusts, massive, granular.

Twinning: Twin plane {100}, common as contact twins.
Cleavage: {100} perfect.
Hardness: $3\frac{1}{2}$–4.
Density: 3.97.
Color: Emerald green to blackish green.
Streak: Pale green.
Luster and opacity: Vitreous; transparent to translucent.
Optical properties: $\alpha = 1.728, \beta = 1.771, \gamma = 1.800; \gamma - \alpha = 0.072$; negative, $2V = 77°$.

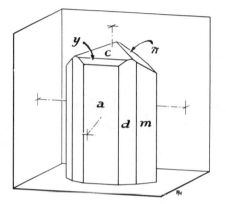

Figure 13-9 Brochantite crystal. Forms: pinacoids $c\{001\}$, $a\{100\}$, $y\{201\}$; prisms $m\{110\}$, $d\{210\}$, $\pi\{\bar{1}11\}$.

Diagnostic features: Fuses; loses water on heating in the closed tube. Brochantite is very similar to antlerite and atacamite in color. All three minerals are found in the secondary oxidized zone of copper deposits, and are difficult to distinguish from one another except by careful study of the crystals and cleavage, combined with chemical or optical tests.

Occurrence: Brochantite is found associated with malachite, limonite, cuprite, and chrysocolla. It occurs abundantly in the copper deposits of Chile (Chuquicamata, Potrerillos, Challacollo) and in the southwestern USA (Bisbee, Clifton-Morenci, Mammoth mine, United Verde mine, in Arizona; and Tintic, Utah).

Antlerite, $Cu_2SO_4(OH)_4$

Crystal system and class: Orthorhombic; $2/m\ 2/m\ 2/m$
Axial elements: $a:b:c = 0.687:1:0.503$.
Cell dimensions and content: $a = 8.24$, $b = 11.99$, $c = 6.03$; $Z = 4$.
Common forms and angles:
    (110) $\wedge$ (1$\bar{1}$0) = 68° 57'   (010) $\wedge$ (011) = 63° 18'
    (010) $\wedge$ (130) = 26° 53'   (110) $\wedge$ (111) = 48° 24'

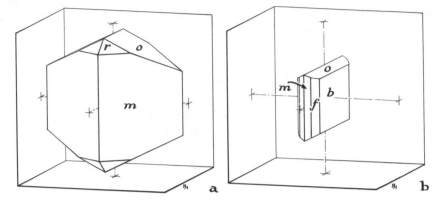

Figure 13-10   Antlerite crystals. Forms: pinacoid $b\{010\}$; rhombic prisms $m\{110\}$, $f\{130\}$, $o\{011\}$; rhombic dipyramid, $r\{111\}$.

Habit: Crystals usually thick tabular on $\{010\}$ or short prismatic along c (Fig. 13-10); as cross-fiber veinlets, friable aggregates, or granular.

Cleavage: $\{010\}$ perfect.

Hardness: $3\frac{1}{2}$.

Density: 3.88.

Color: Emerald green to blackish green.

Streak: Paler green.

Luster and opacity: Vitreous; translucent.

Optical properties: $\alpha = 1.726$, $\beta = 1.738$, $\gamma = 1.789$; $\gamma - \alpha = 0.063$; positive, $2V = 53°$.

Occurrence: Similar to and often associated with brochantite and atacamite, antlerite is the principal ore mineral of copper at the important mine at Chuquicamata in Chile. It is found in copper deposits in arid regions with chalcanthite, gypsum, and other minerals, in addition to the two mentioned above.

Alunite, $KAl_3(SO_4)_2(OH)_6$

Crystal system and class: Trigonal; $\bar{3}\ 2/m$ or $3m$.

Axial elements: $a:c = 1:2.494$.

Cell dimensions and content: $a = 6.97$, $c = 17.38$; $Z = 3$.

Habit: Rarely in small crystals with $\{0001\}$ and the rhombohedron $\{01\bar{1}2\}$, pseudocubic in appearance; usually massive, granular to dense.

Cleavage: $\{0001\}$ distinct.

Fracture and tenacity: Uneven to conchoidal; brittle.

Hardness: $3\frac{1}{2}$–4.

Density: 2.6–2.9.

Color: White, also grayish, yellowish, or reddish.

Streak: White.

Luster and opacity: Vitreous to pearly on {0001}; transparent to translucent.

Optical properties: $\omega = 1.572$, $\epsilon = 1.592$; $\epsilon - \omega = 0.020$; positive.

Diagnostic features: Infusible, and decrepitates in the blowpipe flame, giving a potassium flame; moistened with cobalt nitrate solution it turns blue on heating (aluminum); it gives acid water on heating in the closed tube; it is soluble in sulfuric acid. In the common massive form it is difficult to distinguish from limestone, dolomite, anhydrite, or magnesite without chemical tests. Alunite has been found to be strongly pyroelectric and piezoelectric, and for this reason it is assigned to the class $3m$. Some investigators have been unable to confirm this, and the true symmetry remains in doubt.

Chemistry: Alunite is a basic sulfate of aluminum and potassium in which Na is often found in substitution for K and may extend to Na:K = 7:4. Where Na exceeds K the mineral is called *natroalunite*. The alunite group also includes jarosite and the several compositional variants of jarosite.

Occurrence: It is a mineral of widespread occurrence in near-surface rocks of volcanic regions, which have been altered by solutions containing sulfuric acid. This process of alteration is called "alunitization." This is a widespread feature of the altered or mineralized volcanic rocks of the western USA at Rosita Hills, Custer County, Colorado; Goldfield district, Nye County, Nevada; and Marysvale, Utah.

Jarosite, $KFe_3(SO_4)_2(OH)_6$

Crystal system and class: Trigonal; $\bar{3}\ 2/m$ or $3m$.

Axial elements: $a:c = 1:2.362$.

Cell dimensions and content: $a = 7.21$, $c = 17.03$; $Z = 3$.

Habit: Crystals minute, tabular on {0001} or pseudocubic with {01$\bar{1}$2}, as crusts or coatings of minute crystals; granular massive, fibrous, nodular also pulverulant, earthy.

Cleavage and tenacity: {0001} distinct; brittle.

Hardness: $2\frac{1}{2}$–$3\frac{1}{2}$.

Density: 2.91–3.26.

Color: Ocherous, amber yellow to dark brown.

Streak: Pale yellow.

Luster: Vitreous to resinous.

Optical properties: $\epsilon = 1.715$, $\omega = 1.820$; $\omega - \epsilon = 0.105$; negative.

Chemistry: Jarosite is isostructural with alunite; ordinarily there is no appreciable substitution of Al for Fe^3 or vice versa, but there are a few occurrences which indicate the possibility of a continuous series between alunite and jarosite. Na commonly substitutes for K in jarosite, and a series extends toward natrojarosite. Other members of this group have NH_4, Ag, or Pb in place of K.

Occurrence: It is a widespread secondary mineral in the form of crusts or coatings on ferruginous ores and adjoining rocks; it is commonly a constituent of limonitic gossans and is difficult to distinguish from earthy limonite. It occurs commonly at Chuquicamata, Chile; in the May Day vein, Pioche district, Lincoln County, Nevada; and at many other localities in the USA.

CHROMATES

Crocoite, $PbCrO_4$

Crystal system and class: Monoclinic; $2/m$.
Axial elements: $a:b:c = 0.957:1:0.914$; $\beta = 102° 25'$.
Cell dimension and content: $a = 7.12$, $b = 7.44$, $c = 6.80$; $Z = 4$.
Habit: Crystals usually prismatic and striated along c; also massive, columnar to granular.
Cleavage: {110} distinct.
Tenacity: Sectile.
Hardness: $2\frac{1}{2}$–3.
Density: 6.10.
Color: Hyacinth red, deep orange red, or orange.
Streak: Orange yellow.
Luster and opacity: Adamantine to vitreous; transparent.
Crystal structure: Crocoite is isostructural with monazite.
Diagnostic features: Easily fusible; gives a lead globule when fused with sodium carbonate on charcoal; gives a green chromium bead with borax in the oxidizing flame. The color, luster, and specific gravity are distinctive.
Occurrence: Crocoite is a secondary mineral of rather limited occurrence; it is found in the gossan of ore deposits associated with pyromorphite, cerussite, and other secondary lead minerals. Notable occurrences are in the Beresov district, in the Urals, the USSR; and in the Adelaide, West Comet, and other mines in the Dundas district, Tasmania.

MOLYBDATES, TUNGSTATES

Wolframite, $(Fe,Mn)WO_4$

Crystal system and class: Monoclinic; $2/m$.
Axial elements: $a:b:c = 0.839:1:0.867$; $\beta = 90° 49'$

Cell dimensions and content: $a = 4.81$, $b = 5.73$, $c = 4.97$; $Z = 2$; dimensions
are slightly larger for $MnWO_4$, and slightly smaller with $\beta = 90°$ for $FeWO_4$.
Common forms and angles:

$(100) \wedge (110) = 40° 00'$ $(100) \wedge (102) = 27° 58'$
$(100) \wedge (10\bar{2}) = 26° 41'$ $(011) \wedge (0\bar{1}1) = 81° 55'$

Habit: Crystals commonly short prismatic along c, often flattened on $\{100\}$
(Fig. 13-11); faces striated parallel to c, in subparallel groups, lamellar or
massive granular; huebnerite ($MnWO_4$) is commonly prismatic to long
prismatic along c; ferberite ($FeWO_4$) is commonly elongated along b.

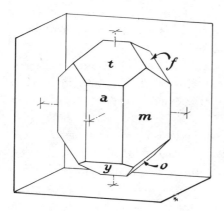

Figure 13-11 Wolframite crystal. Forms: pinacoids $a\{100\}$, $t\{102\}$, $y\{\bar{1}02\}$; prisms $m\{110\}$, $f\{011\}$.

Twinning: Twin plane $\{100\}$, common as simple contact twins.
Cleavage and tenacity: $\{010\}$ perfect; brittle.
Hardness: $4-4\frac{1}{2}$.
Density: 7.1–7.5, increasing with iron content.
Color: Brownish black to iron black.
Streak: Reddish brown to brownish black.
Luster: Submetallic.
Diagnostic features: Huebnerite is reddish brown, with a resinous luster, and
is translucent; ferberite is black, with a nearly black streak and an almost
metallic luster, is virtually opaque, and is weakly magnetic; wolframite is
fusible at 3 to a magnetic globule; it is almost insoluble in acids; after fusion
with sodium carbonate it is soluble in hydrochloric acid, and the solution
turns blue when boiled with a fragment of metallic tin; it gives a bluish-green
color (manganese) to a sodium carbonate bead in the oxidizing flame. Its
color, single direction of perfect cleavage, and high specific gravity are distinctive for wolframite.
Chemistry: The wolframite minerals form a complete series between ferberite
($FeWO_4$) and huebnerite ($MnWO_4$); nearly all specimens show intermediate

composition, and wolframite with 20–80 atom percent iron (and 80–20 atom percent manganese) includes most occurrences.

Occurrence: The minerals of this series constitute the principal ores of tungsten. Wolframite is largely found:

1. In greisen, quartz-rich veins, or pegmatitic veins, in close association with granitic intrusive rocks. There is often evidence of pneumatolytic origin, and the common associated minerals are topaz, cassiterite, arsenopyrite, lithia micas, tourmaline. It is found rarely in normal granite pegmatites.

2. In high-temperature hydrothermal veins with pyrrhotite, pyrite, chalcopyrite, bismuthinite, and some of the minerals in (1). These veins may be genetically associated with the deposits of (1), and in both types wall-rock alteration, accompanied by the development of tourmaline, mica, topaz, fluorite, and chlorite, is usually well-marked.

The minerals of the wolframite group are also found in moderate- and low-temperature veins with sulfides, cassiterite, scheelite, bismuth, quartz, or siderite; in contact metamorphic deposits adjacent to granitic intrusives; also in alluvial deposits.

The major commercial sources of wolframite as an ore of tungsten are in the Nanling Range in southern China, where it is found in both types of occurrence described above. An important producing deposit of ferberite occurs in Boulder County and adjoining parts of Gilpin County, Colorado; here ferberite is the principal metallic mineral in veins of moderate temperature origin. Wolframite is also recovered commercially in Beira Beixa province, Portugal; in Cornwall, England; at Chicote Grande, Llallagua, and Tasna, in Bolivia; in the Malay peninsula and Burma. Huebnerite is of less common occurrence but has been mined as an ore mineral in Ouray County, Colorado; and at Morococha and in the Conchucos district, Peru. Wolframite was an important mineral in the mining districts of Saxony (Schneeberg, Zinnwald, and Altenberg) where it was first recognized.

Scheelite, $CaWO_4$

Crystal system and class: Tetragonal; $4/m$.
Axial elements: $a:c = 1:2.1694$.
Cell dimensions and content: $a = 5.242$, $c = 11.372$; $Z = 4$.
Common forms and angles:
  $(101) \wedge (10\bar{1}) = 49° 30'$ $(0\bar{1}3) \wedge (013) = 71° 44'$
Habit: Crystals usually dipyramidal with $\{101\}$ and $\{1\bar{1}2\}$ predominant Fig. 13-12); commonly massive, granular.
Twinning: Common on $\{110\}$, usually as penetration twins.
Cleavage: $\{101\}$ distinct.

Fracture: Uneven.

Hardness: $4\frac{1}{2}$–5.

Density: 6.12 (pure), decreasing with substitution of Mo for W.

Color: Colorless to white, pale yellow, or brownish; also greenish, gray, or reddish; fluoresces bright bluish white in short ultraviolet radiation.

Streak: White.

Luster: Vitreous.

Optical properties: $\omega = 1.920$, $\epsilon = 1.937$; $\epsilon - \omega = 0.017$; positive.

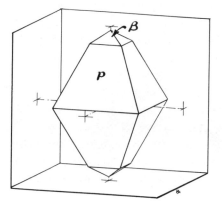

Figure 13-12 Scheelite crystal. Forms: tetragonal dipyramids $p\{101\}$, $\beta\{103\}$.

Diagnostic features: Fusible with difficulty; decomposed by boiling in hydrochloric acid, whereupon it leaves a yellow powder of hydrous tungstic oxide which is soluble in ammonia. The light color, high specific gravity, crystal form, and fluorescence are distinctive, but a test for tungsten may be necessary.

Chemistry: Mo may substitute for W, and a partial series extends toward *powellite* (CaMoO₄), though most scheelite contains only a minor amount of Mo.

Occurrence: Scheelite is often found with wolframite in deposits of high temperature origin. It is widespread and often constitutes an ore of tungsten. It is found in contact metamorphic deposits (including most economic occurrences); in high-temperature hydrothermal veins with wolframite and other associated minerals listed with that mineral, or as a minor mineral in some gold-quartz veins; also in pegmatites and moderate-temperature hydrothermal veins.

Scheelite is often a minor mineral in wolframite veins, particularly in Zinnwald, in Saxony and Bohemia; Cornwall, England; southern China, northern Bolivia. In the USA commercial amounts have been found: near Mill City, Humboldt Range, and other localities in Nevada; at the Boriana mine, Hualpai Mountains, Mohave County, in the Huachuca Mountains, Cochise County, and in the Helvetia district, Pima County, Arizona; near Bishop, Atolia, and Randsburg in California. In Canada it is found: in the Bridge River, Cariboo, and Nelson districts, British Columbia; in a skarn zone just east of the Yukon-Northwest Territories border, where production commenced in 1965.

Wulfenite, PbMoO$_4$

Crystal system and class: Tetragonal; 4/m or 4. Pyramidal symmetry, 4, is indicated by the crystal development, though it is not always clearly displayed. Dipyramidal symmetry, 4/m, is indicated by the etch-figure symmetry, the structure, and the apparent absence of piezoelectricity.

Axial elements: $a:c = 1:2.228$.

Cell dimensions and content: $a = 5.435$, $c = 12.11$; $Z = 4$.

Common forms and angles:

   (001) $\wedge$ (101) = 65° 50'   (001) $\wedge$ (112) = 57° 36'

Habit: Crystals commonly square tabular on {001} bounded by the prism {100} and pyramids {101} and {10$\bar{1}$} (Fig. 13-13), also massive, coarse to fine granular: some crystals are distinctly hemimorphic indicating pyramidal symmetry (4).

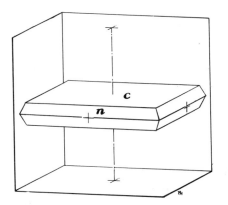

Figure 13-13 Wulfenite crystal. Forms: pinacoid c{001} (or pedions c{001} and $-c${00$\bar{1}$}), tetragonal dipyramid n{101} (or tetragonal pyramids n{101} and $-n${10$\bar{1}$}).

Cleavage: {101} distinct.

Tenacity: Not very brittle.

Hardness: 3.

Density: 6.5–7.0 (6.815 calc.), lower due to substitution of Ca for Pb, or higher due to substitution of W for Mo.

Color: Orange yellow to wax yellow, also yellowish gray, olive green, or brown.

Streak: White.

Luster and opacity: Resinous; transparent or translucent.

Diagnostic features: Decrepitates in the blowpipe flame and fuses readily (at 2 or below); gives lead globule on fusing with sodium carbonate on charcoal; decomposed by evaporation in hydrochloric acid with separation of lead chloride and molybdic oxide. The square tabular crystals, color, and luster are distinctive.

Chemistry: W substitutes for Mo to a small extent, but evidence of a series extending to $PbWO_4$ is lacking. Ca substitutes for Pb, indicating a partial series to powellite $Ca(Mo,W)O_4$. Wulfenite apparently differs from scheelite-powellite in that it lacks a mirror plane normal to the tetragonal axis.

Occurrence: Wulfenite is a secondary mineral formed in the oxidized zone of ore deposits containing lead and molybdenum minerals and associated with pyromorphite, vanadinite, cerussite, limonite, calcite, and other secondary minerals. Next to molybdenite it is the commonest molybdenum mineral. A few of the best known localities are: Pribram, Bohemia; Oudida, Morocco; Sidi Renman, Algeria; Broken Hill district, New South Wales: Mapimi, Durango, Chihuahua, Sonora, and other localities in Mexico; Red Cloud and Hamburg mines, Yuma County, Arizona; Mammoth mine, Pinal County, Arizona; and localities in New Mexico, Nevada and Utah.

Use: Wulfenite is a minor source of molybdenum.

14

Class VII: Phosphates, Arsenates, Vanadates

H																	
Li	*Be*		*C*												O		
Na	Mg	Al	*Si*											P	S		
K	Ca	Sc		V		Mn	Fe	Co	Ni	Cu	Zn			As			
	Sr	Y															
	Ba	La Lu										Pb	*Bi*				
		Th		U													

This class of minerals includes a large number of naturally occurring oxysalts with anionic groups of the type $(XO_4)^{-n}$ in which X is P, As, or V, and n is 3.* There is, commonly, extensive substitution between P and As, and between As and V. These minerals are usually classified on the basis of the $A:(XO_4)$ ratio for anhydrous, acid, and hydrated salts. The commonly occurring members of this class described here are listed below.

* Table above shows elements which form phosphate minerals; those in italics are constituents of complex phosphate minerals.

Anhydrous Normal Phosphates
    Type $A(XO_4)$
        Xenotime, YPO_4
        Monazite, $(Ce,La,Y,Th)PO_4$
Hydrated Normal Phosphates
    Type $A_3(XO_4)_2 \cdot 8H_2O$
        Vivianite, $Fe_3(PO_4)_2 \cdot 8H_2O$
        Erythrite, $Co_3(AsO_4)_2 \cdot 8H_2O$
Anhydrous Phosphates with Hydroxyl or Halogen
    Type $AB(XO_4)Z_q$
        Amblygonite Series: $(Li,Na)Al(PO_4)(F,OH)$
    Type $A_5(XO_4)_3Z_q$
        Apatite Group
            Apatite Series: $Ca_5(PO_4)_3(F,Cl,OH)$
            Pyromorphite Series:
                Pyromorphite, $Pb_5(PO_4)_3Cl$
                Mimetite, $Pb_5(AsO_4)_3Cl$
                Vanadinite, $Pb_5(VO_4)_3(Cl)$
Hydrated Phosphates Containing Hydroxyl
        Turquois, $CuAl_6(PO_4)_4(OH)_8 \cdot 4H_2O$
Uranyl Phosphates
        Torbernite, $Cu(UO_2)_2(PO_4)_2 \cdot 8\text{--}12H_2O$
        Autunite, $Ca(UO_2)_2(PO_4)_2 \cdot 10\text{--}12H_2O$
Vanadium Oxysalts
        Carnotite: $K_2(UO_2)_2(VO_4)_2 \cdot 3H_2O$
        Tyuyamunite: $Ca(UO_2)_2(VO_4)_2 \cdot nH_2O$

Anhydrous Normal Phosphates

Xenotime, YPO_4

Crystal system and class: Tetragonal; $4/m\ 2/m\ 2/m$.
Axial elements: $a:c = 1:0.875$.
Cell dimensions, and content: $a = 6.89$, $c = 6.03$; $Z = 4$; isostructural with zircon.
Common forms and angles:
  $(100) \wedge (101) = 48° 48'$ $(211) \wedge (121) = 32° 43'$
Habit: Short to long prismatic along c; also equant, pyramidal; closely resembling zircon in habit; as radial aggregates of coarse crystals or as rosettes.
Cleavage: {100} perfect.
Hardness: 4–5.

Density: 4.4–5.1 (4.25, calc.).

Color: Commonly yellowish brown to reddish brown, also hair brown, flesh red, grayish white, yellow, greenish.

Streak: Pale brown, yellowish, or reddish.

Luster: Vitreous to resinous, translucent to opaque.

Optical properties: $\omega = 1.720$, $\epsilon = 1.815$; $\epsilon - \omega = 0.095$; positive.

Chemistry: Xenotime is essentially YPO_4; other rare-earth elements, especially erbium and ytterbium, may substitute for Y. Th, U, Zr and Ca also substitute for Y in small amounts. The substitution of Zr^4 and U^4 for Y^3 apparently involves concomitant substitution of SiO_4 for PO_4.

Diagnostic features: Easily confused with zircon, but may be distinguished by its inferior hardness and perfect cleavage; infusible; when moistened with sulfuric acid colors the flame bluish green.

Occurrence: Xenotime is of widespread occurrence as a minor accessory mineral in acidic and alkalic rocks, also in larger crystals in the associated pegmatites. It has also been noted in mica- and quartz-rich gneisses. It is also a common detrital mineral. Xenotime occurs in the granite pegmatites of southern Norway at Hitterö, Tvedestrand, and near Arendal; and at Ytterby in Sweden. It occurs in alluvial deposits in Madagascar; in Brazil (Dattas, Diamantina, and Bahia); in North Carolina, and Georgia, the USA.

Monazite, (Ce,La,Y,Th)PO₄

Crystal system and class: Monoclinic; $2/m$.

Axial elements: $a:b:c = 0.969:1:0.922$; $\beta = 103° 38'$.

Cell dimensions and content: $a = 6.79$, $b = 7.01$, $c = 6.46$; $Z = 4$.

Common forms and angles:

$(010) \wedge (110) = 46° 43'$

$(100) \wedge (101) = 39° 18'$

$(100) \wedge (10\bar{1}) = 53° 38'$

$(101) \wedge (\bar{1}11) = 94° 15'$

Habit: Crystals commonly small but sometimes large, usually equant, often flattened on {100} or somewhat elongated on b (Fig. 14-1); the crystal faces are often rough, striated, or uneven.

Twinning: Common on {100} as contact or cruciform twins.

Cleavage: {100} distinct, {001} parting.

Hardness: 5–5½.

Density: 4.6–5.4.

Color: Yellowish or reddish brown to brown.

Streak: Nearly white.

Luster and opacity: Resinous or waxy; subtranslucent.

Optical properties: $\alpha = 1.77–1.80, \beta = 1.78–1.80; \gamma = 1.83–1.85; \gamma - \alpha = 0.05–$
0.07; positive, $2V = 10–20°$.

Chemistry: Monazite is an anhydrous phosphate of the cerium metals; the lanthanum earths are ordinarily present in a ratio of 1:1 with cerium. Th is usually present to the extent of a few percent, and a series probably extends to at least 30 percent ThO_2. The Y earths and Ca substitute in small amounts for (Ce,La); U has been reported occasionally. The phosphate with the smaller cation, Y, is tetragonal (xenotime), and isostructural with zircon.

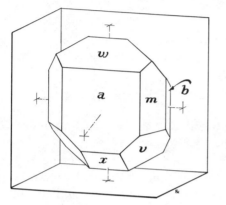

Figure 14-1 Monazite crystal. Forms: pinacoids $b\{010\}$, $a\{100\}$, $w\{101\}$, $x\{\bar{1}01\}$; rhombic prisms $m\{110\}$, $v\{\bar{1}11\}$.

Diagnostic features: Infusible; slowly decomposed by acids and by fusion with sodium carbonate. It is softer than zircon and harder than sphene but may be recognized with certainty only by crystal form and chemical tests, particularly for phosphate, and by its optics.

Occurrence: Monazite is of widespread occurrence as an accessory mineral in granite, syenite, and gneissic rocks of related composition, also as large crystals in associated pegmatite dikes. Detrital sands derived from these rocks may, in some regions, contain commercial quantities of the mineral. It is often associated with zircon, xenotime, magnetite, apatite, columbite, and rare-earth niobate-tantalates. Monazite occurs at numerous localities in southern Norway; commercial quantities of monazite are recovered from detrital deposits, carbonatites, or pegmatites in Australia, India, Brazil, the USA, South Africa, and Malagasay Republic. The mineral is important as a source of cerium and thorium.

Hydrated Normal Phosphates

Vivianite, $Fe_3(PO_4)_2 \cdot 8H_2O$

Crystal system and class: Monoclinic; $2/m$.
Axial elements: $a:b:c = 0.7502:1:0.3502$; $\beta = 104° 18'$.

Cell dimensions and content: $a = 10.06$, $b = 13.41$, $c = 4.696$; $Z = 2$.

Common forms and angles:

(010) $\wedge$ (110) $= 53° 59'$ (010) $\wedge$ ($\bar{2}$21) $= 60° 15'$

(100) $\wedge$ (20$\bar{1}$) $= 54° 41'$

Habit: Crystals usually prismatic along c with pinacoids {010} and {100} dominant; as reniform or tubular masses or concretions, or incrusting with a divergent bladed or fibrous structure; also earthy or pulverulent.

Cleavage: {010} perfect, thin laminae are flexible.

Tenacity: Sectile.

Hardness: $1\frac{1}{2}$–2.

Density: 2.68.

Color: Colorless and transparent when fresh, rapidly becoming pale to dark blue by oxidation.

Streak: White, changing to blue or brown.

Luster: Vitreous.

Optical properties: $\alpha = 1.580$, $\beta = 1.598$, $\gamma = 1.627$; $\gamma - \alpha = 0.047$; positive, $2V = 80°$.

Chemistry: The hydrated arsenates of Co, Ni, and Zn with the composition $A_3(XO_4)_2 \cdot 8H_2O$ are isostructural with vivianite and comprise the vivianite group. Vivianite shows only a slight tendency to form a solid solution with the other members; Mn^2, Mg, and Ca may substitute for Fe^2 to a limited extent. The principal variations in vivianite are due to the partial oxidation of Fe^2 to Fe^3; with this change the mineral becomes dark blue or nearly black.

Diagnostic features: Usually altered to blue or green color, flexible cleavage lamellae are distinctive; fusible to a gray magnetic bead; yields water in the closed tube; soluble in acids; nitric acid solution gives yellow phosphate precipitate with ammonium molybdate solution.

Occurrence: Vivianite, a mineral of secondary origin, occurs commonly in the gossan of metallic ore deposits and in the surface zone of pegmatites containing iron-manganese phosphates. It is also found in clays as concretions, and in recent sedimentary deposits associated with bone or other organic remains. Vivianite occurs with pyrrhotite in the tin veins of Cornwall, England; in the pegmatites of Hagendorf, Bavaria; in the sedimentary iron ores on the Kerch and Taman peninsulas on the Black Sea, the USSR; in the tin veins at Llallagua and Tasna in Bolivia; at Leadville, Lake County, Colorado.

Erythrite, $Co_3(AsO_4)_2 \cdot 8H_2O$

Crystal system and class: Monoclinic; $2/m$.

Axial elements: $a:b:c = 0.7629:1:0.3545$; $\beta = 105° 01'$.

Cell dimensions and content: $a = 10.20$, $b = 13.37$, $c = 4.74$; $Z = 2$.

Habit: Crystals usually prismatic to acicular along c and flattened on $\{010\}$, usually striated parallel to c; often as radial groups or reniform shapes with a drusy surface and columnar or fibrous structure; earthy or pulverulent.

Cleavage: $\{010\}$ perfect.

Tenacity: Sectile.

Hardness: $1\frac{1}{2}-2\frac{1}{2}$.

Density: 3.06.

Color: Crimson-red, becoming paler with increasing content of Ni, and white or gray with Ni > Co; *annabergite*, the nickel analogue, is apple green.

Streak: Paler than color.

Luster: Adamantine to dull.

Optical properties: $\alpha = 1.626$, $\beta = 1.662$, $\gamma = 1.699$; $\gamma - \alpha = 0.073$; $2V$ near 90°.

Chemistry: Erythrite, in contrast to vivianite, shows rather wide variations in composition. Ni commonly substitutes for Co, and a complete series extends to *annabergite*, $Ni_3(AsO_4)_2 \cdot 8H_2O$. Ca, Zn, Mg, and Fe^2 also substitute for Co or Ni, but only Zn forms an isostructural mineral, and a complete series may extend to the mineral *koettigite* $Zn_3(AsO_4)_2 \cdot 8H_2O$.

Diagnostic features: Its color and association with other cobalt minerals are distinctive features. Erythrite is fusible to a gray bead; gives the odor of arsenic when heated on charcoal; yields water in the closed tube; imparts a deep blue cobalt color to the borax bead.

Occurrence: Erythrite and annabergite are secondary minerals formed by surface oxidation of cobalt and nickel arsenides. Their occurrence as a pulverulent coating on the primary minerals led to the common terms "cobalt-bloom" (pink) and "nickel-bloom" (green). Where the primary mineral contains both nickel and cobalt in substantial quantities the resulting bloom is white or gray. The presence of pink cobalt-bloom in the zone of surface weathering is a distinctive indication of the presence of cobalt arsenides; its presence led prospectors to the cobalt-silver veins of Cobalt, Ontario, and to the cobalt-silver-uranium veins of Great Bear Lake, Canada. Erythrite also occurs at Schneeberg, Saxony; Joachimsthal and Pribram, Bohemia; Churchill County, Nevada; the Blackbird district, Lemhi County, Idaho.

Anhydrous Phosphates with Hydroxyl or Halogen

Amblygonite, $(Li,Na)Al(PO_4)(F,OH)$

Crystal system and class: Triclinic; $\bar{1}$.

Axial elements: $a:b:c = 0.729:1:0.708$; $\alpha = 112° 02'$, $\beta = 97° 49'$, $\gamma = 68° 07'$.

Cell dimensions and content: $a = 5.19$, $b = 7.12$, $c = 5.04$; $Z = 2$.

Common forms and angles:

(100) $\wedge$ (110) = 45° 06′	(100) $\wedge$ (0$\bar{1}$1) = 75° 11′
(100) $\wedge$ (001) = 90° 15′	(010) $\wedge$ (021) = 25° 31′

Habit: Crystals usually rough, especially when large; small crystals usually equant; in large cleavable masses, columnar, or compact.

Twinning: Common on $\{\bar{1}11\}$; lamellar on $\{111\}$.

Cleavage: $\{100\}$ perfect, $\{110\}$ good, $\{0\bar{1}1\}$ distinct.

Fracture and tenacity: Uneven; brittle.

Hardness: $5\frac{1}{2}$–6.

Density: 2.98–3.11.

Color: White to milky or creamy, also yellowish, pinkish, greenish, bluish, gray, rarely colorless or transparent.

Streak: White.

Luster: Vitreous to greasy, pearly on well developed cleavages.

Optical properties: α = 1.58–1.60, β = 1.59–1.62, γ = 1.60–1.63; $\gamma - \alpha$ = 0.02–0.03; negative, $2V$ = 50–90°.

Chemistry: Amblygonite shows considerable variation in composition by substitution of Na for Li and substitution of OH for F, probably forming a complete series. Li is normally present much in excess of Na, and F in excess of OH as in amblygonite proper. When OH exceeds F the name *montebrasite* applies; only one occurrence has been noted where Na is in excess of Li.

Diagnostic features: Amblygonite may easily be mistaken for white albite, with which it may be associated. The different orientation, cleavage angles, and less perfect quality of the cleavage, together with the higher density, provide the best means of distinguishing the mineral; more easily fusible than feldspar; gives red flame indicative of lithium; insoluble in acid; gives ammonium molybdate test for phosphate after fusion with sodium carbonate and dissolving in nitric acid.

Occurrence: The members of this series occur chiefly in granite pegmatites of the types rich in Li and PO_4, often as very large crystals. Amblygonite is commonly associated with spodumene, apatite, lepidolite, tourmaline, and other Li minerals. It may also be found in high-temperature veins and greisen, with cassiterite, topaz, and mica. It is found at the famous pegmatite localities of Varuträsk, Sweden; Karibib, South West Africa; Custer district, Black Hills, South Dakota; Pala, San Diego County; Fano mine, Riverside County, California; Newry, Hebron, Buckfield, and other localities in Maine; and Yellowknife-Beaulieu area, Northwest Territories, Canada.

APATITE GROUP

The members of the apatite group fall into two series, the apatite series and the pyromorphite series. These minerals are isostructural, and extensive substitutional solid solution takes place between the members of each series, but only to a limited extent between members of different series. The unit cell volume of the pyromorphites is about one-fifth larger than that of the apatites. The apatite series includes the following varieties of apatite:

Fluorapatite	$Ca_5(PO_4)_3F$
Chlorapatite	$Ca_5(PO_4)_3Cl$
Hydroxylapatite	$Ca_5(PO_4)_3(OH)$
Carbonate-apatite	$Ca_{10}(PO_4)_6(CO_3)H_2O$

Sr, Mn, Ce, and, to a lesser extent, Na, substitute in part for Ca in the apatite series. In the pyromorphite series the principal metal is the larger ion, Pb for which Ca may substitute to a minor extent. This series includes:

Pyromorphite	$Pb_5(PO_4)_3Cl$
Mimetite	$Pb_5(AsO_4)_3Cl$
Vanadinite	$Pb_5(VO_4)_3Cl$

Substitution of V, As, and P for one another in minerals of this series is common, whereas As and V show almost no tendency to substitute for P in apatites. The larger ion, Cl, is least common in apatite, but it is the only significant halogen in pyromorphites.

Apatite, $Ca_5(PO_4)_3(F,Cl,OH)$

Crystal system and class: Hexagonal; $6/m$.
Axial elements: $a:c = 1:0.735$.
Cell dimensions and content: $a = 9.38$, $c = 6.89$; $Z = 2$.
Common forms and angles:
  $(10\bar{1}0) \wedge (10\bar{1}1) = 49° 41'$
  $(0001) \wedge (11\bar{2}1) = 55° 46'$
Habit: Crystals in crystalline limestones and igneous rocks usually short to long prismatic along c with commonest forms, $\{10\bar{1}0\}$, and $\{10\bar{1}1\}$, (Figs. 14-2, 14-3). Crystals of hydrothermal origin in pegmatites and veins are commonly thick tabular on $\{0001\}$ with $\{10\bar{1}0\}$ and dipyramid forms; the hexagonal prism $\{41\bar{5}0\}$ or dipyramids $\{21\bar{3}1\}$, $\{31\bar{4}1\}$ are found very rarely as small faces; etch-pits and x-ray Laue photographs clearly support the $6/m$ symmetry; very commonly massive, coarse granular to compact, sometimes globular or reniform, stalactitic, earthy, oolitic.

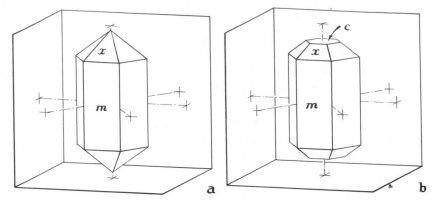

Figure 14-2 Apatite crystals. Forms: pinacoid $c\{0001\}$; hexagonal prism $m\{10\bar{1}0\}$; hexagonal dipyramid $x\{10\bar{1}1\}$.

Cleavage: $\{0001\}$ indistinct.

Fracture and tenacity: Conchoidal to uneven; brittle.

Hardness: 5.

Density: 3.1–3.2.

Color: Green, bluish green, grayish green, also blue, violet, colorless, brown, red.

Streak: White.

Luster and opacity: Vitreous; transparent to opaque.

Optical properties: $\epsilon = 1.630$, $\omega = 1.633$; $\omega - \epsilon = 0.003$; negative (fluorapatite).

Chemistry: Most apatites fit the formula given above, in which F, Cl, and (OH) are present in varying amounts, though the pure end members with F, Cl, or (OH) alone are also found. The specific names, fluorapatite, chlorapatite, and hydroxylapatite, are given to the members in which F, Cl, or (OH) predominate. Carbonate-apatite contains up to 4.4 percent CO_2, which may be present as C in substitution for P or as CO_3 for F, but the mechanism is not fully understood. Mn may substitute for Ca up to about 10 percent MnO, especially in the fluorapatite of granite pegmatites. Sr and rare-earth elements, especially Ce, may substitute for Ca to a minor extent.

Diagnostic features: Usually recognized by its crystal form, color, and hardness; it is distinguished from beryl by inferior hardness, from quartz by color and hardness. Massive, sugary varieties may resemble some diopside or olivine, though they are of inferior hardness and usually of darker color. It is fusible with difficulty. It is soluble in acids; dilute nitric acid solution gives a phosphate test.

Occurrence: Apatite occurs extensively as massive cryptocrystalline material which constitutes the bulk of phosphate rock and fossil bone. This material has usually been called *collophane*, a useful field name. The material is often opaline or horn-like, with a dense, layered or colloform structure, also concretionary or pulverulent. It is usually grayish white, yellowish, or brown, with a dull luster, low H (3—4) and G (2.5—2.9). In composition it usually falls in the hydroxylapatite-fluorapatite series, and it contains minor CO_2. The minerals of the apatite series are, by far, the most abundant phosphorus-

Figure 14-3 Apatite crystal with hexagonal prism $\{10\bar{1}0\}$, hexagonal pyramid $\{10\bar{1}1\}$, and pinacoid $\{0001\}$. (Renfrew County, Ontario.) [Courtesy Royal Ontario Museum.]

bearing minerals in the earth's crust. Fluorapatite, fluor-hydroxylapatite, and carbonatian varieties of these are the most widespread; other members of the series are of rare and restricted occurrence. Fluorapatites are found in almost all igneous rocks as an early-formed accessory mineral, usually as minute crystals not visible to the naked eye. In the Kola Peninsula of the USSR, extensive apatite deposits, formed as late magmatic segregations of alkalic igneous rocks, are mined as a source of phosphate. Crystallized apatite is also found in pegmatites (Mount Apatite, Auburn, Maine) associated with both acidic and basic types of igneous rocks; in magnetite deposits (Durango, Mexico; Adirondack Mountains, New York); and in high temperature hydrothermal veins. It is also very common in metamorphosed rocks of both regional and contact types, especially in crystalline limestones, where it is commonly associated with pyroxene (diopside), amphibole (tremolite or actinolite), idocrase, phlogopite, and spinel.

Apatite is often found as large crystals in crystalline limestones of the Grenville rocks of Precambrian age in eastern Ontario and adjoining areas of southern Quebec. This material was formerly mined as a source of phosphate, and in many localities phlogopite of commercial grade was also obtained.

The more extensive bedded phosphate deposits (collophane) now supply the market for raw phosphate required in the manufacture of fertilizer in North America and most other regions. Some such deposits are of marine origin, others are residual or detrital deposits derived from the marine beds. Extensive "pebble" phosphate deposits of Florida supply most of the United States' requirements, and marine phosphate sediments underlie a large area of Idaho, Wyoming, Montana, and Utah. Weathered sedimentary beds of phosphate rock in Morocco constitute the world's second largest producing deposits. Other deposits of importance occur in Algeria, Tunisia, and Egypt.

Pyromorphite Series

Crystal system and class: Hexagonal; $6/m$.

	PYROMORPHITE	MIMETITE	VANADINITE
Axial elements:	$(a:c) = 0.733$	0.725	0.7108
Cell dimensions:	$a = 10.00$	$a = 10.26$	$a = 10.331$
	$c = 7.33$	$c = 7.44$	$c = 7.343$
Cell content:	$Z = 2$	$Z = 2$	$Z = 2$
Hardness:	$3\frac{1}{2}$–4	$3\frac{1}{2}$–4	$2\frac{1}{2}$–3
Density:	7.04	7.24	6.86
Optical properties:			
ω	2.059	2.144	2.416
ϵ	2.049	2.129	2.350
	negative	negative	negative

Pyromorphite, $Pb_5(PO_4)_3Cl$; Mimetite, $Pb_5(AsO_4)_3Cl$

Habit: Usually prismatic and simple in habit with the commonest forms $\{10\bar{1}0\}$, $\{0001\}$, and $\{10\bar{1}1\}$; also in rounded barrel-shaped forms, or as subparallel groups of prismatic crystals; often globular reniform, or wart-like; granular.

Fracture and tenacity: Uneven; brittle.

Color: Green, yellow, or brown of various shades for pyromorphite, usually pale yellow to yellow brown or colorless for mimetite.

Streak: Nearly white.

Luster: Resinous.

Chemistry: These two minerals form a complete series of solid solutions; the name pyromorphite applies to that half of the series with PO_4 predominant, and the name mimetite to the half with AsO_4 predominant. Ca substitutes to a minor extent for Pb, and V substitutes for (As,P) to a small extent in mimetite.

Diagnostic features: The crystal form, high luster, and high specific gravity are distinctive properties; it is difficult to distinguish pyromorphite and mimetite without chemical tests. Easily fusible; both give lead globule when fused with sodium carbonate on charcoal; soluble in acids; a dilute nitric acid solution gives a phosphate test; mimetite, when heated in the closed tube in contact with a splinter of charcoal, gives a deposit of metallic arsenic on the wall of the tube.

Occurrence: These minerals are of secondary origin and occur in the oxidized zone of ore deposits containing galena. Pyromorphite is more common than mimetite. They are associated with limonite, cerussite, and, to a lesser extent, smithsonite, hemimorphite, anglesite, malachite, vanadinite, and wulfenite. Pyromorphite occurs at Schneeberg, Saxony; at Pribram, Czechoslovakia; at Beresovsk in the Urals, the USSR; at various localities in Cornwall, at Roughton Gill, Cumberland, England; at Leadhills, Scotland; at Broken Hill, New South Wales; in Ojuela mine, Mapimi, Mexico; in mines of the Coeur d'Alene district, Idaho; and at Leadville, Colorado. Mimetite occurs at Johanngeorgenstadt, and Braunsdorf, Saxony; at Pribram, Czechoslovakia; in Cornwall, England; at Leadhills, Scotland; at Långban, Sweden; at Broken Hill, New South Wales; at Mammoth Mine, Arizona; at the Tintic district, Utah; and at the Eureka district, Nevada.

Vanadinite, $Pb_5(VO_4)_3Cl$

Habit: Crystals usually short to long prismatic along c with commonest forms $\{10\bar{1}0\}$, $\{0001\}$, $\{10\bar{1}1\}$, $\{20\bar{2}1\}$; sometimes cavernous as hollow prisms, also in rounded forms and in subparallel groupings, in globules.

Fracture and tenacity: Uneven to conchoidal; brittle.

Hardness: 3.

Density: 6.88.

Color: Orange-red to brownish red.

Streak: White to yellowish.

Luster and opacity: Subresinous; nearly opaque.

Chemistry: In vanadinite P may substitute for V to a small extent, and As may substitute to about As:V = 1:1.

Diagnostic features: Vanadinite is distinguished by its luster, high specific gravity, crystal form, and, from pyromorphite-mimetite, by its color. It is easily fusible; gives a globule of lead when fused with sodium carbonate on charcoal; in the oxidizing flame it gives an amber color (vanadium) to the salt of phosphorus bead; a dilute nitric acid solution with silver nitrate gives a precipitate of silver chloride. Arsenian vanadinite gives a test for arsenic as for mimetite.

Occurrence: Vanadinite is a secondary mineral found in the surface oxidized zone of ore deposits containing galena and other sulfides. It is found associated with pyromorphite, wulfenite, cerussite, anglesite, and limonite. The vanadium in the mineral represents an enrichment from the sparse vanadium content of primary sulfide and gangue minerals. Vanadinite is a minor source of vanadium, particularly from the oxidized ores of Broken Hill (Zambia), and the Otavi district (South West Africa).

Hydrated Phosphates Containing Hydroxyl

Turquois, $CuAl_6(PO_4)_4(OH)_8 \cdot 4H_2O$

Crystal system and class: Triclinic; $\bar{1}$.

Axial elements: $a:b:c = 0.752:1:0.770$; $\alpha = 111°\,39'$, $\beta = 115°\,23'$, $\gamma = 69°\,26'$.

Cell dimensions and content: $a = 7.48$, $b = 9.95$, $c = 7.68$; $Z = 1$.

Habit: Crystals very rare, usually massive, dense and cryptocrystalline to fine-grained, also as veinlets or crusts, concretionary.

Cleavage: {001} perfect; {010} good, but rarely seen.

Fracture: Conchoidal to smooth in massive material.

Hardness: 5–6.

Density: 2.6–2.8.

Color: Sky blue, bluish green to apple green.

Streak: White or greenish to pale green.

Luster: Waxy, vitreous in crystals.

Optical properties: $\alpha = 1.61$, $\beta = 1.62$, $\gamma = 1.65$; $\gamma - \alpha = 0.04$; positive, $2V = 40°$.

Chemistry: Turquois is a member of a mineral series which by substitution of Fe for Al probably extends to the rare mineral *chalcosiderite*. Most turquois falls near the high Al end of the series.

Diagnostic features: Turquois is characterized by its distinctive blue color. It is harder than chrysocolla. It is infusible but turns brown before the blowpipe; readily soluble in hydrochloric acid after ignition; solution gives phosphate test.

Occurrence: It is a secondary mineral found with limonite, chalcedony, and kaolin, and formed by the action of surface waters on aluminous igneous or sedimentary rocks. It is usually found in arid regions. The fine blue crypto-crystalline material has been valued as an ornamental or semiprecious stone since earliest times. Fine quality turquois has been obtained from a deposit near Nishapur, Khorosan, Persia; also at Jordansmühl, Silesia, Germany; at Montebras, France; at Lynch Station, Campbell County, Virginia, as tiny crystals; in the Los Cerillos Mountains, 20 miles southwest of Santa Fe, New Mexico; and at numerous other localities in the southwestern USA.

Uranyl Phosphates

TORBERNITE AND METATORBERNITE GROUPS

The minerals of these two groups conform to the general formula $A(UO_2)_2(XO_4)_2 \cdot nH_2O$, where A is Cu, Ca, Ba or Mg, X is P or As, and n may have various values from 2 to probably 12. The two groups have closely related layer structures. The structure of the Torbernite Group is stable for values of nH_2O over 8 and ranging up to 12, and the metatorbernite structure is stable for $n = 6-8$ down to 5 or $2\frac{1}{2}$. The transition temperature between the two structures varies with the nature of A and X and with vapor pressure, but is in general close to room temperature. The transition between the two structures is reversible for autunite but probably is not reversible for torbernite. Isostructural compounds in which A is $Na_2, K_2, (NH_4)_2$, or H_2 have been prepared artificially.

	TORBERNITE	META-TORBERNITE	AUTUNITE	META-AUTUNITE
Crystal system:	Tetragonal	Tetragonal	Tetragonal	Tetragonal
Crystal class:	$4/m\ 2/m\ 2/m$	$4/m\ 2/m\ 2/m$	$4/m\ 2/m\ 2/m$	$4/m\ 2/m\ 2/m$
Axial elements:				
$(a{:}c)$	$1{:}2.90$	$1{:}1.238$	$1{:}2.952$	$1{:}1.206$
Cell dimensions:	$a = 7.06$	$a = 6.96$	$a = 7.003$	$a = 7.00$
	$c = 20.5$	$c = 8.62$	$c = 20.67$	$c = 8.44$
Cell content:	$2[Cu(UO_2)_2(PO_4)_2$	$Cu(UO_2)(PO_4)_2$	$2[Ca(UO_2)_2(PO_4)_2$	$Ca(UO_2)_2(PO_4)_2$
	$\cdot 8{-}12H_2O]$	$\cdot 8H_2O$	$\cdot 10{-}12H_2O]$	$\cdot 2\frac{1}{2}{-}6\frac{1}{2}H_2O$
Hardness:	$2-2\frac{1}{2}$	$2\frac{1}{2}$	$2-2\frac{1}{2}$	—
Density:	3.22	3.5–3.7	3.1–3.2	—
Optical properties:				
ω	1.592	1.624	1.577	1.613
ϵ	1.582	1.626	1.553	1.600

Torbernite, $Cu(UO_2)_2(PO_4)_2 \cdot 8{-}12H_2O$

Habit: Thin to thick tabular on {001}, usually square in outline, rarely pyramidal; subparallel, foliated, micaceous, or scaly aggregates.

Cleavage: {001} perfect, laminae brittle.

Color: Emerald green, grass green to apple green; not fluorescent in ultra-violet light.

Streak: Pale green.

Luster and opacity: Vitreous, pearly on {001}; transparent to translucent.

Chemistry: Small amounts of As substitute for P; there is no evidence for substitution of Ca for Cu.

Diagnostic features: Color and copper content distinguish torbernite from autunite and other secondary uranium minerals except metatorbernite and zeunerite, $Cu(UO_2)_2(AsO_4)_2 \cdot 8\text{–}12H_2O$.

Occurrence: Torbernite is a secondary mineral formed as an oxidation of uraninite, especially in veins which also contain copper sulfides. It is found especially in Cornwall, England, in the mines near Redruth, and St. Just; at Joachimsthal and Zinnwald in Bohemia; at Schneeberg in Germany; and in the Katanga district in the Congo.

Autunite, $Ca(UO_2)_2(PO_4)_2 \cdot 10\text{–}12H_2O$

Habit: Thin tabular on {001}, closely similar to torbernite in form and angles; common as subparallel growths, as scaly aggregates or thick crusts with a serrated surface due to crystals standing on edge.

Cleavage: {001} perfect, not brittle.

Color: Lemon yellow, sometimes greenish yellow, strongly fluorescent to yellowish green in ultraviolet light.

Streak: Yellowish.

Luster: Vitreous, pearly on {001}.

Chemistry: Autunite may include small amounts of Mg or Ba substituting for Ca. The water content varies from 10 to 12 in autunite; on drying or slight heating autunite changes reversibly to meta-autunite with $2\frac{1}{2}\text{–}6\frac{1}{2}$ H_2O. On further heating to about 80° meta-autunite changes irreversibly to an ortho-rhombic phase which has not been found in nature.

Diagnostic features: Autunite is difficult to distinguish from many other yellow secondary uranium minerals except by chemical and optical tests or by x-ray diffraction.

Occurrence: It is a secondary mineral usually found in the zone of oxidation and weathering of veins or pegmatites containing uranium. It is found at several localities near Autun, France; at Sabugal, Portugal; in Cornwall, England; in the Johanngeorgenstadt district, Germany; in the Katanga district, the Congo; at Spruce Pine, North Carolina; and near Spokane, Washington.

VANADIUM OXYSALTS

Carnotite, $K_2(UO_2)_2(VO_4)_2 \cdot 3H_2O$

Crystal system and class: Monoclinic; $2/m$.
Axial elements: $a:b:c = 1.245:1:0.822$; $\beta = 103° 40'$.
Cell dimensions and content: $a = 10.47$, $b = 8.41$, $c = 6.91$; $Z = 2$.
Habit: Occurs usually as a powder, or as loosely coherent fine crystalline aggregates, compact or disseminated, rarely as crusts or minute platy crystals flattened on $\{001\}$.
Cleavage: $\{001\}$ perfect.
Hardness: about 2.
Density: 4–5.
Color: Bright yellow to lemon yellow.
Luster: Dull or earthy.
Optical properties: $\alpha = 1.76$, $\beta = 1.90$, $\gamma = 1.92$; $\gamma - \alpha = 0.16$; negative, $2V$ small.

Tyuyamunite, $Ca(UO_2)_2(VO_4)_2 \cdot nH_2O$

Crystal system and class: Orthorhombic; $2/m\ 2/m\ 2/m$?
Axial elements: $a:b:c = 1.26:1:2.35$.
Cell dimensions and content: $a = 10.40$, $b = 8.26$, $c = 19.41$; $Z = 4$.
Habit: Occurs as scales and laths flattened on $\{001\}$ and elongate along a, and as radial aggregates; commonly massive, compact, or pulverulent, to cryptocrystalline.
Cleavage: $\{001\}$ perfect, micaceous.
Hardness: about 2.
Density: 3.6–4.4, increasing with lower water content.
Color: Canary yellow to lemon yellow or greenish.
Luster: Waxy on massive material, otherwise earthy.
Optical properties: $\alpha = 1.67$, $\beta = 1.87$, $\gamma = 1.90$; $\gamma - \alpha = 0.23$; negative, $2V$ small.
Diagnostic features: Carnotite and tyuyamunite are not easily distinguished except by chemical, optical, or x-ray diffraction tests.
Occurrence: Carnotite and tyuyamunite are commonly found together as secondary minerals and were formed by the action of meteoric waters on uranium and vanadium minerals. They are widespread in sandstones of the Colorado plateau area in Colorado and adjoining parts of Utah, New Mexico, and Arizona. They are found disseminated, or locally as relatively pure masses around petrified tree trunks or other vegetal matter.

15

Class VIII: Silicates

H																	
Li	Be	*B*													O	F	
Na	Mg	Al	Si												S	Cl	
K	Ca	Sc	Ti	V	*Cr*	Mn	Fe		Ni	Cu	Zn						
		Y	Zr									Sn					
Cs	Ba	La Lu	Hf									Pb	Bi				
			Th		U												

The silicates include a large number of minerals, about a third of all species.* Many of these are quite rare, but others make up a large part of the earth's crust. The crust has been estimated to be about 95 percent silicate minerals, of which some 60 percent is feldspar and 12 percent quartz. The predominance of silicates and aluminosilicates reflects the abundance of oxygen, silicon, and aluminum, which are the commonest elements in the crust (O = 47 percent, Si = 28 percent, Al = 8 percent).

The number of elements which form silicate minerals is comparatively few,

* Table above shows elements which form silicate minerals; those in italics are minor constituents.

as can be seen in the above table. Some elements, such as Rb and Sr, do not form specific silicate minerals but are dispersed in small amounts replacing more abundant elements in common minerals. Other elements do not form independent silicates but are associated with other cations (e.g., B, which forms a number of borosilicates). The multiplicity of silicate minerals is due to the variety of silicates that may be formed from the same elements. They often have complex and variable compositions, and these complexities hindered the development of a satisfactory classification of silicate minerals. This problem was not finally solved until x-ray investigations provided the means for determining the crystal structure of such compounds.

Prior to the elucidation of the crystal structure of the silicates their composition was generally interpreted in terms of hypothetical silicic acids, all of which were derived from a theoretical orthosilicic acid, H_4SiO_4. Some of the hypothetical silicic acids are:

Orthosilicic acid	H_4SiO_4
Metasilicic acid	$H_2SiO_3 = (H_4SiO_4 - H_2O)$
Orthodisilicic acid	$H_6Si_2O_7 = (2H_4SiO_4 - H_2O)$
Metadisilicic acid	$H_2Si_2O_5 = (2H_4SiO_4 - 3H_2O)$
Trisilicic acid	$H_4Si_3O_8 = (3H_4SiO_4 - 4H_2O)$

In this way ratios of silicon to oxygen could be derived to fit any composition. The silicic acid theory had some success in the interpretation of the simpler compounds, such as the orthosilicates and metasilicates, but led to manifest absurdities when applied to more complex minerals. Thus the isomorphous compounds albite, $NaAlSi_3O_8$, and anorthite, $CaAl_2Si_2O_8$, were placed in different groups. The silicic acid classification of the silicate minerals has now been superseded by one based on the crystal structures of these compounds.

THE STRUCTURE AND CLASSIFICATION OF THE SILICATES

In all silicate structures investigated so far the silicon atoms are in fourfold coordination with oxygen. This arrangement appears to be universal in these compounds, and the bonds between silicon and oxygen are so strong that the four oxygens are always found at the corners of a tetrahedron of nearly constant dimensions and regular shape, whatever the rest of the structure may be like. The different silicate types arise from the various ways in which these silicon-oxygen tetrahedra are related to one another (Figs. 15-1 through 15-4); they may exist as separate and distinct units, or they may be linked by sharing corners (i.e., oxygens). Silicate classification is based on the following types of linkages.

1. *Independent tetrahedral groups:*   In this type the silicon-oxygen tetrahedra are present as separate entities (Fig. 15-1a). The resultant composition is SiO_4, and a typical mineral is forsterite, Mg_2SiO_4. This division of the silicates is known as the *nesosilicates*.

2. *Double tetrahedra structures:*   Two silicon-oxygen tetrahedra are linked by the sharing of one oxygen between them; the resulting composition is Si_2O_7 (Fig. 15-1b), and a typical mineral is hemimorphite, $Zn_4Si_2O_7(OH)_2 \cdot H_2O$. Such silicates are known as *sorosilicates*.

3. *Ring structures:*   In this type two of the oxygens of each tetrahedron are shared with neighboring tetrahedra, and the angular positions of the tetrahedra are such that closed units of a ring-like structure result. Rings

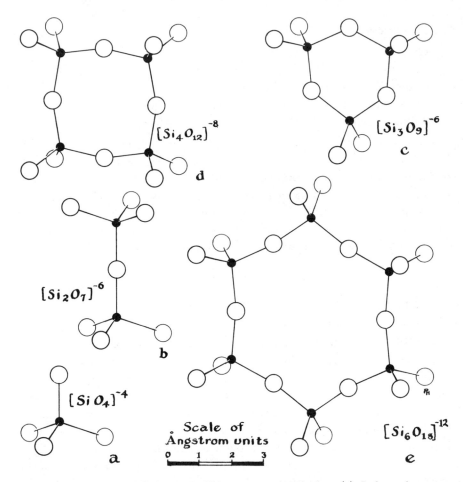

Figure 15-1   Types of linkage of silicon-oxygen tetrahedra. (a) Independent tetrahedra. (b) Double tetrahedra. (c, d, e) Ring structures.

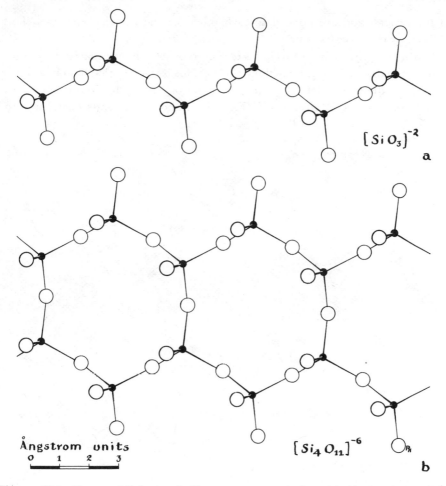

Figure 15-2 Types of linkage of silicon-oxygen tetrahedra. (a) Single chains. (b) Double chains.

of three, four, and six tetrahedra are known (Figs. 15-1c,d,e). Typical examples are benitoite, $BaTiSi_3O_9$, with three linked tetrahedra, axinite, $(Ca,Mn,Fe)_3Al_2(BO_3)Si_4O_{12}(OH)$, with four, and beryl, $Be_3Al_2Si_6O_{18}$, with six. This subclass of the silicates is known as the *cyclosilicates*.

4. *Chain structures:* Tetrahedra are joined together to produce chains of indefinite extent. There are two principal modifications of this structure yielding somewhat different composition: (a) single chains, in which Si:O is 1:3 (Fig. 15-2a), characterized by the pyroxenes and pyroxenoids; and (b) double chains, in which alternate tetrahedra in two parallel single chains are crosslinked and the Si:O ratio is 4:11 (Fig. 15-2b), characterized by the amphiboles. These chains are indefinite in extent, are elongated usually in the *c* direction

of the crystal, and are bonded to each other by the metallic elements. This division of the silicates is known as the *inosilicates*.

5. *Sheet structures:*   Three oxygens of each tetrahedron are shared with adjacent tetrahedra to form extended flat sheets. This is the double-chain inosilicate structure extended indefinitely in two directions instead of only one. This linkage gives an Si:O ratio of 2:5 (Fig. 15-3), and is the fundamental unit in

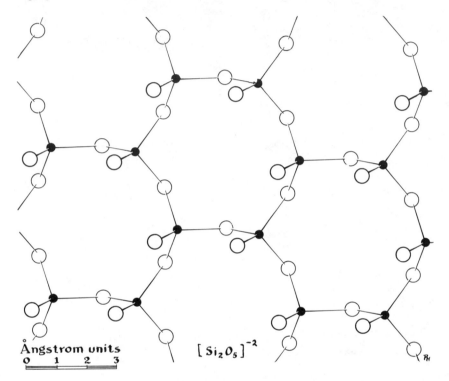

Ångstrom units
0   1   2   3

$[Si_2O_5]^{-2}$

Figure 15-3   Types of linkage of silicon-oxygen tetrahedra. Sheet structure.

all mica and clay structures. The sheets form a hexagonal planar network responsible for the principal characteristics of minerals of this type—their pronounced pseudohexagonal habit and perfect basal cleavage parallel to the plane of the sheet. This division of the silicates is known as *phyllosilicates*.

6. *Three-dimensional networks:*   Every SiO_4 tetrahedron shares all its corners with other tetrahedra, giving a three-dimensional network in which the Si:O ratio is 1:2 (Fig. 15-4). The various forms of silica—quartz, tridymite, cristobalite—have this arrangement. In SiO_2 the positive and negative charges balance. In silicates of this type the silicon is partly replaced by aluminum, and the composition is $(Si,Al)O_2$. The substitution of Al^3 for Si^4 requires additional positive ions in order to restore electrical neutrality [e.g., nepheline,

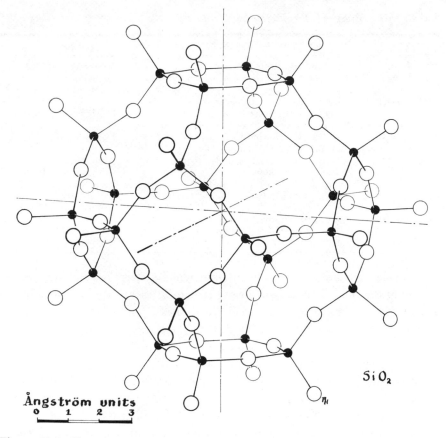

Ångström units
0   1   2   3

SiO₂

Figure 15-4 Types of linkage of silicon-oxygen tetrahedra. Three-dimensional network.

Na(AlSiO₄)]. The feldspars and zeolites are examples of this division of the silicates, which is known as the *tektosilicates*.

The silicate minerals can thus be classified as in Table 15-I, although some contain groupings of more than one type.

The other constituents of a silicate structure, such as additional oxygen atoms, hydroxyl groups, water molecules, and cations, are arranged with the silicate groups in such a way as to produce a mechanically stable and electrically neutral structure. Aluminum, after silicon the most abundant cation in the earth's crust, plays a unique role. As discussed earlier, it is stable both in fourfold and in sixfold coordination. It can replace silicon in the SiO_4 groups, and also the common six-coordination cations: Mg^2, Fe^2, Fe^3, etc. In many mineral groups (e.g., the garnets, the feldspars) aluminum is present entirely in a single coordination; in others (e.g., the amphiboles, the pyroxenes, the micas) it may be present in both coordinations.

Table 15-I *Structural classification of the silicates*

CLASSIFICATION	STRUCTURAL ARRANGEMENT	Si:O	EXAMPLES
Nesosilicates	Independent tetrahedra	1:4	Forsterite, Mg_2SiO_4
Sorosilicates	Two tetrahedra sharing one oxygen	2:7	Hemimorphite, $Zn_4Si_2O_7(OH)_2 \cdot H_2O$
Cyclosilicates	Closed rings of tetrahedra each sharing two oxygens	1:3	Beryl, $Be_3Al_2Si_6O_{18}$
Inosilicates	Continuous single chains of tetrahedra each sharing two oxygens	1:3	Enstatite, $MgSiO_3$
	Continuous double chains of tetrahedra sharing alternately two and three oxygens	4:11	Anthophyllite, $Mg_7(Si_4O_{11})_2(OH)_2$
Phyllosilicates	Continuous sheets of tetrahedra each sharing three oxygens	2:5	Talc, $Mg_3Si_4O_{10}(OH)_2$ Phlogopite, $KMg_3(AlSi_3O_{10})(OH)_2$
Tektosilicates	Continuous framework of tetrahedra each sharing all four oxygens	1:2	Quartz, SiO_2 Nepheline, $NaAlSiO_4$

The valence charge on the silicate unit, which determines the number and charge of the other ions present in the structure, can easily be calculated for any unit if it is remembered that each silicon has a positive charge of four and each oxygen a negative charge of two. Thus the charge on a single SiO_4 unit is $[4 + 4(-2)] = -4$; on an Si_2O_7 unit, -6; on an SiO_3 unit, -2; on an Si_4O_{11} unit, -6; on an Si_2O_5 unit, -2; and on SiO_2, 0.

Corresponding to different linkages of the tetrahedra are different compositions, habits, and physical properties in the various types of silicates. The bonding within the Si–O framework is much stronger than the bonding between the metal cations and the framework, and consequently the cleavage planes of the silicates are parallel to Si–O chains or sheets. The sheet structure, for instance, produces the platy form of the micas, the chlorites, the kaolins, and other minerals. The chain structures produce prismatic or fibrous crystals, as exemplified in the pyroxenes and amphiboles. The three-dimensional network structures usually produce equidimensional crystals. There are characteristic density ranges and refractive index limits in the different types corresponding to differences of structure. In general, increased complexity of silicate linkage results in diminished packing of the ions, manifested by a trend toward lower densities and refractive indices in going from nesosilicate to tektosilicate structures (for comparable compositions; see Table 15-II).

The silicate minerals can thus be grouped into six subclasses according to the type of linkage of the silicon-oxygen tetrahedra. Within each subclass different chemical compositions produce different structural types and hence crystallographic variations, leading to a division into groups. In some groups (e.g., the feldspars, the micas) the individual species are isomorphous with each other, whereas other groups comprise a number of species related chem-

Table 15-II *Relationship between structure type and density in the magnesium silicate minerals*

STRUCTURE TYPE	MINERAL	DENSITY
Nesosilicate	Forsterite, Mg_2SiO_4	3.22
Inosilicate (single chain)	Enstatite, $MgSiO_3$	3.18
Inosilicate (double chain)	Anthophyllite, $Mg_7(Si_4O_{11})_2(OH)_2$	2.96
Phyllosilicate	Talc, $Mg_3(Si_4O_{10})(OH)_2$	2.82

ically or paragenetically (e.g., the feldspathoids, the zeolites). Groups may be divided into series (minerals showing continuous variation in properties with continuous change in composition; e.g., the olivine series, the hornblende series) or individual species (when the mineral shows little or no variation in composition).

In discussing the composition and structure of the silicate minerals it is often convenient to use a general formula which will fit all species within the group. The general formulas in the mineral descriptions use the following symbols:

W = large cations having coordination number greater than six; principally Ca, Na, K.

X = medium-sized bivalent cations (and lithium) in sixfold coordination; principally Mg and Fe^2.

Y = medium-sized trivalent and quadrivalent cations in sixfold coordination; these include Al, Fe^3, and Ti^4.

Z = small cations in four-coordination; principally Si, often partly replaced by Al; also boron in a few minerals.

The silicate minerals that are described in detail in this book are listed below:

SUBCLASS TEKTOSILICATES

Silica Group:

Quartz	SiO_2	Trigonal
Tridymite	SiO_2	Hexagonal
Cristobalite	SiO_2	Isometric
Opal	$SiO_2 \cdot nH_2O$	Amorphous
Feldspar Group:	WZ_4O_8	
Sanidine	$KAlSi_3O_8$	Monoclinic
Orthoclase	$KAlSi_3O_8$	Monoclinic
Microcline	$KAlSi_3O_8$	Triclinic
Plagioclase Series:		Triclinic
Albite	$Ab_{100}An_0-Ab_{90}An_{10}$	$Ab = NaAlSi_3O_8$
Oligoclase	$Ab_{90}An_{10}-Ab_{70}An_{30}$	
Andesine	$Ab_{70}An_{30}-Ab_{50}An_{50}$	
Labradorite	$Ab_{50}An_{50}-Ab_{30}An_{70}$	
Bytownite	$Ab_{30}An_{70}-Ab_{10}An_{90}$	
Anorthite	$Ab_{10}An_{90}-Ab_0An_{100}$	$An = CaAl_2Si_2O_8$
Scapolite Series:	$(Na,Ca)_4[(Al,Si)_4O_8]_3(Cl,CO_3)$	Tetragonal

Feldspathoid Group:
Leucite	$KAlSi_2O_6$	Isometric
Nepheline	$NaAlSiO_4$	Hexagonal
Sodalite	$Na_8(AlSiO_4)_6Cl_2$	Isometric
Cancrinite	$Na_8(AlSiO_4)_6(HCO_3)_2$	Hexagonal

Zeolite Group:
	$W_mZ_rO_{2r}\cdot sH_2O$	
Heulandite	$CaAl_2Si_7O_{18}\cdot 6H_2O$	Monoclinic
Stilbite	$CaAl_2Si_7O_{18}\cdot 7H_2O$	Monoclinic
Laumontite	$CaAl_2Si_4O_{12}\cdot 4H_2O$	Monoclinic
Chabazite	$CaAl_2Si_4O_{12}\cdot 6H_2O$	Trigonal
Analcime	$NaAlSi_2O_6\cdot H_2O$	Isometric
Natrolite	$Na_2Al_2Si_3O_{10}\cdot 2H_2O$	Orthorhombic

SUBCLASS PHYLLOSILICATES

Kaolinite	$Al_4Si_4O_{10}(OH)_8$	Triclinic
Serpentine	$Mg_6Si_4O_{10}(OH)_8$	Monoclinic
Pyrophyllite	$Al_2Si_4O_{10}(OH)_2$	Monoclinic
Talc	$Mg_3Si_4O_{10}(OH)_2$	Monoclinic
Montmorillonite	$Al_2Si_4O_{10}(OH)_2\cdot xH_2O$	Monoclinic
Vermiculite	$Mg_3Si_4O_{10}(OH)_2\cdot xH_2O$	Monoclinic

Mica Group:
	$W(X,Y)_{2-3}(Z_4O_{10})(OH)_2$	
Muscovite	$KAl_2(AlSi_3O_{10})(OH)_2$	Monoclinic
Phlogopite	$KMg_3(AlSi_3O_{10})(OH)_2$	Monoclinic
Biotite	$K(Mg,Fe)_3(AlSi_3O_{10})(OH)_2$	Monoclinic
Lepidolite	$KLi_2Al(Si_4O_{10})(OH)_2$	Monoclinic
Glauconite	$K(Fe,Mg,Al)_2(Si_4O_{10})(OH)_2$	Monoclinic

Chlorite Series:	$(Mg,Fe,Al)_6(Al,Si)_4O_{10}(OH)_8$	Monoclinic
Apophyllite	$KCa_4(Si_4O_{10})_2F\cdot 8H_2O$	Tetragonal
Prehnite	$Ca_2Al(AlSi_3O_{10})(OH)_2$	Orthorhombic

SUBCLASS INOSILICATES

Amphibole Group:
	$(W,X,Y)_{7-8}Z_8O_{22}(OH)_2$	
Anthophyllite Series:	$(Mg,Fe)_7Si_8O_{22}(OH)_2$	Orthorhombic
Cummingtonite Series:	$(Fe,Mg)_7Si_8O_{22}(OH)_2$	Monoclinic
Tremolite-Actinolite Series:	$Ca_2(Mg,Fe)_5Si_8O_{22}(OH)_2$	Monoclinic
Hornblende Series:	$NaCa_2(Mg,Fe,Al)_5(Si,Al)_8O_{22}(OH)_2$	Monoclinic
Alkali-Amphibole Series:	$Na_2(Mg,Fe,Al)_5Si_8O_{22}(OH)_2$	Monoclinic

Pyroxene Group:
	$(W,X,Y)_2Z_2O_6$	
Enstatite-Hypersthene Series:	$(Mg,Fe)SiO_3$	Orthorhombic
Diopside-Hedenbergite Series:	$Ca(Mg,Fe)Si_2O_6$	Monoclinic
Augite	$Ca(Mg,Fe,Al)(Al,Si)_2O_6$	Monoclinic
Aegirine	$NaFeSi_2O_6$	Monoclinic
Jadeite	$NaAlSi_2O_6$	Monoclinic
Spodumene	$LiAlSi_2O_6$	Monoclinic

Pyroxenoid Group:
Wollastonite	$CaSiO_3$	Triclinic
Pectolite	$Ca_2NaHSi_3O_9$	Triclinic
Rhodonite	$MnSiO_3$	Triclinic

SUBCLASS CYCLOSILICATES

Axinite	$(Ca,Mn,Fe)_3Al_2(BO_3)Si_4O_{12}(OH)$	Triclinic
Beryl	$Be_3Al_2Si_6O_{18}$	Hexagonal
Cordierite	$(Mg,Fe)_2Al_4Si_5O_{18}$	Orthorhombic
Tourmaline	$Na(Mg,Fe)_3Al_6(BO_3)_3(Si_6O_{18})(OH)_4$	Trigonal

SUBCLASS SOROSILICATES

Lawsonite	$CaAl_2Si_2O_7(OH)_2 \cdot H_2O$	Orthorhombic
Hemimorphite	$Zn_4Si_2O_7(OH)_2 \cdot H_2O$	Orthorhombic
Idocrase	$Ca_{10}Mg_2Al_4(Si_2O_7)_2(SiO_4)_5(OH)_4$	Tetragonal
Epidote Group:	$W_2(X,Y)_3Z_3O_{12}(OH)$	
Zoisite	$Ca_2Al_3Si_3O_{12}(OH)$	Orthorhombic
Clinozoisite	$Ca_2Al_3Si_3O_{12}(OH)$	Monoclinic
Epidote	$Ca_2(Al,Fe)_3Si_3O_{12}(OH)$	Monoclinic
Allanite	$(Ca,R^*)_2(Al,Fe,Mg)_3Si_3O_{12}(OH)$	Monoclinic

SUBCLASS NESOSILICATES

Olivine Series:	$(Mg,Fe)_2SiO_4$	Orthorhombic
Willemite	Zn_2SiO_4	Trigonal
Aluminum Silicate Group:		
Andalusite	Al_2SiO_5	Orthorhombic
Sillimanite	Al_2SiO_5	Orthorhombic
Kyanite	Al_2SiO_5	Triclinic
Staurolite	$Al_9Fe_2O_6(SiO_4)_4(O,OH)_2$	Monoclinic
Topaz	$Al_2SiO_4(OH,F)_2$	Orthorhombic
Garnet Group:	$X_3Y_2(ZO_4)_3$	
Almandite	$Fe_3Al_2(SiO_4)_3$	Isometric
Pyrope	$Mg_3Al_2(SiO_4)_3$	Isometric
Spessartine	$Mn_3Al_2(SiO_4)_3$	Isometric
Grossular	$Ca_3Al_2(SiO_4)_3$	Isometric
Andradite	$Ca_3Fe_2(SiO_4)_3$	Isometric
Uvarovite	$Ca_3Cr_2(SiO_4)_3$	Isometric
Zircon	$ZrSiO_4$	Tetragonal
Thorite	$ThSiO_4$	Tetragonal
Sphene	$CaTiSiO_5$	Monoclinic
Datolite	$Ca(OH)BSiO_4$	Monoclinic
Dumortierite	$Al_7O_3(BO_3)(SiO_4)_3$	Orthorhombic

Subclass Tektosilicates

The tektosilicates include some of the most important rock-forming minerals. The simplest is quartz, SiO_2, in composition an oxide, but included here because its structure consists of a framework of SiO_4 tetrahedra with each oxygen linked to a silicon atom in a neighboring tetrahedron. Except for quartz and its polymorphs all the minerals in this subclass are aluminosilicates, since in order to produce a net negative charge on the framework some of the Si^4 must be replaced by Al^3. Up to half the Si^4 may be replaced by Al^3; the commonest $Si:Al$ ratio is either $1:1$ or $3:1$, corresponding to formula units of the type $AlSiO_4^{-1}$ and $AlSi_3O_8^{-1}$.

The cations which balance the negative charge on these formula units are always large ions (radius approximately 1 Å or greater) with coordination num-

* R = rare-earth elements.

ber eight or greater. Smaller six-coordination cations such as magnesium and iron do not form tektosilicates. This is understandable when we consider the net negative charge on each oxygen in the tektosilicate structure; for $AlSiO_4^{-1}$ this is one-fourth, for $AlSi_3O_8^{-1}$ it is one-eighth. A bivalent ion in six-coordination, such as Mg^2, requires a contribution of one-third of a negative charge from each of the surrounding oxygens, which is not available in a tektosilicate structure. Only univalent or bivalent cations with coordination eight or more can compensate the negative charge on the aluminosilicate framework. Hence the tektosilicates are aluminosilicates of sodium, potassium, calcium, and barium.

All the tektosilicates are colorless, white, or pale gray when free from inclusions. They also have rather low densities, as a result of the comparatively open nature of the tektosilicate structure. The hardness throughout is rather uniform, about 4–6. On the whole the tektosilicates form a remarkably homogeneous subclass, with many similarities in composition and properties common to all minerals within it.

SILICA GROUP

Silica occurs in nature as seven distinct minerals: quartz, tridymite, cristobalite, opal, coesite, stishovite, and lechatelierite. Of these, quartz is very common; tridymite and cristobalite are widely distributed in volcanic rocks and can hardly be called rare; opal is not uncommon; coesite and stishovite are found only in meteorite craters, where they have been formed from quartz by the high pressure produced by the meteorite impact; lechatelierite (silica glass) is very rare.

Opal and lechatelierite are amorphous. The common crystalline forms exhibit the phenomenon of enantiotropism. Each has its own stability field; at atmospheric pressure quartz is the stable form up to 867°C, tridymite is stable between 867°C and 1470°C, and cristobalite from 1470°C up to the melting point at 1713°C; from 1713°C to the boiling point liquid silica is the stable phase (Fig. 15-5). The relatively great difference in density between quartz and the other two polymorphs means that pressure has a marked effect on inversion temperatures, hence the steep slope of the boundaries between the fields of these polymorphs. It also means that tridymite is not stable above about 3,000 kg/cm², nor cristobalite above about 5,000 kg/cm². The absence of tridymite and cristobalite from plutonic rocks, even such as may have crystallized at high temperatures, is thus readily explicable, since pressures of the above order are attained at comparatively shallow depths in the crust.

Laboratory experiments on the effect of water vapor under pressure on the relationship between quartz, tridymite, and cristobalite have produced most

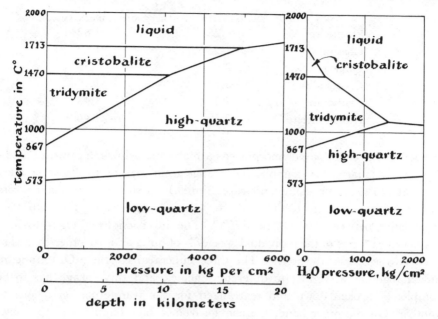

Figure 15-5 Stability relations of the different forms of SiO₂; dry system on the left, system under water vapor pressure on the right.

interesting results. In these experiments quartz, tridymite, and silica glass have been melted in a heated pressure vessel which permitted water vapor pressures up to 2,000 kg/cm². Under these conditions tridymite melted directly to a liquid at pressures between 400 and 1,400 kg/cm², and above 1,400 kg/cm² quartz melted directly (Fig. 15-5). More remarkable still is the lowering of melting point produced. Whereas silica in the "dry" state melts at over 1700°C, quartz under a water vapor pressure of 1,400 kg/cm² melts at about 1100°C. Small amounts of water (the melt contains about 2.3 percent H_2O) thus have an enormous fluxing effect. Results of this kind emphasize the great influence that the content of water and other volatiles must have on the crystallization of natural melts and may help to explain the formation of large masses of quartz in pegmatites and veins.

All the silica polymorphs except stishovite are built of tetrahedral groups of four oxygen atoms surrounding a central silicon atom. The silicon-oxygen tetrahedra are linked together to form a three-dimensional network, but the pattern of linkage is different for each form, hence the difference in their crystal structures and their properties. Cristobalite and tridymite have comparatively open structures, whereas the atoms in quartz are more closely packed; in coesite and stishovite the packing of the atoms is closer again. This is reflected in the densities and the refractive indices:

	DENSITY	REFRACTIVE INDEX (MEAN)
Stishovite	4.28	1.82
Coesite	2.93	1.60
Quartz	2.65	1.55
Cristobalite	2.32	1.49
Tridymite	2.26	1.47
Lechatelierite	2.20	1.46

Each of the common polymorphs has a high- and a low-temperature modification. In quartz, for example, the change from the one to the other takes place at 573°C at atmospheric pressure. Similarly, high-tridymite changes into low-tridymite between 120°C and 160°C, and high-cristobalite into low-cristobalite between 200°C and 275°C. The inversion from high- to low-temperature forms of the individual species is of quite another order than that between the species themselves. The three minerals have the SiO₄ tetrahedra linked together according to different schemes, and this linkage has to be completely broken down and rearranged for the transformation of one to another. On the other hand, the change from a high-temperature to a low-temperature form does not alter the way in which the tetrahedra are linked. They undergo a displacement and rotation which alters the symmetry of the structure without breaking any links. The high-temperature modification is always more symmetrical than the low-temperature modification.

The high-low transformation of each mineral takes place rapidly at the transition temperature and is reversible. The changes from one polymorphic form to another are extremely sluggish, and the existence of tridymite and cristobalite as minerals shows that they can remain unchanged indefinitely at ordinary temperatures. Once formed, the type of linking in tridymite and cristobalite is too firm to be easily broken, and it is possible to study the high-low inversions of tridymite and cristobalite at temperatures where they are really metastable forms. As pointed out in Chapter 3, the presence of foreign elements in the structure may have a stabilizing effect on tridymite and cristobalite. The few comprehensive analyses of these minerals show the presence of Na and Al, suggesting a substitution of NaAl for Si in the open structures; quartz, however, is generally very pure SiO₂. Two other phenomena of great significance should be mentioned here:

1. Even at temperatures below 867°C, especially when crystallization takes place rapidly (for example, in the presence of mineralizers such as hot gases), cristobalite and/or tridymite may crystallize, although quartz is the stable phase.

2. High-quartz and low-quartz are formed only within their stability fields, never at higher temperatures. From these facts the following conclusions can

be drawn. Quartz in an igneous rock signifies that its crystallization from the magma took place below 867°C (with due regard for the effect of pressure); the presence of cristobalite or tridymite, on the other hand, proves nothing about the temperature of crystallization.

As already pointed out, at ordinary temperatures quartz is always present as low-quartz. By the crystal form, nature of the twinning, and other properties of less diagnostic importance, it is often possible to determine the original form. In this way it has been shown that in the quartz-bearing igneous rocks this mineral crystallized originally as high-quartz, i.e., above 573°C. In quartz veins and some pegmatites it crystallized originally as low-quartz. It may, therefore, be concluded that the crystallization of the magma corresponding to the commonest quartz-bearing igneous rocks took place above 573°C, and the residual crystallization, at least in part, at lower temperatures.

Quartz, SiO_2

Crystal system and class: Trigonal; 32 (low-quartz).
Axial elements: $a:c = 1:1.100$.
Cell dimensions, and content: $a = 4.913$, $c = 5.405$; $Z = 3$.
Common forms and angles: (Figs. 15-6, 15-7, 15-8):

$(10\bar{1}1) \wedge (\bar{1}101) = 85° 46'$	$(10\bar{1}0) \wedge (01\bar{1}1) = 66° 52'$
$(10\bar{1}1) \wedge (01\bar{1}1) = 46° 16'$	$(10\bar{1}0) \wedge (11\bar{2}1) = 37° 58'$
$(10\bar{1}0) \wedge (10\bar{1}1) = 38° 13'$	$(10\bar{1}0) \wedge (51\bar{6}1) = 12° 1'$

Habit: Crystals commonly prismatic, terminated by two sets of rhombohedrons. When these two sets are equally developed the appearance is that of a hexagonal dipyramid, but usually inequality in the rate of growth causes one set to be better developed than the other; consequently, alternate rhombohedron faces are similar, but adjacent ones are different (Fig. 15-6c). Prism faces are commonly horizontally striated, these striations being due to the incipient development of several rhombohedron faces. The wide variety in shape of quartz crystals is the result of unequal development of the prisms and rhombohedrons, and unequal development of like faces through differences in rate of growth. The trigonal pyramids appear as small faces truncating the corners between rhombohedron and prism faces; the trapezohedron faces generally appear in combination with the trigonal pyramids, beveling the edges between the pyramids and the prism faces (Fig. 15-6c,d).

Twinning: General, but seldom observable in the external form of the crystals. Common twins are the Dauphiné type, a penetration twin with the *c* axis as twin axis (Figs. 2-43a,c, and Fig. 15-7c) and the Brazil type, a penetration twin with $\{11\bar{2}0\}$ as the twin plane (Fig. 2-43b). Contact twins with $\{11\bar{2}2\}$

as twin plane (Japanese law) (Fig. 15-7a) are of less common occurrence. Twinning is the cause of a large amount of rejected and wasted material in the manufacture of piezoelectric plates, since these plates must be free of twinning.

Cleavage and fracture: Generally none; occasionally an indistinct rhombohedral parting; conchoidal.

Hardness: 7.

Density: 2.65 in macrocrystalline varieties; lower ($\sim$2.60) in cryptocrystalline varieties.

Color: Usually colorless or white, but can occur in practically any shade.

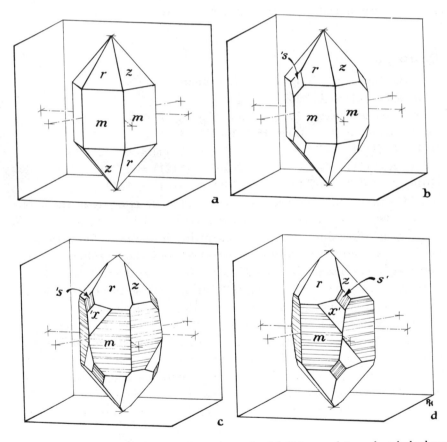

Figure 15-6   Common forms in quartz crystals. (a) Prism and two rhombohedrons. (b) Prism, two rhombohedrons, and left trigonal pyramids. (c) Prism, two rhombohedrons, left trigonal pyramid, and left trigonal trapezohedron. (d) Prism, two rhombohedrons, right trigonal pyramid, and right trigonal trapezohedron.

$$m = \{10\bar{1}0\} \qquad z = \{01\bar{1}1\} \qquad x' = \{51\bar{6}1\} \qquad 'x = \{6\bar{1}\bar{5}1\}$$
$$r = \{10\bar{1}1\} \qquad s' = \{11\bar{2}1\} \qquad 's = \{2\bar{1}\bar{1}1\}$$

Figure 15-7 Quartz crystals. (a) Twinned crystal, Japanese law, twin plane (11$\bar{2}$2). Forms: hexagonal prism {10$\bar{1}$0} striated parallel to *a* horizontal in larger individual, positive and negative rhombohedrons {10$\bar{1}$1}, {01$\bar{1}$1}. (Kai Province, Japan.) [Courtesy Royal Ontario Museum.] (b) Clear crystals, with hexagonal prism {10$\bar{1}$0} and positive rhombohedron {10$\bar{1}$1}. [Courtesy Queen's University.] (c) Smoky quartz, with hexagonal prism {10$\bar{1}$0}, positive and negative rhombohedrons {10$\bar{1}$1}, {01$\bar{1}$1}, left trigonal trapezohedron {6$\bar{1}$51}, twinned according to the Dauphiné law with *c* as twin axis, giving repetition of the left trigonal trapezohedron faces with a rotation of 60° about *c*. (Switzerland.) [Courtesy Royal Ontario Museum.]

Streak: White.

Luster: Vitreous in macrocrystalline varieties, often waxy or dull in cryptocrystalline varieties.

Optical properties: $\omega = 1.544$, $\epsilon = 1.553$; $\epsilon - \omega = 0.009$; positive.

Other properties: Piezoelectric and pyroelectric.

Chemistry: The composition of pure quartz is always close to 100 percent SiO_2;
the structure is such that there is very limited space for accommodating
extra atoms, nor can silicon be readily replaced by other quadrivalent cations. It is insoluble in all acids except HF.

Varieties: Quartz occurs in two distinct types, the macrocrystalline and the
cryptocrystalline varieties, the latter often being classed as the subspecies
chalcedony. Chalcedony consists of quartz crystallites, often fibrous in form
and with submicroscopic pores, which cause it to have a somewhat lower
density than macrocrystalline quartz; it has formed at low to moderate
temperatures often by crystallization of originally colloidal material.

Many names are current for different varieties of quartz; some of the commoner names and their characteristics are given below.

Macrocrystalline Varieties

Rock Crystal: transparent colorless material.

Milky: milk white and almost opaque, luster often somewhat greasy; occurs
in large masses in veins and pegmatites.

Amethyst: transparent purple material, much used as an ornamental and
semiprecious stone; usually occurs lining cavities in volcanic rocks.

Rose: rose red or pink, rarely in crystals.

Yellow (citrine):   yellow and transparent; cut stones are sometimes sold as
topaz. Amethyst on heating often turns yellow, and much citrine is actually heat treated amethyst.

Smoky: transparent and semitransparent brown, gray to nearly black
varieties; sometimes called cairngorm, material from Cairngorm in Scotland having been widely used as an ornamental stone. Radiation from
radioactive material often develops a smoky appearance in colorless quartz,
and this may be responsible for much smoky quartz in nature.

Cryptocrystalline Varieties (chalcedony)

Carnelian and Sard: red and reddish-brown chalcedony; heliotrope and
bloodstone are green chalcedony with red spots.

Agate: banded forms of chalcedony, the banding being due to intermittent
deposition in cavities, the bands being parallel to the walls of the cavity
(Fig. 15-8). Some natural agate shows attractive color banding, but most
of the colored agate used for ornaments has been artificially dyed, the

slightly porous nature of the chalcedony allowing it to absorb material from solution. *Onyx* is plane-banded alternately light and dark agate, used for carved cameos.

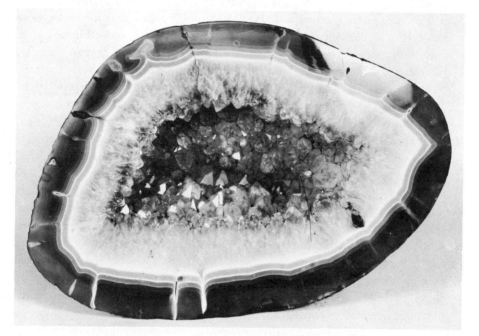

Figure 15-8 Quartz geode. A cavity in basalt has been lined first with structureless chalcedony, then with banded agate, and finally with coarsely crystalline quartz. (Uruguay, South America.) [Courtesy American Museum of Natural History.]

Moss Agate: white or cream-colored chalcedony enclosing brown or black dendritic moss-like aggregates of manganese oxides.

Jasper: opaque chalcedony, generally red, yellow, or brown, the color being due to colloidal particles of iron oxides.

Chert and Flint: massive opaque chalcedony, usually white, pale yellow, gray, or black, and occurring as nodules or extensive beds in sedimentary rocks. The wide occurrence, the hardness and toughness, the conchoidal fracture and sharp cutting edge, caused flint and chert to be widely used by primitive peoples for tools of all kinds.

Silicified Wood: generally consists of reddish or brown chalcedony.

Tiger-eye: Quartz pseudomorphous after asbestos and retaining the fibrous structure of that mineral, and with a yellow, brown, or blue color; it is extensively used as an ornamental stone.

Diagnostic features: Quartz is generally easily recognized by its crystal form, hardness, and lack of cleavage; chalcedonic varieties by their hardness and typical dense structure.

Occurrence: Quartz is stable over practically the whole range of geological conditions, and because SiO_2 is the most abundant oxide in the earth's crust it is a very common mineral. It is present in silica-rich igneous rocks, both volcanic and plutonic, and makes up a large part of hydrothermal veins and granite pegmatites. Since it is hard and extremely resistant to chemical weathering it is the most abundant detrital mineral and the basic material of sandstones. It is an important constituent in most metamorphic rocks. The different varieties of chalcedony are found in sedimentary rocks, and in veins and cavities.

Uses: The uses of quartz are extremely varied. Enormous tonnages are utilized in the construction industry: quartzite and sandstone are used as building stone, as aggregate in concrete, and as sand in mortar and cement. Large amounts are used as a flux in metallurgy, in the manufacture of glass, ceramics, and refractories, as an abrasive, and as a filler. Many of the colored varieties are cut and polished as ornamental and semiprecious stones. Fused silica is a useful material because of its chemical inertness and its low coefficient of expansion—dishes and crucibles of fused silica can be heated and then chilled rapidly without danger of breakage. The most interesting application of quartz crystals makes use of their piezoelectric properties to measure pressures and to control the frequency of electrical impulses; although the total amount used annually for these purposes is comparatively small, the supply of satisfactory natural crystals (mainly from Brazil) is not always adequate, and so laboratory methods of growing suitable crystals have been developed.

Tridymite, SiO_2

Crystal system and class: Hexagonal; $6/m\ 2/m\ 2/m$ (high-tridymite).

Axial elements: $a:c = 1:1.635$.

Cell dimensions and content: $a = 5.04$, $c = 8.24$; $Z = 4$.

Habit: Small (usually 1 mm or less) hexagonal platy crystals, and aggregates of wedge-shaped twins.

Cleavage and fracture: Prismatic; poor.

Hardness: $6\frac{1}{2}$.

Density: 2.26.

Color: Colorless or white.

Streak: White.

Luster: Vitreous.

Optical properties: $\alpha = 1.471\text{--}1.479$, $\beta = 1.472\text{--}1.480$, $\gamma = 1.474\text{--}1.483$; $\gamma - \alpha = 0.002\text{--}0.004$; positive, $2V = 70°\text{--}90°$.

Chemistry: A small amount (2–3 percent) of Si^4 may be replaced by Al^3, and electrical neutrality in the structure is maintained by the introduction of cations such as Na^1 and Ca^2; analyses of natural tridymite generally show about 95 percent SiO_2. Although its stability field is above 867°C, tridymite is probably generally formed in rocks at lower temperatures; this may be conditioned, in part at least, by the presence of foreign ions in the structure extending its stability field to lower temperatures.

Diagnostic features: Shape and size of crystals, and mode of occurrence.

Occurrence: Found in crevices and cavities of volcanic rocks.

Cristobalite, SiO₂

Crystal system and class: Isometric; $4/m\ \bar{3}\ 2/m$ (high-cristobalite).

Cell dimensions and content: $a = 7.28$; $Z = 8$.

Habit: Small octahedral crystals (usually 1 mm or less), often twinned on {111} and aggregated in rounded forms.

Cleavage: None.

Hardness: $6\frac{1}{2}$.

Density: 2.32.

Color: White.

Streak: White.

Luster: Vitreous.

Optical properties: $n = 1.484\text{--}1.487$.

Chemistry: Varies in composition as does tridymite, and for the same reasons.

Diagnostic features: Shape and size of crystals, and mode of occurrence.

Occurrence: Found in crevices and cavities in volcanic rocks.

Opal, SiO₂·nH₂O

Crystal system: None, amorphous.

Habit: Massive, often in rounded and botryoidal forms, sometimes pisolitic (Fig. 15-9).

Hardness: $5\frac{1}{2}\text{--}6\frac{1}{2}$.

Density: 2.0–2.2.

Color: Colorless (var. *hyalite*) or white, but also gray, brown, or red, the color being usually due to fine-grained impurities; also with a rich iridescence and play of colors (var. precious opal).

Streak: White.

Luster: Vitreous or waxy, colored varieties commonly somewhat resinous.

Optical properties: $n = 1.44$–1.46; refractive index (and density) increase as water content decreases.

Chemistry: The water content varies from specimen to specimen, but is generally between 3 and 10 percent. Opal is an example of a solidified colloidal gel, and is essentially amorphous, although x-ray powder photographs sometimes show weak and diffuse reflections, indicating cristobalite-like groupings in the material.

Figure 15-9   Opal, var. hyalite, showing pisolitic structure. (Tateyama, Japan.) [Courtesy American Museum of Natural History.]

Diagnostic features: Mode of occurrence, form, and its low density (which distinguishes it from chalcedony, which it may resemble, and into which it often transforms by crystallization).

Occurrence: Opal is deposited at low temperatures from silica-bearing waters and can occur in fissures and cavities in any rock type. It is often associated with igneous rocks, having been deposited by thermal waters; extensive deposits (siliceous sinter or geyserite) are formed around hot springs, as at Yellowstone National Park, Wyoming. Opal is the form of silica secreted by sponges, radiolaria, and diatoms; diatomite occurs in extensive beds in many parts of the world.

Production and use: Diatomite is extensively used as a filtering medium, as a mild abrasive, and as an insulator against heat, cold, or sound. In the United States large amounts are mined at Lompoc, California (annual production is about 500,000 tons). Precious opal is prized as a gemstone; the finest opals come from deposits in sandstone in Australia, but good gem quality material has been mined in Nevada.

FELDSPAR GROUP

The feldspars owe their importance to the fact that they are the most abundant of all minerals. They are closely related in form and physical properties, but they fall into two subgroups: the potassium and barium feldspars, which are monoclinic or very nearly monoclinic in symmetry, and the sodium and calcium feldspars (the plagioclases), which are definitely triclinic. A point of great interest is the isomorphism between albite, $NaAlSi_3O_8$, and anorthite, $CaAl_2Si_2O_8$. The theory that feldspars of intermediate composition were mixed crystals of these two components was proposed by Tschermak in 1869. It is now known that NaSi often substitutes for CaAl, but Tschermak's theory is of historic importance as a first suggestion that so radical a substitution is possible.

The general formula for the feldspars can be written WZ_4O_8, in which W may be Na, K, Ca, and Ba, and Z is Si and Al, the Si:Al ratio varying from 3:1 to 1:1. Since all feldspars contain a certain minimum amount of Al, the general formula may be somewhat more specifically stated as $WAl(Al,Si)Si_2O_8$, the variable (Al,Si) being balanced by variation in the proportions of univalent and bivalent cations.

The structure of the feldspars is a continuous three-dimensional network of SiO_4 and AlO_4 tetrahedra, with the positively charged sodium, potassium, calcium, and barium situated in the interstices of the negatively charged network. The network of SiO_4 and AlO_4 tetrahedra is elastic to some degree and can adjust itself to the sizes of the cations; when the cations are relatively large (K,Ba) the symmetry is monoclinic or pseudomonoclinic; with the smaller cations (Na,Ca) the structure is slightly distorted, and the symmetry becomes triclinic.

The barium feldspars are rare and of no importance as rock-forming minerals, although minor amounts of barium may be present in potash feldspars. We will omit them from further consideration, and discuss the feldspars as a three-component system, the components being $KAlSi_3O_8$ (Or), $NaAlSi_3O_8$ (Ab), and $CaAl_2Si_2O_8$ (An). Complexities are introduced by the solid solution relations existing among these three components, and the occurrence of polymorphic forms. The following species are recognized:

Potash feldspar

Sanidine	$KAlSi_3O_8$	Monoclinic
Orthoclase	$KAlSi_3O_8$	Monoclinic
Microcline	$KAlSi_3O_8$	Triclinic

Soda-lime feldspar

Plagioclase Series	$(Na,Ca)(Al,Si)AlSi_2O_8$	Triclinic

The potash feldspar minerals occur in several distinct forms having different but intergradational optical and physical properties. Sanidine, the monoclinic high-temperature polymorph, occurs in volcanic rocks. Common orthoclase, another monoclinic variety, and microcline (triclinic) are found in a wide variety of igneous and metamorphic rocks which have crystallized at intermediate to low temperatures. *Adularia* is the name given to a form (which may be either monoclinic or triclinic) with a distinctive crystal habit found in low-temperature hydrothermal veins.

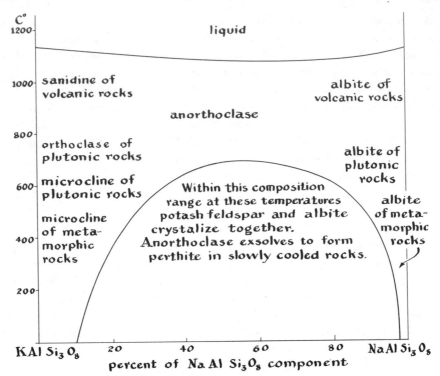

Figure 15-10 Relationship between the alkali feldspars and the influence of temperature of crystallization and rate of cooling.

Recent research has clarified the relationship between these forms. Microcline and sanidine are polymorphs with an order-disorder relationship, the Si and Al atoms being randomly distributed over their lattice positions in sanidine, but being ordered in microcline. The disordered form is the more stable polymorph above about 700°C, and microcline has been transformed into sanidine by hydrothermal treatment at this temperature; the reverse transformation has not been achieved in the laboratory, evidently because of the high activation energy required for the ordering of the Si and Al atoms. Orthoclase and adularia are structurally intermediate between sanidine and microcline. Much orthoclase probably crystallized originally as sanidine. Adularia is evidently a metastable form which develops under conditions of rapid crystallization within the stability field of microcline; the rapid crystallization prevents the attainment of an ordered arrangement of Si and Al.

At temperatures above about 660°C complete solid solution exists between $NaAlSi_3O_8$ and $KAlSi_3O_8$ (Fig. 15-10). The more potassic members, sanidine and orthoclase, are monoclinic, sanidine being characteristic of volcanic rocks and orthoclase being characteristic of plutonic rocks. The more sodic members of the series are triclinic and are called *anorthoclase*. At lower temperatures solid solutions intermediate between potash feldspar and albite are unstable and under conditions of slow cooling break down into an oriented intergrowth of subparallel lamellae, alternately sodic and potassic in composition. Such an intergrowth is called *perthite* (Fig. 15-15) or *antiperthite*. In the perthites the plagioclase occurs as uniformly oriented films, veins, and patches within the orthoclase (or microcline); in the antiperthites this relation is reversed. Perthite, when heated for a long time at 1000°C, becomes homogeneous once more. Not all perthites have been formed by exsolution; some are the product of partial metasomatic replacement of an originally homogeneous potash feldspar by sodium-bearing solutions.

X-ray examination of potash feldspar and of albite provides the following explanation of the perthite lamellar intergrowth. The framework of linked SiO_4 and AlO_4 tetrahedra, being similar for the monoclinic and triclinic forms, is continuous throughout the structure. At high temperatures the K and Na ions

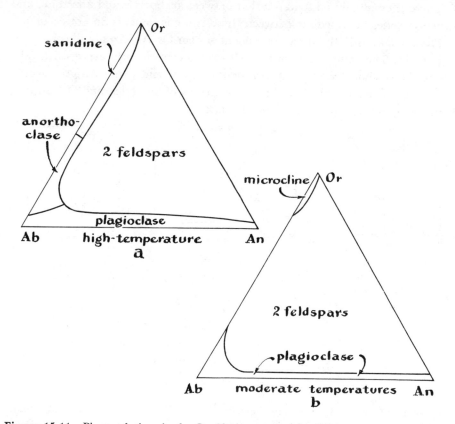

Figure 15-11 Phase relations in the Or-Ab-An system (a) at high temperatures, (b) at moderate temperatures.

are randomly distributed in the framework, producing a homogeneous crystal. At lower temperatures ordering may occur with the formation of potassium-rich and sodium-rich lamellae, producing alternate sheets with monoclinic or pseudomonoclinic and triclinic symmetry, respectively. The a unit length of potash feldspar and albite is markedly different (8.56 Å and 8.14 Å), whereas the b and c unit lengths are almost identical (12.99 Å, 7.19 Å and 12.79 Å, 7.16 Å). This accounts for the lamellae being approximately parallel to {100}, since the b and c lengths coincide in the {100} plane, whereas the a length running through the lamellae shortens in the albite regions and lengthens in the potash feldspar regions.

The system Ab–An shows complete solid solution at high temperatures, but complications enter at low temperatures. The plagioclase feldspars exist in both high- and low-temperature forms, which differ somewhat in structure and in optical properties, and there is incomplete solid solution between the low-temperature forms.

There is essentially no solid solution between orthoclase and anorthite, and hence varieties intermediate between these two components do not occur.

Phase relations in the three-component system Or–Ab–An are illustrated in Fig. 15-11. The composition of any feldspar is conveniently expressed by the use of the symbols Or, Ab, and An for the components; thus $Ab_{32}An_{68}$ would be a plagioclase falling in the labradorite section of the series, and $Or_{26}Ab_{66}An_{8}$ is the composition of a possible anorthoclase.

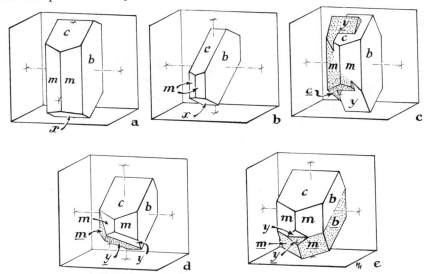

Figure 15-12 Forms in orthoclase (monoclinic). (a, b) Simple crystals. (c) Carlsbad twin. (d) Baveno twin. (e) Manebach twin. Forms lettered as follows: $b\{010\}$, $c\{001\}$, $m\{110\}$, $x\{\bar{1}01\}$, $y\{\bar{2}01\}$.

Sanidine, Orthoclase, KAlSi₃O₈

Crystal system and class: Monoclinic, 2/m.
Axial elements: $a:b:c = 0.6588:1:0.5535$; $\beta = 116°\,01'$.
Cell dimensions and content: $a = 8.562$, $b = 12.996$, $c = 7.193$; $Z = 4$.
Common forms and angles:

$(110) \wedge (1\bar{1}0) = 61°\,16'$	$(001) \wedge (\bar{2}01) = 80°\,07'$
$(130) \wedge (\bar{1}30) = 58°\,46'$	$(001) \wedge (021) = 44°\,51'$
$(001) \wedge (\bar{1}01) = 50°\,06'$	$(021) \wedge (0\bar{2}1) = 89°\,42'$

Habit: Crystals usually short prismatic, somewhat flattened parallel to {010}, or elongated parallel to *a* with {010} and {001} prominent (Fig. 15-12a,b, Fig. 2-15e).

Twinning: Common on the Carlsbad law (composition plane {010}, Figs. 15-12c, 15-13), less common on the Baveno (composition plane {021}, Fig. 15-12d) and Manebach (composition plane {001}, Fig. 15-12e) laws.

Figure 15-13 Orthoclase. Carlsbad twin, twin axis *c* (vertical). Larger individual shows forms {010} (parallel to page), {110}, {001} sloping from top to left, {$\bar{2}$01} sloping from the top to right. Smaller twinned individual has {001} sloping from top to right; {$\bar{2}$01} is very small. Cream-colored; dark spots are adhering mica crystals. (Carlsbad, Czechoslovakia.) [Courtesy Queen's University.]

Cleavage: {001}, perfect; {010}, good.
Hardness: 6.
Density: 2.56.
Color: Sanidine commonly colorless, orthoclase usually white or pink.
Streak: White.
Luster: Vitreous, sometimes pearly on cleavage surfaces.
Optical properties: Sanidine—$\alpha = 1.518$, $\beta = 1.522$, $\gamma = 1.522$ (for pure KAlSi₃O₈; the indices and birefringence increase with replacement of K by Na); negative, $2V$ small. Orthoclase—$\alpha = 1.518$, $\beta = 1.522$, $\gamma = 1.523–1.524$; negative, $2V = 60°–85°$.

Diagnostic features: Sanidine and orthoclase can be distinguished from the plagioclase feldspars by the absence of twinning striations; the distinction from microcline is not easily made, but occurrence is a useful guide, since much of the potash feldspar of igneous rocks is orthoclase but practically all that of pegmatites and hydrothermal veins is microcline. Sanidine has a glassy appearance and is often tabular parallel to {010} (Fig. 15-14).

Figure 15-14   Sanidine. Large pheno-crysts in trachyte. (Drachenfels, Rhineland, Germany.) [Courtesy American Museum of Natural History.]

Occurrence: Sanidine occurs in potash-rich volcanic rocks such as rhyolite and trachyte. Orthoclase is the characteristic potash feldspar of most plutonic rocks, occurring both alone and in perthitic intergrowth with albite; it also occurs in metamorphic rocks. *Adularia* is a variety of potash feldspar formed at comparatively low temperatures in veins and characterized by simple crystals (combinations of {110}, {001}, and {$\bar{1}$01}) which often have the appearance of rhombohedrons.

Microcline, KAlSi$_3$O$_8$

Crystal system and class: Triclinic; $\bar{1}$.

Axial elements: $a:b:c = 0.6614:1:0.5570$; $\alpha = 90°39'$, $\beta = 115°56'$, $\gamma = 87°47'$.

Cell dimensions and content: $a = 8.577$, $b = 12.967$, $c = 7.223$; $Z = 4$.

Common forms and angles: Similar to orthoclase.

Habit: Similar to orthoclase.

Twinning: Usually polysynthetically twinned on both the albite and pericline laws, giving a characteristic grating or gridiron structure on all sections except those in the prism zone [001].

Cleavage: {001}, perfect; {010} good.

Hardness: 6.

Density: 2.56.

Color: White, cream, pink, sometimes bright green (var. amazonstone).

Streak: White.

Luster: Vitreous, sometimes pearly on cleavage surfaces.

Optical properties: $\alpha = 1.518$, $\beta = 1.522$, $\gamma = 1.525$; $\gamma - \alpha = 0.007$; negative, $2V = 80°$.

Diagnostic features: Microcline can be easily distinguished from orthoclase only by its optical properties, although a bright green color is diagnostic. Occurrence is a useful guide (see *Orthoclase*).

Occurrence: Microcline, being an ordered polymorph, is usually formed at lower temperatures than is orthoclase, the disordered polymorph, and microcline is the common potash feldspar of pegmatites and hydrothermal veins. Crystals several feet long are not uncommon in pegmatites. Microcline also occurs in metamorphic rocks.

Figure 15-15 Perthite. Intergrowth of pink microcline (dark) with colorless plagioclase (light); large surface is {001} cleavage, smaller is {010} cleavage; *a* axis horizontal. (Range 6, Lot 3, North Burgess Township, Lanark County, Ontario.) [Courtesy Queen's University.]

Production and use: Large quantities of microcline and microcline perthite (Fig. 15-15) are mined from granite pegmatites and used in the manufacture of glass, porcelain, and enamel. Some 500,000 tons is produced in the United States annually, about half of this coming from pegmatites in North Carolina.

Amazonstone is sometimes cut and polished as an ornamental stone.

Plagioclase Series: $(Ca,Na)(Al,Si)AlSi_2O_8$

Crystal system and class: Triclinic; $\bar{1}$.

Axial elements: $a:b:c = 0.6369:1:0.5599$; $\alpha = 94° 16'$, $\beta = 116° 35'$, $\gamma = 87° 40'$ (albite); $a:b:c = 0.6350:1:1.1003$; $\alpha = 93° 10'$, $\beta = 115° 51'$, $\gamma = 91° 13'$ (anorthite).

Cell dimensions and content: $a = 8.144$, $b = 12.787$, $c = 7.160$; $Z = 4$ (albite) $a = 8.177$, $b = 12.877$, $c = 14.169$; $Z = 8$ (anorthite).

Habit: Crystals usually tabular parallel to {010}; (Fig. 15-16a); commonly as irregular grains and cleavable masses.

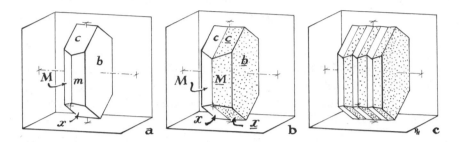

Figure 15-16   Forms in plagioclase feldspars (triclinic). (a) Single crystal. (b) Albite twin. (c) Polysynthetic albite twin. Forms lettered as follows: $b\{010\}$, $c\{001\}$, $m\{110\}$, $M\{1\bar{1}0\}$, $x\{\bar{1}01\}$.

Twinning: Nearly always polysynthetically twinned according to one or both of the albite (Fig. 15-16b,c) and pericline laws; Carlsbad, Baveno, and Manebach twins also common.

Cleavage: {001}, perfect; {010}, good.

Hardness: 6.

Density: 2.62–2.76 (Fig. 15-17).

Color: White or gray, sometimes reddish or reddish brown.

Streak: White.

Luster: Vitreous, sometimes pearly on cleavage surfaces.

Optical properties: These are given in Fig. 15-17 for low-temperature plagioclase (from plutonic and most metamorphic rocks); those of high-temperature plagioclase (i.e., in volcanic rocks) are slightly different, α being approximately the same and γ somewhat lower.

Chemistry: A continuous series is known, both in minerals and in artificial crystals, from pure albite, $NaAlSi_3O_8$ (Ab) to pure anorthite, $CaAl_2Si_2O_8$ (An). The series is arbitrarily divided into six species or subspecies, as follows:

Albite	Ab_{100}–Ab_{90}
Oligoclase	Ab_{90}–Ab_{70}
Andesine	Ab_{70}–Ab_{50}
Labradorite	Ab_{50}–Ab_{30}
Bytownite	Ab_{30}–Ab_{10}
Anorthite	Ab_{10}–Ab_0

It will be noticed that the c dimension of the unit cell of anorthite is twice that of albite. This doubling appears to characterize that part of the series from labradorite to anorthite; the plagioclases can perhaps be considered as a double series in which sodium is replaced by calcium in the albite structure, and at some intermediate point about Ab_{50} the structure changes to one in which the calcium in anorthite is replaced by sodium. This change may also be related to an ordering of the Al and Si atoms, the 1:1 proportion in anorthite favoring an ordered arrangement.

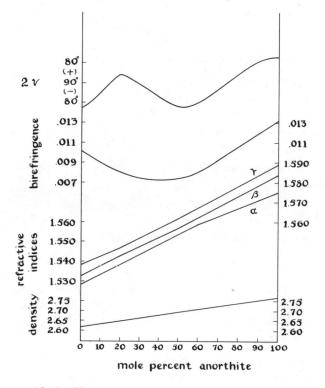

Figure 15-17 The density and optical properties of the plagioclase series (low-temperature).

Diagnostic features: The plagioclases can be distinguished from the potash feldspars by the twinning striations on basal cleavage surfaces. Differentiation between the individual species or subspecies within the plagioclase series is best done optically, but careful density determinations give a good indication of plagioclase composition. Rock type is often a useful guide (see Occurrence).

Labradorite in coarse cleavages commonly shows a play of colors in shades of blue and green.

Occurrence: Anorthite is usually found in contact metamorphosed limestones. Bytownite and labradorite are characteristic of igneous rocks of gabbroic composition and of the anorthosites, andesine of andesites and diorites, oligoclase of monzonites and granodiorites; albite occurs in some igneous rocks of normal texture, but is more common in pegmatites. The plagioclases are common in metamorphic rocks; in low-grade schists and gneisses the plagioclase is usually albite, in medium-grade rocks usually oligoclase or andesine.

Uses: Albite and oligoclase are sometimes mined from pegmatites and used in the manufacture of ceramics.

Scapolite Series: $(Na,Ca)_4[(Al,Si)_4O_8]_3(Cl,CO_3)$

Crystal system and class: Tetragonal; $4/m$.

Axial elements: $a:c = 1:0.623-0.629$.

Cell dimensions and content: $a = 12.05-12.17$, $c = 7.58$; $Z = 2$.

Habit: Crystals usually combinations of prisms and pyramids, often large, with uneven faces; commonly massive, granular.

Cleavage: $\{100\}$, $\{110\}$, distinct, often giving a splintery woody appearance to massive specimens.

Hardness: 6.

Density: 2.56–2.77, increasing with Ca content.

Color: Usually white, gray, or pale brown; sometimes pink, yellow, or blue.

Streak: White.

Luster: Vitreous.

Optical properties: $\omega = 1.545-1.610$, $\epsilon = 1.540-1.570$; $\omega - \epsilon = 0.005-0.040$; negative; refractive indices and birefringence increase with Ca content.

Chemistry: The scapolite series ranges in composition from sodium-rich (var. *marialite*) to calcium-rich (var. *meionite*). In composition scapolite is closely related to plagioclase, the only difference being the presence of small amounts of Cl^{-1} and CO_3^{-2} (and sometimes SO_4^{-2}, F^{-1}, OH^{-1}, also) and additional sodium and calcium to neutralize the additional negative charge.

Diagnostic features: Fuses with intumescence to give a white blebby glass. Sometimes fluoresces orange or yellow in ultraviolet light.

Occurrence: Typically occurs in metamorphosed limestones, often in considerable quantities, especially around intrusive igneous rocks; it also occurs in schists and gneisses, sometimes replacing plagioclase.

Name: The name *wernerite* was proposed for this mineral at about the same time as was *scapolite*, but the latter name is generally used.

FELDSPATHOID GROUP

The feldspathoids are a group of sodium and potassium aluminosilicates which appear in place of the feldspars when an alkali-rich magma is deficient in silica. They never occur together with primary quartz, since they will react with free silica to give feldspar. Although they are all tektosilicates the structural relationship between the different species within the group is less close than within the feldspars, and they crystallize in several systems, although most species are either isometric or hexagonal. The commonest species of the feldspathoid group are:

Leucite	$KAlSi_2O_6$
Nepheline	$NaAlSiO_4$
Cancrinite	$Na_8(AlSiO_4)_6(HCO_3)_2$
Sodalite	$Na_8(AlSiO_4)_6Cl_2$

Analcime, $NaAlSi_2O_6 \cdot H_2O$, is sometimes included with the feldspathoids, since it appears occasionally as a primary mineral of silica-deficient igneous rocks, but here it is classed with other hydrated tektosilicates in the zeolite group.

The feldspathoids are readily attacked by acids. This characteristic is evidently due to the comparatively high Al:Si ratio; the Al is removed in solution and the lattice then collapses, often with the formation of gelatinous silica. This formation of gelatinous silica is the basis of a useful test for the presence of feldspathoids in an igneous rock. Shand recommends spreading a film of syrupy (85 percent) phosphoric acid on a smooth surface of the rock, allowing it to stand for three minutes, dipping it in water to remove the acid, then immersing it in a 0.25 percent solution of methylene blue for one minute. This treatment stains nepheline, sodalite, and analcime deep blue; leucite is not affected.

Leucite, $KAlSi_2O_6$

Crystal system and class: Isometric; $4/m \bar{3} 2/m$ (high temperature form).
Cell dimensions and content: $a = 13.43$; $Z = 16$.
Habit: Usually occurs in trapezohedral crystals.
Cleavage: {110}, very imperfect.
Hardness: 6.
Density: 2.47.
Color: Colorless, white, or gray.
Streak: White.
Luster: Vitreous.
Optical properties: $n = 1.508–1.511$; can be weakly birefringent.
Chemistry: A small amount of the potassium may be replaced by sodium. At 605°C the structure changes from isometric to tetragonal.
Alteration: In many occurrences leucite has altered to *pseudoleucite*, a pseudomorph consisting of a mixture of nepheline, analcime, and orthoclase.
Diagnostic features: Leucite is readily recognized by its crystal form and its

occurrence as embedded crystals in volcanic rocks; analcime resembles it in crystal form, but well-crystallized analcime occurs in cavities and vugs, not as embedded crystals.

Occurrence: Leucite is abundant in the volcanic rocks of a few regions. The best specimens are the fresh crystals found in the lavas of Mount Vesuvius. Leucite does not occur in plutonic rocks, and laboratory investigations have indicated that leucite is not a stable phase in the system $K_2O–Al_2O_3–SiO_2$ at H_2O pressures above 2500 kg/cm^2.

Nepheline, NaAlSiO$_4$

Crystal system and class: Hexagonal; 6.

Axial elements: $a:c = 1:0.840$.

Cell dimensions and content: $a = 10.01$, $c = 8.405$; $Z = 8$.

Habit: Crystals usually simple hexagonal prisms (showing rectangular and hexagonal cross sections in rocks); occurs more commonly as shapeless grains and irregular masses.

Cleavage: $\{10\bar{1}0\}$, poor.

Hardness: 6.

Density: 2.60–2.63.

Color: Usually white, sometimes gray or brown.

Streak: White.

Luster: Vitreous or greasy.

Optical properties: $\omega = 1.530–1.545$, $\epsilon = 1.527–1.541$; $\omega - \epsilon = 0.003–0.005$; negative.

Chemistry: The compound $NaAlSiO_4$ can be made artificially; however, natural nephelines always contain potassium, often approaching the formula $KNa_3(AlSiO_4)_4$. This reflects the space requirements of the nepheline structure, in which one alkali position in four is larger than the other three and preferentially accommodates potassium. A little calcium is generally present in natural nephelines. The structure of nepheline is similar to that of tridynite, half of the silicon being replaced by aluminum, electroneutrality being preserved by the addition of an equivalent number of sodium ions to voids in the structure. The replacement of Si by Al gives the hexagonal axis a polar character which is absent in the parent tridymite structure.

Diagnostic features: Coarsely crystallized nepheline usually has a greasy luster. It is readily decomposed by hydrochloric acid, with the separation of gelatinous silica.

Occurrence: Occurs in both plutonic and volcanic rocks, and in pegmatites associated with nepheline syenites.

Production and use: Pure nepheline is a useful raw material for the manufacture of glass and ceramics. Considerable amounts (about 200,000 tons annually) are mined for this purpose in the Peterborough district, Ontario. The USSR produces nepheline from a large deposit in the Kola Peninsula.

Sodalite, $Na_8(AlSiO_4)_6Cl_2$

Crystal system and class: Isometric; $\bar{4}\,3/m$.
Cell dimensions and content: $a = 8.83-8.91$; $Z = 1$.
Habit: Sometimes as small dodecahedral crystals, but more commonly massive and granular.
Cleavage: {110}, poor.
Hardness: 6.
Density: 2.29.
Color: Often azure blue, but may be white or pink.
Streak: White.
Luster: Vitreous.
Optical properties: $n = 1.483-1.484$.
Diagnostic features: The blue color is often characteristic (it resembles that of azurite, but azurite is readily distinguished by its lower hardness and effervescence in acid); in the absence of the blue color sodalite can be distinguished from the other feldspathoids by decomposing in dilute HNO_3 and testing the solution for chloride.
Occurrence: Sodalite occurs in both volcanic and plutonic rocks of the nepheline syenite family but more commonly in the plutonic types, generally in association with nepheline. Blue sodalite occurs in considerable amount in pegmatitic nepheline syenite near Bancroft, Ontario, and Ice River, British Columbia, and is sometimes cut and polished as an ornamental stone.
Related species: The mineral *lazurite* is isomorphous with sodalite but has sulfide ions in place of chloride ions in the structure; it has a brilliant blue color and is the semiprecious stone known as *lapis lazuli*. The pigment *ultramarine* was formerly made by crushing lazurite, but it is now manufactured synthetically.

Cancrinite, $Na_8(AlSiO_4)_6(HCO_3)_2$

Crystal system and class: Hexagonal; 6.
Axial elements: $a:c = 1:0.405$.
Cell dimensions and content: $a = 12.63-12.78$, $c = 5.11-5.19$; $Z = 1$.
Habit: Rarely in prismatic crystals, usually massive.
Cleavage: {10$\bar{1}$0}, perfect.
Hardness: 6.

Density: 2.4–2.5.

Color: White, pink, sometimes yellow.

Streak: White.

Luster: Vitreous.

Optical properties: $\omega = 1.507$–1.528, $\epsilon = 1.495$–1.503; $\omega - \epsilon = 0.012$–$0.025$; negative.

Chemistry: The composition is somewhat variable; in addition to the elements indicated in the formula cancrinite always contains some calcium, and often small amounts of sulfate and chloride.

Diagnostic features: Cancrinite can be distinguished from other feldspathoids by its effervescence in warm dilute HCl.

Occurrence: Cancrinite occurs solely in plutonic rocks of the nepheline syenite family, evidently because incorporation of carbonate ions in the structure can take place only under considerable pressure of carbon dioxide. It is generally associated with other feldspathoids, especially nepheline and sodalite.

ZEOLITE GROUP

The zeolites may be defined as a group of tektosilicate minerals (1) which are hydrated aluminosilicates of univalent and bivalent bases, in which the $(W_2,W)O:Al_2O_3$ is unity, (2) which can lose a part or the whole of their water without changing crystal structure and can absorb other compounds in place of the water removed, and (3) which are capable of undergoing cation exchange. They form a well-defined group of species closely related to one another in composition, in conditions of formation, and hence in mode of occurrence. They are all aluminosilicates of sodium and calcium chiefly, less commonly potassium, barium, and strontium. The zeolites are not, however, a group of species related in crystal structure, like the feldspars, but comprise a number of independent species diverse in structure and distinct in composition. The name zeolite is derived from a Greek word meaning to boil, and refers to the swelling-up and apparent boiling that occurs when these minerals are heated and water is driven off.

A general formula can be written for the zeolite group: $W_m Z_r O_{2r} \cdot s H_2O$, in which W is chiefly Na and Ca (K, Ba, Sr to a lesser extent), $Z = \mathrm{Si} + \mathrm{Al}$ (Si:Al is 1 or greater), and s is variable. In the zeolites the ratio $Al_2O_3:(CaO + Na_2O)$ is always 1:1, and the (Al + Si):O ratio is always 1:2. There is a limited range of atomic substitution in the individual zeolites, hence the compositional range of a single species is narrow. This probably means a narrow range of stability for most zeolites, and it may explain the large number of species in the group.

The atomic substitutions in the zeolite group are of two types. The first is similar to that in the feldspars:

KSi = BaAl

NaSi = CaAl

The second type of replacement alters the number of cations:

Ba = K_2

Ca = Na_2

For instance, in thomsonite the usual number of cations is represented by the formula $NaCa_2Al_5Si_5O_{20} \cdot 6H_2O$. It is possible, however, to replace half the Ca by Na_2, raising the total number of positive ions to four:

$NaCa_2 = Na_3Ca$

The structure contains a sufficient number of suitable spaces to accommodate the additional ions. Both types of replacement must be borne in mind in considering the composition of the zeolites.

All the zeolites are tektosilicates with particularly open, wide-meshed $(Si,Al)O_2$ frameworks. The cavities in this framework contain the cations which balance the negative charge of the framework, as in the feldspars. A characteristic feature of the zeolites, however, is the ease with which exchange of the cations may take place. In many natural zeolites the cations can be replaced artificially by thallium, potassium, silver, or sodium simply by placing the crystal in a solution containing these ions. The tektosilicate framework is held together by strong bonds and is so rigid that an individual crystal retains its shape during the change. The channels through the framework are sufficiently large to enable the ions to pass freely. Similarly, the water molecules can be removed and the structure will remain stable, in striking contrast with the majority of hydrated inorganic salts, in which the removal of even a part of the water results in the complete breakdown of the structure. The water is given off continuously on heating, not in certain amounts at definite temperatures as is the case with most hydrated compounds. The reason for this difference is easy to see; in zeolites the strongly bonded framework is responsible for the high stability, whereas in the majority of inorganic salts the structure is a simple regular packing of small ions. The removal of water molecules is not achieved without some effect on the zeolite structure; as a part of the water is removed there may be a redistribution of energy within the structure, causing the remaining water to be more firmly held, or the disturbed energy equilibrium may be restored by movements of the atoms of the framework, which has some flexibility; such movements generally result in a decrease in length of one or more sides of the unit cell.

Zeolites may be deposited from magmatic solutions at a late stage of crystallization. More commonly they are the products of hydrothermal activity or diagenesis. They are often the product of alteration of feldspars, feldspathoids, or volcanic glass. They are found especially in amygdaloidal cavities, fractures, and other openings in rocks in many types, especially in igneous rocks on the gabbro and basalt family. They are found also in some metalliferous veins, especially where the gangue is calcite, and in the deposits of hot springs, where they have in some cases been formed within historic time and are still being formed. Many sedimentary rocks, particularly tuffaceous ones, contain considerable amounts of zeolites formed during diagenesis. The zeolites are commonly associated with calcite, prehnite, pectolite, apophyllite, and datolite.

Heulandite, $CaAl_2Si_7O_{18} \cdot 6H_2O$

Crystal system and class: Monoclinic; $2/m$.
Axial elements: $a:b:c = 0.418:1:0.890$; $\beta = 91° 26'$.
Cell dimensions and content: $a = 7.46$, $b = 17.84$, $c = 15.88$; $Z = 4$.
Common forms and angles:
   (010) $\wedge$ (110) $= 67° 19'$ (001) $\wedge$ ($\bar{1}$01) $= 66° 01'$
   (001) $\wedge$ (101) $= 63° 40'$ (001) $\wedge$ (011) $= 41° 40'$
Habit: Trapezoidal crystals tabular parallel to {010} (Fig. 15–18).

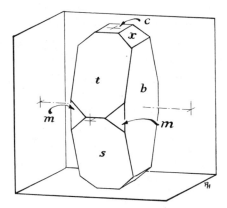

Figure 15-18 Forms in heulandite crystals: $c\{001\}$, $m\{110\}$, $x\{011\}$, $s\{\bar{1}01\}$, $t\{101\}$.

Cleavage: {010}, perfect.
Hardness: 4; 3 on cleavage surfaces.
Density: 2.2.
Color: White, sometimes stained red by iron oxides.
Streak: White.

Luster: Vitreous, pearly on cleavage surfaces.

Optical properties: $\alpha = 1.485$–1.505, $\beta = 1.486$–1.506, $\gamma = 1.488$–1.512; $\gamma - \alpha = 0.003$–0.008; positive, $2V = 10°$–$30°$.

Chemistry: Heulandite commonly contains some sodium substituting for calcium in the structure.

Diagnostic features: The form of heulandite crystals is characteristic and distinctive.

Occurrence: Heulandite is a common mineral associated with other zeolites, especially stilbite, in cavities in basaltic rocks, as in the trap rocks of northern New Jersey. It is also a widespread alteration product of intermediate to acid volcanic glass, and occurs as an authigenic mineral in sedimentary rocks, especially tuffaceous sandstones; this material, although it has the structure of heulandite, differs somewhat in composition and physical properties, and the name *clinoptilolite* has been applied to it.

Stilbite, $CaAl_2Si_7O_{18} \cdot 7H_2O$

Crystal system and class: Monoclinic; $2/m$.

Axial elements: $a:b:c = 0.750:1:0.622$; $\beta = 129° \, 10'$.

Cell dimensions and content: $a = 13.63$, $b = 18.17$, $c = 11.31$; $Z = 4$.

Habit: Rough sheaf-like crystals which are aggregates of cruciform penetration twins (Fig. 15-19).

Figure 15-19 Stilbite, on basalt, showing characteristic sheaf-like form. (Two Islands, Nova Scotia.) [Courtesy American Museum of Natural History.]

Cleavage: {010}, perfect.

Hardness: 4.

Density: 2.1–2.2.

Color: White, cream, or pink.

Streak: White.

Luster: Vitreous, pearly on cleavage surfaces.

Optical properties: $\alpha = 1.484$–1.500, $\beta = 1.492$–1.507, $\gamma = 1.494$–1.513; $\gamma - \alpha = 0.009$–0.012; negative, $2V = 30°$–$50°$.

Chemistry: Stilbite usually contains some sodium replacing calcium.

Diagnostic features: The sheaf-like form of stilbite crystals is characteristic and diagnostic.

Occurrence: Occurs in cavities in basalts, commonly in association with heulandite.

Name: In Germany the name *desmine* is often used for this mineral.

Laumontite, $CaAl_2Si_4O_{12} \cdot 4H_2O$

Crystal system and class: Monoclinic; 2 or *m*.

Axial elements: $a:b:c = 1.131:1:0.573; \beta = 111° 30'$.

Cell dimensions and content: $a = 14.90, b = 13.17; c = 7.55; Z = 4$.

Habit: Commonly in small prismatic crystals with oblique terminations, also massive.

Twinning: Frequently twinned on {100}.

Cleavage: {010}, {110}, perfect.

Hardness: 4.

Density: 2.25–2.30.

Color: White.

Streak: White.

Luster: Vitreous.

Optical properties: $\alpha = 1.505–1.510, \beta = 1.515–1.518, \gamma = 1.516–1.522; \gamma - \alpha = 0.01$; negative, $2V = 20°–40°$.

Other properties: Laumontite is strongly pyroelectric, hence it has no center of symmetry, and its crystal class must therefore be 2 or *m*.

Alteration: Readily loses about an eighth of its water, becoming chalky and friable (var. *leonhardite*); most specimens in collections are actually leonhardite.

Diagnostic features: The crystal form and characteristic alteration usually serve to identify laumontite. It gelatinizes with HCl.

Occurrence: It typically occurs as small, stout prismatic crystals in veins and cavities of igneous rocks; it has also been found in large amount in some tuffaceous sediments as an alteration product of plagioclase and glass, occasionally forming extensive beds of impure laumontite rock.

Chabazite, $CaAl_2Si_4O_{12} \cdot 6H_2O$

Crystal system and class: Trigonal; $\bar{3}\ 2/m$.

Axial elements: $a:c = 1:1.086$.

Cell dimensions and content: $a = 13.78, c = 14.97; Z = 6$.

Habit: Usually in simple rhombohedral crystals which look like cubes, the angles between the rhombohedral faces being close to 90°.

Twinning: Penetration twins common, twin axis *c*.

Cleavage: {$10\bar{1}1$}, poor.

Hardness: 4.

Density: 2.05–2.15.

Color: Colorless or white, sometimes tinted yellow or pink.

Streak: White.

Luster: Vitreous.

Optical properties: $n = 1.48$–1.49; birefringence low, 0.001–0.002.

Chemistry: Often contains a little sodium and potassium replacing calcium.

Diagnostic features: The crystal form serves to distinguish chabazite from the other common zeolites; it is harder than calcite and does not effervesce in acid.

Occurrence: Chabazite typically occurs lining cavities in basalts and andesites, but has also been observed as a major constituent of tuffaceous rocks, where it has evidently formed by the alteration of calcic plagioclase.

Analcime, $NaAlSi_2O_6 \cdot H_2O$

Crystal system and class: Isometric; $4/m\ \bar{3}\ 2/m$.

Cell dimensions and content: $a = 13.71$; $Z = 16$.

Habit: Usually in trapezohedral crystals, also granular and massive as a rock-forming mineral.

Cleavage and fracture: None; uneven.

Hardness: 5.

Density: 2.3.

Color: Colorless or white.

Streak: White.

Luster: Vitreous.

Optical properties: $n = 1.48$–1.49.

Chemistry: Up to about a fourth of the sodium may be replaced by potassium in analcime formed at high temperatures, but there is little atomic substitution at low temperatures. Analyzed analcimes show that in the unit cell Si can range from 31 to 36, Al from 17 to 12; this is balanced by variation in the amount of Na ($+$ K $+$ Ca). Laboratory studies have shown that synthetic analcime is stable up to about 525°C; the presence of analcime as a primary mineral in some igneous rocks suggests that potassium in solid solution may stabilize it to higher temperatures.

Diagnostic features: Resembles leucite, but may be distinguished by its mode of occurrence (see *Leucite*), and by its giving water in the closed tube.

Occurrence: Analcime occurs as well-formed crystals in veins and cavities of igneous rocks; it is rather widespread as a primary mineral in the groundmass of alkali-rich basic igneous rocks; it is widespread and abundant in some sedimentary formations, where it has been formed by the action of alkaline waters rich in sodium on volcanic ash and clay minerals (for example, in the Green River formation of Utah, Colorado, and Wyoming).

Name: The mineral is also known as *analcite;* however, *analcime,* the name originally proposed by Haüy, is the one recommended by committees of both the American and British Mineralogical Societies.

Natrolite, $Na_2Al_2Si_3O_{10} \cdot 2H_2O$

Crystal system and class: Orthorhombic; *mm2.*

Axial elements: $a:b:c = 0.982:1:0.354$.

Cell dimensions and content: $a = 18.38$, $b = 18.71$, $c = 6.63$; $Z = 8$.

Habit: Prismatic crystals, commonly elongated and needle-like, and terminated by pyramid faces; also in radiating nodular forms and as compact masses.

Cleavage: {110}, perfect.

Hardness: 5.

Density: 2.25.

Color: Colorless or white.

Streak: White.

Luster: Vitreous.

Optical properties: $\alpha = 1.479$, $\beta = 1.482$, $\gamma = 1.491$; $\gamma - \alpha = 0.011$; positive, $2V = 60°$.

Diagnostic features: The characteristic crystal form of natrolite distinguishes it from the other common zeolites, and its ready fusibility differentiates it from other acicular silicates such as tremolite.

Occurrence: Typically occurs in association with other zeolites in cavities in basalts, and is sometimes present in igneous rocks as a very late crystallization or as an alteration product of nepheline.

Subclass Phyllosilicates

The basic structural feature of all minerals in this subclass is the presence of SiO_4 tetrahedra linked by sharing three of the four oxygens, and thereby forming sheets with a pseudohexagonal network. This results in a composition for the sheet of $(Si_4O_{10})^{-4}$; Al may replace up to half the Si, giving sheets such as $(AlSi_3O_{10})^{-5}$ and $(Al_2Si_2O_{10})^{-6}$. These sheets are variously referred to as "the silica layer," "the silica sheet," or "the tetrahedral layer."

In all the phyllosilicates except apophyllite and prehnite this tetrahedral

layer is combined with another sheet-like grouping of cations (usually aluminum, magnesium, or iron) in six-coordination with oxygen and hydroxyl anions.
Six-coordination means that the anions are arranged around the cations in an
octahedral pattern, one anion at each solid corner of an octahedron and a cation
at the center. By the sharing of anions between adjacent octahedra a planar
network results, and this is often referred to as "the octahedral layer" (remember that octahedral refers to the arrangement of the anions, not their number).
The minerals $Al(OH)_3$ (gibbsite) and $Mg(OH)_2$ (brucite) (Fig. 10-16) have this
type of structure, and the Al–OH layers and Mg–OH layers in the phyllosilicates are often known as gibbsite layers and brucite layers, respectively. The
gibbsite layer has a dioctahedral arrangement; that is, there are two cations
for each six OH anions, whereas a brucite layer is trioctahedral, there being
three cations for each six OH anions.

The dimensions of the tetrahedral and the octahedral layers are similar,
and, consequently, composite tetrahedral-octahedral layers are readily formed,
either one of each layer (a two-layer structure), or an octahedral layer sand-
wiched between two tetrahedral layers (a three-layer structure). In each tetrahedral layer the free oxygen ion at the apex of each SiO_4 tetrahedron is located
above and at the center of a triangle formed by the other three oxygens. Thus
the free oxygens fall in hexagonal rings with the same spacings as the silicons.
This pattern corresponds to that of the hydroxyl ions on the surface of an

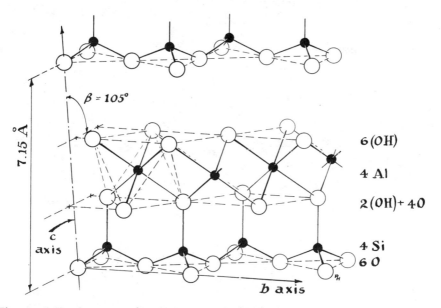

Figure 15-20 Structure of kaolinite, $Al_4Si_4O_{10}(OH)_8$. A tetrahedral sheet Si_4O_{10} linked
to octahedral $Al_4O_4(OH)_8$.

octahedral layer. It is thus possible for octahedral and tetrahedral layers to be fused by sharing in common the oxygen and hydroxyl ions of the individual sheets (Figs. 15-20, 15-21, 15-22, 15-23). If only one surface of an octahedral layer is shared with a tetrahedral layer, a two-layer mineral results (e.g., kaolinite); if both surfaces are shared, a three-layer mineral is obtained (e.g., pyrophyllite, muscovite, chlorite).

The composite octahedral-tetrahedral layers are always stacked in the direction of the *c* axis in the crystal; in the *ab* plane the crystals are pseudohexagonal (most phyllosilicates are monoclinic or triclinic), reflecting the hexagonal nature of the layers (Fig. 15-24). All the phyllosilicates have perfect basal cleavage, which takes place between the composite layers.

We thus obtain a structural classification of the phyllosilicates into dioctahedral and trioctahedral types, according to the nature of the octahedral layer, and into two-layer and three-layer structures, according to whether the octahedral layer is linked on one or both sides with a tetrahedral layer. The resultant classification is illustrated in Table 15-III.

It can be seen that in kaolinite, a two-layer structure, the ratio Si:Al is 1:1;

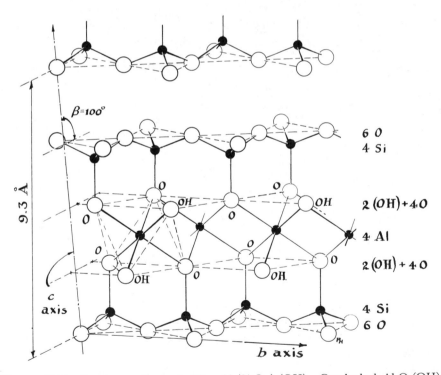

Figure 15-21 Structure of pyrophyllite, $Al_4(Si_4O_{10})_2(OH)_4$. Octahedral $Al_4O_8(OH)_2$ linked to two tetrahedral Si_4O_{10} sheets.

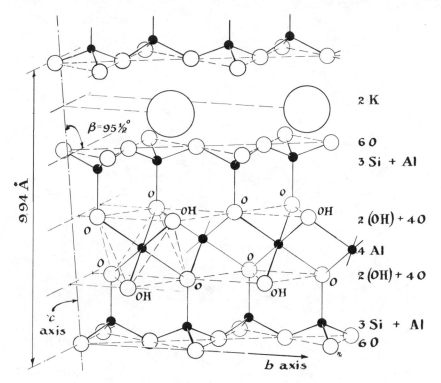

Figure 15-22 Structure of muscovite, $K_2Al_4(Si_6Al_2)O_{20}(OH)_4$. Pyrophyllite layers with one aluminum substituted for one out of four silicons in each tetrahedral layer, linked together by potassium atoms in twelvefold coordination with oxygen.

in pyrophyllite, a three-layer structure with one gibbsite-type layer between two Si_4O_{10} layers, this ratio is 2:1. The muscovite structure is a derivative of the pyrophyllite structure by the substitution of a fourth of the Si by Al, the resultant excess negative charge on the layers being compensated by the introduction of potassium ions between the layers. Phlogopite is a similar derivative of talc. If half the Si is replaced by Al, the excess negative charge on the layers is twice as great, and can be compensated by the introduction of a large bivalent cation such as calcium, giving the rarer minerals margarite and clintonite. The three-layer section of chlorite is similar to that of phlogopite, but the excess negative charge is compensated by a positively charged brucite-type layer instead of by potassium ions.

Although some of the phyllosilicates are stable to quite high temperatures, it is noticeable that many of them are formed at ordinary temperatures during sedimentary processes (the clay minerals). It seems that sheet structures are more readily formed under such conditions than are the other silicate types.

The term *clay* implies an earthy, fine-grained material which develops plas-

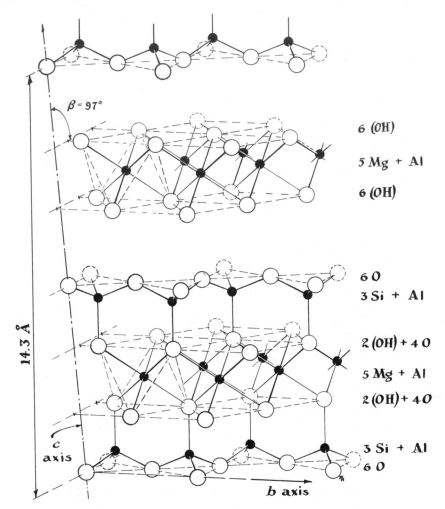

$\beta = 97°$

6 (OH)

5 Mg + Al

6 (OH)

6 O
3 Si + Al

2 (OH) + 4 O

5 Mg + Al

2 (OH) + 4 O

3 Si + Al
6 O

14.3 Å

c axis

b axis

Figure 15-23 Structure of chlorite, $Mg_{10}Al_2(Si_6Al_2)O_{20}(OH)_{16}$. A three-layer structure similar to pyrophyllite but with five Mg and one Al in the octahedral positions, with a further octahedral layer of $Mg_5Al(OH)_{12}$ between pyrophyllite layers.

ticity when mixed with a limited amount of water. Chemical analysis of clays shows that they are made up of hydrous aluminosilicates, frequently with appreciable amounts of iron, magnesium, calcium, sodium, and potassium. They are always very fine-grained, frequently forming colloidal solutions; the upper limit of clay size is generally placed at a particle diameter of 0.004 mm. Genetically most clays are the product of weathering and sedimentation, but they may also be formed by hydrothermal activity.

The characteristic minerals of most clays are phyllosilicates, and they belong

Table 15-III *Composition and classification*
of the important phyllosilicates

DIOCTAHEDRAL (WITH GIBBSITE-TYPE LAYERS)	TRIOCTAHEDRAL (WITH BRUCITE-TYPE LAYERS)
Two-layer Structures	
Kaolinite, Nacrite, Dickite $Al_4Si_4O_{10}(OH)_8$	Antigorite (platy serpentine) $Mg_6Si_4O_{10}(OH)_8$
Halloysite $Al_4Si_4O_{10}(OH)_8$	Chrysotile (fibrous serpentine) $Mg_6Si_4O_{10}(OH)_8$
Three-layer Structures	
Pyrophyllite $Al_2Si_4O_{10}(OH)_2$	Talc $Mg_3Si_4O_{10}(OH)_2$
Montmorillonite $Al_2Si_4O_{10}(OH)_2 \cdot xH_2O^*$	Vermiculite $Mg_3Si_4O_{10}(OH)_2 \cdot xH_2O^*$
Muscovite $KAl_2(AlS_{13}O_{10})(OH)_2$	Phlogopite $KMg_3(AlSi_3O_{10})(OH)_2$
Margarite $CaAl_2(Al_2Si_2O_{10})(OH)_2$	Clintonite $CaMg_3(Al_2Si_2O_{10})(OH)_2$
	Chlorite $Mg_5Al(AlSi_3O_{10})(OH)_8$

* With adsorbed cations and variable water content.

to four groups: the kaolin group, the montmorillonite group, the clay mica group, and the chlorite group. Minerals of the kaolin group have compositions corresponding to the formula $Al_4Si_4O_{10}(OH)_8$, and structures of Si_4O_{10} sheets alternating with gibbsite-type sheets. The montmorillonite group has structures similar to pyrophyllite, but with exchangeable cations and a variable number of water molecules between the layers, which results in swelling when these minerals are immersed in water. Clay mica is essentially fine-grained muscovite. The material known as hydromica or illite is this fine-grained muscovite, often intimately mixed or interlayered with montmorillonite. Clay mica is the commonest mineral in argillaceous rocks and recent marine sediments, and is also present in many soils. Chlorite in clays is always mixed with other clay minerals, and is often difficult to detect.

The clay minerals have many physical features in common. They do not occur as macroscopic or, except for the kaolin minerals, as microscopic crystals; instead they occur as earthy masses. As a result, hardness is not a diagnostic property, since all of them appear very soft (the true hardness of these minerals is about 2–3). Their water content varies with the humidity of the atmosphere, and their apparent density varies accordingly, decreasing with the water content. It is thus very difficult to distinguish between the clay minerals by their physical properties, and indeed the positive identification of the minerals in the clay fraction of sediments is one of the most exacting problems for a mineralogist, generally requiring a combination of optical, x-ray, and chemical tests.

Kaolinite, Al₄Si₄O₁₀(OH)₈

Crystal system and class: Triclinic; $\bar{1}$.

Axial elements: $a:b:c = 0.576:1:0.830$; $\alpha = 91° 48'$, $\beta = 104° 30'$, $\gamma = 90°$.

Cell dimensions and content: $a = 5.14$, $b = 8.93$, $c = 7.37$; $Z = 1$ (Fig. 15-20).

Habit: Usually in earthy aggregates, pseudohexagonal platy crystals sometimes distinguishable under the microscope (Fig. 15-24).

Figure 15-24 Electron micrograph of well-crystallized kaolinite, showing the thin, flat, pseudohexagonal crystals; individual crystals are approximately 0.001 mm across. [Courtesy H. H. Murray.]

Cleavage: {001}, perfect, but not observable to the unaided eye because of small grain size.

Hardness: 2.

Density: 2.6.

Color: White, often stained brown or gray by impurities.

Streak: White.

Luster: Pearly if coarsely crystalline, but usually dull and earthy.

Optical properties: α = 1.553–1.565, β = 1.559–1.569, γ = 1.560–1.570; for fine-grained material only an average refractive index, 1.56–1.57, can be measured.

Chemistry: The composition of kaolinite corresponds closely to the formula, there being little or no atomic substitution. Kaolinite is the commonest of four polymorphs, the others being dickite, nacrite, and halloysite. Dickite and nacrite are rare, halloysite moderately common. The polymorphs differ in the stacking of the basic structural unit (the kaolin layer), which consists of a tetrahedral layer united with a gibbsite-type layer. Regular sequences of one, two, and six kaolin layers are found in kaolinite, dickite, and nacrite, respectively. Halloysite is made up of an irregular sequence of kaolin layers.

Occurrence: Kaolinite is formed by the decomposition of other aluminosilicates, especially the feldspars, either by weathering or by hydrothermal activity. Large deposits have been formed by the hydrothermal alteration of feldspar in granites and granite pegmatites, as in Cornwall (England), Czechoslovakia, and China. Some deposits have been formed by erosion of kaolinized granite and redeposition of the kaolinite, as in Georgia and North Carolina.

Production and uses: Kaolinite is an important industrial mineral. The largest single user in the USA is the paper industry, which consumed about 1.8 million tons in 1964; nearly one-third the weight of today's slick-paper magazines is kaolinite! Other important uses are as raw material for ceramics, and as a filler in rubber, paint, and plastics. World production exceeds 9 million tons annually, some 3 million tons in the USA.

Serpentine, $Mg_6Si_4O_{10}(OH)_8$

Serpentine as a mineral name applies to material containing one or both of the minerals chrysotile and antigorite. Their structures are similar to those of the kaolin minerals, antigorite being the analogue of kaolinite, and chrysotile the analogue of halloysite. The physical properties of chrysotile and antigorite are very similar, but antigorite generally has a lamellar or platy structure, whereas chrysotile is fibrous (serpentine asbestos is chrysotile). A third form, named lizardite, has also been described.

Crystal system: Monoclinic (antigorite has a hexagonal polymorph, chrysotile two orthorhombic polymorphs).

Habit: Crystals unknown; the serpentine minerals usually occur in structureless masses, except when asbestiform.

Cleavage: None observable.

Hardness: Variable, 4–6.

Density: 2.5–2.6.

Color: Usually green, also yellow, brown, reddish brown, and gray.

Streak: White.

Luster: Waxy or greasy in massive varieties, silky in fibrous material.

Optical properties: $\alpha = 1.53$–1.56, $\gamma = 1.54$–1.57; because of the fine-grained nature of most serpentine only an aggregate refractive index, usually 1.54–1.56, can be measured. Birefringence is low, 0.01 or less, and much serpentine is practically isotropic.

Chemistry: The composition of serpentine generally corresponds closely to the above formula; some Fe may be present replacing Mg, and some Al, probably through substitution of Al–Al for Mg–Si. Serpentine decomposes at about 500°C to forsterite, talc, and H_2O, and its presence in a rock indicates an origin below this temperature.

Diagnostic features: The physical properties, especially the color and the luster, usually serve to identify serpentine. Serpentine asbestos can be distinguished from amphibole asbestos by the large amount of water the former gives off at comparatively low temperatures ($\sim$500°C).

Occurrence: Serpentine is formed by the alteration of olivine and enstatite under conditions of low- and medium-grade metamorphism. It sometimes occurs as large rock masses.

Production and uses: Chrysotile provides the greater part of the asbestos of commerce, the combination of fibrous structure, low heat conductivity, high electrical resistance, and chemical inertness being of wide industrial application. The greatest consumption is in asbestos-cement products such as roofing shingles and sheets. The Thetford Mines, Black Lake, and Asbestos districts of southeastern Quebec are the largest producers of chrysotile in the world, with an output of about 1.4 million tons annually.

Massive serpentine is sometimes cut and polished as an ornamental stone.

Varieties and related species: A nickel-bearing variety of serpentine, with an apple-green color, is known as *garnierite*. It is an alteration product of nickel-bearing peridotites, and is mined as an ore of nickel in New Caledonia. *Greenalite* is the ferrous silicate isomorphous with serpentine; it is an important constituent of the iron ores of the Mesabi district in Minnesota.

Pyrophyllite, Al₂Si₄O₁₀(OH)₂

Crystal system and class: Monoclinic; $2/m$.

Axial elements: $a:b:c = 0.577:1:2.084$; $\beta = 99° 55'$.

Cell dimensions and content: $a = 5.15$, $b = 8.92$, $c = 18.59$; $Z = 4$.

Habit: Sometimes in radiating spherulitic aggregates of small platy crystals, otherwise in compact fine-grained masses.

Cleavage: {001}, perfect; cleavage laminae flexible but not elastic.

Hardness: $1–1\frac{1}{2}$.

Density: 2.84.

Color: White or pale yellow, frequently stained reddish or brown by iron oxides.

Streak: White.

Luster: Pearly on cleavage surfaces, otherwise greasy or dull.

Optical properties: $\alpha = 1.566$, $\beta = 1.589$, $\gamma = 1.601$; $\gamma - \alpha = 0.045$; negative, $2V = 60°$; fine-grained pyrophyllite will give refractive index 1.59–1.60.

Chemistry: Composition very uniform, there being little or no atomic substitution.

Diagnostic features: The physical properties of pyrophyllite are practically identical to those of talc, because of the extreme similarity of crystal structure; it can be distinguished from talc by a positive test for Al.

Occurrence: In low- and medium-grade metamorphic rocks rich in aluminum.

Production and uses: The industrial uses of pyrophyllite are similar to those of talc, and in mineral trade statistics the two minerals are seldom reported separately. Pyrophyllite, however, has the distinction of being the best material as a carrier of insecticides, such as DDT, and insecticidal dusting powders generally contain this mineral. There is considerable production of pyrophyllite in the central Piedmont area of North Carolina. The material *agalmatolite*, used extensively in Chinese carvings, is fine-grained pyrophyllite.

Talc, Mg₃Si₄O₁₀(OH)₂

Crystal system and class: Monoclinic; $2/m$.

Axial elements: $a:b:c = 0.578:1:2.067$; $\beta = 100° 00'$.

Cell dimensions and content: $a = 5.27$, $b = 9.12$, $c = 18.85$; $Z = 4$.

Habit: Foliated masses or compact fine-grained aggregates.

Cleavage: {001}, perfect; cleavage laminae flexible but not elastic.

Hardness: 1.

Density: 2.82.

Color: Characteristically pale green, also white or gray, sometimes stained reddish or brown by iron oxides.

Streak: White.

Luster: Pearly on cleavage surfaces, otherwise greasy or dull.

Optical properties: $\alpha = 1.54-1.55$, $\beta = 1.59$, $\gamma = 1.59-1.60$; $\gamma - \alpha = 0.05$; negative, $2V = 0°-30°$.

Chemistry: The composition of talc is very uniform, there being little or no atomic substitution. On heating, talc gives off water and decomposes at 780°C into enstatite and quartz. The extreme softness and cleavability of talc result from the absence of bonding except by van der Waals forces between the layers in the structure.

Diagnostic features: Talc can be distinguished from other minerals (except pyrophyllite) by its extreme softness and its color; certain discrimination from pyrophyllite requires a chemical test for magnesium (although a pale green color suggests talc rather than pyrophyllite).

Occurrence: Talc is a mineral of low- and medium-grade metamorphic rocks rich in magnesium, being often derived from ultrabasic igneous rocks made up of enstatite and olivine. It may be a major rock-forming mineral, such rocks being known as *soapstone* or *steatite.*

Production and uses: Talc has many commercial uses; the most familiar, although one that consumes comparatively little of the mineral, is as talcum powder and face powder. Talc has low conductivity for heat and electricity, is fire resistant, hardens when heated to a high temperature, and is not attacked by acids. These properties are valuable in many industrial applications. It is used in ceramics, especially for electrical porcelain and for refractories, and as a filler in paint, paper, and rubber. Slabs of soapstone are used for electrical switchboards, acid proof table tops and sinks, and laundry tubs. World production (including pyrophyllite) is about 3 million tons annually; the US is a major producer, averaging about 800,000 tons annually, principally from California, North Carolina (largely pyrophyllite), Georgia, Texas, and Montana.

Related species: *Minnesotaite* is the ferrous silicate isomorphous with talc; it occurs as a fine-grained constituent in some of the Minnesota iron ores.

Montmorillonite, $Al_2Si_4O_{10}(OH)_2 \cdot xH_2O$

Crystal system: Monoclinic.

Habit: Always in earthy masses, crystals not distinguishable even in the electron microscope.

Hardness: $2-2\frac{1}{2}$.

Density: 2.0–2.7, decreasing with increasing water content.

Color: Usually gray or greenish gray, but may be white, yellow, yellow green, pink, or brown.

Streak: White.

Luster: Greasy or dull.

Optical properties: Extremely variable, because of the variation in composition, especially water content; the fine-grained nature makes it difficult to determine detailed optical properties, and usually only an aggregate refractive index, which may range from 1.50 to 1.64, can be measured.

Chemistry: The name *montmorillonite* is applied to the mineral approximating the composition given above, and to a group of minerals of similar structure and properties derived from it by atomic substitution. The composition of montmorillonite always deviates from the ideal formula through substitutions in the structure, such as Mg for Al, and Al for Si. This leads to a net negative charge on the layers, which is compensated by cations such as Ca^2, Na^1, and H_3O^1 (i.e., $H^1 + H_2O$) adsorbed between the layers. This explains the swelling of montmorillonite when immersed in water and its property of *cation exchange*, whereby cations in solution can be exchanged for cations adsorbed in the mineral.

Montmorillonite as an individual species is the aluminum-rich form approximating the above formula; *nontronite* is an iron-rich species, greenish yellow when pure, in which Al is largely replaced by Fe^3; in *saponite* the Al is replaced by Mg, in *sauconite* Al is replaced by Zn; in *beidellite* some Si is replaced by Al.

Diagnostic features: Its clay-like character and soapy feel, and the property of swelling and forming a gel-like mass in water.

Occurrence: A clay mineral, commonly formed by the alteration of beds of volcanic ash. *Bentonite* is a rock consisting largely of montmorillonite.

Production and uses: The physical properties of montmorillonite make it an important industrial mineral. It is much used in drilling muds, because of the gel-like suspension it forms in water. A small amount will give plasticity to a large quantity of inert material, and on this account it is used as a binder and plasticizer in molding sands for foundries. Its swelling properties, so useful in industry, often cause serious complications in civil engineering operations, where montmorillonite-bearing material is cut into during excavation.

Most of the world's supply of bentonite is mined in the United States, from deposits in Wyoming, Texas, and Mississippi. Annual production is about 1,800,000 tons.

Vermiculite, $Mg_3Si_4O_{10}(OH)_2 \cdot xH_2O$

Crystal system: Monoclinic.

Habit: Usually as pseudomorphs after phlogopite and biotite.

Cleavage: {001}, perfect.

Hardness: $1\frac{1}{2}$.

Density: 2.4.

Color: Yellow to brown.

Streak: White.

Luster: Pearly, sometimes submetallic.

Optical properties: $\alpha = 1.52–1.56$, $\beta = 1.54–1.58$, $\gamma = 1.54–1.58$; $\gamma - \alpha = 0.02–0.03$; negative, $2V$ near zero.

Chemistry: Chemically and structurally vermiculite is related to talc in the same way that montmorillonite is related to pyrophyllite. Vermiculite always contains some Al replacing Si, the resulting charge deficiency being compensated by Fe^3 replacing Mg, and by adsorbed cations. Vermiculite shows many of the properties of montmorillonite, such as cation exchange and a (001) spacing which varies with the water content and the exchangeable cations present. A unique property of vermiculite is the rapid and large (up to thirty times original volume) expansion in a direction parallel to the c axis when it is heated quickly to 250–300°C.

Diagnostic features: Its habit, and expansion on heating, characterize vermiculite.

Occurrence: Commonly as the product of hydrothermal alteration of phlogopite and biotite. Vermiculite is also found in the clay fraction of some soils.

Production and uses: Large quantities of vermiculite are mined and are expanded by heating for use as an insulator in building construction. It is also mixed with cement and plaster to make lightweight concrete and plaster. The largest and most productive deposits are in Montana, and the USA is the greatest producer (and user) of vermiculite, about 200,000 tons being mined annually.

MICA GROUP

The micas constitute an isomorphous group within the phyllosilicates. Their structures are closely related to those of pyrophyllite and talc (Fig. 15-22). Two sheets of linked SiO_4 tetrahedra are placed together with the vertices of the tetrahedra pointing inward; these vertices are cross-linked either with Al, as in muscovite, or with Mg, as in phlogopite; hydroxyl groups are also incorporated, completing the six-coordination of Al or Mg. A firmly bound layer is thus produced with the bases of the tetrahedra on each outer side. This layer has a negative charge, which is compensated by potassium ions between the layers and linking them together. The ionic linkage between the layers explains the greater hardness of the micas in comparison with talc and pyrophyllite, in which the individual layers are uncharged and hence have practically no cohesive forces.

The individual species in the group include a number of polymorphs, which differ in the number and orientation of the layers in the unit cell. The geometry of the structure shows that six distinct polymorphs are possible. However, since these polymorphs can only be distinguished by x-ray diffraction techniques, or sometimes by their optical properties, they will not be further discussed here.

Compositions are variable within the mica group, but a general formula $W(X,Y)_{2-3}Z_4O_{10}(OH,F)_2$ can be written for the group as a whole. In this formula W is generally K (Na in paragonite); X and Y represent Al, Mg, Fe^2, Fe^3, and Li; and Z is Si and Al, the Si:Al ratio being generally about 3:1. In the following list of species the formulas have been simplified to an ideal type, neglecting variations due to atomic substitution:

Muscovite	$KAl_2(AlSi_3O_{10})(OH)_2$
Paragonite	$NaAl_2(AlSi_3O_{10})(OH)_2$
Phlogopite	$KMg_3(AlSi_3O_{10})(OH)_2$
Biotite	$K(Mg,Fe)_3(AlSi_3O_{10})(OH)_2$
Lepidolite	$KLi_2Al(Si_4O_{10})(OH)_2$

As a group the micas are characterized by their perfect basal cleavage, giving thin, flexible, and elastic cleavage plates. Crystals are usually tabular with prominent basal planes, and hexagonal in outline, the angles (110) $\wedge$ ($1\bar{1}0$) and (110) $\wedge$ (010) being close to 60°. A blow with a dull-pointed instrument on a cleavage plate of any of the micas develops a six-rayed percussion figure, one line, parallel to {010}, being more distinct than the other two, which are parallel to the prism faces.

Paragonite is a rare mineral found in a few schists and gneisses. It is very similar to muscovite in all respects, and can be distinguished only by chemical analysis or x-ray tests; it will not be described in detail. Phlogopite and biotite are described together, since they form a continuous series and the boundary dividing them is arbitrary.

Muscovite, $KAl_2(AlSi_3O_{10})(OH)_2$

Crystal system and class: Monoclinic; $2/m$.
Axial elements: $a:b:c = 0.575:1:2.220$; $\beta = 95° 30'$.
Cell dimensions and content: $a = 5.19$, $b = 9.03$, $c = 20.05$; $Z = 4$.
Habit: Usually in lamellar masses or small flakes; crystals tabular with a hexagonal outline.

Cleavage: {001}, perfect; cleavage laminae flexible and elastic; sometimes shows parting parallel to {010} and {110}.

Hardness: $2\frac{1}{2}$ (on cleavage), 4 (across cleavage).

Density: 2.8–2.9.

Color: Colorless or pale shades of green, gray, or brown in thin sheets.

Streak: White.

Luster: Vitreous, sometimes pearly.

Optical properties: $\alpha = 1.55$–1.57, $\beta = 1.58$–1.61, $\gamma = 1.59$–1.62; $\gamma - \alpha = 0.04$–0.05; negative, $2V = 30°$–$50°$.

Chemistry: Muscovite can vary considerably in composition, as a result of atomic substitutions. Some Na is always present replacing K (replacement is greater the higher the temperature of formation, and the sodium content of muscovite may thus be a useful geological thermometer). Aluminum in six-coordination is often partly replaced by Mg and Fe^2, or less commonly by Cr^3 (giving a bright green color, var. *fuchsite* or *mariposite*), or V^3 (var. *roscoelite*). The ratio Si:Al in four-coordination varies from 3:1 to about 7:1. Some F may be present replacing OH.

Muscovite decomposes between 600°C and 700°C into orthoclase, corundum, and H_2O.

Diagnostic features: The perfect cleavage and the flexible and elastic foliae identify muscovite as belonging to the mica group; brown varieties resemble phlogopite closely, and chemical tests may be necessary to distinguish them, although in the field their associations are usually quite different.

Occurrence: Muscovite occurs in a variety of associations. In igneous rocks it is confined to some granites, but it is a common mineral in granite pegmatites (which provide the material used in industry). It is a common and abundant mineral in schists and gneisses of low- and medium-grade metamorphism; fine-grained muscovite in these rocks is known as *sericite*. It is fairly resistant to weathering, and thus occurs as a detrital mineral in clastic sediments. It evidently forms readily in the sedimentary environment, and clay mica (*illite*) often makes up a large part of argillaceous sediments.

Production and uses: The combination of perfect cleavage, flexibility, elasticity, low thermal conductivity, infusibility, and high dielectric strength makes muscovite a unique mineral and one which is essential in industry. About 90 percent of the production of sheet muscovite is used in the electrical industry for condensers, as insulating material between commutator segments, and in heating elements. Ground mica is used as a filler and as a dusting medium to prevent such materials as rubber goods, asphalt roofing, and asphalt tiles from sticking together. The chief producer of sheet muscovite is India, but considerable amounts have been mined in New Hampshire and North Carolina, especially during wartime.

The average annual world production of muscovite is about 400,000 tons; of this only a small percentage is the valuable sheet mica, the remainder being scrap or ground mica.

Phlogopite, $KMg_3(AlSi_3O_{10})(OH)_2$; Biotite, $K(Mg,Fe)_3(AlSi_3O_{10})(OH)_2$

Crystal system and class: Monoclinic; $2/m$ (one polymorph is trigonal).
Axial elements: $a:b:c = 0.575:1:2.206$; $\beta = 99° 18'$.
Cell dimensions: $a = 5.31$, $b = 9.23$, $c = 20.36$; $Z = 4$.
Habit: Crystals are pseudohexagonal prisms (Fig. 15-25) but habit is more commonly lamellar plates without crystal outline.

Figure 15-25 Phlogopite. Cleavage slab from a single crystal. Perfect basal cleavage and typical rough bounding faces nearly perpendicular to the cleavage. (Frontenac County, Ontario.) [Courtesy Queen's University.]

Cleavage: {001}, perfect, giving flexible and elastic laminae.
Hardness: $2\frac{1}{2}$ (on cleavage).
Density: 2.8–3.4, increasing with iron content.
Color: Pale yellow to brown in phlogopite, dark green, brown, or black in biotite.
Streak: White to gray.
Luster: Vitreous or pearly, sometimes submetallic.
Optical properties: $\alpha = 1.53$–1.63, $\beta = 1.56$–1.70, $\gamma = 1.56$–1.70; $\gamma - \alpha = 0.03$–0.08; negative, $2V$ small; pleochroic in yellow and brown, sometimes green; refractive indices and birefringence increase with the Fe content.
Other properties: Phlogopite sometimes shows asterism—a point of light viewed through a cleavage sheet shows a six-rayed star (caused by regularly arranged microscopic inclusions).
Chemistry: Phlogopite and biotite are arbitrary divisions of a single phase of variable composition, phlogopite being the pale-colored, low-iron part of the series. The composition of the series is exceedingly variable: K can be re-

placed in part by Na, Ca, Ba, Rb, or Cs; Mg can be completely replaced by Fe^2 and Fe^3, and in part by Ti and Mn; the Si:Al ratio is somewhat variable, and some of the OH is replaceable by F. A marked correlation exists between composition and geological environment (Fig. 15-26); in igneous rocks the

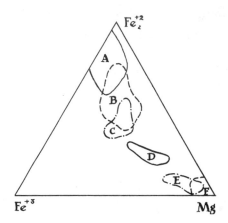

Figure 15-26 Compositional variation in terms of relative amounts of magnesium, ferrous iron, and ferric iron in biotites and phlogopites from different associations.

A   Biotites from granite pegmatites.
B   Biotites from granites and granodiorites.
C   Biotites from tonalites and diorites.
D   Biotites from gabbros.
E   Biotites and phlogopites from ultrabasic igneous rocks.
F   Phlogopites from metamorphosed magnesian limestones.

iron content of biotite increases with silica content of the rock: ultrabasic rocks contain phlogopite, and granites and granite pegmatites contain iron-rich biotites.

Occurrence: Phlogopite occurs in ultrabasic igneous rocks, in some marbles, and in magnesium-rich pegmatites. Biotite is found in many igneous rocks but most commonly in the intermediate and acidic families; it is also an important constituent of metamorphic rocks, especially schists, gneisses, and hornfels.

Production and uses: Phlogopite is used industrially in the same way as is muscovite, and for some purposes is superior to it. It is preferred for electrical commutators, since it wears at about the same rate as the copper segments. It also has greater heat resistance and can withstand temperatures up to 1000°C. In recent years phlogopite (a variety in which all the OH is replaced by F) in sufficiently large crystals for industrial use has been made in the laboratory, but under normal circumstances the natural mineral is cheaper.

Canada and Madagascar are the principal producers of phlogopite for industrial use.

Lepidolite, $KLi_2Al(Si_4O_{10})(OH)_2$

Crystal system and class: Monoclinic; 2/m.
Axial elements: $a:b:c = 0.581:1:2.248$; $\beta = 100° 48'$.
Cell dimensions and content: $a = 5.21$, $b = 8.97$, $c = 20.16$; $Z = 4$.

Habit: Usually in medium- to fine-grained crystalline aggregates (good crystals and large plates are much rarer in lepidolite than in the other micas).
Cleavage: {001}, perfect, giving flexible and elastic laminae.
Hardness: $2\frac{1}{2}$ (on cleavage).
Density: 2.8–2.9.
Color: Commonly pale lilac, but can be colorless, pale yellow, or pale gray.
Luster: Vitreous to pearly.
Optical properties: $\alpha = 1.53$–1.55, $\beta = 1.55$–1.58, $\gamma = 1.55$–1.59; $\gamma - \alpha = 0.02$–0.04; negative, $2V = 0°$–$60°$.
Chemistry: Lepidolite varies considerably in composition; K is often partly replaced by Na, Rb, and Cs; the relative amounts of Li and Al in six-coordination vary widely; some Si may be replaced by Al, and some OH by F. As a result of these variations the amount of Li_2O can vary widely from less than 3 percent to a maximum of about 7 percent; this is an important consideration in the value of a lepidolite as an ore of Li.
Diagnostic features: Lepidolite can be distinguished from the other micas by its lithium content (flame test); its occurrence in complex granite pegmatites and its pale lilac color are often diagnostic.
Occurrence: Lepidolite occurs in a few granites, and in complex granite pegmatites.
Production and uses: Lepidolite is mined from complex granite pegmatites as an ore of lithium; because of the variable lithium content, careful assaying is necessary to determine whether a deposit is commercially valuable. Lepidolite is also used as a raw material in glass and ceramics.

Glauconite, $K(Fe,Mg,Al)_2(Si_4O_{10})(OH)_2$

Crystal system: Monoclinic.
Habit: Small granules in marine sedimentary rocks.
Hardness: 2.
Density: 2.5–2.8.
Color: Green to black.
Streak: Green.
Luster: Earthy and dull.
Optical properties: $\alpha = 1.59$–1.61, $\beta = \gamma = 1.61$–1.64; $\gamma - \alpha = 0.02$–0.03; negative, $2V$ small; pleochroic in green.
Chemistry: Structurally glauconite is a dioctahedral mica, and can be considered as muscovite in which part of the Al in six-coordination is replaced by Mg and Fe^2, thereby reducing the positive charge on the octahedral layer; this is compensated by the replacement of Al in four-coordination by Si; much of the remaining Al in six-coordination may be replaced by Fe^3.

Diagnostic features: Color and habit are characteristic.

Occurrence: Glauconite is an authigenic mineral of marine sedimentary rocks. At the present time it is forming on the sea floor in areas where clastic sedimentation is small or lacking. Its occurrence in the geological column is often linked with disconformities. Glauconite has been recorded as forming from detrital biotite, but most glauconite probably crystallizes directly from an aluminosilicate gel.

Related species: The mineral *celadonite* is identical in structure and composition to glauconite, but its mode of occurrence is quite different; it is found as blue-green earthy material in vesicular cavities in basalt, having been formed by hot solutions during the late cooling stages of the rock.

OTHER PHYLLOSILICATES

Chlorite Series: $(Mg,Fe,Al)_6(Al,Si)_4O_{10}(OH)_8$

Crystal system and class: Monoclinic; $2/m$ (a triclinic polymorph is also known).

Axial elements: $a:b:c = 0.58:1:1.53$; $\beta = 96°\ 17'$.

Cell dimensions and content: $a = 5.3$, $b = 9.3$, $c = 14.3$; $Z = 4$.

Habit: Pseudohexagonal crystals, tabular parallel to $\{001\}$; usually as scaly aggregates and as fine-grained and earthy masses.

Cleavage: $\{001\}$, perfect, cleavage flakes flexible but not elastic.

Hardness: $2\frac{1}{2}$ (on cleavage).

Density: 2.6–3.3, increasing with iron content (Fig. 15-27); 2.7–2.9 in the common varieties.

Color: Characteristically green (hence the name); manganese-bearing varieties orange to brown, chromium-bearing varieties violet.

Streak: White, pale green.

Luster: Vitreous to earthy.

Optical properties: $\alpha = 1.56–1.67$, $\beta = 1.57–1.68$, $\gamma = 1.57–1.68$; $\gamma - \alpha = 0.00–0.01$; iron-poor chlorites positive, iron-rich ones negative; $2V$ usually small. Refractive indices increase with the Fe content (Fig. 15-27).

Chemistry: The chlorites form an extensive isomorphous series with a high degree of atomic substitution. In the general formula Mg and Fe are mutually replaceable, Al in six-coordination ranges from zero to 2, Al in four-coordination from 0.5 to 2, and Si from 2 to 3.5 (Fig. 15-27). Ferric iron is often present. Some varieties of chlorite contain appreciable amounts of chromium, nickel, or manganese. The chemical composition and physical properties of minerals of the chlorite series thus vary widely, and a large number of names are current for different varieties.

The structure of chlorite is illustrated in Fig. 15-23. It consists of alternate

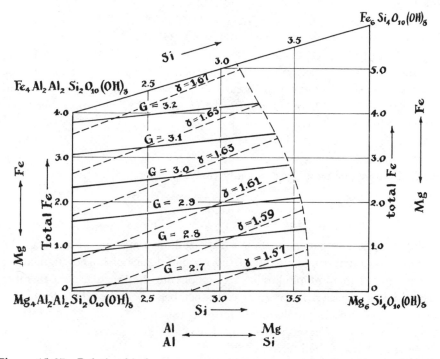

Figure 15-27 Relationship between composition, gamma refractive index, and density in the chlorites. The solid lines are lines of equal density; the broken line indicates the approximate boundary of the composition field of chlorite (compositions to the left of the curve crystallize as chlorite, to the right of the curve as serpentine or greenalite).

mica and brucite layers. The mica layer is negatively charged, and this charge is neutralized by a positive charge on the brucite layer, produced by the partial substitution of Al for Mg. The brucite layer in the chlorites thus corresponds to the potassium ions in the micas.

Diagnostic features: The green color, the micaceous cleavage, and the nonelastic nature of cleavage flakes are characteristic.

Occurrence: Chlorite is an important mineral in low grade schists, and is also common in igneous rocks as an alteration product of biotite and other ferromagnesian minerals. Hydrothermal alteration of pre-existing rocks often results in the formation of chlorite in large amounts. Chlorite is also present in the clay mineral fraction of many sediments. *Chamosite*, an iron-rich chlorite, is an important constituent of some sedimentary iron ores.

Apophyllite, $KCa_4(Si_4O_{10})_2F \cdot 8H_2O$

Crystal system and class: Tetragonal; $4/m\ 2/m\ 2/m$.
Axial elements: $a:c = 1:1.761$.

Cell dimensions and content: $a = 8.960$, $c = 15.78$; $Z = 2$.

Common forms and angles: $(001) \wedge (101) = 60° 25'$.

Habit: Usually in good crystals, combinations of $\{110\}$, $\{101\}$, and $\{001\}$; crystals sometimes apparently isometric combinations (Fig. 2-26b) of cube and octahedron, but the true symmetry is revealed by the unidirectional cleavage, and the difference in luster between basal planes and other faces.

Cleavage: $\{001\}$, perfect.

Hardness: 5.

Density: 2.35.

Color: Colorless or white, occasionally tinted pink, green, or yellow.

Streak: White.

Luster: Pearly on $\{001\}$, otherwise vitreous.

Optical properties: $\omega = 1.535$, $\epsilon = 1.537$; $\epsilon - \omega = 0.002$; positive.

Chemistry: Analyses generally show a little sodium replacing potassium. The structure is interesting: instead of the regular planes of linked six-member rings present in most phyllosilicates, apophyllite has regular four-member rings linked together by irregular eight-member rings, the whole forming a sheet with the usual phyllosilicate formula $(Si_4O_{10})^{-4}$.

Diagnostic features: The crystal form, cleavage, and associations are usually diagnostic for apophyllite.

Occurrence: Apophyllite occurs in hydrothermal veins, and in cavities in basalts, often in association with zeolites, as in northern New Jersey and Nova Scotia.

Prehnite, $Ca_2Al(AlSi_3O_{10})(OH)_2$

Crystal system and class: Orthorhombic; *mm*2.

Axial elements: $a:b:c = 0.843:1:3.378$.

Cell dimensions and content: $a = 4.61$, $b = 5.47$, $c = 18.48$; $Z = 2$.

Habit: Crystals not common, generally pseudocubic or tabular parallel to $\{001\}$; usually massive, globular, or stalactitic with a crystalline surface.

Cleavage: $\{001\}$, good.

Hardness: $6\frac{1}{2}$.

Density: 2.90–2.93.

Color: Characteristically pale green, sometimes white or gray.

Streak: White.

Luster: Vitreous.

Optical properties: $\alpha = 1.615$, $\beta = 1.624$, $\gamma = 1.644$; $\gamma - \alpha = 0.029$; positive, $2V = 68°$.

Diagnostic features: Color and habit are usually diagnostic. Prehnite fuses with intumescence to a blebby enamel-like glass.

Occurrence: Prehnite occurs chiefly in cavities in basic igneous rocks, often associated with zeolites; also in low grade metamorphic rocks, and secondary after calcic plagioclase in altered igneous rocks. Magnificent specimens of prehnite have been found in the trap-rock quarries of northern New Jersey.

Subclass Inosilicates

The inosilicates are a subclass in which the SiO_4 tetrahedra are linked to form chains of indefinite extent. There are two main types of chain structure: single chains, in which each tetrahedron is linked to the next by a common oxygen, producing an overall composition $(SiO_3)^{-2}$; and double chains, in which two single chains placed side by side are cross-linked through alternate tetrahedra, producing an overall composition $(Si_4O_{11})^{-6}$. Each type is represented by an important group of rock-forming minerals, the pyroxene group being single-chain structures and the amphibole group double-chain structures.

The two groups have marked similarities in structure and in physical and chemical properties. In each group the chains are aligned with the elongation in the c direction of the crystals, and are bonded by ions in six-coordination (mainly Ca, Mg, Fe) placed between the chains. The size and shape of the unit cells in each group are closely related:

	a	b	c	β
Tremolite (amphibole)	9.84 Å	18.05 Å	5.28 Å	104° 42′
Diopside (pyroxene)	9.73 Å	8.91 Å	5.25 Å	105° 50′

The unit cells are evidently identical except for a doubling of the period in the b direction in the amphiboles. The orthorhombic members of both groups have about double the spacing of {100} in the monoclinic members.

The bonding between the chains is considerably weaker than the bonding within the chains themselves, and as a result both the pyroxenes and the amphiboles have good prismatic cleavage and occur in crystals in which {110} is a prominent form ({210} in orthorhombic species). The form and the cleavage

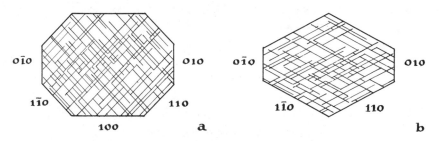

Figure 15-28 Cross sections of (a) pyroxene and (b) amphibole crystals, showing characteristic crystal forms and traces of the prismatic cleavages {110}.

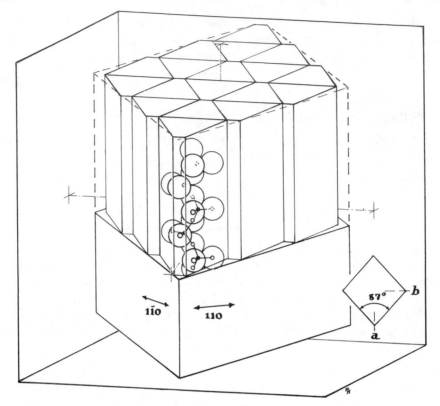

Figure 15-29   The relationship between atomic structure and cleavage in the pyroxenes (this page) and amphiboles (facing page). The linked Si—O chains are shown in cross

are important distinguishing features between the two groups; in fact they often are the only features that can be readily used to tell amphibole from pyroxene in hand specimens. In amphiboles (110) ∧ (1$\bar{1}$0) is about 56°, in pyroxenes about 93°; as a result the cross section of amphibole crystals is approximately diamond-shaped, of pyroxene crystals it is square, and the cleavages intersect at about 60° in the amphiboles and at about 90° in the pyroxenes (Fig. 15-28). The reason for the difference in prism and cleavage angles can be seen from a diagram of the structures (Fig. 15-29). The chains have trapezium-shaped cross sections, the length of which in the *b* direction is twice as great in the amphiboles as in the pyroxenes. Cleavage takes place in a diagonal manner, without cutting through the chains, as shown by the broken lines in the figure. This produces approximately square outlines in the pyroxenes (angles 87° and 93°), and rhombic outlines in the amphiboles (angles 56° and 124°).

In addition to the similarities in structure between the pyroxenes and the amphiboles there are many similarities in chemical composition and physical

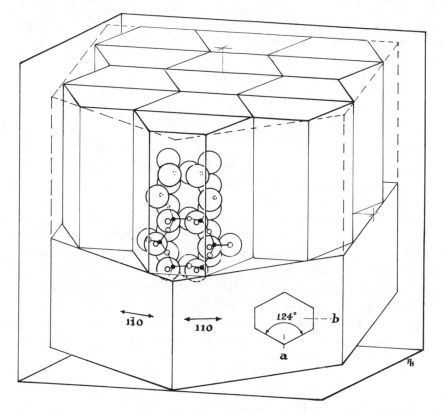

section. The planes of weakness are shown by heavy solid lines, and resulting cleavage directions {110} by broken lines.

properties. Many of the analogous species in the two groups have similar compositions, so similar that it was often argued that such species were actually polymorphs of the same composition. This argument was finally settled in 1930, when the structures were elucidated and the presence of (SiO_3) chains in the pyroxenes and (Si_4O_{11}) chains in the amphiboles was established. The pyroxenes and the amphiboles contain the same elements, so qualitative chemical tests [except for the presence of (OH) groups in the amphiboles] fail to distinguish them. Physical properties such as color, luster, hardness, and density are very similar. Under these circumstances, and in view of the importance of correct identification in determining rock types, the distinguishing features must be borne in mind. Most useful is the difference in cleavage angle, usually most easily observed in coarsely crushed fragments under a hand lens. The form of the crystals is often significant—pyroxene crystals are generally short prismatic, amphibole crystals long prismatic, sometimes acicular or fibrous. Mode of occurrence differs somewhat—pyroxenes are characteristic of basic and ultrabasic igneous rocks, hornblende is more common in intermediate igneous rocks; amphiboles are very common in metamorphic rocks, especially those of medium grade (many species occur only in these rocks), whereas pyroxenes are much less abundant, being largely confined to high-grade types.

AMPHIBOLE GROUP

The amphibole group comprises a number of species, which, although falling in both the orthorhombic and monoclinic systems, are closely related in crystallography and other physical properties, as well as in chemical composition. They can be grouped into a number of series, within which extensive replacement of one ion by others of similar size can take place, giving very complex chemical compositions. The species of the amphibole group form a sequence parallel to those of the allied pyroxene group; they were originally looked upon as complex metasilicates dimorphous with the corresponding pyroxenes. However, this is not so. The amphiboles contain essential (OH) groups in the structure, and the Si:O ratio is 4:11, not 1:3 as in the pyroxenes. An interesting feature of the amphibole structure is the presence of open spaces into which an extra sodium atom for each two Si_4O_{11} groups will fit.

A general formula for all members of the amphibole group can be written $(W,X,Y)_{7-8}(Z_4O_{10})_2(OH)_2$, in which the symbols W, X, Y, Z indicate elements having similar ionic radii and being capable of replacing each other in the structure. The letter W stands for the large cations Ca and Na (K is sometimes present in small amounts); X for the smaller cations Mg and Fe^2 (sometimes Mn); Y for Fe^3, Ti, and Al in six-coordination; and Z for Si, and Al in four-coordination. In the general formula the degree of atomic substitution is as follows:

1. Al may replace up to one Si in the Si_4O_{11} chains (i.e., to the extent of $AlSi_3O_{11}$).

2. Mg and Fe^2 are completely interchangeable.

3. The total (Ca,Na,K) may be zero or near zero, or may vary from 2 to 3; however, total Ca never exceeds 2, and K is present only in minor amounts.

4. OH may be replaced by F; sometimes total (OH,F) is less than 2, presumably by partial substitution of O. With these possibilities it is clear that the composition of the amphiboles may be very complex. However, on the basis of chemical composition the individual amphiboles can be grouped into five series.

Orthorhombic
    Anthophyllite Series $(Mg,Fe)_7Si_8O_{22}(OH)_2$

Monoclinic

Cummingtonite Series	$(Fe,Mg)_7Si_8O_{22}(OH)_2; (Fe > Mg)$
Tremolite–Actinolite Series	$Ca_2(Mg,Fe)_5(Si_8O_{22})(OH)_2$
Hornblende Series	$NaCa_2(Mg,Fe,Al)_5(Al,Si)_8O_{22}(OH)_2$
Alkali Amphibole Series	$Na_2(Mg,Fe,Al)_5Si_8O_{22}(OH)_2$

In the anthophyllite series the ratio $Mg:(Mg + Fe)$ is usually greater than 0.5; Al is often present and can replace Si in (Si_4O_{11}) up to $(AlSi_3O_{11})$ with concomitant replacement of Mg by Al (Fig. 15-30). Other substitutions are minor. Members of the anthophyllite series have been found only in metamorphic rocks.

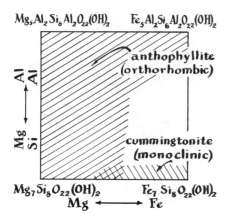

Figure 15-30 Composition fields of the anthophyllite series and the cummingtonite series. Note the extensive replacement of MgSi by AlAl in anthophyllite, and the overlap of composition between the two series.

In the cummingtonite series the ratio $Fe:(Fe + Mg)$ ranges from 1 to about 0.4. Thus the cummingtonite series and the anthophyllite series overlap in the middle composition range (Fig. 15-30). Rocks are known which contain these two amphiboles side by side. Members of the cummingtonite series are also confined to metamorphic rocks.

In the tremolite-actinolite series magnesium is replaceable by ferrous iron and also in part by aluminum and ferric iron, silicon in part by aluminum; titanium and fluorine may be present; and an additional sodium ion for each two (Si_4O_{11}) groups may enter the structure. The product of all these substitutions is the hornblende series. Thus the mineral known as hornblende has a very wide range of composition (Fig. 15-31) and a correspondingly wide range in physical properties. Most hornblende is green, but there is a dark brown variety, known as basaltic hornblende, which is sometimes considered a separate series because of its distinctive properties.

The alkali amphiboles can be considered as being derived from the horn-

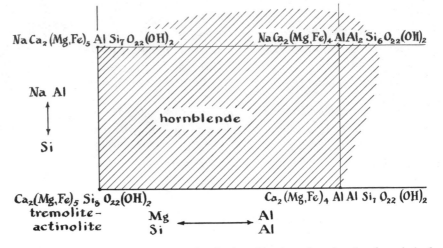

Figure 15-31 Composition diagram for the hornblende series, showing how it is derived from the tremolite-actinolite series by atomic substitution. Most analyses of common hornblende fall in the shaded area.

blende series by the partial or complete substitution of Na for Ca. The best known of these soda amphiboles are glaucophane and the related riebeckite and arfvedsonite, though other varieties have been described.

Anthophyllite, $(Mg,Fe)_7Si_8O_{22}(OH)_2$

Crystal system and class: Orthorhombic; $2/m\ 2/m\ 2/m$.

Axial elements: $a:b:c = 1.027:1:0.292$.

Cell dimensions and content: $a = 18.56$, $b = 18.08$, $c = 5.28$; $Z = 4$.

Habit: Usually in aggregates of prismatic crystals, sometimes fibrous and asbestiform.

Cleavage: $\{210\}$, perfect.

Hardness: 6.

Density: 2.9–3.3, increasing with iron content.

Color: White, gray, brown.

Streak: White.

Luster: Vitreous, somewhat silky in fibrous varieties.

Optical properties: $\alpha = 1.60$–1.68, $\beta = 1.61$–1.68, $\gamma = 1.62$–1.70, refractive indices increasing with the Fe content; $\gamma - \alpha = 0.013$–0.025; can be positive or negative, $2V$ near 90°; sometimes weakly pleochroic in yellow and brown.

Diagnostic features: Anthophyllite is rather distinctive in color and appearance from other amphiboles, except cummingtonite, from which it can be readily distinguished only by careful density determination or by optical or x-ray tests.

Occurrence: Anthophyllite occurs in magnesium-rich metamorphic rocks of medium grade, often associated with talc or cordierite.

Use: Asbestiform anthophyllite is used in industry, but usually the fibers are brittle and of relatively low tensile strength; consequently it is unsuitable for spinning and is used mainly in asbestos cement and as an insulating material.

Related species: The name *gedrite* has been applied to aluminian anthophyllites.

Cummingtonite, $(Fe,Mg)_7Si_8O_{22}(OH)_2$

Crystal system and class: Monoclinic; $2/m$.

Axial elements: $a:b:c = 0.523:1:0.292$; $\beta = 101° 55'$.

Cell dimensions and content: $a = 9.51$, $b = 18.19$, $c = 5.34$; $Z = 2$.

Habit: Usually in aggregates of fibrous crystals, often radiating.

Cleavage: {110}, perfect.

Hardness: 6.

Density: 3.2–3.6, increasing with iron content.

Color: Pale to dark brown.

Streak: White.

Luster: Vitreous to silky.

Optical properties: $\alpha = 1.64–1.68$, $\beta = 1.65–1.71$, $\gamma = 1.66–1.73$; $\gamma - \alpha = 0.02–0.04$; positive, $2V = 60°–90°$; usually polysynthetically twinned on {100}. Refractive indices and birefringence increase with Fe content.

Chemistry: The $Fe:(Fe + Mg)$ ratio ranges from 1 to about 0.4; iron-rich varieties of cummingtonite are called *grunerite*. Manganese sometimes replaces part of the iron and magnesium.

Diagnostic features: The color of cummingtonite is rather characteristic and serves to distinguish it from the other amphiboles, except some anthophyllite. Its density is generally higher than that of anthophyllite.

Occurrence: Cummingtonite occurs in calcium-poor, iron-rich metamorphic rocks of medium grade, often in association with ore deposits.

Tremolite-Actinolite, $Ca_2(Mg,Fe)_5Si_8O_{22}(OH)_2$

Crystal system and class: Monoclinic; $2/m$.

Axial elements: $a:b:c = 0.545:1:0.293$; $\beta = 104° 42'$.

Cell dimensions and content: $a = 9.84$, $b = 18.1$, $c = 5.28$; $Z = 2$.

Habit: Usually in aggregates of long prismatic crystals (Fig. 15-32), sometimes fibrous and asbestiform; occasionally dense and fine-grained (var. *nephrite*).

Cleavage: {110}, perfect.

Density: 2.98–3.46, increasing with iron content (Fig. 15-33).

Color: White in tremolite, becoming green with increasing iron content (actinolite); manganiferous varieties of tremolite pink or pale violet.

Streak: White.

Luster: Vitreous.

Figure   15-32   Tremolite-actinolite. Grayish green radiating groups in cream-colored crystalline limestone. (Verona, Ontario.) [Courtesy Queen's University.]

Optical properties: Refractive indices increase regularly with Fe content, as shown in Fig. 15-33; negative, $2V = 70°–80°$; extinction angle $\gamma:c = 10°–20°$; actinolite pleochroic in green.

Chemistry: The division between tremolite and actinolite is an arbitrary one, tremolite being the magnesian end of the series, which is white, and actinolite comprising the iron-bearing members, which are green.

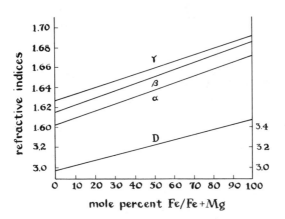

Figure 15-33   Relationship between density, refractive indices, and composition in the tremolite-actinolite series.

Diagnostic features: Color and habit are distinctive. Tremolite resembles wollastonite and sillimanite, but wollastonite is decomposed by HCl, and sillimanite is infusible and has higher refractive indices.

Occurrence: Tremolite and actinolite are common minerals of low- and medium-grade metamorphic rocks, tremolite being characteristic of metamorphosed dolomitic limestones, and actinolite occurring in rocks richer in iron.

Uses: Asbestiform varieties of tremolite are mined to some extent, but are less valuable industrially than serpentine asbestos. Nephrite is one form of jade (the other being the pyroxene jadeite), and it is carved into ornaments. Before the discovery of metals nephrite was extensively used for stone tools.

Hornblende, $NaCa_2(Mg,Fe,Al)_5(Si,Al)_8O_{22}(OH)_2$

Crystal system and class: Monoclinic; $2/m$.

Axial elements: $a:b:c = 0.547:1:0.295$; $\beta = 105° 31'$.

Cell dimensions and content: $a = 9.87$, $b = 18.0$, $c = 5.33$; $Z = 2$.

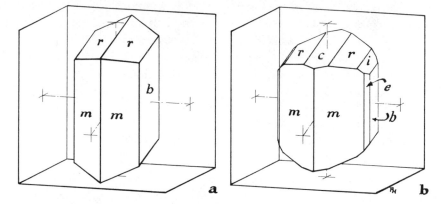

Figure 15-34 Typical crystal forms of hornblende: $b\{010\}$, $c\{001\}$, $r\{011\}$, $i\{031\}$, $m\{110\}$, $e\{130\}$.

Common forms and angles: (Fig. 15-34):

$(110) \wedge (1\bar{1}0) = 55° 35'$ $(001) \wedge (\bar{1}01) = 31° 37'$

$(011) \wedge (0\bar{1}1) = 31° 44'$ $(031) \wedge (0\bar{3}1) = 80° 56'$

$(001) \wedge (100) = 74° 29'$ $(\bar{1}01) \wedge (011) = 34° 41'$

Habit: Sometimes in prismatic crystals, usually with hexagonal cross section (combination of $\{110\}$ and $\{010\}$); also as irregular grains and massive.

Twinning: Common, twin plane $\{100\}$.

Cleavage: $\{110\}$, perfect.

Hardness: 6.

Density: 3.0–3.4, increasing with iron content.

Color: Dark green, dark brown, black.

Streak: White to gray.

Luster: Vitreous.

Optical properties: $\alpha = 1.62$–1.70, $\beta = 1.62$–1.71, $\gamma = 1.63$–1.73; $\gamma - \alpha = 0.014$–0.026 (can be much greater in basaltic hornblendes); negative, $2V = 30°$–$80°$; extinction angle $\gamma{:}c = 10°$–$20°$; usually pleochroic in green and brown.

Chemistry: Hornblende can vary greatly in composition (Fig. 15-31), and many subspecies have been recognized; the dark brown varieties occurring in basic igneous rocks often contain considerable amounts of titanium (up to 10 percent TiO_2 in analyses) and are deficient in hydroxyl, and are known as *basaltic hornblende* or *kaersutite.*

Diagnostic features: Distinguished from pyroxene by its cleavage angles of $56°$ and $124°$, and the six-sided cross section of crystals. Color generally darker than that of other species of amphibole.

Occurrence: Hornblende is an important and widespread rock-forming mineral. It is especially common in medium-grade metamorphic rocks, being an essential constituent of hornblende schists and amphibolites. In igneous rocks it is more common in plutonic rocks than in volcanic rocks, evidently because incorporation of OH in the structure is favored by crystallization under pressure; it occurs especially in diorites and syenites, and in pegamatites associated with these rocks.

Glaucophane-Riebeckite, $Na_2(Mg,Fe,Al)_5Si_8O_{22}(OH)_2$

Crystal system and class: Monoclinic; $2/m$.

Axial elements: $a{:}b{:}c = 0.54{:}1{:}0.30$; $\beta = 104°$.

Habit: Prismatic to acicular crystals, sometimes fibrous or asbestiform; in igneous rocks riebeckite often occurs as moss-like aggregates of tiny grains.

Cleavage: $\{110\}$, perfect.

Hardness: 6.

Density: 3.0–3.4, increasing with iron content.

Color: Pale blue, lavender blue, dark blue to black, deepening with iron content.

Streak: White to blue gray.

Luster: Vitreous; silky in fibrous varieties.

Optical properties: $\alpha = 1.61$–1.70, $\beta = 1.62$–1.71, $\gamma = 1.63$–1.72; $\gamma - \alpha = 0.006$–0.022; negative, $2V$ small to large; pleochroic in blue.

Chemistry: Glaucophane and riebeckite form a series varying in composition mainly in the replacement of magnesium and aluminum by ferrous and ferric iron; the formulas of the end-members can be written $Na_2Mg_3Al_2Si_8O_{22}(OH)_2$ (glaucophane) and $Na_2Fe_3^2Fe_2^3Si_8O_{22}(OH)_2$ (riebeckite).

Diagnostic features: Glaucophane and riebeckite are characterized among the amphiboles by their blue color (in dark-colored varieties it may be necessary to crush the mineral to observe the blue tint).

Occurrence: Glaucophane is confined to low and medium grade metamorphic rocks, being an essential mineral in the glaucophane schists, which are widely distributed in California. Riebeckite is found both in metamorphic and igneous rocks; in the latter it occurs especially in soda-rich granites, rhyolites, and granite pegmatites.

Uses: Riebeckite occurs in an asbestiform variety known as *crocidolite* or blue asbestos, which is mined in South Africa (100,000 tons in 1964) and Western Australia (12,000 tons in 1964). The South African crocidolite is often partly or completely replaced by quartz, and is then used as an ornamental stone under the name *tiger-eye;* tiger-eye may retain the blue of the original mineral, but more commonly the iron has oxidized to a golden brown color.

PYROXENE GROUP

The pyroxenes are a group of minerals closely related structurally and in physical properties, as well as in chemical composition, though they crystallize in two different systems, orthorhombic and monoclinic. In all species within the group the fundamental and common form is the prism, {110} in monoclinic, {210} in orthorhombic, with interfacial angles of about 87° and 93°, and parallel to these prism faces there are good cleavages.

The chemical composition of the pyroxenes can be expressed by the general formula $(W,X,Y)_2Z_2O_6$, in which W, X, Y, Z indicate elements having similar ionic radii and capable of replacing each other in the structure. In the pyroxenes these elements may be:

$W = $ Ca, Na
$X = $ Mg, Fe^2, Mn^2, Li
$Y = $ Al, Fe^3, Ti
$Z = $ Si, Al

The proportion of W atoms in the above formula is generally close to 1 or 0. Of the X group manganese is generally present in minor amounts except in the rare pyroxene *johannsenite*, $CaMnSi_2O_6$; Li occurs as a major constituent in spodumene, $LiAlSi_2O_6$. Of the Y group Ti is only present in minor amounts replacing Al and Fe^3. Z is usually Si, but can be partly replaced by Al up to an Si:Al ratio of 3:1.

The following list gives the principal species names that have been applied to members of the group:

Orthorhombic

Enstatite	$MgSiO_3$
Hypersthene	$(Mg,Fe)SiO_3$

Monoclinic

Clinoenstatite	$MgSiO_3$
Clinohypersthene	$(Mg,Fe)SiO_3$
Diopside	$CaMgSi_2O_6$
Hedenbergite	$CaFeSi_2O_6$
Augite	Intermediate between diopside and hedenbergite, with some Al
Pigeonite	Intermediate between clinoenstatite and augite
Aegirine	$NaFeSi_2O_6$
Jadeite	$NaAlSi_2O_6$
Spodumene	$LiAlSi_2O_6$
Johannsenite	$CaMnSi_2O_6$

The common rock-forming pyroxenes can be considered as phases in the ternary system $MgSiO_3$–$FeSiO_3$–$CaSiO_3$, with $CaSiO_3$ 50 percent or less (Fig. 15-35).

Enstatite, $MgSiO_3$; Hypersthene, $(Mg,Fe)SiO_3$

Crystal system and class: Orthorhombic; $2/m\ 2/m\ 2/m$.
Axial elements: $a:b:c = 2.0636:1:0.5881$.

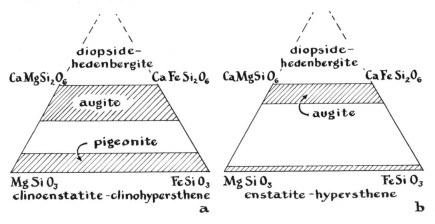

Figure 15-35 Composition and phase relations in the common pyroxenes (a) at high temperatures, (b) at medium temperatures. The unshaded areas represent the miscibility gap between high-calcic and low-calcic pyroxenes, which increases with decreasing temperature.

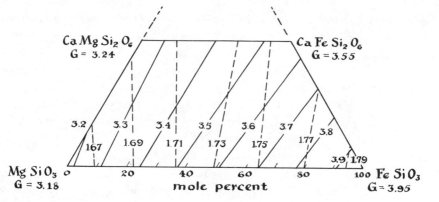

Figure 15-36 Relationship between density, gamma refractive index, and composition in the pyroxenes.

Cell dimensions and content: $a = 18.22$, $b = 8.829$, $c = 5.19$; $Z = 16$.

Habit: Well-formed crystals rare; usually as irregular grains or coarse cleavable masses.

Cleavage: {210}, good; (210) $\wedge$ (2$\bar{1}$0) = 91° 48′.

Hardness: 6.

Density. 3.2–3.9, increasing with Fe content (Fig. 15-36).

Color: White or pale green in low-iron varieties (enstatite), brownish green, dark brown, to black with greater Fe content (hypersthene), sometimes bronzy (var. *bronzite*).

Streak: White to gray.

Luster: Vitreous.

Optical properties: $\alpha = 1.655$–1.755, $\beta = 1.658$–1.765, $\gamma = 1.663$–1.775; $\gamma - \alpha = 0.008$–0.020; refractive indices and birefringence increase with Fe content; enstatite is positive, $2V = 55°$–$90°$, hypersthene negative, $2V = 90°$–$50°$

Chemistry: Enstatite and hypersthene form a continuous series, and the boundary between them is an arbitrary one generally fixed at about 10 percent of the $FeSiO_3$ component. The series does not extend to pure $FeSiO_3$, the orthorhombic pyroxenes ranging in composition from $MgSiO_3$ to about 90 percent $FeSiO_3$; the compound $FeSiO_3$ is not a stable one at high temperatures, since from a melt of its composition tridymite and fayalite (Fe_2SiO_4) will crystallize.

When heated above 1050°C enstatite changes to protoenstatite, another orthorhombic phase with a different structure. Protoenstatite has not been found as a mineral, but on rapid cooling it changes to clinoenstatite, a monoclinic phase with the same structure as diopside. Hypersthene has a corresponding monoclinic polymorph known as clinohypersthene. Clinoenstatite and clinohypersthene are practically unknown in terrestrial rocks but have been recognized in some meteorites. A mineral intermediate in composition between these substances and augite is found in basic volcanic rocks and is known as *pigeonite*. In its properties it is very similar to augite and can be distinguished from it only by microscopic examination.

Diagnostic features: Dark-colored varieties are not easily distinguished from augite. The pyroxene cleavage and the color serve to characterize enstatite and bronzite.

Occurrence: The orthorhombic pyroxenes typically occur in basic and ultra-basic rocks low in calcium, such as pyroxenites, peridotites, norites, and some basalts and andesites. They also occur in some high-grade metamorphic rocks. Most orthorhombic pyroxenes of igneous rocks contain low to moderate amounts of iron. Compositions closest to $MgSiO_3$ are recorded for enstatite from meteorites.

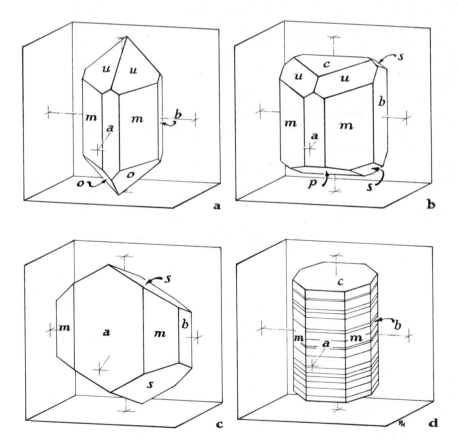

Figure 15-37   (a–c) Typical crystal forms of augite: $a\{100\}$, $b\{010\}$, $c\{001\}$, $m\{110\}$, $p\{\bar{1}01\}$, $u\{111\}$, $s\{\bar{1}11\}$, $o\{\bar{2}21\}$. (d) Typical crystal showing the basal parting.

Diopside, CaMgSi$_2$O$_6$; Hedenbergite, CaFeSi$_2$O$_6$; Augite, Ca(Mg,Fe,Al)(Al,Si)$_2$O$_6$

Crystal system and class: Monoclinic; 2/m.

Axial elements: a:b:c = 1.092:1:0.589; β = 105° 50′.

Cell dimensions and content: a = 9.73, b = 8.91, c = 5.25; Z = 4.

Figure 15-38 Diopside. Dark green crystal. Forms: {010}, {110}, {111}, {$\bar{1}$01}; *c*-axis vertical, (010) parallel to page, bottom of crystal broken off along (00$\bar{1}$) parting plane; sloping face at upper left is (111). (Wakefield, Quebec.) [Courtesy Queen's University.]

Common forms and angles: (Figs. 15-37, 15-38; Fig. 2-15d):

(110) ∧ (1$\bar{1}$0) = 92° 50′   (021) ∧ (0$\bar{2}$1) = 97° 10′

(001) ∧ (100) = 74° 10′   (001) ∧ (111) = 33° 49′

(001) ∧ ($\bar{1}$01) = 31° 19′   (011) ∧ (0$\bar{1}$1) = 59° 04′

Habit: Short prismatic crystals, often square in cross section as a result of the predominance of {110}; also granular and massive.

Twinning: Common, twin plane {100} (Fig. 2-45a).

Cleavage: {110}, good; sometimes a well-developed parting parallel to {100} (var. *diallage*), or parallel to {001}.

Hardness: 6.

Density: 3.25–3.55, increasing with Fe content (Fig. 15-36).

Color: Colorless or white in rare iron-free varieties, usually dark green to black.

Streak: White to gray.

Luster: Vitreous.

Optical properties: $\alpha = 1.66-1.73$, $\beta = 1.67-1.74$, $\gamma = 1.70-1.76$, increasing with Fe content (Fig. 15–36); $\gamma - \alpha = 0.03$; positive, $2V = 50°-60°$; extinction angle $\gamma\!:\!c = 40°-50°$; iron-rich varieties pleochroic in green, titaniferous augite pleochroic in violet.

Chemistry: Diopside, hedenbergite, and augite are convenient species designations for parts of a continuous series in chemical composition and physical properties. Augite is the commonest pyroxene; in composition it is intermediate between diopside and hedenbergite, with some substitution of Al–Al for Mg–Si; some Ti is often present.

Diagnostic features: Characterized by crystal form, cleavage, and color. Diopside is white to pale green, in contrast to the darker color of augite and hedenbergite.

Occurrence: Augite is the most important ferromagnesian mineral of igneous rocks; it is especially abundant in basic and ultrabasic rocks, being characteristic of gabbros and basalts, and also occurs in many andesites. Diopside and hedenbergite occur in medium- and high-grade metamorphic rocks, especially those rich in calcium. White and light-green diopsides characteristically occur in metamorphosed dolomitic limestones; hedenbergite is often found associated with ore deposits formed at high temperatures.

Aegirine, $NaFeSi_2O_6$

Crystal system and class: Monoclinic; $2/m$.

Axial elements: $a\!:\!b\!:\!c = 1.099\!:\!1\!:\!0.601$; $\beta = 106° \, 49'$.

Habit: Prismatic crystals, sometimes elongated and terminated with steep pyramids; also as irregular grains.

Twinning: Often twinned, twin plane $\{100\}$.

Cleavage: $\{110\}$, good.

Hardness: 6.

Density: 3.5–3.6.

Color: Dark green, green brown, or nearly black.

Streak: Gray.

Luster: Vitreous.

Optical properties: $\alpha = 1.77$, $\beta = 1.81$, $\gamma = 1.83$ (for pure aegirine, refractive indices decrease in aegirine-augite); $\gamma - \alpha = 0.06$; negative, $2V = 60°-70°$; extinction angle small; pleochroic in green and brown.

Chemistry: Solid solution can produce a complete series of compositions between aegirine and augite, intermediate varieties being known as aegirine-augite.

Diagnostic features: Resembles augite, but is more apt to occur in long prismatic crystals. Fuses rather easily, coloring the flame yellow.

Occurrence: Aegirine and aegirine-augite occur in sodium-rich igneous rocks, especially those of the nepheline syenite family, but also in some granites and syenites, and in pegmatites associated with these rocks.

Name: The name *acmite* is sometimes used as a synonym of aegirine, or for a variety occurring in long slender crystals, often brown in color. Acmite was described in 1821 as an independent species, and its identity with aegirine was not established until 1871.

Jadeite, $NaAlSi_2O_6$

Crystal system and class: Monoclinic; $2/m$.

Axial elements: $a:b:c = 1.104:1:0.610$; $\beta = 107° 20'$.

Cell dimensions and content: $a = 9.48$, $b = 8.59$, $c = 5.24$; $Z = 4$.

Habit: Crystals very rare; usually fine-grained granular or dense masses.

Cleavage: $\{110\}$, good, but seldom visible in hand specimen; fine-grained material exceedingly tough.

Hardness: $6\frac{1}{2}$.

Density: 3.25–3.35.

Color: Usually green, sometimes white, rarely violet or brown.

Streak: White.

Luster: Vitreous, often somewhat waxy in polished specimens.

Optical properties: $\alpha = 1.658$, $\beta = 1.663$, $\gamma = 1.673$; $\gamma - \alpha = 0.015$; positive, $2V = 70°$; extinction angle $\gamma:c = 40°$.

Chemistry: The Na and Al may be partly replaced by Ca and Mg, giving the variety *diopside-jadeite*. The Al is sometimes partly replaced by Fe^3, giving a dark green variety *chloromelanite*.

Jadeite is a mineral of much significance in petrogenesis, since it is intermediate in composition between albite and nepheline, yet it is exceedingly rare, whereas they are common. Analcime, a common zeolite, has the same composition as jadeite, with the addition of one molecule of combined water. A significant feature is the high density (3.3) of jadeite as compared with that of the other sodium aluminum silicates (albite, 2.6; nepheline, 2.6; analcime, 2.3), indicating that its formation is promoted by high pressure. Jadeite cannot be made by melting together its component oxides in the correct proportions; such a melt crystallizes as a mixture of albite and nepheline. For a long time all efforts at synthesis failed, but it has been made at high pressures and at temperatures between 300°C and 600°C, confirming the theoretical and geological evidence for its conditions of stability.

Diagnostic features: Jadeite resembles nephrite, from which it can be distinguished by a determination of density. Jadeite fuses more readily than nephrite and gives a strong sodium flame.

Occurrence: The mode of occurrence of jadeite has been somewhat of a mystery, since it was known for a long time only as stream-worn boulders brought from Burma, and as carved objects from Mayan ruins in Mexico and Guatemala. It has now been found *in situ* in Burma, Japan, California, and Guatemala, in metamorphic rocks associated with serpentine, and has also been recognized as microscopic grains in low-grade metamorphosed graywackes in California.

Spodumene, $LiAlSi_2O_6$

Crystal system and class: Monoclinic; $2/m$.

Axial elements: $a:b:c = 1.144:1:0.631$; $\beta = 110° 20'$.

Cell dimensions and content: $a = 9.52$, $b = 8.32$, $c = 5.25$; $Z = 4$.

Habit: Long prismatic crystals, and platy masses; very large crystals have been recorded—some from the Etta pegmatite, near Keystone, South Dakota, were 40 feet long and 2–6 feet wide.

Cleavage: $\{110\}$, perfect; often also a well developed parting parallel to $\{100\}$, giving a marked splintery fracture.

Hardness: $6\frac{1}{2}$.

Density: 3.1–3.2.

Color: Usually white or grayish white, some varieties transparent pink and violet (*kunzite*), or green (*hiddenite*).

Streak: White.

Luster: Vitreous.

Optical properties: $\alpha = 1.653–1.663$, $\beta = 1.659–1.668$, $\gamma = 1.676–1.679$; $\gamma - \alpha = 0.015–0.024$; positive, $2V = 60°–70°$; extinction angle $\gamma:c = 25°$.

Chemistry: Spodumene, unlike the other pyroxenes, is not subject to wide variations in composition as a result of atomic substitutions; however, lithium may be partly replaced by sodium.

Alteration: Alters readily to clay minerals, and is often partly replaced by albite and muscovite by the action of hydrothermal solutions carrying sodium and potassium.

Diagnostic features: Readily distinguished by its habit and cleavage, the presence of lithium (flame test), and its typical occurrence in complex granite pegmatites.

Occurrence: Spodumene occurs in complex granite pegmatites, associated with other lithium aluminosilicates such as lepidolite, and with beryl and colored tourmalines.

Uses: Spodumene is mined as a raw material for lithium compounds and for ceramics, the principal producer in the USA being a large pegmatite at Kings Mountain, North Carolina. Kunzite and hiddenite are cut as gemstones; kunzite is obtained from pegmatites at Pala, California, and in Madagascar and Brazil, and the finest hiddenite came from a small deposit in Alexander County, North Carolina.

PYROXENOID GROUP

The compound $CaSiO_3$, though analogous in formula to the pyroxenes, does not crystallize in the pyroxene structure. Wollastonite, $CaSiO_3$, and some other nonpyroxene minerals with a Si:O ratio of 1:3, such as rhodonite, $MnSiO_3$, pectolite, $Ca_2NaHSi_3O_9$, and bustamite, $CaMnSi_2O_6$, have been classed together in a pyroxenoid group. All these minerals are inosilicates, but they differ from the pyroxenes in the nature and arrangement of the SiO_3 chains. In the pyroxenes the chains are parallel to the c axis, and the unit cell is 5.3 Å high, corresponding to a "repeat" after every second SiO_4 tetrahedron; in wollastonite, pectolite, and bustamite the chains are parallel to the b axis, and the unit cell is 7.3 Å long in this direction, corresponding to a repeat after every third SiO_4 tetrahedron. This is shown in Fig. 15-39, which also illustrates how

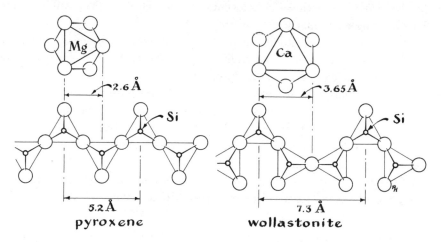

Figure 15-39 The SiO_3 chains in pyroxene and in wollastonite. Note the relationship of the repeat period in the chains to the sizes of Mg—O and Ca—O coordination octahedra.

this type of chain is conditioned by the larger size of a six-coordinated calcium-oxygen group, whereas the arrangement of the pyroxene chain is controlled by the smaller size of the six-coordinated magnesium-oxygen group. In rhodonite the chains are parallel to the c axis, as in the pyroxenes, but the unit cell is 12.2 Å high, corresponding to a repeat after every fifth SiO_4 tetrahedron.

The presence of SiO_3 chains in the pyroxenoids is manifested by their often fibrous nature.

Wollastonite, $CaSiO_3$

Crystal system and class: Triclinic; $\bar{1}$.
Axial elements: $a:b:c = 1.085:1:0.966; \alpha = 90° 02', \beta = 95° 22', \gamma = 103° 26'$.
Cell dimensions and content: $a = 7.94, b = 7.32, c = 7.07; Z = 6$.
Habit: Usually in cleavable or fibrous masses, sometimes granular and compact.
Cleavage: $\{100\}$, perfect; $\{\bar{1}02\}$ and $\{001\}$, good, giving a splintery fracture.
Hardness: 5.
Density: 2.9.
Color: Usually white, sometimes grayish.
Streak: White.
Luster: Vitreous, somewhat silky in fibrous varieties.
Optical properties: $\alpha = 1.618, \beta = 1.628, \gamma = 1.631; \gamma - \alpha = 0.013$; negative, $2V = 40°$.
Other properties: Sometimes fluorescent, the material from Franklin, New Jersey, being particularly noteworthy in this respect.
Diagnostic features: Wollastonite can be distinguished from other white fibrous silicates, such as tremolite and sillimanite, by its solubility in HCl with the separation of silica and its optical properties.
Occurrence: Wollastonite is formed by the metamorphism of siliceous limestones at temperatures of about 450°C and higher, and thus occurs in igneous contact zones and in high-grade regionally metamorphosed rocks.
Uses: A large deposit of wollastonite at Willsboro, New York, is mined for use in ceramics and paints.
Related species: A rare monoclinic polymorph, *parawollastonite*, occurs in limestone blocks ejected from volcanoes, as at Monte Somma, Italy.

Pectolite, $Ca_2NaHSi_3O_9$

Crystal system and class: Triclinic; $\bar{1}$.
Axial elements: $a:b:c = 1.135:1:0.997; \alpha = 90° 03', \beta = 95° 17', \gamma = 102° 28'$.
Cell dimensions and content: $a = 7.99, b = 7.04, c = 7.02; Z = 2$.
Habit: Aggregates of needle-like crystals elongated parallel to b, often radiating and forming globular masses (specimens should be handled with care, as the needle-like crystals readily penetrate the skin).
Cleavage: $\{001\}, \{100\}$, perfect.
Hardness: 5.

Density: 2.86.

Color: White.

Streak: White.

Luster: Vitreous or silky.

Optical properties: $\alpha = 1.600$, $\beta = 1.605$, $\gamma = 1.636$; $\gamma - \alpha = 0.036$; positive, $2V = 50°$.

Chemistry: Most pectolite has a composition close to the above formula, but in some specimens the calcium is extensively replaced by manganese, which raises the density and refractive indices.

Diagnostic features: Easily fusible to a white enamel, and gives a bright yellow flame. Decomposed by warm dilute HCl with the separation of silica.

Occurrence: Pectolite typically occurs associated with zeolites in cavities in basalts, as in northern New Jersey.

Rhodonite, $MnSiO_3$

Crystal system and class: Triclinic; $\bar{1}$.

Axial elements: $a:b:c = 0.874:1:1.592$; $\alpha = 111° \, 06'$, $\beta = 86° \, 00'$, $\gamma = 93° \, 12'$.

Cell dimensions and content: $a = 6.68$, $b = 7.66$, $c = 12.22$; $Z = 10$.

Habit: Crystals uncommon, usually tabular parallel to $\{001\}$ (Fig. 2-12b); usually massive, coarse to fine granular or dense aggregates.

Cleavage: $\{110\}$, $\{1\bar{1}0\}$, perfect; fine-grained material very tough, with a conchoidal fracture.

Hardness: 6.

Density: 3.5–3.7.

Color: Pink to rose red, often veined by black alteration products.

Streak: White.

Luster: Vitreous.

Optical properties: $\alpha = 1.720$–1.729, $\beta = 1.725$–1.734, $\gamma = 1.733$–1.741; $\gamma - \alpha = 0.011$–0.014; positive, $2V = 60°$–$80°$.

Chemistry: Composition variable, part of the Mn being replaceable by Fe and Ca (in the rhodonite structure there are five positions for the Mn and Ca atoms, one of these being preferentially filled by Ca); the variation in composition is reflected in the variable density and refractive indices.

Alteration: Alters readily by oxidation to black manganese oxides.

Diagnostic features: The rose red color is characteristic; rhodonite resembles rhodochrosite but is distinguished by greater hardness and no effervescence in warm HCl.

Occurrence: In veins and irregular masses formed by hydrothermal or metasomatic processes, and by metamorphism of sedimentary manganese ores.

Excellent crystals have been found at Franklin and Sterling Hill, New Jersey. Other notable localities are Långban, Sweden; Broken Hill, Australia; and Ouro Preto, Brazil.

Uses: Massive rhodonite is sometimes cut and polished as an ornamental stone; a deposit near Sverdlovsk, in the Ural Mountains, has provided a large amount of rhodonite which has been used to make vases, tabletops, and other decorative objects; one of the stations of the Moscow subway is extensively decorated with this rhodonite.

Subclass Cyclosilicates

The cyclosilicates are so named because they contain rings of linked SiO_4 tetrahedra. The tetrahedra are linked in the same manner as in the single-chain inosilicates, each SiO_4 group sharing two oxygens with adjoining tetrahedra on either side, giving the same overall formula $(SiO_3^{-2})_n$. However, instead of forming straight chains the tetrahedra are joined at angles which result in the formation of rings. The minimum number of tetrahedra to form a ring is of course three, and a few three-member-ring minerals are known, the best-known being benitoite, $BaTiSi_3O_9$. A few minerals have four-member rings in the structure (e.g., axinite) and several important minerals (beryl, cordierite, and tourmaline) have six-member rings. Five-member rings and single rings with more than six tetrahedra are unknown.

Axinite, $(Ca,Mn,Fe)_3Al_2(BO_3)Si_4O_{12}(OH)$

Crystal system and class: Triclinic; $\bar{1}$.
Axial elements: $a:b:c = 0.779:1:0.978$; $\alpha = 91° 52'$, $\beta = 98° 09'$, $\gamma = 77° 19'$.
Cell dimensions and content: $a = 7.151$, $b = 9.184$, $c = 8.935$; $Z = 2$.
Common forms and angles:

    $(010) \wedge (1\bar{1}0) = 135° 25'$ $(010) \wedge (011) = 45° 12'$
    $(0\bar{1}0) \wedge (1\bar{2}0) = 28° 58'$ $(010) \wedge (\bar{1}21) = 33° 20'$

Habit· Commonly as wedge-shaped crystals; also massive, lamellar, or granular.
Cleavage: {100}, good.
Hardness: 7.
Density: 3.2–3.3.
Color: Usually violet to brown, but sometimes yellow, greenish yellow, or pink.
Streak: White.
Luster: Vitreous.

Optical properties: $\alpha = 1.674$–1.693, $\beta = 1.681$–1.701, $\gamma = 1.684$–$1,704$; $\gamma - \alpha = 0.01$; negative, $2V = 60°$–$80°$.

Chemistry: The composition of axinite varies considerably within the formula, as a result of varying proportions of Ca, Mn, and Fe. The proportion of Ca to Mn + Fe is commonly about 2:1. A little Mg may be present (up to 2.4 percent MgO in analyses).

Diagnostic features: The crystal form and color are useful diagnostic properties. Axinite is easily fusible with intumescence, giving a dark glass.

Occurrence: Axinite is generally formed by metasomatic reactions at comparatively high temperatures, the boron being derived from magmatic emanations; thus it is commonly found in contact-altered calcareous rocks, in cavities in granite, and in hydrothermal veins. Fine crystals are known from Riverside, California; Bourg d'Oisans, France; Obira, Japan; and St. Just, Cornwall, England.

Beryl, $Be_3Al_2Si_6O_{18}$

Crystal system and class: Hexagonal; $6/m\ 2/m\ 2/m$.

Axial elements: $a:c = 1:0.9975$.

Cell dimensions and content: $a = 9.215$, $c = 9.192$; $Z = 2$.

Habit: Prismatic crystals (Fig. 2-31), combinations of $\{10\bar{1}0\}$ and $\{0001\}$ (sometimes up to several feet long); also massive.

Cleavage: $\{0001\}$, poor.

Hardness: 8.

Density: 2.66–2.92.

Color: Commonly pale green, also white or yellow; gem varieties transparent, dark green (*emerald*), pale blue or green (*aquamarine*), yellow (*heliodor*), pink (*morganite*).

Streak: White.

Luster: Vitreous.

Optical properties: $\omega = 1.566$–1.608, $\epsilon = 1.562$–1.600; $\omega - \epsilon = 0.004\ 0.008$; negative.

Chemistry: In addition to the elements indicated in the formula, beryl often contains appreciable amounts of the alkali elements and large amounts of occluded gases. Until the structure was elucidated the reason for this was not understood. The structure of beryl contains tubular channels defined by the Si_6O_{18} rings. In the structure Li can substitute for Al in six-coordination and Al for Be in four-coordination; charge imbalance resulting from these substitutions is satisfied by the introduction of Na, K, and Cs ions into these tubular channels, which also serve to accommodate occluded gases. These substitutions result in an increase in refractive indices and density, and a

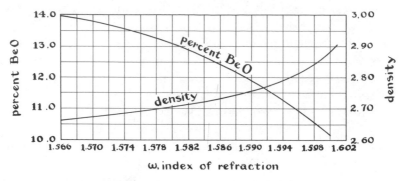

Figure 15-40   Relationship between omega index of refraction, density, and BeO content of beryl.

decrease in the beryllium content (Fig. 15-40). Since the BeO content of beryl ranges from 10 percent to 14 percent, the nature of the beryl must be considered when evaluating a beryl deposit.

Diagnostic features: When in crystals beryl is easily recognized, the only similar mineral being apatite, which is much softer. Massive white beryl, however, is much like milky quartz, and differentiation of the two is important in prospecting and mining operations. A simple field test uses acetylene tetra-bromide diluted with benzene, in which beryl sinks and quartz floats, the density of beryl being generally somewhat greater than that of quartz.

Occurrence: The usual mode of occurrence is in granite pegmatites, but beryl is occasionally found in druses in granite; the fine emeralds from Muso in Colombia occur in cavities in a bituminous limestone, and emeralds have also been found in mica schist (generally, however, in the wall rock of pegmatite veins).

Production and uses: Beryl is the commonest beryllium mineral, and the only commercial source of this element. It is mined from granite pegmatites in New Mexico and South Dakota, Brazil, Argentina, India, South Africa, and Australia. World production is about 9,000 tons annually.

Cordierite, $(Mg,Fe)_2Al_4Si_5O_{18}$

Crystal system and class: Orthorhombic; $2/m\ 2/m\ 2/m$.

Axial elements: $a:b:c = 1.748:1:0.954$.

Cell dimensions and content: $a = 17.13$, $b = 9.80$, $c = 9.35$; $Z = 4$.

Habit: Rarely as prismatic or pseudohexagonal twinned crystals; generally massive or as irregular grains.

Twinning: Twin plane $\{110\}$, usually repeated, giving pseudohexagonal crystals.

Cleavage: {010}, poor; parting on {001} develops with the onset of alteration.
Hardness: 7.
Density: 2.55–2.75, increasing with Fe content.
Color: Characteristically pale to dark blue or violet, also colorless, gray, yellow, or brown.
Streak: White.
Luster: Vitreous.
Optical properties: α = 1.522–1.558, β = 1.524–1.574, γ = 1.527–1.578 (increasing with the Fe content); $\gamma - \alpha$ = 0.005–0.018; usually negative, $2V$ = 65°–90°.
Chemistry: The structure of cordierite is similar to that of beryl; three Al atoms are in six-coordination, the fourth substitutes for one Si in the ring structure, giving an $AlSi_5O_{18}$ group. As in beryl there are channels within the ring structure, in which the H_2O molecules and the alkali metals recorded in some analyses are evidently accommodated.
Alteration: Cordierite alters so readily that it is more commonly found in the altered condition rather than fresh. The commonest type of alteration is to chlorite or muscovite.
Diagnostic features: When it shows the typical blue or blue-gray color cordierite is easy to recognize, but when it occurs as colorless or gray grains in a rock it is very like quartz and can be easily distinguished only by optical or chemical tests.
Occurrence: Cordierite is formed by the medium- to high-grade metamorphism of aluminum-rich rocks, and is found in schists, gneisses, and hornfels. It has been recorded from volcanic rocks, but an investigation of one such occurrence showed it to be a mineral very similar but distinct, containing units of two six-member rings joined through the vertices of the SiO_4 tetrahedra, giving $(Si,Al)_{12}O_{30}$ groups in the structure. This new mineral has been called *osumilite*, and probably most if not all volcanic cordierite is actually this mineral.
Use: Clear blue varieties of cordierite are occasionally cut as gemstones.
Name: Cordierite is also known as *iolite*, for its violet color, or *dichroite*, for its variation in color with crystallographic direction (dichroism).

Tourmaline, $Na(Mg,Fe)_3Al_6(BO_3)_3(Si_6O_{18})(OH)_4$

Crystal system and class: Trigonal; $3m$.
Axial elements: $a:c$ = 1:0.448.
Cell dimensions and content: a = 15.8–16.0, c = 7.1–7.2; Z = 3.
Common forms and angles: (Figs. 15-41, 15-42):

(0001) $\wedge$ (10$\bar{1}$1) = 27° 20′   (02$\bar{2}$1) $\wedge$ ($\bar{2}$021) = 77° 0′

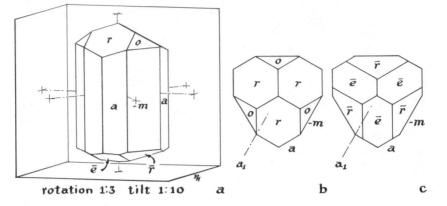

rotation 1:3   tilt 1:10   a   b   c

Figure 15-41 Forms in a crystal of tourmaline showing hemimorphism. (a) General view. (b) Terminal faces, upper end. (c) Terminal faces, lower end; $r\{10\bar{1}1\}$, $r\{0\bar{1}\bar{1}1\}$, $-m\{01\bar{1}0\}$, $a\{11\bar{2}0\}$, $o\{20\bar{2}1\}$, $e\{10\bar{1}2\}$.

$$(0001) \wedge (02\bar{2}1) = 45° 57' \qquad (32\bar{5}1) \wedge (\bar{3}5\bar{2}1) = 66° 1'$$
$$(10\bar{1}1) \wedge (\bar{1}101) = 46° 52' \qquad (32\bar{5}1) \wedge (5\bar{3}\bar{2}1) = 42° 36'$$

Habit: Generally in prismatic crystals, often with rounded triangular cross-section due to the predominance of the triangular prism $\{10\bar{1}0\}$ or $\{01\bar{1}0\}$; prism faces usually vertically striated; doubly terminated crystals often hemimorphic; commonly in parallel or radiating groups of columnar to acicular crystals.

Cleavage: $\{11\bar{2}0\}$, $\{10\bar{1}1\}$, poor.

Hardness: $7\frac{1}{2}$.

Density: 3.0–3.2, increasing with Fe content.

Color: Usually black; also brown, dark blue, colorless (iron-free varieties); lithium-bearing varieties pink, green, and blue, commonly zoned, also transparent and of gem quality.

Streak: White.

Luster: Vitreous.

Optical properties: $\omega = 1.635–1.675$, $\epsilon = 1.610–1.650$; $\omega - \epsilon = 0.02–0.03$; negative; strongly pleochroic in shades of blue, gray, and brown, except in iron-free varieties.

Other properties: Strongly pyroelectric and piezoelectric.

Chemistry: Tourmaline varies greatly in composition: Mg and Fe are completely interchangeable, and Mn, Ti, and Cr may also be present; Al may be partly replaced by Fe^3; Na may be replaced by Ca; Li is often present; and some of the OH may be replaced by F.

Diagnostic features: The color, triangular cross section, and striation of the prism faces are useful diagnostic features of tourmaline.

Occurrence: Tourmaline, both the common black variety and the colored lithium-bearing types, occurs in granite pegmatites; it is also common as an accessory mineral in metamorphic rocks, especially schists and gneisses; brown, Mg-rich tourmaline is found in metamorphosed limestones. It also occurs in high-temperature metalliferous veins.

Figure 15-42 Tourmaline crystal, black, showing typical striations parallel to *c*, trigonal outline, and trigonal pyramid {10$\bar{1}$1} termination. (Tongafero, Madagascar.) [Courtesy Royal Ontario Museum.]

Uses: The piezoelectric property of tourmaline is utilized industrially in the construction of pressure gauges to measure blast pressures in air and under water. Gem tourmalines have been mined in the USA from pegmatites in New England and southern California; the major producer is the state of Minas Gerais in Brazil.

Subclass Sorosilicates

In the sorosilicates two SiO_4 tetrahedra are linked by sharing one oxygen, thereby forming discrete $(Si_2O_7)^{-6}$ groups in the structure. A few simple sorosilicates exist as minerals. Structures containing both Si_2O_7 and independent SiO_4 groups also occur, the most important being idocrase and the minerals of the epidote group.

Lawsonite, $CaAl_2Si_2O_7(OH)_2 \cdot H_2O$

Crystal system and class: Orthorhombic; 222.
Axial elements: $a:b:c = 1.545:1:2.314$.
Cell dimensions and content: $a = 8.90$, $b = 5.76$, $c = 13.33$; $Z = 4$.

Habit: Prismatic crystals, sometimes tabular parallel to c; also massive, granular.

Cleavage: {100}, {001}, perfect; {110}, poor.

Hardness: 8.

Density: 3.09.

Color: White, pale blue, pale gray.

Streak: White.

Luster: Vitreous to greasy.

Optical properties: $\alpha = 1.665$, $\beta = 1.675$, $\gamma = 1.686$; $\gamma - \alpha = 0.021$; positive, $2V = 80°$.

Chemistry: The composition of lawsonite is similar to that of anorthite, but the structure is quite different, being much more closely packed (note the higher density and superior hardness of lawsonite). The structure consists of chains of six-coordinated aluminum-oxygen (and hydroxyl) groups, linked sideways by the Si_2O_7 groups; in the framework so formed there are "holes" occupied by the calcium ions and water molecules.

Diagnostic features: The hardness, comparatively high density, and common association with glaucophane are diagnostic for lawsonite.

Occurrence: Lawsonite is a mineral of metamorphic rocks, and typically occurs in glaucophane schists. It is widely distributed in these rocks in California (it was first recognized in 1895 in the glaucophane schists of the Tiburon Peninsula, in San Francisco Bay). It occurs in similar rocks in Italy, Corsica, and New Caledonia.

Hemimorphite, $Zn_4Si_2O_7(OH)_2 \cdot H_2O$

Crystal system and class: Orthorhombic; *mm2*.

Axial elements: $a:b:c = 0.7808:1:0.4776$.

Cell dimensions and content: $a = 8.370$, $b = 10.719$, $c = 5.120$; $Z = 2$.

Habit: Crystals usually thin tabular parallel to {010}; also massive, often in stalactitic or mammillary forms.

Cleavage: {110}, perfect.

Hardness: 5.

Density: 3.4–3.5.

Color: White, sometimes stained brown (with iron), or blue or green (with copper).

Streak: White.

Luster: Vitreous.

Optical properties: $\alpha = 1.614$, $\beta = 1.617$, $\gamma = 1.636$; $\gamma - \alpha = 0.022$; positive, $2V = 46°$.

Other properties: Strongly pyroelectric and piezoelectric.

Chemistry: On heating it decomposes into willemite at about 240°C, which fixes an upper temperature limit in hemimorphite deposits.

Diagnostic features: Soluble in HCl, giving gelatinous silica on partial evaporation. In the closed tube it decrepitates and gives off water.

Occurrence: In the oxidized zone of zinc deposits, often associated with smithsonite. Fine specimens have been found in Franklin, New Jersey; Leadville, Colorado; Santa Eulalia and Mapimi, Mexico; and Broken Hill, Zambia.

Use: A minor ore of zinc; the pure mineral contains 54 percent Zn.

Name: This mineral was originally known as *calamine,* which name however has also been used for zinc carbonate. The name hemimorphite was proposed in 1853 from the hemimorphic nature of the crystals, and this name has been adopted by international agreement to eliminate the confusion caused by the dual application of calamine.

Idocrase, $Ca_{10}Mg_2Al_4(Si_2O_7)_2(SiO_4)_5(OH)_4$

Crystal system and class: Tetragonal, $4/m\ 2/m\ 2/m$.
Axial elements: $a:c = 1:0.757$.
Cell dimensions and content: $a = 15.66$, $c = 11.85$; $Z = 4$.

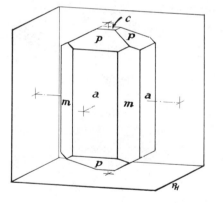

Figure 15-43 Typical forms in idocrase crystals: $c\{001\}$, $a\{100\}$, $m\{110\}$, $p\{101\}$.

Common forms and angles: (Figs. 15-43, 15-44):

(001) ∧ (112) = 28° 11′	(101) ∧ (011) = 50° 32′
(001) ∧ (101) = 37° 08′	(211) ∧ (121) = 31° 36′
(001) ∧ (301) = 66° 14′	

Habit: Prismatic or pyramidal crystals; also massive, granular, or compact.
Cleavage: {110}, poor.
Hardness: 7.
Density: 3.3–3.5.

Color: Usually brown or green, sometimes yellow, occasionally blue (var. *cyprine*).

Streak: White.

Luster: Vitreous, often somewhat resinous.

Optical properties: ω = 1.703–1.752, ϵ = 1.700–1.746; $\omega - \epsilon$ = 0.001–0.008; negative; may be weakly pleochroic in yellow and brown.

Chemistry: Idocrase shows considerable variation in composition; some Ca may be replaced by Na and K, Mg by Fe^2 and Mn, and Al by Fe^3 and Ti; some varieties contain Be.

Figure 15-44   Idocrase crystals, greenish gray, showing difference in habit from one locality. Forms: {100}, {110}, {101}, {001}, {211} (one small face). (Wilui River, Siberia, the USSR.) [Courtesy Royal Ontario Museum.]

Diagnostic features: When in crystals idocrase is easily recognized, but massive varieties resemble closely some forms of garnet; it can be distinguished by its ready fusibility and intumescence in the blowpipe flame, and its optics.

Occurrence: Idocrase is formed by the contact metamorphism of impure limestones, and is commonly associated with calcite, lime garnet (grossularite or andradite), and wollastonite. Fine crystals have been found at Eden Mills, Vermont, and near Olmsteadville, New York; it also occurs at Magnet Cove, Arkansas; Crestmore, California; and at many localities abroad.

Name: The name *vesuvianite* is also used for this mineral; it alludes to the original discovery of the mineral in ejected limestone blocks on Mount Vesuvius.

EPIDOTE GROUP

The epidote group comprises several minerals with the general formula $X_2Y_3Si_3O_{12}(OH)$; X is Ca, partly replaced by rare-earth elements (**R**) in allanite; Y is Al and Fe^3, partly replaced by Mg and Fe^2 in allanite, and by Mn^3 in piemontite. The commoner species are:

Zoisite	$Ca_2Al_3Si_3O_{12}(OH)$	Orthorhombic
Clinozoisite	$Ca_2Al_3Si_3O_{12}(OH)$	Monoclinic

Epidote $Ca_2(Al,Fe)_3Si_3O_{12}(OH)$ Monoclinic

Allanite $(Ca,\mathbf{R})_2(Al,Fe,Mg)_3Si_3O_{12}(OH)$ Monoclinic

The simplest composition $Ca_2Al_3Si_3O_{12}(OH)$ is thus dimorphous, with the orthorhombic zoisite and the monoclinic clinozoisite; all other species in the group are monoclinic.

The structure of minerals of the epidote group is a mixed type, containing both Si_2O_7 and SiO_4 groups; the formula of clinozoisite can be written $Ca_2Al_3O(SiO_4)(Si_2O_7)(OH)$, the odd oxygen and the OH group being attached to the Al atoms. The structure is best described in relation to the continuous chains of AlO_6 and $AlO_4(OH)_2$, which are parallel to the b axis (hence the elongation of the crystals in this direction). These chains are formed by the arrangement of oxygens and hydroxyls in octahedra around the Al ions, these octahedra having edges in common. Single SiO_4 and double Si_2O_7 tetrahedra bridge the (Al–O,OH) chains and share some of the oxygens. Outside the chains and completing the structure are Al and Fe ions (also in six-fold coordination), and Ca ions, each of which is surrounded irregularly by eight oxygens.

Minerals of the epidote group characteristically occur in metamorphic rocks (except allanite, which is found in granites and granite pegmatites). In this connection it is interesting to note that the composition of zoisite and clinozoisite is very similar to that of anorthite. However, the densities are very different, reflecting the different atomic packing in the different structures (zoisite and clinozoisite, $G = 3.2$–3.3; anorthite, $G = 2.76$). This is also reflected in the general absence of anorthite-rich plagioclase in low- and medium-grade schists and gneisses; instead we find albite-rich plagioclase and an epidote-group mineral. Evidently under the conditions of moderate temperature and high pressure of low- and medium-grade metamorphism the epidote-group minerals are more stable than anorthite.

Zoisite, $Ca_2Al_3Si_3O_{12}(OH)$

Crystal system and class: Orthorhombic; $2/m\ 2/m\ 2/m$.

Axial elements: $a:b:c = 2.879:1:1.791$.

Cell dimensions and content: $a = 16.24$, $b = 5.58$, $c = 10.10$; $Z = 4$.

Habit: Aggregates of long prismatic crystals parallel to b, striated along b, seldom with terminations.

Cleavage: {001}, perfect.

Hardness: $6\frac{1}{2}$.

Density: 3.3.

Color: Usually gray, sometimes pink (var. *thulite*) or apple green.

Streak: White.

Luster: Vitreous, pearly on cleavage surfaces.

Optical properties: $\alpha = 1.685-1.705$, $\beta = 1.688-1.710$, $\gamma = 1.697-1.725$; $\gamma - \alpha = 0.004-0.008$; positive, $2V = 0°-60°$.

Chemistry: Its composition is usually close to the formula, the only substitution being a small amount (up to about 10 atom percent) of Fe^3 for Al.

Diagnostic features: Zoisite may resemble some of the amphiboles in appearance and mode of occurrence, but it can be distinguished by the single perfect cleavage, and its optics.

Occurrence: Zoisite occurs as a constituent of schists and gneisses, but less commonly than its dimorph clinozoisite, and in quartz veins; the material known as *saussurite* (altered plagioclase in basic igneous rocks) is a mixture of albite and zoisite or clinozoisite.

Clinozoisite, $Ca_2Al_3Si_3O_{12}(OH)$; Epidote, $Ca_2(Al,Fe)_3Si_3O_{12}(OH)$

Crystal system and class: Monoclinic; $2/m$.
Axial elements: $a:b:c = 1.592:1:1.812$; $\beta = 115° 24'$.
Cell dimensions and content: $a = 8.98$, $b = 5.64$, $c = 10.22$; $Z = 2$.

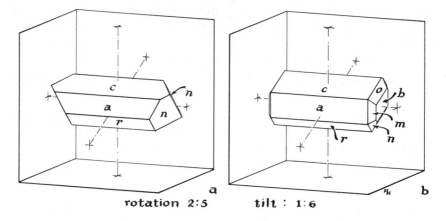

rotation 2:5   tilt : 1:6

Figure 15-45   Typical forms in epidote crystals: $c\{001\}$, $a\{100\}$, $r\{\bar{1}01\}$, $n\{\bar{1}11\}$, $m\{110\}$, $o\{011\}$.

Common forms and angles: (Fig. 15-45):

(100) $\wedge$ (110) = 55° 11'   (001) $\wedge$ (011) = 58° 35'
(001) $\wedge$ (100) = 64° 36'   (001) $\wedge$ ($\bar{1}$11) = 75° 10'
(001) $\wedge$ ($\bar{1}$01) = 63° 32'   (100) $\wedge$ (10$\bar{1}$) = 51° 52'

Habit: Crystals elongated parallel to b; also massive, fibrous or granular.
Cleavage: $\{001\}$, perfect.
Hardness: 7.
Density: 3.3–3.6, increasing with Fe content.

Color: Pale green or greenish gray in clinozoisite, yellowish to brownish green to black in epidote.

Streak: White or grayish white.

Luster: Vitreous.

Optical properties: $\alpha = 1.67–1.75$, $\beta = 1.68–1.78$, $\gamma = 1.69–1.80$; $\gamma - \alpha = 0.005–0.05$ (birefringence and refractive indices increase with the Fe content); clinozoisite is positive, $2V = 15°–90°$, epidote negative, $2V = 60°–90°$; epidote is pleochroic, yellow to green.

Chemistry: Clinozoisite and epidote are in effect a single species, the boundary between them being arbitrarily drawn at a replacement of 10 atom percent Al by Fe; the limit of replacement of Al by Fe is about one-third, equivalent to a formula $Ca_2Al_2FeSi_3O_{12}(OH)$.

Diagnostic features: Epidote can usually be readily recognized by its characteristic yellowish green (pistachio) color, seldom seen in other minerals; before the blowpipe it fuses with intumescence to a black scoriaceous glass, usually magnetic.

Occurrence: Clinozoisite and epidote are important constituents of low and medium grade metamorphic rocks, often being the principal calcium aluminum silicates in these rocks. Epidote frequently occurs in large amount in contact-metamorphosed limestones, especially in association with iron ores; it is also associated with zeolites in cavities in basalts. Very fine crystals of epidote have come from the Untersulzbachtal, in the Austrian Tyrol, and from Prince of Wales Island, Alaska.

Related species: Piemontite is a variety of epidote in which some of the Al is replaced by Mn^3; trivalent manganese is a strong chromophore, and piemontite has a violet-red color easily recognized even when the mineral is present in very small amounts, as in some schists and quartzites.

Allanite, $(Ca,R)_2(Al,Fe,Mg)_3Si_3O_{12}(OH)$

Crystal system and class: Monoclinic; $2/m$.

Axial elements: $a:b:c = 1.562:1:1.779$; $\beta = 115° 0'$.

Cell dimensions and content: $a = 8.98$, $b = 5.75$, $c = 10.23$; $Z = 2$.

Habit: Crystals tabular parallel to $\{100\}$, or long and slender, elongated parallel to b; also massive and as embedded grains.

Cleavage and fracture: $\{001\}$, seldom observable; conchoidal.

Hardness: $5\frac{1}{2}–6\frac{1}{2}$.

Density: 3.3–4.0.

Color: Usually black, sometimes dark brown.

Streak: Gray brown.

Luster: Vitreous, sometimes pitchy.

Optical properties: α = 1.69–1.79, β = 1.70–1.81, γ = 1.71–1.83; $\gamma - \alpha$ = 0.01–0.04; usually negative, $2V$ = 40°–90°; for metamict allanite n ranges from 1.54 to 1.72.

Chemistry: The composition is variable; in addition to the elements indicated in the formula it usually contains some thorium, up to 3 percent having been recorded; beryllium, sodium, and potassium have also been recorded in some analyses. Allanite is commonly metamict, presumably as a result of radiation damage caused by the radioactive decay of thorium. The wide variation in density is mainly due to this loss of crystallinity and an accompanying adsorption of water.

Diagnostic features: Allanite can usually be recognized by its mode of occurrence and by its weak radioactivity.

Occurrence: Allanite is not uncommon as an accessory mineral in small grains in granitic rocks; good crystals and large masses are found in some granite pegmatites.

Name: The name *orthite* is commonly used for this mineral in Europe, although allanite has priority (allanite was applied to this mineral in 1811, orthite in 1818).

Subclass Nesosilicates

Nesosilicates are those silicates with isolated SiO_4^{-4} groups in the structure. Under the obsolete silicic acid theory they were considered to be salts of orthosilicic acid, H_4SiO_4, and called "orthosilicates." It is noteworthy that in the nesosilicates Al seldom substitutes for Si, so their formulas generally correspond strictly to (SiO_4) or a multiple thereof. Another feature of the nesosilicates is the absence of compounds with the alkali elements.

All the nesosilicates have notably dense atomic packing, and their physical properties reflect this. They are relatively harder and have a higher density than corresponding compounds of other structure types. The absence of chains and sheets is reflected in the generally equidimensional nature of their crystals.

Olivine, $(Mg,Fe)_2SiO_4$

Crystal system and class: Orthorhombic; $2/m\ 2/m\ 2/m$.

Axial elements: $a:b:c$ = 0.467:1:0.586 (Mg_2SiO_4).

Cell dimensions and content: a = 4.76, b = 10.20, c = 5.98; (Mg_2SiO_4) Z = 4. $(a$ = 4.82, b = 10.48, c = 6.11, fayalite, $Fe_2SiO_4)$.

Common forms and angles:

  (110) $\wedge$ (1$\bar{1}$0) = 50° 04'   (021) $\wedge$ (0$\bar{2}$1) = 99° 4'

$$(120) \wedge (\bar{1}20) = 93° 54' \qquad (111) \wedge (1\bar{1}1) = 40° 8'$$
$$(101) \wedge (\bar{1}01) = 102° 54' \qquad (121) \wedge (1\bar{2}1) = 72° 18'$$

Habit: Rarely as crystals; usually in granular masses, and as rounded grains in igneous rocks.

Cleavage: {010}, indistinct.

Hardness: $6\frac{1}{2}$.

Density: 3.22 (Mg_2SiO_4) to 4.39 (Fe_2SiO_4) (Fig. 3-1); common olivine about 3.3–3.4.

Color: Usually olive-green (hence the name), also white (forsterite) and brown to black (fayalite).

Streak: White or gray.

Luster: Vitreous.

Optical properties: $\alpha = 1.635–1.827$, $\beta = 1.651–1.869$, $\gamma = 1.670–1.879$ (Fig. 15-46); $\gamma - \alpha = 0.035–0.052$; $2V$ ranges from positive, 82°, to negative, 44°. Common olivine has $\alpha \sim 1.64$, $\beta \sim 1.67$, $\gamma \sim 1.69$, $2V$ near 90°.

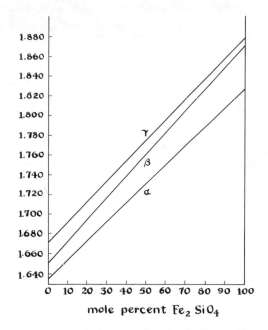

mole percent Fe_2SiO_4

Figure 15-46 Variation of refractive indices with composition in the olivine series.

Chemistry: The olivine series is an example of continuous solid solution of two components, Mg_2SiO_4 and Fe_2SiO_4; three names are currently used: *forsterite* for pure or nearly pure Mg_2SiO_4, *fayalite* for pure or nearly pure Fe_2SiO_4, and *olivine* for the common intermediate varieties. Forsterite and olivine are incompatible with free silica, reacting with it to give pyroxene, and therefore olivine and quartz cannot crystallize together in a rock; fayalite, however, does not react in this way, and consequently occurs occasionally in granites and rhyolites.

Alteration: Olivine alters readily; hydrothermal alteration generally results in the formation of serpentine, surface or near-surface alteration in the oxidation of the iron, and the removal of the magnesium and silica, leaving a brown or red-brown pseudomorph consisting of goethite or hematite.

Diagnostic features: The color and generally granular nature serve to distinguish olivine from other rock-forming silicates. Olivine dissolves easily in hot 1:1 HCl, with the separation of gelatinous silica.

Occurrence: Olivine is typically a mineral of basic and ultrabasic igneous rocks; sometimes it forms major rock masses (dunite); basalts occasionally contain nodules of granular olivine, probably inclusions derived from dunite masses at depth. Olivine is a common mineral of stony and stony-iron meteorites. Forsterite is formed by the metamorphism of dolomitic limestone.

Uses: Fayalite melts at 1205°C, forsterite at 1890°C; magnesium-rich olivine thus has a very high melting point, and is used in the manufacture of refractory bricks. Transparent olivine of good color can be cut into attractive gemstones (*peridot*); good material of this kind comes from St. John's Island in the Red Sea, and from Burma.

Related species: There are a number of minerals isomorphous with olivine, all silicates of bivalent elements; they are of rare occurrence except for *monticellite*, $CaMgSiO_4$, which is sometimes found in metamorphosed limestones.

Willemite, Zn_2SiO_4

Crystal system and class: Trigonal; $\bar{3}$.

Axial elements: $a:c = 1:0.668$.

Cell dimensions and content: $a = 13.94$, $c = 9.31$; $Z = 18$.

Habit: Small prismatic or rhombohedral crystals; usually massive, coarse to fine granular.

Cleavage: {0001}, good.

Hardness: $5\frac{1}{2}$.

Density: 3.9–4.2.

Color: Variable; white, yellow, green, reddish brown, sometimes black.

Streak: White.

Luster: Vitreous, occasionally resinous.

Optical properties: $\omega = 1.69$–1.72, $\epsilon = 1.72$–1.73 (indices increase with the Mn content); $\epsilon - \omega = 0.02$–0.03; positive.

Chemistry: A considerable part of the zinc may be replaced by manganese, and a small amount of iron is often present.

Other properties: Willemite from Franklin, New Jersey, shows a magnificent green fluorescence in ultraviolet light; this fluorescence is activated by the partial substitution of Mn for Zn, and Mn-free willemite is not fluorescent.

Diagnostic properties: The associations and fluorescence characterize the willemite from Franklin. Willemite is soluble in HCl with the separation of gelatinous silica.

Occurrence: Willemite occurs abundantly in the unique metamorphic ore deposit at Franklin, New Jersey, where it is a major zinc ore. Here it occurs with franklinite and zincite in a green, black, and red mixture, and also as large coarse crystals embedded in crystalline limestone. It is not uncommon in the oxidized zone of zinc-bearing deposits, especially in arid regions, but is often inconspicuous and easily overlooked.

Use: A minor ore of zinc.

ALUMINUM SILICATE GROUP

This group can be considered as including the three polymorphs of Al_2SiO_5 (andalusite, sillimanite, and kyanite), together with topaz, $Al_2SiO_4(OH,F)_2$, and staurolite, whose formula can be written $2Al_2SiO_5 \cdot Fe(OH)_2$ and whose structure resembles that of kyanite.

The structures of the three polymorphs of Al_2SiO_5 are rather closely related. All of them have aluminum-oxygen chains parallel to the c axis, these chains consisting of octahedral groups of six oxygens around each Al, each group sharing an edge with the group on either side. The c repeat represents the length occupied by two octahedral groups, and is thus nearly the same (5.6–5.7 Å) in all three minerals. In sillimanite the Al–O chains are crosslinked by Si and Al, both in four-coordination. In kyanite the lateral bonding is by Si in four-coordination and Al in six-coordination. In andalusite, the cross-linking is by Si in four-coordination and Al in five-coordination. This is the only mineral in which this peculiar coordination of Al has been found; three O atoms surround an Al in one plane, and the other two oxygens are, respectively, above and below; i.e., a trigonal dipyramid of O around the Al. Thus the structural difference between the three minerals consists in the coordination around one of the Al atoms of Al_2SiO_5; it is four in sillimanite, five in andalusite, and six in kyanite. The other Al is always in six-coordination, the Si in four-coordination.

Until recently andalusite, sillimanite, and kyanite had not been made in the laboratory, and experimental data on their stability fields have been lacking. Kyanite is much denser than the other two, and hence its formation is favored by high pressures. Andalusite is typically formed by the contact meta-

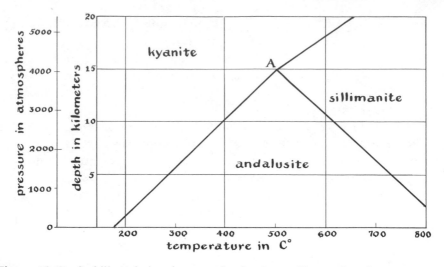

Figure 15-47   Stability relations between the aluminum silicate minerals.

morphism of argillaceous rocks. Sillimanite is characteristic of high-grade metamorphic rocks. On the basis of laboratory data an equilibrium diagram for the different polymorphs has been drawn up and is shown in Fig. 15-47. The triple point, *A*, at which kyanite, sillimanite, and andalusite can exist together in equilibrium, is at about 500°C.

On heating to 1300°C kyanite, sillimanite, andalusite, and topaz decompose into *mullite*, $Al_6Si_2O_{13}$, and a silica-rich glass. Mullite is occasionally found as a mineral where aluminum-rich rocks have been subjected to very high temperatures, such as where fragments of shale have been picked up by a lava flow. Mullite is used in industry as a refractory, for which purpose it is made by heating kyanite or andalusite.

Andalusite, Al_2SiO_5

Crystal system and class: Orthorhombic; $2/m\ 2/m\ 2/m$.
Axial elements: $a:b:c = 0.982:1:0.703$.
Cell dimensions: $a = 7.78$, $b = 7.92$, $c = 5.57$; $Z = 4$.
Habit: Coarse prismatic crystals with a nearly square cross section, the prism angle being 89° 12′; also massive. Crystals often have carbonaceous inclusions, arranged in shapes conforming to the symmetry of the mineral (var. *chiastolite*, Fig. 15-48).
Cleavage: {110}, distinct.
Hardness: $7\frac{1}{2}$.
Density: 3.15.

Color: Variable; white, gray, rose red, brown, sometimes green (var. *viridine*).

Streak: White.

Luster: Vitreous, often dull.

Optical properties: $\alpha = 1.629-1.640$, $\beta = 1.633-1.644$, $\gamma = 1.638-1.650$; $\gamma - \alpha = 0.01$; positive, $2V = 73°-86°$; sometimes weakly pleochroic in yellow and pink.

Diagnostic features: The crystal form and mode of occurrence in metamorphosed shales are characteristic.

Alteration: Paramorphs of kyanite and sillimanite after andalusite have been recorded, and andalusite also alters readily to sericite.

Occurrence: Andalusite occurs mainly in contact-metamorphosed shales; it is occasionally found in regionally metamorphosed rocks, and has been recorded from pegmatites.

Production and uses: Andalusite is used in the manufacture of mullite refractories, especially spark plugs; it has been mined at White Mountain, Mono

Figure 15-48 Andalusite, var. chiastolite, showing characteristic regular arrangement of inclusions. (Madera County, California.) [Courtesy American Museum of Natural History.]

County, California, and considerable tonnages are also produced from andalusite-bearing sands in western Transvaal, South Africa.

Sillimanite, Al_2SiO_5

Crystal system and class: Orthorhombic; $2/m\ 2/m\ 2/m$.

Axial elements: $a:b:c = 0.975:1:0.752$.

Cell dimensions and content: $a = 7.47$, $b = 7.66$, $c = 5.76$; $Z = 4$.

Habit: Usually in finely fibrous or coarse prismatic masses. The fibrous nature is due to the presence of chains of aluminum-oxygen groups parallel to the c axis.

Cleavage: {010}, perfect.

Hardness: 7.

Density: 3.24.

Color: White, sometimes brownish or greenish.

Streak: White.

Luster: Vitreous, often silky in fibrous material.

Optical properties: $\alpha = 1.657$, $\beta = 1.660$, $\gamma = 1.677$; $\gamma - \alpha = 0.020$; positive. $2V = 25°$.

Diagnostic features: Sillimanite resembles other fibrous silicates such as wollastonite and tremolite, but may be distinguished by its infusibility and insolubility in acids, and its refractive indices.

Occurrence: Sillimanite is formed in aluminum-rich rocks under conditions of high grade regional metamorphism; and is found in schists and gneisses. Workable deposits of sillimanite are rarer than andalusite or kyanite, and hence it has been little used in the manufacture of mullite.

Kyanite, Al_2SiO_5

Crystal system and class: Triclinic; $\bar{1}$.

Axial elements: $a:b:c = 0.917:1:0.720$; $\alpha = 90°\ 6'$, $\beta = 101°\ 2'$, $\gamma = 105°\ 45'$.

Cell dimensions and content: $a = 7.10$, $b = 7.74$, $c = 5.57$; $Z = 4$.

Habit: Crystals {100} tablets elongated in the c direction, seldom terminated; also as bladed masses. Crystals are distinctly flexible and often bent or twisted.

Cleavage: {100}, perfect; {010}, good; parting on {001}.

Hardness: Variable; on {100} 4–5 parallel to c, 6–7 parallel to b.

Density: 3.63.

Color: Characteristically patchy blue, also green, white, or gray.

Streak: White.

Luster: Vitreous, sometimes pearly on cleavage surfaces.

Optical properties: $\alpha = 1.713$, $\beta = 1.720$, $\gamma = 1.728$; $\gamma - \alpha = 0.015$; negative, $2V = 83°$; may be weakly pleochroic in shades of blue.

Diagnostic features: The color and habit usually serve to characterize kyanite.

Occurrence: Kyanite is the product of medium-grade regional metamorphism of aluminum-rich rocks, and it occurs in schists and gneisses and in quartz veins and pegmatites cutting these rocks.

Production and uses: Kyanite is important commercially for the manufacture of mullite refractories, and is mined in India and in Kenya; there is some production from the metamorphic belt in Virginia, South Carolina, and Georgia.

Name: Kyanite is sometimes spelled *cyanite;* it is also known as *disthene,* a name derived from its variable hardness.

Staurolite, $Fe_2Al_9O_6(SiO_4)_4(O,OH)_2$

Crystal system and class: Monoclinic (pseudo-orthorhombic); $2/m$.

Axial elements: $a:b:c = 0.471:1:0.340$; $\beta = 90°$.

Cell dimensions and content: $a = 7.83$, $b = 16.62$, $c = 5.65$; $Z = 4$.

Common forms and angles:

   $(110) \wedge (1\bar{1}0) = 50° 26'$ $(001) \wedge (201) = 55° 17'$

Habit: Prismatic crystals with $\{001\}$ and $\{010\}$, commonly twinned (Figs. 15-49, 15-50; Fig. 2-44).

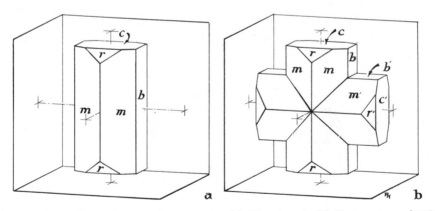

Figure 15-49 Forms in staurolite crystals. (a) Untwinned. (b) Twinned on $\{031\}$, $c\{001\}$, $b\{010\}$, $m\{110\}$, $r\{201\}$.

Twinning: Cruciform twins common; twin plane $\{031\}$, giving a rectangular cross, and twin plane $\{231\}$, giving a cross at 60°.

Cleavage: $\{010\}$, distinct.

Figure 15-50 Staurolite, grayish brown. Forms: {110}, {010}, {201}. (a) Single crystal, {110} parallel to page, *c* vertical. (b) Twin crystal, {031} twin plane. (c) Twin crystal with {231} twin plane. (Blue Ridge, Georgia.) [Courtesy Queen's University.]

Hardness: 7.

Density: 3.7–3.8.

Color: Brown.

Streak: Gray.

Luster: Vitreous or somewhat resinous.

Optical properties: $\alpha = 1.739–1.747$, $\beta = 1.745–1.753$, $\gamma = 1.752–1.761$; $\gamma - \alpha = 0.012–0.014$; positive, $2V$ near $90°$; pleochroic from colorless to golden yellow.

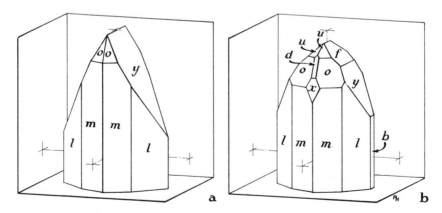

Figure 15-51 Typical forms in topaz crystals: $c\{001\}$, $b\{010\}$, $m\{110\}$, $l\{120\}$, $y\{021\}$, $d\{101\}$, $x\{201\}$, $f\{011\}$, $o\{111\}$, $u\{112\}$.

Chemistry: Some of the ferrous iron may be replaced by magnesium and manganese, some of the aluminum by ferric iron. The structure of staurolite is closely related to that of kyanite, and consists of layers with the kyanite structure alternating with layers containing the iron atoms and the hydroxyl groups. Parallel crystallization of kyanite and staurolite is sometimes observed.

Diagnostic features: The color and habit usually serve to identify staurolite.

Occurrence: Staurolite is characteristic of medium grade metamorphic rocks, and is found in aluminum-rich schists and gneisses.

Use: The crossed twins found in Virginia and elsewhere are sometimes known as fairy crosses and are sold as good luck charms; imitations made of baked brown clay are occasionally sold.

Topaz, $Al_2SiO_4(OH,F)_2$

Crystal system and class: Orthorhombic; *mm*2 or 222.

Axial elements: $a:b:c = 0.528:1:0.955$.

Cell dimensions and content: $a = 4.65$, $b = 8.80$, $c = 8.40$; $Z = 4$.

Common forms and angles: (Fig. 15-51):

$(010) \wedge (120) = 43° 26'$	$(001) \wedge (011) = 43° 41'$
$(010) \wedge (110) = 62° 10'$	$(001) \wedge (021) = 62° 22'$
$(001) \wedge (121) = 69° 12'$	$(001) \wedge (111) = 63° 56'$

Habit: In well-developed prismatic crystals, occasionally very large (up to several hundred pounds weight); also massive, coarse to fine granular.

Cleavage: {001}, perfect.

Hardness: 8.

Density: 3.5–3.6, increasing with F content.

Color: Colorless or white, also transparent pale blue, yellow, yellow-brown; rarely pink.

Streak: White.

Luster: Vitreous.

Optical properties: $\alpha = 1.606–1.629$, $\beta = 1.609–1.631$, $\gamma = 1.616–1.638$ (the refractive indices increase with the OH content); $\gamma - \alpha = 0.01$; positive, $2V = 50°–70°$.

Other properties: Slightly pyroelectric and piezoelectric.

Diagnostic features: Topaz in crystals is readily recognizable by its habit, its hardness, and its single perfect cleavage; massive topaz is known by its density, and by tests for aluminum and fluorine, and by its optical properties.

Occurrence: Topaz occurs in pegmatites and high temperature quartz veins, and in cavities in granites and rhyolites. Clear crystals a foot and more across are found in Minas Gerais, Brazil. In spite of its good cleavage topaz sometimes survives a moderate degree of transportation and occurs as rounded pebbles in alluvial deposits.

Use: Transparent topaz is cut as a gemstone, but a great deal of brown and yellow quartz is also sold under this name.

GARNET GROUP

Crystal system and class: Isometric; $4/m\ \bar{3}\ 2/m$.

Habit: Commonly in crystals, the crystal forms being characteristic, usually dodecahedrons {110} or trapezohedrons {211}, or combinations of these, rarely combinations of dodecahedron and hexoctahedron {321} (Figs. 15-52, 15-53); also massive, coarse to fine granular.

Structure: The unit cell of garnet contains eight formula units. The [SiO$_4$] tetrahedra are present as independent groups linked to [AlO$_6$] octahedra,

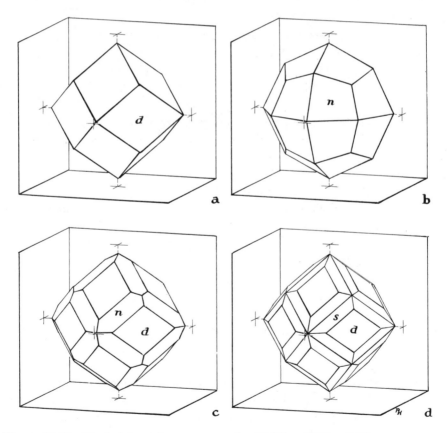

Figure 15-52   Typical forms in garnet crystals: $d\{110\}$, $n\{211\}$, $s\{321\}$.

Figure 15-53 Garnet, dark red and brownish black. (a) Dodecahedron {110}. (b) Dodecahedron beveled by trapezohedron {211}. (Fort Wrangel, Alaska.) (c) Trapezohedron {211}. (Burma.) (d) Trapezohedron {211} with minor dodecahedron faces. (Burma.) In contrast to (a) and (b), (c) and (d) have a_1 turned 45° to the right. [(a) and (b) courtesy Queen's University. (c) and (d) courtesy Royal Ontario Museum.]

and the divalent ions are situated in the interstices of this structure, each divalent ion being surrounded by eight oxygens; i.e., in 8-fold coordination. This is an unusual coordination for Fe^2 and Mg, and probably explains why pyrope can only be synthesized at high pressure, which favors high coordination.

Cleavage: None.

Hardness: $7-7\frac{1}{2}$.

Density: 3.6–4.3, depending on the composition.

Color: Variable, but dark red and reddish brown are common. Specifically, almandite is generally red or reddish brown; pyrope, red to nearly black; spessartine, orange to dark red or brown; grossular, white when pure, otherwise yellow, pink, green, or brown; andradite, yellow, greenish yellow, greenish brown, or black; uvarovite, a characteristic emerald green.

Streak: White, or a pale shade of the color of the specimen

Luster: Vitreous to resinous.

Optical properties: The refractive index for each of the pure components is given below; the index of a naturally occurring garnet will depend upon the relative amounts of the different components in the specimen. Grossular, spessartine, andradite, and uvarovite may show weak birefringence.

Chemistry: The garnets comprise a group of isomorphous minerals with the general formula $X_3Y_2(SiO_4)_3$ in which X may be Ca, Mg, Fe^2, or Mn^2, and Y may be Al, Fe^3, or Cr^3, sometimes in part Ti or Mn^3. The following names are used for the minerals and the components:

SPECIES	FORMULA	DENSITY	n	UNIT CELL EDGE
Almandite	$Fe_3Al_2(SiO_4)_3$	4.32	1.832	11.53
Pyrope	$Mg_3Al_2(SiO_4)_3$	3.56	1.723	11.46
Spessartine	$Mn_3Al_2(SiO_4)_3$	4.19	1.807	11.62
Grossular	$Ca_3Al_2(SiO_4)_3$	3.59	1.731	11.85
Andradite	$Ca_3Fe_2(SiO_4)_3$	3.86	1.886	12.05
Uvarovite	$Ca_3Cr_2(SiO_4)_3$	3.85	1.868	12.01

The composition of naturally occurring garnets seldom approaches the formulas given above, as a result of extensive atomic substitution, and the specific name applied is that of the component present in largest amount. Ferrous iron and magnesium are interchangeable, and a series of intermediate compositions is known between almandite and pyrope; similarly, a series of intermediate compositions exists between almandite and spessartine, and between grossular and andradite (Fig. 15-54). However, the difference in ionic size between calcium on the one hand and ferrous iron and magnesium

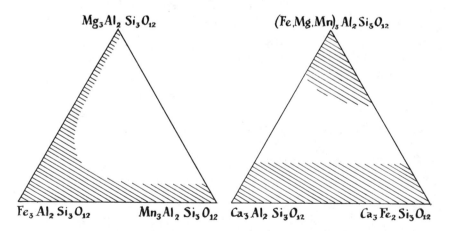

Figure 15-54 Mineral compositions in the garnet group. Shaded areas indicate compositional regions for natural garnets; unshaded areas indicate compositional regions unrepresented by natural garnets.

on the other results in rather limited substitution and a wide composition gap between grossular—andradite on the one hand, and almandite, pyrope, and spessartine on the other. There is also a wide composition gap between spessartine and pyrope, due, at least in part, to the difference in size of the magnesium and manganese ions, and possibly to the different geological environments in which spessartine and pyrope form.

Diagnostic features: The crystal form, color, and hardness usually serve to identify garnet; the color and mode of occurrence will often indicate which of the garnet species is present.

Occurrence: Garnets are typically minerals of metamorphic rocks, although they have been found in some igneous rocks. Specifically, they differ somewhat in their mode of occurrence, as follows (Fig. 15-55):

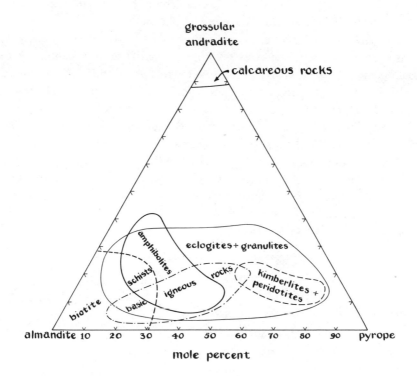

Figure 15-55 The limits of composition of garnet from various rock types.

Almandite: The common garnet of schists and gneisses is usually almandite. It is also recorded from granites, rhyolites, and pegmatites.

Pyrope: Less common than the other garnets (except uvarovite), pyrope occurs in ultrabasic igneous rocks and serpentines derived from them, and in high-grade magnesium-rich metamorphic rocks.

Spessartine: Garnets from granite pegmatites are often spessartine or intermediate between spessartine and almandite; spessartine also occurs in metamorphosed manganese-bearing rocks.

Grossular: Grossular is typically formed by the contact or regional metamorphism of impure limestones, and is thus often associated with calcite, wollastonite, and idocrase. Grossular commonly contains combined water, as a result of the partial substitution of $(OH)_4^{-4}$ for $(SiO_4)^{-4}$ in the structure.

Andradite: Andradite is formed by the metasomatic alteration of limestones by iron-bearing solutions, and commonly occurs associated with ore deposits in calcareous rocks.

Uvarovite: Uvarovite is rare, occurs in association with chromite in serpentine.

Garnet, being resistant to both mechanical and chemical breakdown, occurs as a detrital mineral in sands and sandstones.

Production and uses: Garnet has some value as an abrasive, being fairly hard and without cleavage, hence breaking into irregular grains. Although garnet is a common mineral, material suitable for this use is seldom found in workable quantity. The requirements are for large isolated crystals that are crushed to provide the garnet "sand" used to make sandpaper. Several thousand tons of such garnet are produced annually at Gore Mountain, in the Adirondack Mountains of New York.

Transparent unflawed garnet of good color can be cut into attractive gemstones. Much of the red garnet jewelry consists of pyrope from Czechoslovakia. Uvarovite would make a magnificent gemstone but does not occur in sufficiently large pieces; the green garnet gemstones are cut from a variety of andradite known as *demantoid*.

OTHER NESOSILICATES

Zircon, ZrSiO₄

Crystal system and class: Tetragonal; $4/m\ 2/m\ 2/m$.
Axial elements: $a\!:\!c = 1\!:\!0.9054$.
Cell dimensions and content: $a = 6.604$, $c = 5.979$; $Z = 4$.
Common forms and angles: (Fig. 15-56; Fig. 2-26a):

$(101) \wedge (011) = 56° 40'$	$(100) \wedge (101) = 47° 51'$
$(301) \wedge (031) = 83° 08'$	$(100) \wedge (301) = 20° 13'$
$(110) \wedge (211) = 31° 43'$	$(211) \wedge (121) = 32° 56'$

Habit: Prismatic crystals terminated by pyramids.
Twinning: Occasionally twinned, twin plane {112}, giving knee-shaped twins.
Cleavage: {110}, indistinct.

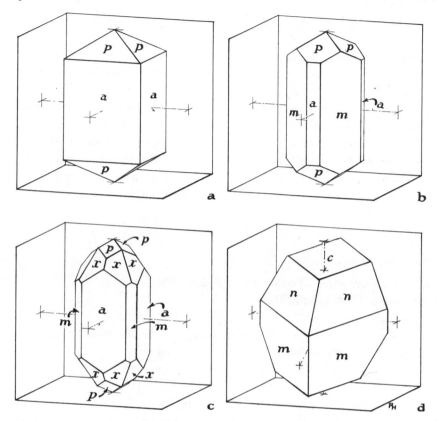

Figure 15-56   (a–c) Typical forms in zircon crystals: $a\{100\}$, $m\{110\}$, $p\{101\}$, $x\{211\}$.
(d) Typical form of sphene crystals: $c\{001\}$, $m\{110\}$, $n\{111\}$.

Hardness: $7\frac{1}{2}$.

Density: 4.6–4.7 when crystalline, decreasing to 3.9 when metamict.

Color: Usually brown or reddish brown, but can be colorless, gray, green, or violet; the transparent blue varieties used as gemstones are produced by heat treatment of natural zircon.

Streak: White.

Luster: Vitreous to adamantine.

Optical properties: $\omega = 1.92$–1.96, $\epsilon = 1.97$–2.00; $\epsilon - \omega = 0.04$–0.06 (non-metamict); in metamict zircons n ranges from 1.78 to 1.87.

Chemistry: In zircon some of the Zr is always replaced by Hf (generally about 1 percent, but 4 percent and more has been recorded). Part of the Zr can also be replaced by rare earths, coupled with the replacement of silicon by phosphorus. Zircon is frequently radioactive due to the presence of Th and U replacing Zr in the structure; as a result of radiation damage from these radioactive elements, these zircons are often metamict.

Diagnostic features: Habit, hardness, color, and density are useful distinguishing features of zircon.

Occurrence: Zircon is a common accessory mineral of igneous rocks and pegmatites of the granite, syenite, and nepheline syenite families. The presence of uranium and thorium makes it a useful mineral for age determination of such rocks. Because zircon is resistant to mechanical and chemical disintegration it appears as a detrital mineral in river and beach sands.

Production and uses: Zircon is the principal source of zirconium and hafnium for industry. It is extracted from sands, the most important source being beach deposits on the coast of Queensland in Australia. About 250,000 tons are produced annually, 200,000 tons in Australia, and nearly 30,000 tons from deposits in Florida.

Thorite, $ThSiO_4$

Crystal system and class: Tetragonal; $4/m\ 2/m\ 2/m$.

Axial elements: $a:c = 1:0.885$.

Cell dimensions and content: $a = 7.117$, $c = 6.295$; $Z = 4$.

Habit: Crystals rare, squat pyramids or square prisms with pyramidal terminations; usually as irregular grains and masses.

Cleavage and fracture: $\{110\}$, seldom observed; conchoidal.

Hardness: 5.

Density: 6.7 for pure crystalline $ThSiO_4$, variable and generally considerably lower (down to 4 and less) due to varying degree of metamictness and accompanying hydration.

Color: Reddish brown to black, sometimes orange (var. *orangite*).

Streak: Light orange to brown.

Luster: Resinous.

Optical properties: $\omega = 1.78\text{--}1.82$, $\epsilon = 1.79\text{--}1.84$; $\epsilon - \omega = 0.01\text{--}0.02$; usually metamict, $n = 1.66\text{--}1.86$; color brown, yellow, or green.

Chemistry: Some of the thorium may be replaced by uranium (var. *uranorthorite*) and some of the $(SiO_4)^{-4}$ by $(OH)_4^{-4}$ (var. *thorogummite*). Thorite is nearly always metamict. It is strongly radioactive.

Diagnostic features: Color, luster, habit, and strong radioactivity characterize thorite.

Occurrence: An accessory mineral in granites, syenites, and nepheline syenites, and in pegmatites associated with these rocks; also as a detrital mineral in sands. Uranothorite is partially responsible for the uranium content of some ores in the Bancroft district of Ontario.

Uses: Thorite is the commonest thorium mineral, and it might become an important raw material if thorium is used for the production of atomic energy. The present demand for thorium is met by its extraction from mònazite, which often contains a considerable amount of this element.

Sphene, CaTiSiO₅

Crystal system and class: Monoclinic; 2/m.

Axial elements: $a:b:c = 0.753:1:0.854$; $\beta = 119° 43'$.

Cell dimensions and content: $a = 6.55$, $b = 8.70$, $c = 7.43$; $Z = 4$.

Habit: Generally as wedge-shaped crystals (hence the name) with large {001} and {111} faces (Figs. 15-56d, 15-57); sometimes massive or granular.

Figure 15-57 Sphene, brownish black. Forms: {110}, {111}, {001}. (Frontenac County, Ontario.) [Courtesy Queen's University.]

Twinning: Twin plane {100}, giving contact twins or cruciform penetration twins.

Cleavage: {110}, distinct.

Hardness: 6.

Density: 3.5.

Color: Usually brown, sometimes yellow, green, or gray.

Streak: White.

Luster: Adamantine.

Optical properties: $\alpha = 1.84–1.95$, $\beta = 1.87–2.03$, $\gamma = 1.94–2.11$; $\gamma - \alpha = 0.10–0.15$; positive, $2V = 20°–40°$.

Chemistry: The composition is somewhat variable; calcium may be partly substituted by sodium or rare earths, titanium by aluminum and sometimes by niobium.

Diagnostic features: The color, the typical wedge-shaped or diamond-shaped cross sections, and the adamantine luster serve to identify sphene.

Occurrence: Sphene is widely distributed as an accessory mineral in intermediate and acid igneous rocks and their associated pegmatites, and in metamorphic rocks.

Name: The alternative name *titanite* is derived from the chemical composition.

Datolite, Ca(OH)BSiO₄

Crystal system and class: Monoclinic; $2/m$.

Axial elements: $a:b:c = 1.264:1:0.632$; $\beta = 90°\,09'$.

Cell dimensions and content: $a = 9.66$, $b = 7.64$, $c = 4.83$; $Z = 4$.

Habit: Usually in short prismatic crystals, often with a variety of forms; also as granular or compact porcelain-like masses.

Cleavage: None.

Hardness: $6\frac{1}{2}$.

Density: 3.01.

Color: Transparent and colorless, or pale shades of yellow and green, also white and opaque.

Streak: White.

Luster: Vitreous.

Optical properties: $\alpha = 1.626$, $\beta = 1.653$, $\gamma = 1.670$; $\gamma - \alpha = 0.044$; negative, $2V = 74°$.

Diagnostic features: The color and crystal form, the ease of fusibility, and the boron flame serve to identify datolite.

Occurrence: A secondary mineral, usually found in cavities in basalts associated with zeolites, as in the trap rocks of New Jersey and Massachusetts. White porcellanous masses are found associated with native copper in the Lake Superior deposits.

Dumortierite, Al₇O₃(BO₃)(SiO₄)₃

Crystal system and class: Orthorhombic; $2/m\ 2/m\ 2/m$.

Axial elements: $a:b:c = 0.583:1:0.234$.

Cell dimensions and content: $a = 11.76$, $b = 20.18$, $c = 4.72$; $Z = 4$.

Habit: Prismatic crystals rare; usually in columnar or fibrous masses.

Cleavage: {100}, good; {110}, imperfect.

Hardness: $8\frac{1}{2}$.

Density: 3.3–3.4.

Color: Blue, violet, pink, brown.

Streak: White.

Luster: Vitreous or dull.

Optical properties: $\alpha = 1.66$–1.68, $\beta = 1.68$–1.72, $\gamma = 1.69$–1.72; $\gamma - \alpha =$

0.02–0.04; negative, $2V$ small; strongly pleochroic in blue, violet, or red.

Chemistry: Some of the aluminum may be replaced by ferric iron.

Diagnostic features: The color and habit often serve to identify dumortierite. It is infusible and gives a test for boron.

Occurrence: Dumortierite is a rare mineral which occurs in considerable amounts in a few localities. It occurs in aluminum-rich metamorphic rocks and occasionally in pegmatites. Dumortierite has been mined in Nevada for use in the manufacture of porcelain for spark plugs, since on firing at a high temperature it is converted to the refractory aluminum silicate mullite.

Selected Readings (Chapters 7–15)

Bates, R. L., 1960, *Geology of the industrial rocks and minerals:* New York, Harper & Row.

Bragg, W. L., and G. F. Claringbull, 1965, *The crystal structures of minerals:* London, Bell & Sons.

Dana, J. D., *A system of mineralogy* (7th ed.), vol. 1, 1944; vol. 2, 1951; vol. 3, 1962; vol. 4, in preparation: New York, Wiley. Rewritten by C. Palache, H. Berman, and C. Frondel.

Deer, W. A., R. A. Howie, and J. Zussman, 1966, *An introduction to the rock-forming minerals:* New York, Wiley.

Eitel, W., 1954, *The physical chemistry of the silicates:* Chicago, Univ. Chicago Press.

Geology and economic minerals of Canada: Economic Geology Series No. 1 (4th ed.), 1957, Department of Mines and Technical Surveys, Ottawa.

Gillson, J. L., ed., 1960, *Industrial minerals and rocks* (3rd ed.): New York, American Institute of Mining and Metallurgical Engineers.

Grim, R. E., 1953, *Clay mineralogy:* New York, McGraw-Hill.

Hey, M. H., 1955, *Chemical index of minerals* (2nd ed.): London, British Museum. (Appendix, 1963.)

Johnstone, S. J., 1961, *Minerals for the chemical and allied industries* (2nd ed.): London, Chapman and Hall.

Jones, W. R., 1963, *Minerals in industry* (4th ed.): Harmondsworth, England, Penguin Books.

Minerals, Canada and the world: Mines Branch Report No. 860 (with statistical supplement), 1957, Ottawa, Department of Mines and Technical Surveys.

Strunz, H., 1966, *Mineralogische Tabellen* (4th ed.) Leipzig, Akademische Verlagsgesellschaft.

U. S. Bureau of Mines, *Minerals yearbook:* Published annually by the Government Printing Office, Washington, D. C.

———, 1965, *Mineral facts and problems* (Bull. 630): Washington, Government Printing Office.

Part III

DETERMINATIONS

16

Determinative Tables

In the identification of minerals a set of determinative tables is of great assistance when it is used systematically by a student with keen observation and retentive memory. The whole detective process is stimulating, and running an unknown mineral to earth is a satisfying triumph. No set of tables does away with the necessity of individual enterprise. The student who follows the general precepts and special advice of determinative tables will certainly be helped to avoid waste of time, but in the domain of mineral identification there is no foolproof system. Expressed in two words, the best key to mineral identification is "Be smart"!

The primary basis of discrimination used in the tables in this book is *luster*, whether metallic or nonmetallic. A few opaque or nearly opaque minerals with a dark brown to black color are sometimes described as having submetallic luster; they are grouped with the minerals with a metallic luster, but some of them are also listed in the nonmetallic group.

The secondary basis of discrimination is *hardness*. Relative hardness is easily determined with the simplest of tools, and is one of the few such properties capable of being expressed in numerical terms. The tables are divided into three groups: minerals with hardness less than $2\frac{1}{2}$ (can be scratched with the fingernail); minerals with hardness between $2\frac{1}{2}$ and $5\frac{1}{2}$ (can be scratched with a knife blade); minerals with hardness greater than $5\frac{1}{2}$ (cannot be scratched with a knife blade). With a little practice a student should be able to place a figure on the hardness of a mineral in the $2\frac{1}{2}$–$5\frac{1}{2}$ group from its relative ease of scratching. The hardness given for each mineral is that of a clean smooth surface; minerals in earthy varieties, such as some forms of hematite, are much softer than the same mineral in good crystals.

Minerals with a nonmetallic luster far exceed in numbers those with a metallic luster, and it is therefore convenient to find a further criterion for the division of this nonmetallic group. In this book the tables for the nonmetallic group are subdivided into two sections, based on color: one section includes those minerals which are usually colorless or white, unless tinted by impurities; the other section includes those minerals which are colored.

By using luster and hardness, and dividing the nonmetallic minerals into two sections on the basis of color, we thus obtain a set of tables comprising nine groups, as follows:

LUSTER	HARDNESS	NUMBER OF MINERALS	PAGE REF.
Metallic			
	$H < 2\frac{1}{2}$	4	516
	$H = 2\frac{1}{2}-5\frac{1}{2}$	29	516
	$H > 5\frac{1}{2}$	19	520
Nonmetallic			
Colorless or white			
	$H < 2\frac{1}{2}$	8	522
	$H = 2\frac{1}{2}-5\frac{1}{2}$	38	523
	$H > 5\frac{1}{2}$	19	527
Colored			
	$H < 2\frac{1}{2}$	13	529
	$H = 2\frac{1}{2}-5\frac{1}{2}$	32	531
	$H > 5\frac{1}{2}$	44	534

Within each group the minerals are described in order of increasing *density*. Density is, of course, not so easily estimated as hardness; however, practice in handling minerals of known density should develop the ability to distinguish densities differing by one unit, and thus you should eventually be able to say whether a specimen has a density near 3, or 4, or 5, or greater than 5. In this connection the availability of one or more of the heavy liquids mentioned on p. 105 is a useful determinative aid. With a small jar of tetrabromoethane (acetylene tetrabromide), which has a density of 2.96, it takes only a moment to determine whether a fragment of an unknown mineral sinks or floats. By using tetrabromoethane in this way the larger groups can each be divided into two smaller ones.

Having determined luster and hardness, and made a reasonable estimate of density, you will then have reduced the number of possible choices to a comparatively few species, and at this stage the other physical properties, particularly crystal form, cleavage, and color, should guide you to the correct identification. The study of physical properties may be extended, if necessary, by one or two simple chemical tests or, for the nonmetallic minerals, by determining the refractive index. In the following tables the mean refractive index for each of the nonmetallic minerals is given in a column headed *n*; birefringence and other optical properties are given in the descriptive section of the book. For the metallic minerals the relative fusibility, which can be a useful diagnostic property, is given in a column headed F.

These determinative tables give a very brief statement of the more significant diagnostic properties. The page number under the mineral name refers to the full description of the mineral given in Chapters 8–15, where, for most minerals, a section headed *Diagnostic features* outlines a few special tests which will be found useful in confirming an identification.

Luster metallic; H < $2\frac{1}{2}$

NAME FORMULA	COLOR	STREAK	G	H	SYSTEM AND HABIT	CLEAVAGE	F	OBSERVATIONS
Graphite C (p. 221)	Black	Black; black on glazed porcelain	2.23	1	Hexagonal; hexagonal plates or massive, foliated or earthy	{0001} perfect	7	Greasy feel; plates flexible, inelastic; marks paper; often occurs with calcite in metamorphic rocks
Stibnite Sb_2S_3 (p. 252)	Black	Gray	4.6	2	Orthorhombic; prismatic crystals elongated ∥ c. Bladed masses	{010} perfect	1	Brittle; occurs in quartz veins
Molybdenite MoS_2 (p. 264)	Blue-gray	Blue-gray; greenish black on glazed porcelain	4.73	1	Hexagonal; hexagonal plates or massive, foliated	{0001} perfect	7	Like graphite, but heavier and more metallic; marks paper; usually associated with quartz or feldspar
Sylvanite $(Ag,Au)Te_2$ (p. 267)	White to gray	Gray	8.1	2	Monoclinic; prismatic and tabular crystals, also massive bladed	{010} perfect	1	Occurs in quartz veins with gold or telluride minerals

Luster metallic; H = $2\frac{1}{2}$–$5\frac{1}{2}$

NAME FORMULA	COLOR	STREAK	G	H	SYSTEM AND HABIT	CLEAVAGE	F	OBSERVATIONS
Goethite $HFeO_2$ (p. 314)	Brown, black	Yellow to brown	4.2	$5\frac{1}{2}$	Orthorhombic; long prismatic crystals; usually massive, botryoidal or stalactitic	{010} perfect	7	Earthy varieties known as *limonite*; cement and coloring matter in sedimentary rocks; forms pseudomorphs after pyrite
Chalcopyrite $CuFeS_2$ (p. 239)	Brass yellow	Greenish, black	4.3	4	Tetragonal; sphenoidal crystals; usually massive, compact	{011} imperfect	2	Often tarnished blue or gray; occurs with pyrite, galena, sphalerite; yellower and softer than pyrite

Mineral	Color	Streak	G	H	Crystal system and habit	Cleavage	No.	Remarks
Manganite $MnO(OH)$ (p. 312)	Black	Brown	4.3	4	Monoclinic (pseudo-orthorhombic); prismatic crystals, often elongated $\parallel c$	{010} perfect	7	Occurs in hydrothermal deposits; alters to pyrolusite, the streak then becoming black
Enargite Cu_3AsS_4 (p. 271)	Black	Black	4.4	3	Orthorhombic; tabular of prismatic crystals; also massive, compact or granular	{110} perfect; {100}, {010} distinct	1	Often associated with chalcocite and bornite
Psilomelane $(Ba,H_2O)_2Mn_5O_{10}$ (p. 297)	Black	Black	4.7	5½	Monoclinic; always massive, often botryoidal	None visible	7	Secondary mineral, weathering product of other Mn minerals
Covellite CuS (p. 247)	Indigo blue to black	Black	4.7	2	Hexagonal; hexagonal plates; also massive, foliated or granular	{0001} perfect	3	Uncommon copper mineral; generally associated with chalcocite or enargite
Pyrrhotite $Fe_{1-x}S$ (p. 243)	Bronze yellow	Gray-black	4.6–4.65	4	Hexagonal; usually massive, granular or compact	None	3	Magnetic; surface often tarnished brown
Pentlandite $(Fe,Ni)_9S_8$ (p. 246)	Brass yellow	Brown	4.6–5.0	4	Isometric; always massive	None; parting on {111}	2	Occurs intermixed with pyrrhotite
Tetrahedrite $Cu_{12}Sb_4S_{13}$ Tennantite $Cu_{12}As_4S_{13}$ (p. 269)	Black	Brown to black	4.6–5.1	3–4	Isometric; tetrahedral crystals; also massive, granular or compact	None	2	Widely distributed in hydrothermal Cu deposits; often contains Ag
Bornite Cu_5FeS_4 (p. 228)	Bronze brown, purple tarnish	Gray-black	5.07	3	Isometric; usually massive, granular or compact	None	2	Often associated with chalcopyrite or chalcocite; purple tarnish characteristic
Pyrolusite MnO_2 (p. 301)	Gray to black	Black	5.0–5.2	to 5½	Tetragonal; usually massive, columnar, fibrous, or powdery	{110} perfect	7	Secondary after other Mn minerals; often soft and earthy

Luster metallic; H = $2\frac{1}{2}$–$5\frac{1}{2}$ (*Continued*)

NAME FORMULA	COLOR	STREAK	G	H	SYSTEM AND HABIT	CLEAVAGE	F	OBSERVATIONS
Millerite NiS (p. 245)	Pale brass yellow	Greenish black	5.5	3–$3\frac{1}{2}$	Trigonal; usually as hair-like crystals, often in radiating groups	$\{10\bar{1}1\}$, $\{01\bar{1}2\}$ perfect	2	Commonly occurs in cavities associated with calcite or siderite
Pyrargyrite Ag_3SbS_3 Proustite Ag_3AsS_3 (p. 268)	Deep red to black	Red	5.6–5.8	$2\frac{1}{2}$	Trigonal; prismatic crystals; also massive, granular	$\{10\bar{1}1\}$ good	1	Often associated with galena in lead-silver ores
Arsenic As (p. 213)	Gray white, tarnishing to dark gray	Gray	5.7	$3\frac{1}{2}$	Hexagonal; usually massive, often botyroidal	$\{0001\}$ good	—	Volatilizes without melting
Chalcocite Cu_2S (p. 226)	Gray to black	Gray to black	5.8	3	Orthorhombic; usually massive, compact or granular	$\{110\}$ imperfect	2	Sectile; often has sooty coating
Bournonite $PbCuSbS_3$ (p. 272)	Black	Gray black	5.8–5.9	3	Orthorhombic; short prismatic or tabular crystals; also massive, granular or compact	$\{010\}$ imperfect	1	Occurs in veins with other sulfides, often with galena and tetrahedrite
Boulangerite $Pb_5Sb_4S_{11}$ (p. 272)	Lead gray	Brownish gray	6.23	$2\frac{1}{2}$–3	Monoclinic; long prismatic crystals, also fibrous and hair-like	$\{100\}$ good	1	Occurs in vein deposits associated with other lead minerals
Antimony Sb (p. 214)	White to gray	Gray	6.7	3	Hexagonal; usually massive, lamellar	$\{0001\}$ perfect	1	Specimen often coated with white antimony oxide
Argentite Ag_2S (p. 225)	Black	Dark gray	7.0–7.4	$2\frac{1}{2}$	Isometric; cubes or octahedrons; also massive	$\{100\}$, $\{110\}$ imperfect	2	Sectile; occurs with other silver minerals

Mineral	Color	Streak	G	H	Crystal system / habit	Cleavage	Fusibility	Remarks
Wolframite (Fe,Mn)WO$_4$ (p. 374)	Black	Brown black	7.1–7.5	4½	Monoclinic; crystals often tabular ‖ (100); also massive, bladed or compact	{010} perfect	4	Resembles columbite-tantalite, but distinguished by perfect cleavage; occurs in quartz veins
Galena PbS (p. 229)	Gray	Gray	7.6	2½	Isometric; cubes or cubo-octahedrons; generally massive, granular	{100} perfect	3	The commonest lead mineral; often occurs with sphalerite
Iron Fe (p. 212)	Gray	Gray	7.8	4	Isometric; always massive	{100} not prominent	6	Generally as meteorites which always contain Ni; magnetic
Niccolite NiAs (p. 245)	Copper red	Brown black	7.8	5½	Hexagonal; usually massive, compact or granular	None	2	Often coated with green annabergite
Copper Cu (p. 209)	Copper red, often tarnished	Red; metallic	8.9	3	Isometric; massive and in dendritic or wire-like forms	None	3	Often in cavities in basic igneous rocks; malleable
Calaverite AuTe$_2$ (p. 266)	Brass yellow	Green gray	9.3	3	Monoclinic; bladed crystals or massive	None	1	Fuses easily on charcoal giving globules of Au
Bismuth Bi (p. 215)	Silver white, cream	Silver white	9.7–9.8	2½	Hexagonal; indistinct crystals, often hopper-shaped; usually in platy masses	{0001} perfect	1	Occurs in quartz veins and some pegmatites; often has brassy tarnish
Silver Ag (p. 208)	Silver white, often tarnished black	Silver	10.1–11.1	3	Isometric; usually in dendritic or wire-like forms	None	3	Often occurs with barite and calcite in veins; tarnishes gray or black; malleable
Platinum Pt (p. 211)	Steel gray	Steel gray	14–19	4	Isometric; usually in nuggets or grains	None	7	Usually as grains in alluvial deposits; may be magnetic if it contains iron
Gold Au (p. 207)	Yellow	Yellow	15–19	3	Isometric; usually in dendritic forms, or as alluvial grains	None	2	Easily recognized by its color, softness, and malleability

Luster metallic; H > $5\frac{1}{2}$

NAME FORMULA	COLOR	STREAK	G	H	SYSTEM AND HABIT	CLEAVAGE	F	OBSERVATIONS
Anatase TiO_2 (p. 305)	Usually brown	White	3.9	6	Tetragonal; usually in acute pyramidal crystals	{001}, {011} perfect	7	Occurs in small crystals in joint planes and veins in schists and gneisses
Brookite TiO_2 (p. 305)	Brown to black	White to gray	4.1	6	Orthorhombic; tabular and prismatic crystals	None	7	Occurs as anatase; $n > 2.0$
Rutile TiO_2 (p. 299)	Red brown	Yellow to gray	4.2–4.3	$6\frac{1}{2}$	Tetragonal; usually in long prismatic crystals; also massive, granular	{110} imperfect	7	Common accessory mineral of igneous and metamorphic rocks
Chromite $(Mg,Fe)Cr_2O_4$ (p. 282)	Black	Brown	4.5–4.8	6	Isometric; octahedral crystals rare; usually massive, granular or compact	None	7	Occurs in ultrabasic rocks and serpentines
Ilmenite $FeTiO_3$ (p. 294)	Black	Black	4.6–4.8	6	Trigonal; tabular crystals; usually massive, compact or granular	None, parting on {0001}, {01$\bar{1}$2}	7	Occurs associated with basic igneous rocks, and in sands derived therefrom; often with intergrown magnetite
Braunite $3Mn_2O_3 \cdot MnSiO_3$ (p. 295)	Black	Black	4.7–4.8	6	Tetragonal; pyramidal crystals, or massive granular	{112} perfect	7	Ore mineral of manganese; occurs in veins and in metamorphic deposits
Hausmannite Mn_3O_4 (p. 285)	Black	Brown	4.84	6	Tetragonal; pyramidal crystals, or massive granular	{001} good	7	Occurs as braunite; distinguished from this mineral by brown streak
Marcasite FeS_2 (p. 260)	Light brass yellow	Gray black	4.88	$6\frac{1}{2}$	Orthorhombic; complex twinned crystals; also massive, stalactitic or nodular	{101} imperfect	3	Occurs in low-temperature surface and near-surface deposits
Pyrite FeS_2 (p. 256)	Light brass yellow	Brown black	5.0	$6\frac{1}{2}$	Isometric; pyritohedrons, cubes, or octahedrons; also massive, compact or granular	None	3	The commonest sulfide mineral; alters to limonite

Mineral	Color	Streak	G	H	Crystal form / habit	Cleavage		Remarks
Franklinite $ZnFe_2O_4$ (p. 285)	Black	Brown	5.1–5.3	6	Isometric; rounded octahedrons or massive granular	None	7	Occurs in metamorphosed limestone at Franklin, N. J.; weakly magnetic
Magnetite Fe_3O_4 (p. 281)	Black	Black	5.2	6	Isometric; octahedrons or massive granular	None; parting on {111}	7	Accessory mineral of igneous and metamorphic rock, and in sands; strongly magnetic
Columbite-tantalite $(Fe,Mn)(Nb,Ta)_2O_6$ (p. 306)	Red brown to black	Brown to black	5.2–8.0	6	Orthorhombic; tabular or prismatic crystals, also massive	{010} imperfect	7	Occurs in granite pegmatites
Hematite Fe_2O_3 (p. 291)	Red to black	Red brown	5.26	6	Trigonal; tabular crystals and massive	None; parting on {0001}, {0112}	7	Common iron mineral of sediments and metamorphosed sediments; often earthy and soft
Arsenopyrite FeAsS (p. 262)	Silver white	Gray black	6.1	6	Monoclinic; prismatic crystals; usually massive, columnar or granular	{101} imperfect	2	Generally occurs in metalliferous veins; the commonest arsenic mineral
Cobaltite CoAsS (p. 259)	Silver white	Gray black	6.3	6	Isometric; pyritohedrons, cubes, or massive, granular or compact	{100} perfect	3	Occurs in vein deposits; sometimes coated with pink erythrite
Skutterudite Smaltite Chloanthite $(Co,Ni)As_{3-x}$ (p. 267)	Silver white	Black	6.5	6½	Isometric; cubes and octahedrons, also massive, granular or compact	{100}, {111} imperfect	3	Occurs in vein deposits; sometimes coated with pink erythrite
Cassiterite SnO_2 (p. 304)	Usually brown to black	White	7.0	6½	Tetragonal; prismatic crystals and massive, sometimes botryoidal or concretionary	{110} imperfect	7	Occurs in quartz veins and granite pegmatites; also detrital

Luster metallic; H > 5½ (Continued)

NAME FORMULA	COLOR	STREAK	G	H	SYSTEM AND HABIT	CLEAVAGE	F	OBSERVATIONS
Uraninite (pitchblende) UO_2 (p. 308)	Black	Brown black	8–10	6½	Isometric; cubes; also massive, sometimes botryoidal	None	7	Occurs in granite pegmatites, hydrothermal veins, and disseminated in sandstones and limestones; yellow alteration products sometimes present
Sperrylite $PtAs_2$ (p. 259)	Silver white	Black	10.6	6½	Isometric; cubic crystals and rounded grains	{001} imperfect	3	Occurs in basic igneous rock, and in sands derived therefrom

Luster nonmetallic; usually colorless or white unless tinted by impurities; streak white; H < 2½

NAME FORMULA	COLOR	G	H	SYSTEM AND HABIT	CLEAVAGE	n	OBSERVATIONS
Epsomite $MgSO_4 \cdot 7H_2O$ (p. 369)	White	1.67	2	Orthorhombic; in acicular crystals, elongated ∥ c, or powdery crusts	{010} good	1.45	Secondary mineral; soluble in H_2O; bitter, salty taste
Soda-niter $NaNO_3$ (p. 352)	Colorless, white	2.25	2	Trigonal; usually in granular masses	{10$\bar{1}$1} perfect	1.58	Soluble in H_2O, cooling and saline taste; occurs only in rainless deserts
Gypsum $CaSO_4 \cdot 2H_2O$ (p. 364)	Colorless, white; sometimes tinted pink, yellow, gray	2.32	2	Monoclinic; tabular prismatic crystals, elongated ∥ c, tabular ∥ {010}; swallowtail twins; often granular, massive or fibrous	{010} perfect	1.52	Cleavage laminae flexible but not elastic
Montmorillonite $Al_2Si_4O_{10}(OH)_2 \cdot xH_2O$ (p. 446)	White, gray, green gray	2.0–2.7	2	Monoclinic; always in clayey masses	{001} perfect, not megascopically observable	1.50 –1.64	Clay mineral; absorbs water with swelling; often has soapy feel
Brucite $Mg(OH)_2$ (p. 309)	Colorless, white, pale green	2.4	2	Trigonal; in hexagonal plates or massive compact; sometimes fibrous	{0001} perfect	1.57	Occurs in association with dolomite or serpentine, or in metamorphosed dolomite

NAME FORMULA	COLOR	G	H	SYSTEM AND HABIT	CLEAVAGE	n	OBSERVATIONS
Kaolinite $Al_4Si_4O_{10}(OH)_8$ (p. 442)	White, gray, brown	2.6	2	Triclinic; always in clayey masses	{001} perfect, not usually observable	1.56	Clay mineral; clayey odor when breathed upon; does not swell in water
Talc $Mg_3Si_4O_{10}(OH)_2$ (p. 445)	Green	2.82	1	Monoclinic; in foliated or fine-grained masses	{001} perfect	1.58	Greasy feel; occurs in magnesium-rich metamorphic rocks
Pyrophyllite $Al_2Si_4O_{10}(OH)_2$ (p. 445)	White, yellow, brown	2.84	1	Monoclinic; radiated fibrous, or fine-grained	{001} perfect	1.59	Greasy feel; occurs in aluminum-rich metamorphic rocks

Luster nonmetallic; usually colorless or white unless tinted by impurities; streak white; $H = 2\frac{1}{2}-5\frac{1}{2}$

NAME FORMULA	COLOR	G	H	SYSTEM AND HABIT	CLEAVAGE	n	OBSERVATIONS
Carnallite $KMgCl_3 \cdot 6H_2O$ (p. 327)	Colorless or white	1.60	$2\frac{1}{2}$	Orthorhombic; usually in coarsely granular masses	None	1.48	Bitter taste, occurs in salt deposits; deliquescent
Borax $Na_2B_4O_5(OH)_4 \cdot 8H_2O$ (p. 353)	White	1.71	$2\frac{1}{2}$	Monoclinic; short prismatic crystals	{100} indistinct	1.47	Sweetish alkaline taste; deposited from salt lakes; effloresces in dry air
Kernite $Na_2B_4O_6(OH)_2 \cdot 3H_2O$ (p. 352)	Colorless	1.91	$2\frac{1}{2}$	Monoclinic; massive, large cleavages	{100} perfect; {001} good	1.47	Fibrous fracture; sometimes coated with white tincalconite, $Na_2B_4O_7 \cdot 5H_2O$
Sylvite KCl (p. 321)	Colorless, sometimes white or gray	1.99	$2\frac{1}{2}$	Isometric; cubic crystals; usually massive, granular	{100} perfect	1.49	Salty taste like halite, but somewhat bitter, sectile
Opal $SiO_2 \cdot nH_2O$ (p. 415)	Colorless, white	2.0–2.2	$5-5\frac{1}{2}$	Amorphous; massive compact, sometimes mammillary	None	1.45	Secondary silica deposited at low temperatures
Chabazite $CaAl_2Si_4O_{12} \cdot 6H_2O$ (p. 434)	Colorless, white, pink	2.1	4	Trigonal; usually in cube-like rhombohedral crystals	{10$\bar{1}$1} imperfect	1.48	Occurs in cavities in basalts; crystal form distinguishes it from other zeolites

Luster nonmetallic; usually colorless or white unless tinted by impurities; streak white; $H = 2\frac{1}{2}-5\frac{1}{2}$ (*Continued*)

NAME FORMULA	COLOR	G	H	SYSTEM AND HABIT	CLEAVAGE	n	OBSERVATIONS
Stilbite $CaAl_2Si_7O_{18}\cdot 7H_2O$ (p. 433)	White, pale yellow	2.1–2.2	4	Monoclinic; sheaf-like aggregates of twinned crystals	{010} perfect	1.50	Occurs in cavities in basalts; crystal form distinguishes it from other zeolites
Halite $NaCl$ (p. 320)	Colorless, white, red	2.16	$2\frac{1}{2}$	Isometric; cubic crystals; usually massive, granular	{100} perfect	1.54	Characteristic salty taste
Heulandite $CaAl_2Si_7O_{18}\cdot 6H_2O$ (p. 432)	White	2.2	3–4	Monoclinic; trapezoidal crystals tabular ∥ {010}	{010} perfect	1.49	Pearly luster on cleavage; occurs in cavities in basalts; crystal form distinguishes it from other zeolites
Natrolite $Na_2Al_2Si_3O_{10}\cdot 2H_2O$ (p. 436)	Colorless, white	2.25	5	Orthorhombic; long prismatic crystals and radiating nodular forms	{110} perfect	1.48	Occurs in cavities in basalts; crystal form distinguishes it from other zeolites
Laumontite $CaAl_2Si_4O_{12}\cdot 4H_2O$ (p. 434)	White	2.25–2.30	4	Monoclinic; small prismatic crystals; also massive, often powdery	{010}, {110} perfect	1.52	Occurs in cavities in basalts; crystal form distinguishes it from other zeolites
Analcime $NaAlSi_2O_6\cdot H_2O$ (p. 435)	Colorless, white	2.3	5	Isometric; trapezohedral crystals; also massive, compact or granular	None	1.48	Occurs in cavities in basalts with other zeolites, also in some igneous and sedimentary rocks
Apophyllite $KCa_4(Si_4O_{10})_2F\cdot 8H_2O$ (p. 455)	Colorless, white	2.35	5	Tetragonal; prismatic crystals with basal plane and pyramids	{001} perfect	1.54	Occurs in cavities in basalts and in hydrothermal veins, often with zeolites
Gibbsite $Al(OH)_3$ (p. 316)	White, red, brown	2.3–2.4	3	Monoclinic; usually massive, compact, pisolitic, or earthy	{001} perfect	1.58	A major component of bauxite
Colemanite $CaB_3O_4(OH)_3\cdot H_2O$ (p. 354)	Colorless, white	2.42	$4\frac{1}{2}$	Monoclinic; short prismatic crystals, often with complex terminations; also massive	{010} perfect; {001} imperfect	1.51	Occurs as geode-like masses in sedimentary rocks
Calcite $CaCO_3$ (p. 337)	Colorless, white	2.71	3	Trigonal; scalenohedral and rhombohedral crystals; also massive, granular or compact	{1011} perfect	1.66	Often colored brown, gray, or black with impurities; effervesces in cold dilute HCl

Mineral	Color	G	H	Crystal system and habit	Cleavage	n	Remarks
Dolomite $CaMg(CO_3)_2$ (p. 342)	White, yellow, pink	2.85	3½–4	Trigonal; small curved rhombohedral crystals; also massive, granular	$\{10\bar{1}1\}$ perfect	1.68	Effervesces in warm dilute HCl
Alunite $KAl_3(SO_4)_2(OH)_6$ (p. 372)	White, pink	2.6–2.9	3½–4	Trigonal; usually massive, granular or compact	$\{0001\}$ imperfect	1.58	Often occurs as hydrothermal alteration of volcanic rocks
Muscovite $KAl_2(AlSi_3O_{10})(OH)_2$ (p. 449)	Colorless or pale tints	2.8–2.9	2½–3	Monoclinic; usually in irregular platy crystals; also massive, sometimes compact	$\{001\}$ perfect	1.59	Large crystals in granite pegmatites; small flakes in many metamorphic rocks
Pectolite $Ca_2NaHSi_3O_9$ (p. 476)	White	2.86	5	Triclinic; usually in radiating aggregates of needle-like crystals	$\{001\}$, $\{100\}$ perfect	1.60	Typically occurs associated with zeolites in cavities in basalt
Wollastonite $CaSiO_3$ (p. 476)	White	2.9	5	Triclinic; usually in columnar or fibrous masses	$\{100\}$ perfect; $\{\bar{1}02\}$, $\{001\}$ good	1.63	Occurs in metamorphosed limestones
Aragonite $CaCO_3$ (p. 344)	White	2.93	3½–4	Orthorhombic; hexagonal twinned crystals; also massive, sometimes coralloid	$\{010\}$ imperfect	1.68	Effervesces in cold dilute HCl; much less common than calcite
Cryolite Na_3AlF_6 (p. 327)	White	2.96	2½	Monoclinic; usually massive, coarsely granular	None; parting on $\{001\}$ and $\{110\}$	1.34	Wet-snow luster; occurs in large amounts at Ivigtut, Greenland, associated with siderite
Anhydrite $CaSO_4$ (p. 363)	White, gray, pale blue	3.0	3½	Orthorhombic; usually massive, granular	$\{010\}$ perfect; $\{100\}$, $\{001\}$ good	1.58	Forms extensive beds in sedimentary rocks; alters to gypsum
Magnesite $MgCO_3$ (p. 337)	White	3.0	4	Trigonal; usually massive compact	$\{10\bar{1}1\}$ perfect	1.70	Cleavage not seen in compact forms, which have conchoidal fracture; effervesces in warm dilute HCl

Luster nonmetallic; usually colorless or white unless tinted by impurities; streak white; $H = 2\frac{1}{2}$-$5\frac{1}{2}$ (Continued)

NAME FORMULA	COLOR	G	H	SYSTEM AND HABIT	CLEAVAGE	n	OBSERVATIONS
Boehmite $AlO(OH)$ (p. 312)	White, yellow, brown	3.0–3.1	3	Orthorhombic; always massive, compact; often pisolitic	$\{010\}$ not visible	1.65	Important constituent of bauxite
Amblygonite $LiAlPO_4(F,OH)$ (p. 385)	White	3.1	$5\frac{1}{2}$	Triclinic; usually in cleavable masses	$\{100\}$ perfect; $\{110\}$, $\{0\bar{1}1\}$ distinct	1.61	Often shows twinning striations on cleavages; occurs in granite pegmatites
Hemimorphite $Zn_4Si_2O_7(OH)_2\cdot H_2O$ (p. 484)	White	3.4–3.5	5	Orthorhombic; usually massive, often in mammillary forms	$\{110\}$ perfect	1.67	Alteration product of sphalerite; often associated with smithsonite
Periclase MgO (p. 277)	White	3.56	5	Isometric; usually in small rounded grains	$\{100\}$ perfect	1.74	Occurs as small grains often altered to brucite in metamorphosed dolomite
Strontianite $SrCO_3$ (p. 347)	White	3.78	$3\frac{1}{2}$	Orthorhombic; usually massive, columnar to fibrous	$\{110\}$ good	1.66	Red flame coloration; effervesces in cold dilute HCl
Celestite $SrSO_4$ (p. 361)	White, pale blue	3.9–4.0	3	Orthorhombic; crystals tabular ∥ $\{001\}$, also massive, often fibrous	$\{001\}$ perfect; $\{210\}$ good	1.62	Red flame coloration
Witherite $BaCO_3$ (p. 346)	White	4.31	3	Orthorhombic; hexagonal twinned crystals, also massive, columnar or granular	$\{010\}$ distinct	1.67	Green flame coloration; effervesces in cold dilute HCl
Smithsonite $ZnCO_3$ (p. 340)	White, yellow, blue, green	4.4	4	Trigonal; usually massive, compact, often concretionary	$\{10\bar{1}1\}$ good	1.85	Effervesces in warm dilute HCl
Barite $BaSO_4$ (p. 358)	Colorless, white, gray	4.5	3	Orthorhombic; crystals usually tabular ∥ $\{001\}$, also massive, granular	$\{001\}$, $\{210\}$ perfect	1.64	Green flame coloration

NAME FORMULA	COLOR	G	H	SYSTEM AND HABIT	CLEAVAGE	n	OBSERVATIONS
Scheelite CaWO$_4$ (p. 376)	White, yellow, brown	6.1	5	Tetragonal; usually massive, granular	{101} distinct	1.92	Fluoresces white or yellow in ultraviolet light
Anglesite PbSO$_4$ (p. 362)	White, gray	6.3–6.4	3	Orthorhombic; crystals prismatic or tabular ∥ {001}; also massive, compact	{001}, {210} distinct	1.88	Often has a nucleus of unchanged galena
Cerussite PbCO$_3$ (p. 347)	White, gray	6.5–6.6	3	Orthorhombic; in lattice-like aggregates of twinned crystals; also massive, granular	{110}, {021} distinct	2.07	Effervesces weakly in cold dilute HCl, strongly in warm
Mimetite Pb$_5$(AsO$_4$)$_3$Cl (p. 390)	White, yellow	7.2–7.3	4	Hexagonal; in small prismatic crystals; also massive, often globular and botryoidal	None	2.14	Found in the oxibized zone of lead deposits

Luster nonmetallic; usually colorless or white unless tinted by impurities; streak white; H > 5½

NAME FORMULA	COLOR	G	H	SYSTEM AND HABIT	CLEAVAGE	n	OBSERVATIONS
Tridymite SiO$_2$ (p. 414)	Colorless, white	2.26	6½	Hexagonal; small thin hexagonal plates, often twinned	{10$\bar{1}$0} imperfect	1.48	Occurs in cavities in volcanic rocks
Cristobalite SiO$_2$ (p. 415)	White	2.32	6½	Isometric; minute octahedrons or small rounded aggregates	None	1.49	Occurs in cavities in volcanic rocks
Leucite KAlSi$_2$O$_6$ (p. 427)	White, gray	2.45	6	Isometric; trapezohedral crystals	{110} imperfect	1.51	Occurs as isolated crystals in basic volcanic rocks
Nepheline NaAlSiO$_4$ (p. 428)	White, brown, gray	2.5–2.6	6	Hexagonal; hexagonal prisms, and massive granular	{10$\bar{1}$0} imperfect	1.53	Gelatinizes in HCl
Orthoclase KAlSi$_3$O$_8$ (p. 421)	White, pink	2.56	6	Monoclinic; prismatic crystals, flattened ∥ {010} or elongated ∥ a; also massive, granular	{001} perfect, {010}; good	1.52	Occurs in potassium-rich igneous and metamorphic rocks

Luster nonmetallic; usually colorless or white unless tinted by impurities; streak white; $H > 5\frac{1}{2}$ (Continued)

NAME FORMULA	COLOR	G	H	SYSTEM AND HABIT	CLEAVAGE	n	OBSERVATIONS
Microcline $KAlSi_3O_8$ (p. 422)	White, pink, green	2.56	6	Triclinic; like orthoclase	{001} perfect; {010} good	1.52	Like orthoclase, but may show imperfect polysynthetic twinning striations, or grid structure on {001}; green color sometimes characteristic
Scapolite $(Na,Ca)_4[(Al,Si)_4O_8]_3(Cl,CO_3)$ (p. 426)	White, yellow, pink, gray	2.56–2.77	6	Tetragonal; prismatic crystals terminated by pyramids; also massive granular or columnar	{100}, {110} good	1.54–1.61	Occurs in lime-rich metamorphic rocks
Plagioclase $(Na,Ca)(Al,Si)_4O_8$ (p. 423)	White, gray, brown	2.62–2.76	6	Triclinic; prismatic crystals, flattened ∥ {010}; also massive, granular	{001} perfect; {010} good	1.53–1.58	Polysynthetic twinning usually visible on {001} cleavage surfaces
Quartz SiO_2 (p. 409)	Colorless, white, gray, pink, green	2.65	7	Trigonal; prismatic crystals terminated by rhombohedrons; also massive, granular or compact (*chalcedony*)	None	1.55	Occurs in silica-rich igneous rocks, granite pegmatites, sedimentary and metamorphic rocks
Anthophyllite $(Mg,Fe)_7Si_8O_{22}(OH)_2$ (p. 462)	White, gray, pink	2.9–3.3	6	Orthorhombic; massive, bladed or fibrous aggregates	{210} perfect	1.61–1.68	Occurs in magnesium-rich metamorphic rocks
Tremolite $Ca_2Mg_5Si_8O_{22}(OH)_2$ (p. 463)	White	3.0	6	Monoclinic; usually in columnar or fibrous aggregates	{110} perfect	1.61	Occurs in calcareous and magnesian metamorphic rocks, associated with calcite, dolomite, talc
Datolite $Ca(OH)BSiO_4$ (p. 508)	Colorless, pale yellow, pale green, white	3.0	$6\frac{1}{2}$	Monoclinic; short prismatic crystals; also massive, granular or compact	None	1.65	Occurs in cavities in basic igneous rocks associated with zeolites
Lawsonite $CaAl_2Si_2O_7(OH)_2 \cdot H_2O$ (p. 483)	White, pink, gray	3.1	8	Orthorhombic; prismatic and tabular crystals; also massive granular	{010}, {001} perfect	1.68	Occurs in schists and gneisses with glaucophane

NAME FORMULA	COLOR	G	H	SYSTEM AND HABIT	CLEAVAGE	n	OBSERVATIONS
Spodumene LiAlSi$_2$O$_6$ (p. 474)	Colorless, white, pink	3.1–3.2	6½	Monoclinic; long prismatic crystals, also platy masses	{110} perfect; parting on {100}	1.66	Red flame coloration; occurs in granite pegmatites
Andalusite Al$_2$SiO$_5$ (p. 494)	White, gray, pink, brown	3.15	7½	Orthorhombic; coarse prismatic crystals, square cross-section; also massive, columnar or granular	{110} good	1.64	Often contains inclusions symmetrically arranged (var. *chiastolite*); typically occurs in hornfels
Sillimanite Al$_2$SiO$_5$ (p. 496)	White, brown, gray	3.24	7	Orthorhombic; usually in fibrous aggregates	{010} perfect	1.66	Occurs in high-grade schists and gneisses
Diaspore AlO(OH) (p. 313)	White, gray, brown	3.3–3.4	6½	Orthorhombic; usually massive, bladed, foliated, or compact	{010} perfect	1.72	Occurs in aluminum-rich metamorphic rocks
Diamond C (p. 220)	Colorless, yellow, pink, gray, black	3.5	10	Isometric; rounded octahedral crystals, often twinned	{111} perfect	2.42	Occurs in ultrabasic igneous rocks or sands derived from them
Topaz Al$_2$SiO$_4$(OH,F)$_2$ (p. 499)	Colorless, white, yellow, blue	3.5–3.6	8	Orthorhombic; prismatic crystals terminated by pyramids; also massive, granular	{001} perfect	1.61 –1.63	Occurs in veins and cavities in silica-rich igneous rocks and pegmatites

Luster nonmetallic; usually colored; streak white unless otherwise stated; H < 2½

NAME FORMULA	COLOR	G	H	SYSTEM AND HABIT	CLEAVAGE	n	OBSERVATIONS
Melanterite FeSO$_4$·7H$_2$O (p. 368)	Pale green	1.90	2	Monoclinic; usually stalactitic or powdery crusts	{001} perfect	1.48	Secondary mineral after iron sulfides; soluble in H$_2$O; tastes like ink
Vermiculite Mg$_3$Si$_4$O$_{10}$(OH)$_2$·xH$_2$O (p.447)	Yellow, brown, green	2.4	1½	Monoclinic; pseudohexagonal plates, pseudomorphs after biotite	{001} perfect	1.54 –1.58	Expands enormously when heated; occurs in magnesium-rich rocks

Luster nonmetallic; usually colored; streak white unless otherwise stated; H < 2½ *(Continued)*

NAME FORMULA	COLOR	G	H	SYSTEM AND HABIT	CLEAVAGE	n	OBSERVATIONS
Glauconite $K(Fe,Mg,Al)_2Si_4O_{10}(OH)_2$ (p. 453)	Dark green to black	2.5–2.8	2	Monoclinic; occurs as rounded grains	{001} perfect; not megascopically observable	1.61 –1.64	Streak pale green; occurs in marine sedimentary rocks as pure beds or as grains in limestone, sandstone, mudstone
Chlorite $(Mg,Fe,Al)_6(Al,Si)_4O_{10}(OH)_8$ (p. 454)	Green	2.6–2.4	2	Monoclinic; in pseudohexagonal plates or micaceous masses	{001} perfect	1.57 –1.67	Cleavage laminae flexible but not elastic; common in schists
Vivianite $Fe_3(PO_4)\cdot 8H_2O$ (p. 383)	Blue, blue green	2.7	1½	Monoclinic; in prismatic crystals, elongated ∥ c, or earthy masses	{010} prominent in crystals	1.60	Streak blue; earthy masses in clay and fossil bones; crystals associated with iron sulfides in ore deposits
Erythrite $Co_3(AsO_4)_2\cdot 8H_2O$ (p. 384)	Carmine red	3.06	1½	Monoclinic; prismatic crystals and powdery masses	{010} perfect	1.66	Streak pink; occurs as coatings or crusts on cobalt arsenides
Autunite $Ca(UO_2)_2(PO_4)_2\cdot 10-12H_2O$ (p. 394)	Lemon yellow	3.1–3.2	2–2½	Tetragonal; thin tabular on {001}; scaly aggregates	{001} perfect	1.58	Streak yellow; occurs as an oxidation product from veins or pegmatites containing uraninite
Torbernite, Metatorbernite $Cu(UO_2)(PO_4)_2\cdot 8-12H_2O$ (p. 393)	Emerald green to apple green	3.2–3.7	2–2½	Tetragonal; tabular on {001} foliated or micaceous	{001} perfect	1.59	Streak green; occurs as an oxidation product over veins containing uraninite and copper sulfides
Orpiment As_2S_3 (p. 251)	Yellow	3.48	1½	Orthorhombic; acicular crystals, elongated ∥ c; usually in foliated or granular masses	{010} perfect	2.81	Streak yellow; often occurs with realgar
Realgar AsS (p. 250)	Red	3.56	2	Monoclinic; short prismatic crystals; usually in granular masses	{010} distinct	2.62	Streak orange; often occurs with orpiment

NAME FORMULA	COLOR	G	H	SYSTEM AND HABIT	CLEAVAGE	n	OBSERVATIONS
Tyuyamunite Ca(UO$_2$)$_2$(VO$_4$)$_2$·6H$_2$O (p. 395)	Yellow, greenish yellow	3.6–4.4	2	Orthorhombic; usually massive, powdery	{001} perfect, seldom visible	1.87	Streak yellow; occurs disseminated in sandstone, and as crusts on limestone
Carnotite K$_2$(UO$_2$)$_2$(VO$_4$)$_2$·3H$_2$O (p. 395)	Yellow	4–5	2	Monoclinic; usually massive, powdery	{001} perfect, seldom visible	1.90	Streak yellow; occurs disseminated in sandstone; like tyuyamunite, but rarer
Cerargyrite AgCl (p. 322)	Colorless, yellow, gray	5.6	2	Isometric; usually massive, wax-like	None	2.10	Sectile; occurs in oxidized zone of silver deposits

Luster nonmetallic; usually colored; streak white unless otherwise stated; H $= 2\frac{1}{2}$–$5\frac{1}{2}$

NAME FORMULA	COLOR	G	H	SYSTEM AND HABIT	CLEAVAGE	n	OBSERVATIONS
Sulfur S (p. 215)	Yellow	2.07	$2\frac{1}{2}$	Orthorhombic; pyramidal crystals; usually massive, granular or compact	{001}, {110}, {111} imperfect	1.96	Burns, giving choking fumes of SO$_2$; deposited in fumaroles, and occurs in limestones and gypsum
Chalcanthite CuSO$_4$·5H$_2$O (p. 367)	Blue	2.28	$2\frac{1}{2}$	Triclinic; short prismatic crystals; also massive, granular	{1$\bar{1}$0} imperfect	1.53	Streak pale blue; metallic nauseous taste; occurs in oxidized zone of sulfide copper ores
Serpentine Mg$_6$Si$_4$O$_{10}$(OH)$_8$ (p. 443)	Green, yellow, brown, gray	2.5–2.6	$2\frac{1}{2}$	Monoclinic; always massive, compact or fibrous (asbestos)	None visible	1.54 –1.57	Alteration product of ultrabasic igneous rocks
Garnierite (Mg,Ni)$_6$Si$_4$O$_{10}$(OH)$_8$ (p. 444)	Apple green	2.5–2.6	$2\frac{1}{2}$	Monoclinic; always massive, often as porous crusts	None visible	1.63	A nickel-bearing serpentine
Lepidolite KLi$_2$Al(Si$_4$O$_{10}$)(OH)$_2$ (p. 452)	Colorless, mauve, gray	2.8–2.9	$2\frac{1}{2}$–3	Monoclinic; usually massive, coarse to fine granular	{001} perfect	1.55 –1.59	Pale mauve color often diagnostic; occurs in granite pegmatites

Luster nonmetallic; usually colored; streak white unless otherwise stated; $H = 2\frac{1}{2}-5\frac{1}{2}$ (Continued)

NAME FORMULA	COLOR	G	H	SYSTEM AND HABIT	CLEAVAGE	n	OBSERVATIONS
Phlogopite $KMg_3(AlSi_3O_{10})(OH)_2$ (p. 451)	Pale brown	2.8–2.9	$2\frac{1}{2}$–3	Monoclinic; usually in irregular platy crystals	{001} perfect	1.56	Occurs in ultrabasic rocks and metamorphosed dolomites
Biotite $K(Mg,Fe)_3AlSi_3O_{10}(OH)_2$ (p. 451)	Green, brown, black	2.9–3.4	$2\frac{1}{2}$	Monoclinic; usually in irregular platy crystals	{001} perfect	1.57 –1.70	Occurs in acid igneous rocks, pegmatites, and metamorphic rocks
Jarosite $KFe_3(SO_4)_2(OH)_6$ (p. 373)	Brown	2.9–3.3	3	Trigonal; usually massive, compact	{0001} imperfect	1.82	Resembles limonite, but contains sulfate; yellow streak
Fluorite CaF_2 (p. 323)	Purple, green, yellow, colorless	3.18	4	Isometric; cubic crystals, often twinned on (111); also massive, granular	{111} perfect	1.43	Common vein mineral, and replacing limestones and dolomites
Apatite $Ca_5(PO_4)_3F$ (p. 387)	Green, blue, brown, white	3.1–3.2	5	Hexagonal; prismatic crystals, also massive, granular or compact	{0001} poor	1.63	Occurs in pegmatites and metamorphosed limestones; also in sedimentary rocks (phosphorite)
Kyanite Al_2SiO_5 (p. 496)	Blue, green, white	3.63	4–7	Triclinic; usually in bladed crystals	{100} perfect; {010} good	1.72	Occurs in aluminum-rich schists and gneisses
Rhodochrosite $MnCO_3$ (p. 339)	Pink	3.7	$3\frac{1}{2}$–4	Trigonal; usually massive, granular	{10$\bar{1}$1} perfect	1.81	Effervesces in warm dilute HCl; occurs in ore deposits
Atacamite $Cu_2(OH)_3Cl$ (p. 326)	Green	3.76	3	Orthorhombic; slender prismatic crystals, also massive granular	{010} perfect	1.51	Streak green; associated with other secondary copper minerals; gives azure-blue flame coloration
Azurite $Cu_3(CO_3)_2(OH)_2$ (p. 350)	Blue	3.77	4	Monoclinic; crystals of various habit, also massive, spherical aggregates	{011} good	1.76	Streak blue; effervesces in cold dilute HCl

Mineral	Color	G	H	Crystal form and habit	Cleavage	n	Remarks
Siderite $FeCO_3$ (p. 338)	Yellow, brown	3.3–4.0	4	Trigonal; rhombohedral aggregates, also massive, granular to compact	$\{10\bar{1}1\}$ perfect	1.87	Effervesces in warm dilute HCl; occurs in veins and replacement of limestone
Antlerite $Cu_3(SO_4)(OH)_4$ (p. 371)	Dark green	3.88	$3\frac{1}{2}$	Orthorhombic; usually in granular aggregates or cross-fiber veinlets	$\{010\}$ perfect	1.74	Streak green; soluble in HCl without effervescence
Wurtzite ZnS (p. 239)	Brownish black	4.09	$3\frac{1}{2}$–4	Hexagonal; pyramidal or prismatic crystals; usually massive, fibrous, columnar, or as banded crusts	$\{11\bar{2}0\}$ distinct	2.35	Streak brown; much less common than sphalerite
Sphalerite $(Zn,Fe)S$ (p. 233)	Yellow, brown, black	4.0–4.1	4	Isometric; usually massive, coarse to fine granular	$\{110\}$ perfect	2.37 –2.47	Dissolves in warm HCl giving off H_2S
Brochantite $Cu_4(SO_4)(OH)_6$ (p. 370)	Green	4.0	4	Monoclinic; prismatic to acicular crystals; also massive granular	$\{100\}$ perfect	1.77	Streak green; dissolves in cold dilute HCl without effervescence
Malachite $Cu_2(CO_3)(OH)_2$ (p. 349)	Green	4.0	4	Monoclinic; usually massive compact, often stalactitic or botryoidal	$\{\bar{2}01\}$ perfect	1.88	Streak green; effervesces in cold dilute HCl
Goethite (Limonite) $HFeO_2$ (p. 314)	Yellow, brown	4.2	3–$5\frac{1}{2}$	Orthorhombic; usually massive, compact or earthy, also stalactitic	$\{010\}$ perfect	2.41	Streak yellow; dissolves in dilute HCl giving a yellow solution
Pyrochlore-microlite $NaCa(Nb,Ta)_2O_6F$ (p. 296)	Yellow, brown, black	4.2–6.4	5	Isometric; octahedral crystals and rounded grains	Generally absent	1.93 –2.18	Occurs in carbonatites (pyrochlore) and granite pegmatites
Xenotime YPO_4 (p. 381)	Yellowish to reddish brown	4.4–5.1	4–5	Tetragonal prismatic with pyramidal faces resembling zircon	$\{100\}$ perfect	1.72	Accessory mineral in acidic igneous rocks, pegmatites, and gneisses
Monazite $CePO_4$ (p. 382)	Yellow, brown	4.6–5.4	5	Monoclinic; small crystals flattened $\parallel \{100\}$, also as detrital grains	Parting on $\{001\}$	1.79	Accessory mineral in granites and pegmatites; concentrates in sands and sandstones

Luster nonmetallic; usually colored; streak white unless otherwise stated; H = $2\frac{1}{2}$–$5\frac{1}{2}$ (Continued)

NAME FORMULA	COLOR	G	H	SYSTEM AND HABIT	CLEAVAGE	n	OBSERVATIONS
Thorite $ThSiO_4$ (p. 506)	Orange, brown, black	5.2–6.7	5	Tetragonal; pyramidal crystals; also massive, compact	{100} imperfect	1.66 –1.86	Strongly radioactive; occurs in pegmatites and as detrital grains
Zincite ZnO (p. 278)	Red	5.7	4	Hexagonal; usually in platy masses or rounded grains	{0001} perfect	2.01	Streak orange; occurs with franklinite and willemite at Franklin, N.J.
Crocoite $PbCrO_4$ (p. 374)	Orange red	6.1	3	Monoclinic; usually in prismatic crystals	{110} imperfect	2.37	Streak orange yellow; rare secondary mineral in lead veins
Cuprite Cu_2O (p. 276)	Red, often tarnished	6.1	4	Isometric; cubic crystals, also massive, granular or hair-like	{111} imperfect	2.85	Streak red brown; easily reduced to copper on charcoal
Vanadinite $Pb_5(VO_4)_3Cl$ (p. 391)	Yellow, red, brown	6.9	3	Hexagonal; usually in small hexagonal prisms	None	2.42	Streak yellow; found in the oxidation zone of lead deposits
Wulfenite $PbMoO_4$ (p. 378)	Yellow, red, brown	6.5– 7.0	3	Tetragonal; usually in square, platy crystals, tabular $\parallel$ {001}	{101} distinct	2.40	Often associated with vanadinite
Pyromorphite $Pb_5(PO_4)_3Cl$ (p. 390)	Green, yellow, brown	7.0– 7.1	4	Hexagonal; in small prismatic crystals; also massive, often globular and botryoidal	None	2.06	Found in the oxidized zone of lead deposits
Cinnabar HgS (p. 248)	Red	8.1	$2\frac{1}{2}$	Hexagonal; usually massive, granular	{10$\bar{1}$0} perfect	2.91	Streak red; volatilizes on heating

Luster nonmetallic; usually colored; streak white unless otherwise stated; H > $5\frac{1}{2}$

NAME FORMULA	COLOR	G	H	SYSTEM AND HABIT	CLEAVAGE	n	OBSERVATIONS
Sodalite $Na_4(AlSiO_4)_6Cl_2$ (p. 429)	Blue, white, pink	2.3	6	Isometric; usually massive, granular	{110} imperfect	1.48	Blue color characteristic; occurs in nepheline syenites

Mineral	Color	G	H	Crystal system; habit	Cleavage	R.I.	Remarks
Cancrinite $Na_8(AlSiO_4)_6(HCO_3)_2$ (p. 429)	Yellow, white, pink	2.4–2.5	6	Hexagonal; usually massive, granular	{10$\bar{1}$0} perfect	1.51–1.53	Effervesces in warm dilute HCl; its yellow color often characteristic
Cordierite $(Mg,Fe)_2Al_4Si_5O_{18}$ (p. 480)	Colorless, blue, gray, brown	2.55–2.75	7	Orthorhombic; usually massive, granular	{010} imperfect; parting on {001}	1.52–1.57	Occurs in aluminous metamorphic rocks
Turquois $CuAl_6(PO_4)_4(OH)_8 \cdot 4H_2O$ (p. 392)	Blue, green	2.6–2.8	6	Triclinic; usually massive, compact	{001} perfect	1.62	Occurs in veinlets in igneous and sedimentary rocks
Beryl $Be_3Al_2Si_6O_{18}$ (p. 479)	Colorless, white, green, yellow, blue	2.65–2.85	8	Hexagonal; prismatic crystals, or massive granular	{0001} imperfect	1.57–1.60	Occurs in granite pegmatites
Allanite $(Ca,Ce)_2(Al,Fe)_3 Si_3O_{12}OH$ (p. 489)	Black, dark brown	2.7–4.0	6½	Monoclinic; platy or columnar crystals, also massive, granular	{001} seldom observable	1.54–1.81	Streak gray or green gray; often somewhat radioactive
Prehnite $Ca_2Al_2Si_3O_{10}(OH)_2$ (p. 456)	White, pale green	2.9	6½	Orthorhombic; generally massive, encrusting, with crystalline surface	{001} not prominent	1.62	Occurs in cavities in basic igneous rocks associated with zeolites
Tourmaline Na,Mg,Fe,Al borosilicate (p. 481)	Black, brown, pink, green, blue	3.0–3.2	7½	Trigonal; prismatic crystals with rounded trigonal cross-section; also massive, columnar	None	1.63–1.68	Prism faces vertically striated; occurs in granite pegmatites, and in schists and gneisses
Actinolite $Ca_2(Mg,Fe)_5Si_8O_{22}(OH)_2$ (p. 463)	Green	3.0–3.4	6	Monoclinic; usually in columnar or fibrous aggregates	{110} perfect	1.62–1.66	Occurs in schists and gneisses
Hornblende $NaCa_2(Mg,Fe,Al)_5 (Al,Si)_8O_{22}(OH)_2$ (p. 465)	Green, brown, black	3.0–3.4	6	Monoclinic; long prismatic crystals, also columnar, fibrous or granular	{110} perfect	1.62–1.71	A common mineral of both igneous and metamorphic rocks

Luster nonmetallic; usually colored; streak white unless otherwise stated; $H > 5\frac{1}{2}$ (Continued)

NAME FORMULA	COLOR	G	H	SYSTEM AND HABIT	CLEAVAGE	n	OBSERVATIONS
Glaucophane $Na_2Mg_3Al_2Si_8O_{22}(OH)_2$ (p. 466)	Blue, lavender, blue-black	3.1–3.3	6	Monoclinic; in aggregates of bladed or acicular crystals, or as felt or silky needles	{110} perfect	1.62 –1.67	Occurs in schists and gneisses
Dumortierite $Al_7BSi_3O_{18}$ (p. 508)	Blue, pink	3.3–3.4	7	Orthorhombic; usually massive, or fibrous compact	{100} distinct	1.68 –1.72	Occurs in aluminum-rich metamorphic rocks
Cummingtonite $(Fe,Mg)_7Si_8O_{22}(OH)_2$ (p. 463)	Brown	3.2–3.6	6	Monoclinic; usually massive, fibrous aggregates	{110} perfect	1.65 –1.71	Occurs in iron-rich schists and gneisses
Enstatite $(Mg,Fe)SiO_3$ (p. 468)	White, pale green, brown, gray	3.2–3.4	6	Orthorhombic; usually massive, fibrous or granular	{210} good	1.65 –1.67	Often has bronze color (var. *bronzite*)
Augite $(Ca,Mg,Al,Fe)(Al,Si)_2O_6$ (p. 471)	Black, dark green	3.25–3.55	6	Monoclinic; short prismatic crystals; also massive, granular	{110} good	1.70 –1.75	The common pyroxene of igneous rocks
Diopside $Ca(Mg,Fe)Si_2O_6$ (p. 471)	White, green	3.25–3.40	6	Monoclinic; prismatic crystals, usually massive, granular	{110} good	1.70 –1.73	Occurs in metamorphosed dolomites
Axinite Ca,Mn,Fe,Al borosilicate (p. 478)	Brown, yellow	3.3	7	Triclinic; wedge-shaped crystals, also massive granular	{100} good	1.68 –1.70	Occurs in contact zones of igneous intrusions
Jadeite $NaAlSi_2O_6$ (p. 473)	White, green	3.3	$6\frac{1}{2}$	Monoclinic; usually massive, granular to compact	{110} good	1.66	Metamorphic mineral, often associated with serpentine

Mineral	Color	G	H	Crystallography	Cleavage	n	Remarks
Zoisite Ca₂Al₃Si₃O₁₂(OH) (p. 487)	Gray, green, pink	3.3	7	Orthorhombic; usually massive; columnar or bladed	{001} perfect	1.69 –1.71	Occurs in metamorphic rocks rich in lime and alumina
Riebeckite Na₂(Mg,Fe)₅Si₈O₂₂(OH)₂ (p. 466)	Dark blue, black	3.3–3.6	6	Monoclinic; bladed crystals, also massive granular; sometimes asbestiform	{110} perfect	1.67 –1.71	Blue-gray streak; occurs in soda-rich igneous rocks
Olivine (Mg,Fe)₂SiO₄ (p. 490)	Olive green	3.3–3.6	6½	Orthorhombic; usually massive, granular	{010} imperfect	1.65 –1.87	Occurs in basic and ultrabasic igneous rocks
Idocrase Ca,Mg,Al silicate (p. 485)	Brown, green, yellow, blue	3.3–3.5	7	Tetragonal; short prismatic or low pyramidal crystals; also massive granular	{100} imperfect	1.70 –1.75	Occurs in metamorphosed limestones
Clinozoisite, Epidote Ca₂(Al,Fe)₃ Si₃O₁₂(OH) (p. 488)	Green to black	3.3–3.6	7	Monoclinic; prismatic crystals elongated ∥ *b*; also massive, columnar or granular	{001} perfect	1.68 –1.78	Yellow-green color typical; occurs in metamorphic rocks
Hedenbergite Ca(Fe,Mg)Si₂O₆ (p. 471)	Green to black	3.40–3.55	6	Monoclinic; usually massive, granular or fibrous	{110}	1.73 –1.76	Often occurs in skarn associated with ore deposits
Hypersthene (Mg,Fe)SiO₃ (p. 468)	Brown, black	3.4–4.0	6	Orthorhombic; usually massive, platy	{210} good	1.67 –1.79	Occurs in basic igneous rocks
Sphene CaTiSiO₅ (p. 507)	Brown, yellow, green	3.5	6	Monoclinic; usually in small diamond- or wedge-shaped crystals	{110} imperfect	1.87 –2.03	An accessory mineral of igneous and metamorphic rocks
Rhodonite MnSiO₃ (p. 477)	Pink	3.5–3.7	6	Triclinic; tabular crystals; usually massive, granular to compact	{110}, {110} good	1.73	Often veined with black manganese oxides

Luster nonmetallic; usually colored; streak white unless otherwise stated; H > 5½ (Continued)

NAME FORMULA	COLOR	G	H	SYSTEM AND HABIT	CLEAVAGE	n	OBSERVATIONS
Aegirine $NaFeSi_2O_6$ (p. 472)	Dark green, brown, black	3.6	6	Monoclinic; usually in prismatic crystals, often elongated ∥ c	{110} perfect	1.81	Occurs in sodium-rich igneous rocks
Spinel $MgAl_2O_4$ (p. 281)	Black, brown, green, red, blue	3.6	8	Isometric; octahedral crystals, often twinned on {111}	{111} imperfect	1.72	Occurs in metamorphic rocks and ultrabasic igneous rocks, and in sands derived from them
Grossular $Ca_3Al_2Si_3O_{12}$ (p. 502)	White, yellow, pink, brown	3.6–3.7	7	Isometric; dodecahedrons and trapezohedrons, and combinations; also massive, granular or compact	None	1.73–1.80	Occurs in metamorphosed limestones
Pyrope $Mg_3Al_2Si_3O_{12}$ (p. 502)	Red	3.6–3.9	7½	Isometric; usually in anhedral grains	None	1.73–1.76	Occurs in high-grade metamorphic rocks
Andradite $Ca_3Fe_2Si_3O_{12}$ (p. 502)	Yellow, green, brown, black	3.7–3.8	7	Isometric; dodecahedron and trapezohedrons, and combinations; also massive granular	None	1.80–1.88	Occurs in metamorphosed limestones, often associated with ore deposits
Chrysoberyl $BeAl_2O_4$ (p. 286)	Yellow, green, brown	3.7	8½	Orthorhombic; crystals tabular ∥ {001}, and peudohexagonal twins	{110} distinct	1.75	Occurs in granite pegmatites and in alluvial deposits
Staurolite $Fe_2Al_9O_6(SiO_4)_4(O,OH)_2$ (p. 497)	Brown	3.76	7	Orthorhombic; prismatic crystals, often twinned in right-angle and diagonal crosses	{010} distinct	1.75	Occurs in medium-grade schists and gneisses
Uvarovite $Ca_3Cr_2Si_3O_{12}$ (p. 502)	Green	3.8	7	Isometric; usually in dodecahedrons	None	1.86	Generally associated with serpentine and chromite

Mineral	Color	H	G	Crystal system and habit	Cleavage	Index	Remarks
Willemite Zn_2SiO_4 (p. 492)	Yellow, green, brown, black	6	3.9–4.2	Trigonal; usually massive, granular	{0001}, {1120} imperfect	1.69–1.72	Often fluoresces green in ultraviolet light
Zircon $ZrSiO_4$ (p. 504)	Brown, gray	$7\frac{1}{2}$	3.9–4.7	Tetragonal; prismatic crystals terminated by pyramids, and as rounded grains	{100} imperfect	1.78–1.96	An accessory mineral of igneous and metamorphic rocks; a detrital mineral in sands
Anatase TiO_2 (p. 305)	Usually brown	6	3.9	Tetragonal; usually in acute pyramidal crystals	{001}, {011} perfect	2.56	Occurs in small crystals in joint planes and veins in schists and gneisses
Corundum Al_2O_3 (p. 288)	Gray, blue, pink, brown	9	4.0	Trigonal; hexagonal prisms and pyramids, also massive, granular	None; parting on {0001}, {01$\bar{1}$2}	1.77	Occurs in metamorphic rocks, and occasionally in aluminum-rich igneous rocks
Almandite $Fe_3Al_2Si_3O_{12}$ (p. 502)	Red to black	$7\frac{1}{2}$	4.0–4.3	Isometric; dodecahedrons, trapezohedrons, and combinations; also massive, granular	None	1.78–1.83	Occurs in metamorphic rocks; the commonest garnet
Spessartine $Mn_3Al_2Si_3O_{12}$ (p. 502)	Orange, red, brown	$7\frac{1}{2}$	4.0–4.2	Isometric; trapezohedrons, often in combination with dodecahedrons; also massive, granular	None	1.78–1.81	Occurs in granite pegmatites and metamorphosed manganese-rich rocks
Brookite TiO_2 (p. 305)	Brown to black	6	4.1	Orthorhombic; tabular and prismatic crystals	None	2.58	Occurs in joint planes and veins in schists and gneisses
Rutile TiO_2 (p. 299)	Red brown to black	$6\frac{1}{2}$	4.2–4.5	Tetragonal; usually in long prismatic crystals or massive	{110} imperfect	2.61	Common accessory mineral of igneous and metamorphic rocks
Cassiterite SnO_2 (p. 304)	Brown to black	$6\frac{1}{2}$	7.0	Tetragonal; prismatic crystals and massive, sometimes botryoidal or concretionary	{110} imperfect	2.00	Occurs in quartz veins and granite pegmatites; also detrital

Index

Mineral names in **boldface** type are those described in detail; page numbers in **boldface** type refer to the principal description of the mineral.

The periodic table, giving the charge and radius (in Å) of the commoner ions

H																	He
Li^1 0.68	Be^2 0.35	B^3 0.23	C^4 0.16											N^5 0.13	O^{-2} 1.40	F^- 1.36	Ne
Na^1 0.97	Mg^2 0.66	Al^3 0.51	Si^4 0.42											P^5 0.35	S^6 0.30 S^{-2} 1.84	Cl^- 1.81	A
K^1 1.33	Ca^2 0.99	Sc^3 0.81	Ti^4 0.68	V^5 0.59 V^3 0.74	Cr^6 0.52 Cr^3 0.63	Mn^4 0.60 Mn^2 0.80	Fe^3 0.64 Fe^2 0.74	Co^3 0.63 Co^2 0.72	Ni^2 0.69	Cu^2 0.72 Cu^1 0.96	Zn^2 0.74	Ga^2 0.62	Ge^4 0.53	As^5 0.46	Se^6 0.42	Br^- 1.95	Kr
Rb^1 1.47	Sr^2 1.12	Y^3 0.92	Zr^4 0.79	Nb^5 0.69	Mo^6 0.62		Ru^4 0.67	Rh^3 0.68	Pd^2 0.80	Ag^1 1.26	Cd^2 0.97	In^3 0.81	Sn^4 0.71	Sb^5 0.62 Sb^3 0.76	Te^6 0.56	I^- 2.16	Xe
Cs^1 1.67	Ba^2 1.34	La^{3-} 1.14 Lu^3 0.85	Hf^4 0.78	Ta^5 0.68	W^6 0.62	Re^4 0.72	Os^6 0.69	Ir^4 0.68	Pt^2 0.80	Au^1 1.37	Hg^2 1.10	Tl^1 1.47	Pb^4 0.84 Pb^2 1.20	Bi^3 0.96			
			Th^4 1.02		U^4 0.97												